CAMBRIDGE LIBRARY COLLECTION

Books of enduring scholarly value

Physical Sciences

From ancient times, humans have tried to understand the workings of the world around them. The roots of modern physical science go back to the very earliest mechanical devices such as levers and rollers, the mixing of paints and dyes, and the importance of the heavenly bodies in early religious observance and navigation. The physical sciences as we know them today began to emerge as independent academic subjects during the early modern period, in the work of Newton and other 'natural philosophers', and numerous sub-disciplines developed during the centuries that followed. This part of the Cambridge Library Collection is devoted to landmark publications in this area which will be of interest to historians of science concerned with individual scientists, particular discoveries, and advances in scientific method, or with the establishment and development of scientific institutions around the world.

An Account of Experiments to Determine the Figure of the Earth by Means of the Pendulum Vibrating Seconds in Different Latitudes

As early as the seventeenth century, scientists realised that a pendulum swings more slowly at the equator than it would at the North Pole. Newton predicted that gravity increased with latitude, and that the Earth could not be perfectly spherical. Although various experiments were undertaken to determine the exact degree of this ellipticity, none proved successful until physicist Edward Sabine (1788–1883) embarked on a series of expeditions across the world. Based on pendulum measurements from a wide range of latitudes, from Jamaica to Spitsbergen, his results were very different to mathematical predictions, and far more accurate; Charles Babbage would even complain that they were too good to be true. In this account, which first appeared in 1825, Sabine explains his methodology and presents his findings. His book opens a fascinating window into nineteenth-century geodesy for students in the history of science.

An Account of Experiments to Determine the Figure of the Earth by Means of the Pendulum Vibrating Seconds in Different Latitudes

As Well as on Various Other Subjects of Philosophical Inquiry

Edward Sabine

CAMBRIDGE UNIVERSITY PRESS

Cambridge, New York, Melbourne, Madrid, Cape Town,
Singapore, São Paolo, Delhi, Mexico City

Published in the United States of America by Cambridge University Press, New York

www.cambridge.org
Information on this title: www.cambridge.org/9781108062077

This edition first published 1825
This digitally printed version 2013

ISBN 978-1-108-06207-7 Paperback

Some of the tables in this volume may fall very close to the margins of the book. They can also be viewed online at http://www.cambridge.org/9781108062077

AN ACCOUNT

OF

EXPERIMENTS

TO DETERMINE

THE FIGURE OF THE EARTH,

BY MEANS OF THE

PENDULUM VIBRATING SECONDS IN DIFFERENT LATITUDES;

AS WELL AS ON

VARIOUS OTHER SUBJECTS

OF

PHILOSOPHICAL INQUIRY.

By EDWARD SABINE,

CAPTAIN IN THE ROYAL REGIMENT OF ARTILLERY; FELLOW OF THE ROYAL AND LINNÆAN SOCIETIES OF LONDON; MEMBER OF THE ROYAL SOCIETY OF SCIENCES OF NORWAY; CORRESPONDING MEMBER OF THE ROYAL SOCIETY OF SCIENCES AT GOTTINGEN; HONORARY MEMBER OF THE HISTORICAL, AND OF THE LITERARY AND PHILOSOPHICAL, SOCIETIES OF NEW YORK.

PRINTED AT THE EXPENSE OF THE BOARD OF LONGITUDE.

LONDON:
JOHN MURRAY,
BOOKSELLER TO THE BOARD OF LONGITUDE.

MDCCCXXV.

TO

DAVIES GILBERT, Esq., M.P.

TREASURER AND A VICE-PRESIDENT OF THE ROYAL SOCIETY,

AND A COMMISSIONER OF THE BOARD OF LONGITUDE;

WHO, IN THE SESSION OF 1816, MOVED AN ADDRESS OF THE HOUSE OF COMMONS TO THE CROWN, PRAYING, "THAT HIS MAJESTY WOULD BE GRACIOUSLY PLEASED TO GIVE DIRECTIONS FOR DETERMINING THE VARIATION IN THE LENGTH OF THE PENDULUM VIBRATING SECONDS, AT THE PRINCIPAL STATIONS OF THE TRIGONOMETRICAL SURVEY OF GREAT BRITAIN,"

THIS VOLUME,

CONTAINING AN ACCOUNT OF THE EXTENSION OF THE EXPERIMENTS, TO STATIONS INCLUDING THE UTMOST ACCESSIBLE DISTANCE ON THE MERIDIAN OF A HEMISPHERE,

IS VERY RESPECTFULLY INSCRIBED.

CONTENTS.

EXPERIMENTS FOR DETERMINING THE VARIATION IN THE LENGTH OF THE SECONDS' PENDULUM.

SECTION I. *With Detached Invariable Pendulums.*

SECTION II. *With Attached Invariable Pendulums.*

Latitudes of the Pendulum Stations.

Application of the observed Variation in the Length of the Seconds' Pendulum to the Determination of the Figure of the Earth.

GEOGRAPHICAL NOTICES.

Longitudes of the Pendulum Stations.

HYDROGRAPHICAL NOTICES.

EXPERIMENTS FOR DETERMINING THE VARIATION IN THE INTENSITY OF TERRESTRIAL MAGNETISM.

ATMOSPHERICAL NOTICES.

PREFACE.

In the year 1672, a pendulum, conveyed by Richer from Paris to Cayenne, first made known the variation in the force of gravitation in different latitudes; and the progressive increase in its intensity from the Equator to the Pole. The fact, thus evidenced, furnished, in the hands of Newton, an experimental demonstration of the deviation of the figure of the Earth from perfect sphericity, and of its oblateness or compression at the Poles.

In 1743, Clairault demonstrated his celebrated theorem, that the sum of the two fractions, of which the one expresses the Ellipticity of the Earth, and the other the ratio of the force of gravity at the Poles to that at the Equator, is a constant quantity, and is equal to $\frac{5}{2}$ of the fraction expressing the ratio of the centrifugal force at the Equator to that of gravity. From thenceforward, the Ellipticity of the Earth was deducible, whenever the difference between gravity at the Pole and at the Equator should be ascertained; and the pendulum became a means of investigating the precise figure of the Earth, inasmuch as it might be rendered an exact measure of the variation of intensity, of which it had furnished the first general intimation.

Thus, a century and a half has elapsed since the earth has been known to be compressed at the Poles; during eighty years of which period, we have recognised and possessed in the pendulum a means of determining the exact measure of the compression.

The progress which the inquiry had made in the first forty years after the publication of Clairault's theorem, and the importance which, in the judgment of the most eminent philosophers of that period, attached to its further prosecution, cannot be better stated than in the words of the admirable *Memoire redigé par l'Académie Royale des Sciences (de Paris), pour servir aux Savans embarqués sous les ordres de M. de La Perouse,* in which memoir experiments with the pendulum form the first subject recommended to the attention of the navigators: "Une des recherches les plus intéressantes est celle qui concerne la determination de la longueur du pendule à secondes, à différentes latitudes. Les inductions que l'on a tirées jusqu'ici de cet instrument, pour determiner les variations de la Pesanteur, ont eu pour fondement des opérations faites en petit nombre par divers observateurs, et avec des instruments différens; et ce defaut d'uniformite dans les operations a dû influer sur la certitude des conséquences déduites de la comparaison des résultats. *On sent de quel prix serait un ensemble d'opérations da ce genre, faites avec soin, par les mêmes personnes, et avec les mêmes instrumens.*"

The calamity, which terminated prematurely the researches of M. de La Perouse and his associates, in an enterprise preeminent amongst Voyages of Discovery for the just and appropriate regard paid to the advancement of every branch of natural knowledge, deprived the Academy and the public of the immediate and justly-expected fruit of a memoir, in the preparation of which so many eminent philosophers had participated; but the memoir itself has happily survived, to stimulate the exertions and direct the researches of individuals of a succeeding generation, in this as well as in many other respects.

Early in the present century, a series of experiments to determine the intensity of gravitation by means of the pendulum, at the extremities and at some intermediate stations of the arc of the meridian passing through France, and comprised between Dunkirk and Formentera, was undertaken on the recommendation of the Academy of Sciences, and carried into effect at the expense of the French Government, by the Members of the Bureau des Longitudes. Experiments in France were thus in progress, when, in 1816, a corresponding undertaking was originated in Great Britain by an address to the Crown, moved in the House of Commons by Mr. Davies Gilbert, praying that His Majesty (then Prince Regent) "would be graciously pleased to give directions for ascertaining the length of the pendulum vibrating seconds of time in the latitude of London, as compared with the standard measure in the possession of the House of Commons; and *for determining the variations in length of the said pendulum at the principal stations of the trigonometrical survey extended through Great Britain.*" His Majesty's Ministers having requested the assistance of the Royal Society, in carrying into effect the objects of the address, their accomplishment was undertaken by one of the most distinguished members of that Society, and completed in 1819; in which year the account of the experiments for determining the variation in the length of the pendulum vibrating seconds at the principal stations of the trigonometrical survey of Great Britain, was published by Captain Kater in the Philosophical Transactions.

In 1821 the account of the experiments on the length of the seconds pendulum at different points of the arc of the meridian between Formentera and Dunkirk, in which MM. Biot, Arago, Mathieu, Bouvard, and Chaix, had participated, including a subsequent extension of the series by M. Biot to the northern extremity of the British Islands, was published, forming the close of the third volume of the *Base du Système Métrique.*

The suite of experiments thus executed in Great Britain and in France, having, for their ultimate purpose, the highest accomplishment of practical geometry, that of determining the exterior configuration of the Earth, and conducted by the most distinguished experimental philosophers in Europe, was nevertheless found to fail in arriving at a satisfactory conclusion. By the improvements successively introduced in the apparatus and in the methods of procedure, by the eminent mechanical skill and ingenuity of the conductors, the pendulum was indeed rendered an exceedingly precise measure of the relative intensities of gravitation, at the several stations of experiment; but the regularity with which gravitation itself had been supposed to vary in conformity with the general Ellipticity of the meridian, was discovered to be greatly interfered with by inequalities in the density of the materials which form the strata near the surface of the Earth, the effects of which had not been duly appreciated, or anticipated. The diminution of gravity from the pole to the equator is derived theoretically from the decrease observed to take place between any two given latitudes; consequently, if no irregular attraction occurred, the result deduced from the comparison of the intensity at every two latitudes whatsoever should be the same. The discrepancies, however, in the results obtained by combining the lengths of the pendulums observed at the different stations in Great Britain and in France, were so great and so irregular, as to prevent any independent conclusion whatsoever, relatively to the general figure of the Earth, being drawn from the experiments, either of the French philosophers, or of Captain Kater.

The state in which the inquiry into the figure of the Earth by means of the pendulum, was left at the close of the experiments in Great Britain and France, may be best shewn by the following extracts from the respective published accounts. The memoir of Captain Kater concludes with this general remark: "It must be evident that nothing very decisive respecting the general ellipticity of the meridian can be deduced from the present experiments. For that purpose it is requisite that the extreme stations should comprise an arc of sufficient length to render the effect of irregular attraction insensible; and that effect might be diminished, if not wholly prevented, by selecting stations of similar geological character, and which should differ as little as possible in elevation above the level of the sea." Similar in effect is the conclusion of the memoir of the French experiments. "La concordance des résultats du Capitaine Kater avec les nôtres, lorsque les uns et les autres ne peuvent pas se représenter rigoureusement, par une variation de la pesanteur proportionnelle au carré du sinus de la latitude, acheve de mettre en évidence que cette impossibilité est réel, et qu'ainsi l'on ne peut pas se flatter de représenter les longueurs du pendule pour tout le globe par une même formule, qui les reproduise avec une complette rigueur, mais seulement dans les limites des différences que les variations locales de la pesanteur peuvent y occasionner. Alors tout ce que reste à faire consiste à employer toujours des procédés d'observation assez exacts pour que les erreurs propres qu'ils comportent, soient fort inférieurs en étendue aux effets réels des causes accidentelles, afin de pouvoir déduire celle-ci de leur comparaison avec la formule théorique construite sur l'ensemble de toutes les observations."

Such was the state of the inquiry when the present experiments were undertaken: their design was, to give the method of experiment the advantage of being tried under the circumstances most favourable for the production of a conclusive result; to extend the suite of stations, previously confined to Great Britain and France, to the Equator on the one side, and to the highest accessible latitudes of the northern hemisphere on the other; to multiply the stations at both extremities of the meridian, so that by their general combination, the irregular influences of local density might mutually destroy each other, and the variations of gravity due to the Ellipticity alone be eliminated; and to ensure the uniformity of procedure and strict comparability of the results at all the stations, by the unity of the observer, and the identity of the instruments. In effect, to terminate the inquiry with the pendulum,—either by obtaining decisively the result which it might be capable of furnishing,—or by manifesting that no decisive result whatsover was attainable by it, even under the most favourable circumstances of operation

The success which has attended the attempt, and the conclusive nature of the result which it has furnished, will be best seen in the part of the following volume, in which the results at the several stations of experiment are collected and applied, with those at the stations in Great Britain and France, in the deduction of the tota lincrease of gravitation between the Equator and the Pole, and of the corresponding Ellipticity of the Earth.

The Reader is requested to correct the following Errata before perusal.

Page	
4	line 10, and page 187, line 22, *for* resistance *read* buoyancy.
45	line 19, *for* alternate *read* ultimate.
85	heading, *for* Reduced Vibrations at 38°.1 *read* 81°.83.
86	heading, *for* Reduced Vibrations at 82°.1 *read* 82°.02.
115	line 3, *for* the strict relation of the force of gravity *read* the strict relation of the variation of the force of gravity.
148	last line, *for* 11° West *read* 9° East.
296	recapitulation, *for* Arcturus *read* α Centauri.
305, 307,	*for* α Gruris *read* α Gruis.
308	recapitulation, *for* α Crucis *read* α Gruis.
333	line 9 of text, *for* on the principle that the length *read* on the principle that the proportion of the total difference in the length.
345	line 3 of Geological Characters, *for* computed *read* compact.
420	last line, *for* westward *read* eastward.
360	line 22, *for* nearly two centuries *read* nearly a century.

d

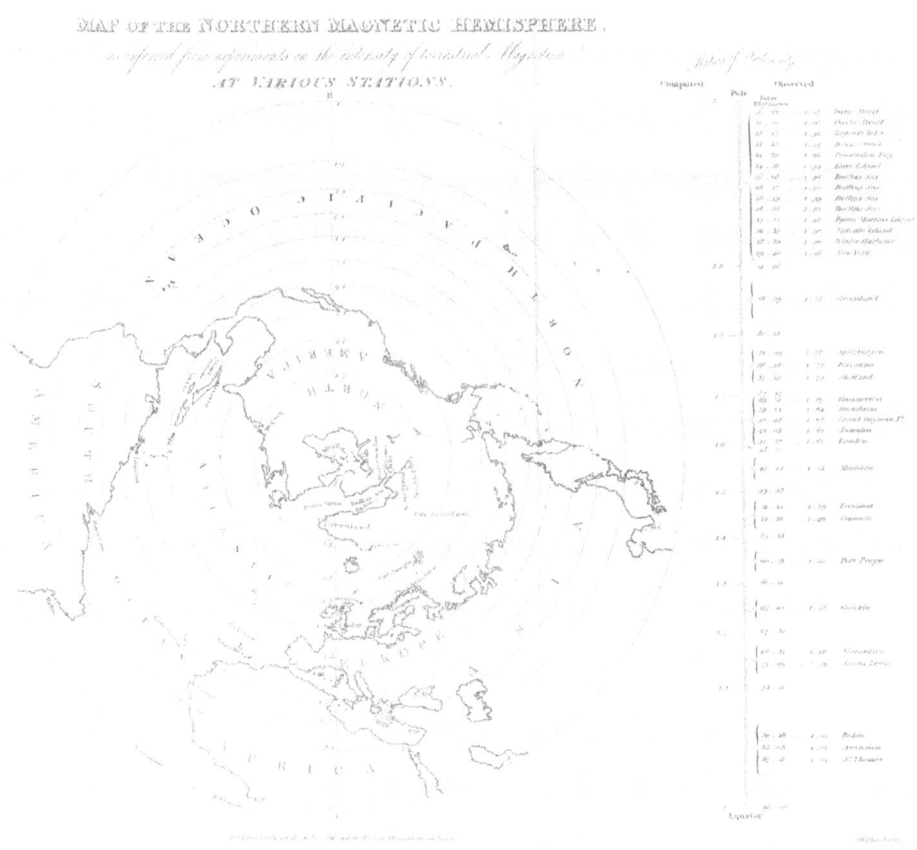

The material originally positioned here is too large for reproduction in this reissue. A PDF can be downloaded from the web address given on page iv of this book, by clicking on 'Resources Available'.

The material originally positioned here is too large for reproduction in this reissue. A PDF can be downloaded from the web address given on page iv of this book, by clicking on 'Resources Available'.

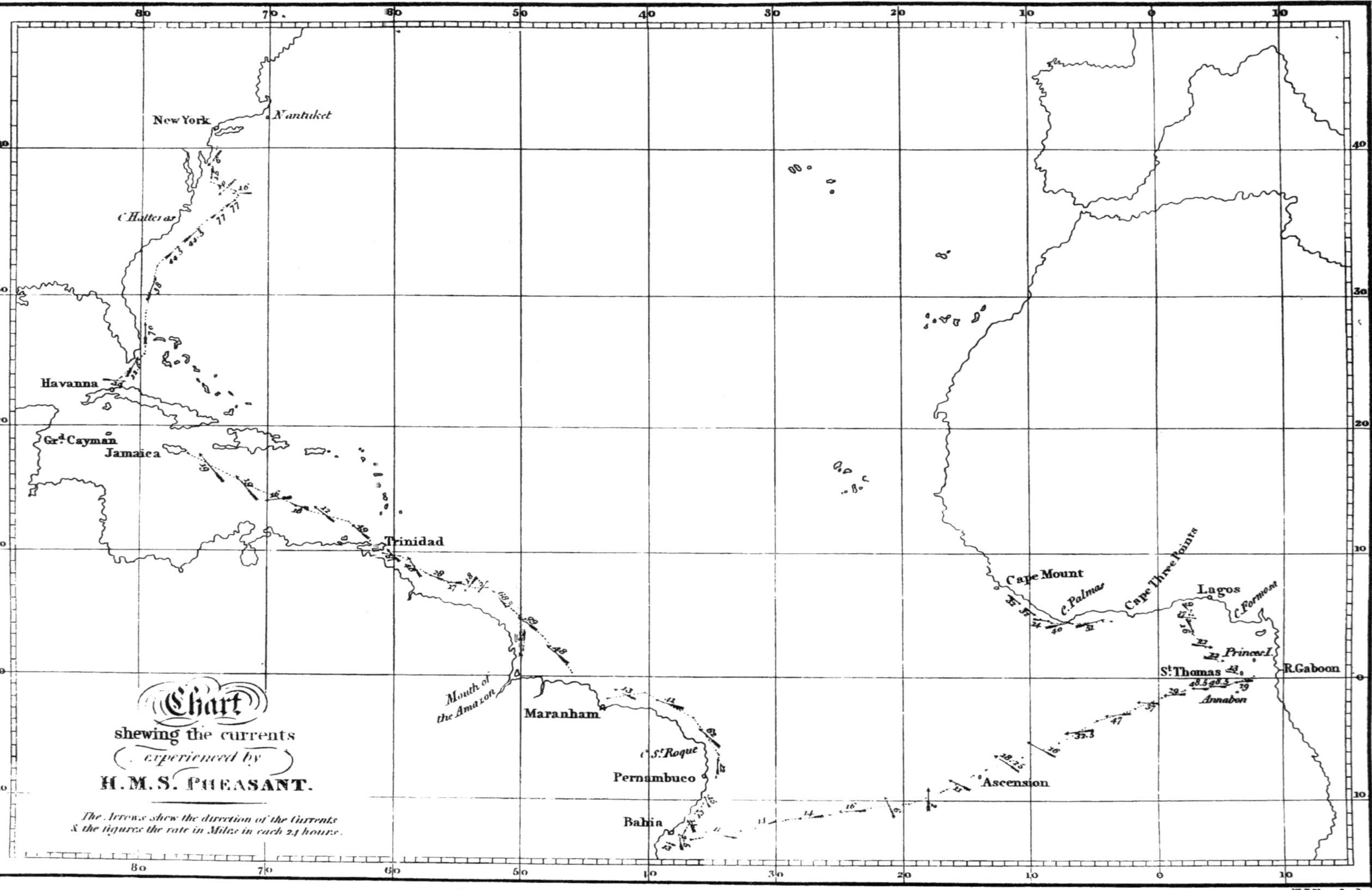

Published as the act directs June 1825 by John Murray, Albemarle Street London.

J. Walker Sculpt.

AN ACCOUNT

OF

EXPERIMENTS FOR DETERMINING THE VARIATION IN THE LENGTH OF THE PENDULUM VIBRATING SECONDS,

IN DIFFERENT LATITUDES;

MADE WITH A VIEW TO ASCERTAIN THE TRUE FIGURE OF THE EARTH.

SECTION I.

With detached Invariable Pendulums.

THE pendulums employed in the experiments of the present section were *detached*, or unconnected with machinery of any sort to maintain or register their oscillations; when set in motion, they vibrated by the influence of gravity alone, in arcs progressively diminishing by the resistance of the air, until the pendulum arrived at rest; the continuance of the vibration was sufficient, when commenced in an arc of moderate dimension, to admit of its rate being very accurately determined.

The pendulums were so constructed as to be invariable in their length excepting by the expansion of the metal of which they were composed, due to different degrees of temperature.

The rate of vibration of a pendulum of this description in an uniform temperature in different parts of the globe, being strictly proportioned to the force of gravity, furnishes the means of computing the variation in the length of the pendulum vibrating seconds, at the respective stations of experiment; and, if at one of the stations, the *absolute* length of the seconds' pendulum has been determined, by other processes and with instruments purposely contrived, the true as well as the relative length becomes also known at every other station at which the invariable pendulum is employed.

The length of the seconds' pendulum in London has been determined with great precision by Captain Kater, at Mr. Browne's house in Portland-place; the rates of vibration of the present pendulums were obtained at the same spot; consequently their comparative rates at other stations furnish respectively the true lengths of the seconds'- pendulum, subject to the errors (if any) of the original determination, and to their own possible errors.

The rate of a pendulum is otherwise expressed by the number of vibrations which it would make in any determinate period of time, (the period throughout these experiments being twenty-four hours of mean solar time), independently of the resistance of the air, and in arcs infinitely small; this number, expressive of the rate, is deducible by certain known methods from the number of vibrations *actually* performed in the same period, in arcs of observed magnitude, and in the then existing circumstances of the atmosphere. The ascertainment of these particulars with the necessary exactness constitutes a principal part of the experiment; and as it is impossible to command an uniform temperature at different stations, and especially at those widely apart in latitude, it is also necessary to be very accurately acquainted with the temperature of the pendulums at the time of their observed vibration, in order to remedy the only exception to their rigorous invariability; and by making their expansion within the limits of the extreme differences of temperature which occur in the course of the observations, the subject itself of direct experiment, the value of the several reductions to a general mean temperature is ascertained, and the results are ultimately rendered strictly comparable with each other.

The method of ascertaining the rate of a detached pendulum, by observing the times of coincidence of its vibration with that of the pendulum of an astronomical clock, which is itself duly compared with the heavens, has been so recently practised and so fully described, both in Great

Britain and France, as to render a particular account of the process by which it is accomplished unnecessary on the present occasion. The principle of the method adopted in these experiments was the same as in the instances alluded to; but an alteration was made in the mode of observing the coincidences which it was conceived would tend to greater accuracy; the nature of the alteration, and the reasons which induced its preference, will be explained in a subsequent part of the volume, and in a place more appropriate for their discussion.

The following were the instruments employed in the experiments with the detached pendulums:

An astronomical clock, the property of the Royal Society, which was lent by the President and Council for the occasion; it was one of two clocks made for the society in 1769, by Shelton, for the purpose of accompanying Messrs. Baily and Dixon to the North Cape and Hammerfest, to observe the transit of Venus. It does not appear by any document in the records of the Society, that these clocks were subsequently used, until the expeditions of Arctic Discovery commenced in 1818, when being fitted with invariable pendulums, they were employed in the experiments of which I have given an account in the *Philosophical Transactions* for 1821*; the clock which was distinguished in that paper as No. 2, had now its original compensation pendulum restored, and was thoroughly cleaned and oiled afresh; two iron plates were affixed to the case, one at the top and the other at the bottom; the upper plate projected on each side of the case to receive screws, by which the clock might be attached to pickets of wood driven into a vertical wall; a similar screw in the

* It is there stated, on the authority of Captain Kater, that they were the same clocks which accompanied Captain Cook round the world: I have since been informed by Mr. Lee, Librarian to the Royal Society, that the clocks which were lent to Captain Cook, one of which was also made by Shelton and the other by Ellicot, were sold by the Society, in 1771.

middle of the lower plate secured the clock firmly to the wall when finally adjusted, but admitted of a small motion of the case to the right and left before the screw was tightened, for the purpose of putting the clock in beat; the lower plate was also furnished with screws acting against the wall, to render the agate support of the pendulum truly horizontal; the ball of the pendulum was covered with a black varnish, and had a circular disk of white paper pasted on the surface, near the middle of the ball.

A transit instrument, by Dolland, with a telescope of thirty inches focal length, and an aperture of two inches; the transverse axis, being fifteen inches, was supported on hollow brass cones; the level was furnished with a small cross level, and a graduated scale shewing tenths of seconds of time at the zenith; instead of the usual mode of placing wires or cobwebs in the focus of the object and the eye-glasses, the telescope of this transit was fitted with a piece of finely polished parallel glass, with fine lines drawn upon it with a diamond at the proper intervals, whereby the equality of distance, parallelism, and permanency, were designed to be better secured than by the usual method.

A repeating circle, of six inches diameter, made by Dolland, upon the same principles and construction as the one he made under the direction of Captain Kater for the British Board of Longitude, and which was originally designed for Professor Playfair.

Four chronometers, Nos. 357, 384, 423, and 493, which Messrs. Parkinson and Frodsham were so kind as to place in my charge, and entrust to my entire disposal; as the makers were yet desirous of other opportunities of manifesting the excellency of their chronometers, than those of the Arctic Expeditions, in which they had already obtained so much distinction: No. 423 was fitted for the pocket, and the other three were box chronometers; No. 357 was wound weekly, the others daily.

Two invariable pendulums, one of which was the property of the Board of Longitude, as were also the transit and repeating circle: the other pendulum was my own.

In entering on so extensive a series of experiments, the value of which would depend entirely on their relation to each other being strictly maintained throughout, I could not but feel great anxiety in anticipating the possibility of an injury taking place, in the course of so many embarkations and disembarkations, to some essential part of the apparatus, and pre-eminently to the pendulum; it is unquestionably a great disadvantage attendant on the method of proceeding with invariable pendulums, in such an inquiry as the present, that an accident befalling the pendulum at any period of the experiments, not only interrupts their continuity, but may prossibly render the whole previous labour of no avail; in this respect, and it is one which cannot fail to occasion continual and excessive solicitude, the method of Borda, pursued in France, is much its superior, as each determination obtained by it is in itself complete, and is final; it is a superiority, however, by no means uncompensated even in sources of anxiety: by providing myself with a second pendulum, and by employing both pendulums at every station, I hoped to avoid the inconveniences, whilst I should avail myself of the advantages of the method, which after much consideration I was induced to prefer.

The construction of the pendulums was precisely similar to that of the one employed by Captain Kater, in determining the variation in the length of the pendulum, at the principal stations of the trigonometrical survey of Great Britain; each pendulum was composed of a bar of plate brass 1.6 inch wide, and less than the eighth of an inch thick; a strong cross piece of brass was rivetted and soldered to the top, having a prism of wootz, passing through a triangular hole in the bar, firmly secured to it by screws of which the heads were sunk; the angle of the prism on which the vibrations were performed, and which is usually called a knife edge, was ground to about 120 degrees it was fixed as nearly as could be done at right angles to the bar; the length of the bar from the knife edge to the extremity was about five feet, but a flat circular weight, nicely turned and pierced in the direction of its diameter to receive the bar, was soldered at such a distance from the knife edge which served as the point of suspension, that the pendulum made two vibrations less in about eleven minutes than a clock keeping nearly mean time; the part of the bar which was below the weight was reduced to the width of 0.7 of an inch, and was varnished black to be more distinctly visible, in the observation of coincidences, in contrast with the disk of white paper on the pendulum of the clock; the bar terminated in an angle, the point of which served to indicate the magnitude of the vibration on an arc, divided into degrees and tenths, which fitted into the opening of the door of the clock-case.

The frame on which the pendulum was supported was of cast iron; it was designed to be fixed to the same wall with the clock, and at such a distance above it, as would allow the end of the invariable pendulum to reach a little below the disk on the pendulum of the clock; the upper and horizontal part of the frame was nearly square, and stood out from the wall about twenty inches, in order to carry the pendulum clear of the clock-case, when it should be suspended in front of the clock: the side of the square which joined the wall was pierced with three holes, one in the middle and the others at the extremities, to receive screws of nearly five inches in length, which were intended to work into pickets of wood driven into the wall; the horizontal part of the frame received additional support in brackets firmly screwed to its under side having strips of sheet lead between; the brackets were so contrived as to spread at the bottom to the width of three feet, the more effectually to oppose any lateral motion arising from the vibration of the pendulum; they were fastened to the wall at their lower extremities by screws, similar to those already described; the weight of the frame exceeded ninety pounds.

The planes on which the knife edge immediately rested were of hard Brazil pebble, and were fixed in a brass frame which screwed firmly to the iron support with sheet lead intermediately; the frame was furnished with three adjusting screws, by which the planes could be rendered truly horizontal, as shewn by a moveable level placed on them for that purpose; it had also the usual contrivance of Y's moveable by a screw, to raise the pendulum when not in use, and to ensure its being lowered on exactly the same part of the planes on every occasion.

The pendulums were numbered 3 and 4, and are distinguished accordingly.

A barometer, by Newman, and two thermometers, by Jones, having their scales sufficiently large to admit a fair estimation of tenths of a degree.

A small telescope, for the observation of coincidences, completed the apparatus connected with the detached pendulums which was carried to the stations in the neighbourhood of the Equator, where the accommodation of a house with stone walls for the support of the clock and pendulum frame could be depended on; the provision which was subsequently made for the same purpose in the northern stations, where no such accommodation existed, will be hereafter described.

I shall now proceed to detail the operations with the detached pendulums, at the several stations, as they were successively visited.

SIERRA LEONE.

The Iphigenia arrived at Sierra Leone on the 22d of February; I had had the advantage of being previously known in England to His Excellency Brigadier-General Sir Charles Mac Carthy, Governor of the British possessions in Western Africa, who was resident at Sierra Leone, as the seat of government; and in the expectation of having to accomplish the present purpose at some station within his command, I had received from him the assurance of every assistance in his power; I had now the pleasure of finding this assurance most amply realized, accompanied with the utmost personal kindness.

On examination of the few stone buildings which have been as yet constructed at Freetown, the officers' range of barracks at Fort Thornton appeared on the whole the most eligible situation for the experiments, being convenient and suitable in every respect, excepting in the height above the sea, which was somewhat greater than could have been wished, but was a minor consideration in comparison with the other circumstances that determined the preference. Apartments in the barracks were accordingly vacated and assigned me, being the Captain's quarter at the north end of the range, immediately adjoining to and opening into the north-west bastion: to these apartments the instruments were removed, as soon as the necessary measures could be taken for their disembarkation and conveyance.

Sir Robert Mends having assigned the Pheasant sloop of war for the further prosecution of the service in which I was engaged, it was arranged that Captain Clavering should return from a cruize to the harbour of Sierra Leone, in the first week in April, by which time I engaged to have completed the experiments, and to be ready to embark.

Captain Clavering was so kind as to land two of the marines of the Pheasant's complement, and to leave them with me as orderlies; an arrangement which proved of material service to me.

The end wall of the range of barracks, which I had designed for the pendulum and clock supports, was above three feet thick, composed of irregularly shaped masses of granite imbedded in a strong cement; I had intended to have had a sufficient number of holes bored in the wall, and at the proper distances, to have received separate plugs of wood for each screw; but after much perseverance and labour, the attempt was obliged to be abandoned, in consequence of the extreme hardness of the granite, which occasioned the stones to be loosened with the force necessary to cut them: the supports were finally screwed to four-inch planks, fixed to the wall by eight-inch spikes driven into the cement, the planks being unconnected with the floor; the supports were not less firmly attached by this method, than if they had been fixed to the wall itself, but the several screws of the clock and of the pendulum support, on one side, were fastened into the same plank, an arrangement with which I was hardly so well satisfied at the time, as I had afterwards reason to believe that I might have been.

Glass windows not being in use at Sierra Leone, it was with some difficulty that a sufficient number of panes were obtained, which should give the necessary light without admitting currents of air; the room in other respects was closed, and entered only for the purposes of observation; the telescope, for the determination of coincidences, was placed at nine feet six inches from the front of the clock case, which distance was preserved at all the subsequent stations.

By the kindness of Thomas Stuart Buckle, Esq., Surveyor and Civil Engineer of the Colony, to whose assistance I was materially indebted on this and other similar occasions, a very firm support of masonry was erected for the transit instrument, on the parapet of the fort, adjoining the bastion. On unpacking the transit, I had the mortification to find that the female screw, into which one of the collimating screws worked, had become unsoldered from the stop; its repair occasioned some little delay, and might have been more difficult, and have caused further accidents of more importance, if the system of wires had been of cobweb as is usual; but the lines cut on glass are much less liable to injury: the delay, however, and the inconvenience, might have been avoided altogether, if the female screw had been made in the stop itself, cast in a proper shape for that purpose, instead of being in a detached piece requiring to be soldered to the stop; the repair being completed, and the collimation perfect, the instrument was placed in the meridian, or at least in a sufficient approximation to it, on the evening of the 6th of March, by transits of Capella and Rigel, and proved by those of Castor, Procyon, and Pollux; and marks were taken for its subsequent adjustment and verification.

The rate of the clock appearing by the transits of the 7th, 8th, and 9th, to be sufficiently steady, the observation of coincidences was commenced with pendulum 3 on the morning of the 10th, and continued in the fore and afternoons until the 14th, when the number of observations, and their agreement with each other, being considered sufficient, pendulum 4 was substituted on the support, with which a similar series was completed between the 10th and 25th.

Previously to my quitting England, the rates of the pendulums had been ascertained in London, the details of which, reserved for the present, will be found in a subsequent part of the volume; as it has been deemed a preferable arrangement, to connect together, in one view, the several experiments made with the pendulums in London at different periods.

On examining the rates now obtained at Sierra Leone with those of London, I was much surprised and disturbed to find that the retardation was several seconds less than I had anticipated; as I had left England in the expectation, founded on the inferences drawn by Monsieur La Place, that the variation of gravity would accord, at least nearly, with an Ellipticity of about $\frac{1}{300}$. My first impression was to suspect error in the experiments; but on due consideration of every circumstance connected with them, I could discover no reasonable ground of distrusting the general applicability of the mode of experiment, or of supposing any peculiarity in the present instance, which could render the results doubtful beyond a small fraction of a second; had I been furnished with one pendulum only, I might have supposed that it had received some injury in the intermediate voyage; but the results of the two pendulums so nearly accorded as to render such a supposition inadmissible, and I now felt the value of this double provision, in the additional confidence which it authorised: I had already reason to believe that the thermometers with which I had been supplied were not very accurately graduated for tropical temperatures, but error from this source could produce but a very small portion of the difference existing between the experiments and my previous expectation: register thermometers, in different parts of the room, shewed that the extremes of temperature in twenty-four hours, did not vary so much as three degrees; and pieces of floss silk, suspended in the neighbourhood of the pendulum, manifested the absence of currents of air of sufficient note to influence its motion: the horizontality of the agate supports had been verified continually, and was as perfect at the close as at the commencement: fortunately a few days yet remained before the Pheasant would return, and I determined to employ them in putting up the apparatus afresh, and in a new place, and in effecting an entire repetition of the experiments. For this purpose I availed myself of the eastern wall of the same room, which was sufficiently substantial, and being built of brick, permitted a separate picket of oak to be driven into it for each screw, whereby a communication between the detached pendulum and that of the clock was rendered even more improbable than before: the preparations being completed, the observation of coincidences with No. 3 was commenced on the morning of the 27th, and ended on the evening of the 29th; and commenced with No. 4 on the morning of the 31st., and ended on the evening of the 2d of April.

The results of the second series accorded so closely with those of the preceding as to be in effect identical, a circumstance not only highly satisfactory, in increasing the dependence which may be placed on the experiments at Sierra Leone, but also important, in giving additional confidence to the mode of experiment generally; as identity on repetition renders highly improbable the existence of interfering causes dependant on accident; and strengthens the conclusion, that the rate of a detached pendulum does indeed afford a just relative measure of the force of gravity at the place of observation.

Being now fully persuaded that the true rates were obtained at Sierra Leone, the pendulums were repacked for embarkation; the experiments having produced an impression, either that differences in the density of materials near the surface and in the neighbourhood of the pendulum station, have more influence on the rates than had been anticipated, or that the Compression of the earth was itself greater than was generally supposed; and consequently that, in either case, it would be desirable to repeat the experiments at a greater number of stations than had been previously designed.

Mr. Buckle was kind enough to ascertain the height of the pendulums above half tide to be 190 feet, by levelling to the water side; the slope was gradual towards the sea, and the fort stands on the highest ground in its own immediate neighbourhood, except a small hill on which a martello tower is built, at a distance rather exceeding a quarter of a mile:

the situation of Freetown, however, may be more generally stated to be at the foot, on the northern side, of the range of mountains, which coming from the interior, finds here its termination in the sea, and gives the name to the cape, harbour, and colony of Sierra Leone; the general height of the range, so far as it has been yet explored, is from two to three thousand feet; the principal geological feature, in the neighbourhood of Sierra Leone, is a red granite, of easy and rapid decomposition.

The subjoined tables comprise the detail of the observations of which the history has been thus related: Table I. contains the times of transit of stars, by which the rate of the clock was ascertained; the transits were observed by the chronometer, of which the times are entered in the table, and consequently the record is direct; they were noted by the beats without an assistant; the transit instrument being without shelter, and the sun nearly vertical, the observations were confined to stars, and to such as passed the meridian about sunset; the telescope was removed into the house during the remainder of the twenty-four hours from motives of precaution, the pillars only being stationary. Table II. is annexed, as affording evidence, that by means of the meridian marks, the telescope was adjusted throughout to the same vertical plane. Table III. contains the daily rate of the chronometers, deduced from the preceding transits: and Table IV. the comparisons of the chronometer and clock at exact intervals each of twelve hours of mean time, whereby the clock's rate on mean solar time was also obtained. The stoppage of the clock, on the 15th of March, took place for the purpose of making a small alteration in the position of the telescope for observing coincidences, when the Series with No. 3 was completed and the pendulums changed, as the telescope adjusted for the one pendulum was not precisely in the prolongation of the line, connecting the disk on the clock pendulum and the tail piece of the other detached pendulum; the stoppage of the clock, however, being inconvenient on many accounts, was dispensed with at the subsequent stations, it being found equally easy, with a little practice, to adjust the telescope when both pendulums were in motion, as when they were at rest.

Tables V. and VI. comprise an account of the coincidences in the double series with each pendulum. In these tables the times of disappearance of the disk and also of its re-appearance are registered, and a mean between the two is deemed and entered as the true time of coincidence; the "time of disappearance" is the second which immediately follows the first passage of the disc in entire obscuration; the "time of re-appearance" is the second immediately preceding the re-appearance of the first portion of the disk, when passing the opening in the diaphragm of the telescope. As each result is obtained from a mean of ten successive intervals, the first and eleventh coincidence only are recorded, being those alone which are influencial on the deduction. The two last columns are added for the purpose of shewing the accordance of the particular results, when reduced to a mean temperature; a degree of Fahrenheit's scale being considered as equivalent to 0.42 of a vibration in twenty-four hours.

The correction for the arc which added to the observed number of vibrations in twenty-four hours, shews what they would have been in an arc infinitely small, has been computed agreeably to the formula for that purpose, given and demonstrated by Mr. Watts, in the seventeenth article of the second number of the *Edinburgh Philosophical Journal.* If N be the number of vibrations in twenty-four hours in circular arcs, and a and b the commencing and concluding arcs, the corrected number of vibrations will be $N + \frac{N(a+b)(a-b)}{241886.08 \log\left(\frac{a}{b}\right)}$

Table I. TRANSITS OBSERVED AT SIERRA LEONE.

1822.	STARS.	TIMES OF TRANSIT BY THE CHRONOMETER 423.					Mean by the Chronometer.
		1st Wire.	2nd Wire.	Meridian Wire.	4th Wire.	5th Wire.	
		M. S.	M. S.	H. M. S.	M. S.	M. S.	H. M. S.
Mar. 6	Capella ...	55 25.6	56 03.2	6 56 40.8	57 18.4	57 56	6 56 40.8
,,	Rigel	...	...	6 59 03.6	...	...	6 59 03.6
,,	Castor	14 56.4	15 27.2	9 15 58.4	16 29.6	17 00.8	9 15 58.47
,,	Procyon ...	21 48.4	22 14.8	9 22 41.2	23 07.6	23 34	9 22 41.2
,,	Pollux	26 08.4	26 38	9 27 07.6	27 37.6	28 07.2	9 27 07.73
9.	Capella ...	43 40	44 17.6	6 44 55.2	45 33.2	...	6 44 55.3
,,	Rigel	...	46 52	6 47 18.4	47 44.8	48 11.2	6 47 18 4
,,	ε Orionis....	7 34.4	8 00.4	7 08 26.4	8 53.2	9 19.2	7 07 26.67
,,	ζ Orionis....	12 09.2	12 35.6	7 13 02	13 28.4	13 54.8	7 13 02
,,	κ Orionis....	19 40	20 10.4	7 20 32.8	20 58.8	21 25.2	7 20 32.67
,,	α Orionis....	25 52.8	26 19.2	7 26 45.6	27 12	27 38	7 26 45.53
,,	δ Canis Maj..	...	...	8 42 09.2	42 38.4	43 07.2	8 42 09.2
,,	λ Gemini....	47 57.6	48 24.8	8 48 52	49 20	49 47.2	8 48 52.27
,,	Castor.....	3 10.8	3 41.6	9 04 12.4	4 43.6	5 14.8	9 04 12.6
,,	Procyon ...	10 02.8	10 28.8	9 10 55.2	11 22	11 48	9 10 55.33
,,	Pollux	14 22	14 52	9 15 21.6	15 52	16 21.6	9 15 21.8
,,	ξ Navis	21 44.4	22 13.2	9 22 42	23 10.8	23 39.6	9 22 42
,,	e Navis	29 08.4	29 36.8	9 30 05.2	30 33.6	31 01.6	9 30 05.13
10	Capella ...	39 42.6	40 23.2	6 41 00.8	41 38.8	42 16.4	6 41 00.93
,,	Rigel	...	...	6 43 24	...	...	6 43 24
,,	ε Orionis....	3 40.4	4 06.4	7 04 32.4	4 58.4	5 24.4	7 04 32.4
,,	ζ Orionis....	8 15.2	8 41.6	7 09 07.6	9 34	10 00	7 09 07.67
,,	κ Orionis....	15 45.6	16 12	7 16 38.4	17 04.8	17 31.2	7 16 38.4
,,	α Orionis....	21 58.4	22 24.8	7 22 51.2	23 17.6	23 44	7 22 47.2
11.	Capella ...	35 50.8	36 28.4	6 37 06	37 44	38 21.6	6 37 06.13
,,	Rigel	...	39 02.8	6 39 29.2	39 55.6	40 22	6 39 29.2
,,	ε Orionis....	59 45.2	00 11.2	7 00 37.6	01 04	01 30	7 00 37.6
,,	ζ Orionis....	04 20.4	04 46.8	7 05 12.8	05 39.2	06 05.2	7 05 12.87
,,	κ Orionis....	11 50.4	12 17.2	7 12 43.6	13 10.4	13 36.8	7 12 43.67
,,	α Orionis....	18 03.6	18 30	7 18 56.4	19 22.8	19 49.2	7 18 56.4

TRANSITS OBSERVED AT SIERRA LEONE, *continued.*

1822.	STARS.	TIMES OF TRANSIT BY THE CHRONOMETER 423.					Mean by the Chronometer.
		1st Wire.	2nd Wire.	Meridian Wire.	4th Wire.	5th Wire.	
		M. S.	M. S.	H. M. S.	M. S.	M. S.	H. M. S.
Mar. 14	Capella ...	24 05.6	24 43.6	6 25 20.8	25 58.8	....	6 25 21
,,	Rigel.....	...	27 17.6	6 27 44	28 10.4	...	6 27 44
,,	ε Orionis....	...	48 26	6 48 52 4	49 18.8	....	6 48 52.4
,,	ζ Orionis....	52 35.2	53 01.2	6 53 27.6	53 53.6	54 19.6	6 53 27.47
,,	κ Orionis....	00 05.2	00 32	7 00 58.4	01 24.8	01 51.6	7 00 58.4
,,	α Orionis....	06 18.4	06 44.8	7 07 11.2	07 37.6	08 04.4	7 07 11.27
,,	δ Canis Maj..	21 36.4	22 05.6	8 22 34.4	23 03.2	23 32.4	8 22 34.4
,,	λ Gemini....	28 24	28 51.2	8 29 18	29 45.2	30 12.4	8 29 18.13
,,	Castor	43 36.4	44 07.2	8 44 38	45 08.8	45 39.6	8 44 38
,,	Procyon ...	50 27.6	50 54	8 51 20.8	51 46.8	52 13.2	8 51 20.53
,,	Pollux	54 47.6	55 17.2	8 55 47.2	56 16.8	56 46.4	8 55 47.07
,,	ξ Navis.....	02 10	02 38.4	9 03 07.2	03 36	04 04.8	9 03 07.27
,,	e Navis.....	09 33.2	10 02	9 10 30.4	10 58.8	11 27.2	9 10 30.33
19	Canopus...	20 29	21 11.6	7 21 55	22 38.6	23 20.8	7 21 55
,,	γ Gemini....	28 27.2	28 54.4	7 29 21.6	29 48.8	30 16	7 29 21.6
,,	Sirius.....	38 18	38 44.4	7 39 11.6	39 38.8	40 05.6	7 39 11.67
,,	δ Canis Maj..	...	...	8 02 58.2	...	...	8 02 58.2
,,	λ Gemini....	8 46.8	9 14.4	8 09 41.6	10 08.4	10 36	8 09 41.47
,,	Castor	23 59.6	24 30.8	8 25 01.6	25 32.4	26 04	8 25 01 67
,,	Procyon ...	30 52	31 18	8 31 44	32 10.4	32 36.8	8 31 44.2
,,	Pollux	35 11.2	35 40.8	8 36 10.8	36 40.4	37 10	8 36 10.67
,,	ξ Navis.....	42 34	43 02.8	8 43 31.2	44 00	44 28.4	8 43 31.27
21	Canopus...	12 36.8	13 20.4	7 14 04	14 47.6	15 31.2	7 14 04
,,	γ Gemini....	20 36.8	21 04	7 21 31.2	21 58.4	22 25.6	7 21 31.2
22	Canopus...	8 43.2	9 26.4	7 10 09.6	10 52.8	11 36	7 10 09.6
,,	γ Gemini....	...	16 08.8	7 17 36.4	18 03.8	....	7 17 36.35
24	α Orionis....	27 07.6	27 34	6 28 00.4	28 26.8	28 53.2	6 28 00.4
,,	Canopus...	00 54	01 36.8	7 02 20	03 03.2	03 46.8	7 02 20.17
,,	γ Gemini....	08 53.2	09 20.4	7 09 47.6	10 14.8	10 42	7 09 47.6
,,	Sirius.....	18 42.8	19 10	7 19 37.6	20 04.8	20 32	7 19 37.47

TRANSITS OBSERVED AT SIERRA LEONE, *continued*.

1822.	STARS.	TIMES OF TRANSIT BY THE CHRONOMETER 423.					Mean by the Chronometer.
		1st Wire.	2nd Wire.	Meridian Wire.	4th Wire.	5th Wire.	
		M. S.	M. S.	H. M. S.	M. S.	M. S.	H. M. S.
Mar. 24	δ Canis Maj..	42 25.6	42 54.4	7 43 23.8	43 53.2	44 22	7 43 23.8
,,	λ Gemini....	48 12.4	49 39.6	7 50 07.2	50 34.4	51 01.6	7 50 03.07
,,	Castor	04 25.6	04 56.4	8 05 27	05 58.4	06 29.2	8 05 27.27
,,	Procyon ...	11 18	11 44	8 12 10	12 36	13 02.4	8 12 10.07
,,	Pollux	15 36.8	16 06.8	8 16 36.4	17 06.4	17 36.4	8 16 36.53
,,	ξ Navis	22 59.2	23 28	8 23 56.8	24 25.6	24 54.4	8 23 56.8
26	γ Gemini....	01 03.6	01 30.8	7 01 58	02 25.2	02 52.4	7 01.58
,,	Sirius.....	10 52.8	11 20	7 11 47.2	12 14.4	12 41.6	7 11 43.2
,,	δ Canis Maj..	34 36.4	35 05.2	7 35 34.4	36 03.6	36 32.4	7 35 34.4
,,	λ Gemini....	41 22.4	41 50	7 42 17.2	42 44.8	43 12.4	7 42 17.33
29	α Orionis....	7 31.6	7 58	6 08 24.4	8 50.8	9 17.6	6 08 24.46
,,	Canopus....	41 18	42 00.8	6 42 44	43 27.2	44 10	6 42 44
,,	γ Gemini....	49 17.6	49 44.8	6 50 12	50 39.6	51 06.8	6 50 12.13
,,	Sirius.....	59 06.8	59 34	7 00 01.4	00 28.8	00 56	7 00 01.4
,,	δ Canis Maj..	22 50.4	23 19.6	8 23 48.8	24 18	24 47.2	7 23 48.8
,,	λ Gemini....	29 37.6	30 04.8	7 30 32	30 59.2	31 26.8	7 30 32.07
April 2	Canopus..	25 36.8	26 20	6 27 03.2	27 46.4	28 29.6	6 27 03.2
,,	γ Gemini....	33 25.8	34 03	6 34 30.2	34 57.6	35 24.8	6 34 30.2
,,	Sirius.....	43 25.6	43 52.4	6 44 20	44 47.6	45 15.2	6 44 20.13
,,	δ Canis Maj..	07 08.4	07 37.6	7 08 06.8	08 36	09 05.2	7 08 06.8
,,	λ Gemini....	13 55.2	14 22 4	7 14 50	15 17.2	15 44.8	7 14 49.87

TABLE II.

SIERRA LEONE.

DEVIATION of the **TRANSIT INSTRUMENT** from the **MERIDIAN**, on the under-mentioned Days of March, 1822, as shewn by the Interval between the Transits of Stars, differing considerably in Declination, but having nearly the same Right Ascension.

STARS.	Differences.		INTERVAL BETWEEN THE TRANSITS.							Mean Interval.		South Star too soon.	Deviation of the Instrument from the Meridian.
	In R.A.	In Declin.	6th.	9th.	10th.	11th.	14th.	19th.	24th.	Solar.	Sidereal.		
	M. S.	°	M. S.	M. S.	M. S.	M. S.	M. S.	M. S.	M. S.	M. S.	M. S.	S.	S.
Capella and Rigel.....	2 25.5	54 13.25	2 22.8	2 23.1	2 23.07	2 23.07	2 23	......	......	2 23.01	2 23.4	2.1	1.8
Castor and Procyon ..	6 44.75	26 35.75	6 42.73	6 42.73	......	......	6 42.53	6 42.53	6 42.8	6 42.66	6 43.76	0.99	1.88
Procyon and Pollux....	4 26.35	22 46.5	4 26.53	4 26.47	......	......	4 26.54	4 26.47	4 26.46	4 26.49	4 27.22	0.87	1.98
Mean Deviation of the Transit Instrument to the East, when pointed to the South Horizon......................													1.89

TABLE III. SIERRA LEONE.—DEDUCTION of the RATE of the Chronometer No. 423 from TRANSITS; between the 9th of March and the 2d of April, 1822.

STARS.	9 to 10	10 to 11	11 to 12	12 to 13	13 to 14	14 to 19	19 to 20	20 to 21	21 to 22	22 to 23	23 to 24	24 to 26	26 to 27	27 to 28	28 to 29	29 to 30	30 to 31	31 to Ap. 1	1 to 2
	s.	s.	s.	s.	s.	s.	s.	s.	s.	s.	s.	s.	s.	s.	s.	s.	s.	s.	s.
Capella...	1.54	1.11	0.67	0.87	0.87	..	..	..	..	..	..	..	..	..	..	..	..	..	..
Rigel	1.41	1.11	0.84	0.84	0.84	..	..	..	..	..	..	..	..	..	..	..	..	..	..
ι Orionis...	1.64	1.11	0.84	0.84	0.84	..	..	..	..	..	..	..	..	..	..	..	..	..	..
ζ Orionis...	1.58	1.11	0.78	0.78	0.78	..	..	..	..	..	..	..	..	..	..	..	..	..	..
κ Orionis...	1.64	1.18	0.82	0.82	0.82	..	..	..	..	..	..	..	..	..	..	..	..	..	..
α Orionis...	1.58	1.11	0.87	0.87	0.87	..	..	..	..	..	..	0.72	0.72	0.72	0.72	..	..	..	..
Canopus..	..	..	..	..	..	..	0.77	0.77	0.77	1.2	1.2	0.68	0.68	0.68	0.68	0.71	0.71	0.71	0.71
γ Gemini...	..	..	..	..	..	..	0.71	0.71	1.06	1.53	1.53	1.11	0.62	0.62	0.62	0.43	0.43	0.43	0.43
Sirius	..	..	..	..	..	..	1.06	1.06	1.06	1.06	1.06	1.01	0.64	0.64	0.64	0.59	0.59	0.42	0.42
δ Canis	0.95	0.95	0.95	0.95	0.95	0.67	1.03	1.03	1.03	1.03	1.03	1.21	0.7	0.7	0.7	0.42	0.42	0.59	0.59
λ Gemini...	1.09	1.09	1.09	1.09	1.09	0.56	1.04	1.04	1.04	1.04	1.04	1.04	0.69	0.69	0.69	0.46	0.46	0.46	0.46
Castor....	0.99	0.99	0.99	0.99	0.99	0.64	1.03	1.03	1.03	1.03	1.03	..	..	..	..	..	..	..	..
Procyon..	0.96	0.96	0.96	0.96	0.96	0.64	1.04	1.04	1.04	1.04	1.04	..	..	..	..	..	..	..	..
Pollux....	0.95	0.95	0.95	0.95	0.95	0.63	1.1	1.1	1.1	1.1	1.1	..	..	..	..	..	..	..	..
ξ Navis	0.98	0.98	0.98	0.98	0.98	0.71	1.02	1.02	1.02	1.02	1.02	..	..	..	..	..	..	..	..
e Navis	0.98	0.89	0.98	0.98	0.98	..	..	..	..	..	..	..	..	..	..	..	..	..	..
MEANS— Gaining per Diem......	1.25	1.05	0.92	0.92	0.92	0.64	0.98	0.98	1.02	1.12	1.12	0.96	0.67	0.67	0.67	0.52	0.52	0.52	0.52
	1.01 Sidereal = 1.013 Solar						1.044 Sidereal = 1.047 Solar						0.67 Sid. = 0.673 S.			0.52 Sidereal = 0.523 Sol.			

Table IV. SIERRA LEONE.——Comparisons of the Astronomical Clock with the Chronometer No. 423, from the 9th to the 24th of March, 1822; with the Clock's Rate on Mean Solar Time deduced.

1822.	Chronometer.	Clock.	Clock's Loss on 423.		Daily Rates. Chron.	Daily Rates. Clock.
	H. M. S.	H. M. S.	s.	s.	Gaining. s.	Losing. s.
March 9 P. M.	6 30 00	6 10 21.4				
" 9 P. M.		9 40 03 1				
" 10 A. M.		9 39 00.9	124.7			
" 10 P. M.		9 37 58.4				
" 11 A. M.		9 36 56	125			
" 11 P. M.		9 35 53.4				
" 12 A. M.		9 34 51.2	124.7	124 52	1.01	123.51
" 12 P. M.		9 33 48.7				
" 13 A. M.		9 32 46.8	124.1			
" 13 P. M.		9 31 44.6				
" 14 A. M.		9 30 42.6	124.1			
" 14 P. M.	10 00 00	9 29 40.5				
" 19 P. M.		9 49 43				
" 20 A. M.		9 48 41	124.1			
" 20 P. M.		9 47 38.9				
" 21 A. M.		9 46 37	123.6			
" 21 P. M.		9 45 35.3				
" 22 A. M.		9 44 33.5	123.6	124.12	1.05	123.07
" 22 P. M.		9 43 31.7				
" 23 A. M.		9 42 29.8	124 3			
" 23 P. M.		9 41 27.4				
" 24 A. M.		9 40 24.9	125			
" 24 P. M.		9 39 22.4				

SIERRA LEONE.—Comparisous of the Astronomical Clock with the Chronometer No. 423, from the 26th of March to the 2d of April, 1822; with the Clock's Rate on Mean Solar Time deduced.

1822.	Chronometer.	Clock.	Clock's Loss on 423.		Daily Rates. Chron.	Daily Rates. Clock.
	H. M. S.	H. M. S.	s.	s	Gaining. s.	Losing. s.
March 26 P. M.	10 00 00	9 48 41.6				
" 27 A. M.		9 47 40	123.1			
" 27 P. M.		9 46 38.5				
" 28 A. M.		9 45 37	122.9	123.067	0.673	122.39
" 28 P. M.		9 44 35.6				
" 29 A. M.		9 43 33.8	123.2			
" 29 P. M.		9 42 32.4				
" 30 A. M.		9 41 31	122.6			
" 30 P. M.		9 40 29.8				
" 31 A. M.		9 39 28.5	123.3			
" 31 P. M.		9 38 26.5		123.225	0.523	122.7
April 1 A. M.		9 37 24.4	123.8			
" 1 P. M.		9 36 22.7				
" 2 A. M.		9 35 21	123.2			
" 2 P. M.		9 34 19.5				

TABLE V. SIERRA LEONE.—COINCIDENCES OBSERVED with PENDULUM 3.

1822. MARCH.	Barometer.	No. of Coincidence.	Temperature.	Time of Disappearance.	Time of Re-appearance.	True Time of Coincidence.	Arc of Vibration.	Mean Temperature.	Mean Interval.	Correction for the Arc.	Vibrations in 24 hours.	Reduction to a mean Temperature.	Reduced Vibrations at 81.3° Faht.
First Series; Clock losing 123.51 seconds per diem.													
			°	M. S.	M. S.	H. M. S.	°	°	S.	+ S.			
10. A.M.	29.865	1	80.	19 38	19 40	8 19 39	1.22	80.4	644.85	1.35	86010.25	−0.38	86009.87
		11	80.8	07 03	07 12	10 07 07.5	0.64						
10. P.M.	29 835	1	81.6	35 17	35 20	2 35 18.5	1.2	82.15	643.6	1.33	86009.75	+0.35	86010.16
		11	82 7	22 28	22 41	1 22 34 5	0.63						
10. P.M. (by lamp light.)	29.850	1	81.2	10 39	10 41	8 10 40	1.27	81.2	643.4	1.49	86009.81	−0 04	86009.77
		11	81.2	57 45	58 03	9 57 54	0.67						
11. A.M.	29.885	1	80.2	31 09	31 10	8 31 09.5	1.29	80.6	644.35	1.53	86010.23	−0.29	86009.94
		11	81.	18 28	18 38	10 18 33	0.68						
11. P.M.	29.850	1	80.8	51 34	51 42	3 51 38	1.17	80.8	644.	1.31	86009.85	−0.21	86009.64
		11	80.8	38 49	39 07	5 38 58	0.61						
12. A.M.	29.900	1	80.5	29 45	29 46	8 29 45.5	1 25	81.05	643.7	1.45	86009.89	−0.11	86009.78
		11	81.6	16 57	17 08	10 17 02.5	0.66						
12. P.M.	29.850	1	82.3	00 59	01 02	3 01 00.5	1.22	82.6	642.7	1.35	86009.37	+0.55	86000.92
		11	82.9	48 03	48 12	4 48 07.5	0.64						
13. A.M.	29.760	1	80.6	20 02	20 04	7 20 03	1.22	80.65	644.25	1.35	86010.01	−0.27	86009.74
		11	80.7	07 21	07 30	9 07 25.5	0.64						
13. P.M.	29.880	1	82.9	27 04	27 06	2 27 05	1.2	82.9	642.35	1 33	86009 19	+0.67	86009.86
		11	82.9	14 04	14 13	4 14 08.5	0.63						
14. A.M.	29.870	1	80.8	52 15	52 17	8 52 16	1.18	81.8	643.25	1 31	86009.57	+0.21	86009.78
		11	82.8	39 26	39 30	10 39 28	0.62						
Second Series; Clock losing 122.39 seconds per diem.													
27. A.M.	29.860	1	79.8	49 30	49 31	9 49 30.5	1.25	79.95	642.1	1.45	86010.29	−0.57	86009.72
		11	80.1	36 26	36 37	11 36 31.5	0.66						
27. P.M.	29.800	1	81.2	22 47	22 51	2 22 49	1.22	81.6	640.8	1.35	86009 67	+0.13	86009.80
		11	82.	09 31	09 43	4 09 37	0 64						
29. A.M.	29.800	1	80.4	30 11	30 13	9 30 12	1.29	80.6	641.35	1.53	86010.07	−0.29	86009.78
		11	80.8	17 00	17 11	11 17 05 5	0.68						
29. P.M.	29.750	1	82.	43 05	43 08	2 43 06.5	1.25	82.1	640.15	1.45	86009.51	+0.34	86009.85
		11	82.2	29 43	29 53	1 29 48	0.66						
Means	29.840							81.3			86009.82		86009.82

TABLE VI. SIERRA LEONE.—COINCIDENCES OBSERVED with PENDULUM 4.

1822. March & April.	Barometer.	No. of Coincidence.	Temperature.	Time of Disappearance.	Time of Re-appearance.	True Time of Coincidence.	Arc of Vibration.	Mean Temperature.	Mean Interval between the Coincidences.	Correction for the Arc.	Vibrations in 24 hours.	Reduction to a mean Temperature.	Reduced Vibrations at 81.75 Faht.
First Series; Clock losing 123.07 seconds per diem.													
	IN.		°	M. S.	M. S.	H. M. S.	°	°	S.	+ S.			
Mar. 20 A.M.	29.842	1	82.1	43 48	43 54	8 43 51	1.18	82.1	666.	1.31	86019.13	+0.15	86019.28
		11	82.1	34 44	34 58	10 34 51	0.62						
,, 20 P.M.	29.920	1	80.8	00 03	00 04	3 00 03.5	1.29	81.1	666.85	1.53	86019.67	−0.27	86019.40
		11	81.4	51 06	51 18	4 51 12	0.68						
,, 21 A.M.	29.88	1	81.7	41 16	41 20	8 41 18	1.28	81.8	665.95	1.51	86019.33	+0.02	86019.35
		11	81.9	32 11	32 24	10 32 17.5	0.67						
,, 22 A.M.	29.912	1	79.6	19 59	20 02	6 20 00.5	1.28	79.85	667.9	1.51	86020.09	−0.80	86019.29
		11	81.1	11 13	11 26	8 11 19.5	0.67						
,, 22 P.M.	29.800	1	82.3	42 09	42 14	2 42 11.5	1.27	82.85	665.4	1.49	86019.09	+0.46	86019.55
		11	83.4	31 59	32 12	4 32 05.5	0.66						
,, 23 A.M.	29.820	1	81.3	51 57	52 00	7 51 58.5	1.27	81.75	666.1	1.49	86019.37	. . .	86019.37
		11	82.2	42 53	43 06	9 42 59.5	0.66						
,, 24 P.M.	29.84	1	83.6	7 40	7 46	3 07 43	1.2	83.35	665.	1.33	86018.77	+0.67	86019.44
		11	83.1	58 24	58 42	4 58 33	0.63						
,, 25 A.M.	29.85	1	81.3	47 32	47 38	9 47 35	1.2	81.95	666.35	1.33	86019.17	+0.08	86019.25
		11	82.6	38 31	38 46	11 38 38.5	0.63						
Second Series; Clock losing 122.7 seconds per diem.													
,, 31 A.M.	29.860	1	81.4	06 00	06 03	10 06 01.5	1.28	81.7	665.15	1.51	86019.37	−0.02	86019.35
		11	82	56 46	57 00	11 56 53	0.67						
April 1 P.M.	29.83	1	82	26 18	26 23	2 26 20.5	1.23	82.2	665.15	1.36	86019.22	+0.19	86019.41
		11	82.4	17 05	17 19	4 17 12	0.64						
,, 2 A.M.	29.860	1	80.2	02 54	02 59	7 02 56.5	1.25	80.6	666.55	1.44	86019.88	−0.48	86019.40
		11	81	53 54	54 11	8 54 02.5	0.64						
Means.	29.856							81.75			86019.37		86019.37

ISLAND OF ST. THOMAS.

The Pheasant having occasion to touch at the settlements on the Gold Coast on her passage, did not arrive at St. Thomas's until the 15th of May, on which day she anchored in the harbour of Santa Anna de Chaves. The Island being a Portuguese possession, I had been furnished with letters, obtained by the Admiralty through the Secretary of State for Foreign Affairs, from the Marquess de Souza, Ambassador in London, to the Governor of the Island, Senor Joao Baptiste, explanatory of the purpose for which it was proposed that I should visit his command, and requesting the good offices of the authorities in its promotion. On anchoring, we were attended by Mr. John Fernandez, supercargo of a British merchant ship in the harbour, who had received from Sir Robert Mends a temporary appointment as vice consul; and we were informed by him, that the inhabitants, following the example of the larger Portuguese Colonies, had recently established a provisional government, in a Junta of three members, of which the former governor was the President; his colleagues being, the principal Ecclesiastic entitled the representative of the bishop, and a Colonel André, a native of the Island, of mixed blood, who we further learnt was the efficient member of the administration; notwithstanding these proceedings, however, the island still acknowledged the supremacy of Portugal.

Captain Clavering and myself, accompanied by Mr. Fernandez, waited the same day on Senor Baptiste to present the Ambassador's letter, and also one which Sir Charles Mac Carthy had been kind enough to give us; we were courteously and kindly received, as we had been led to expect from the President's European birth and general character; he assured us of his personal readiness to comply with the request which the Ambassador's letter contained, but added, that it would be necessary to be submitted in the first instance to the consideration of the Junta, and concluded by desiring me to make known to Colonel André the particulars of the assistance which I should require. Our reception by that gentleman was civil, but not so cordial as by the President; we understood that he had been a considerable sufferer, in the capture, by British cruisers, of vessels which he had employed in the illicit trade in slaves, which circumstance might have induced perhaps an indisposition to forward the views of individuals of that nation; we left him, however, with permission to examine the town and neighbourhood for a suitable situation for the experiments.

The town of Santa Anna, which is much more extensive than it appears when viewed from the harbour, is built of wood with the exception only of two stone houses; one of these was occupied by Mr. Fernandez, who was so obliging as to offer to resign it; but the situation was too public, and the view of the heavens too confined, for the purposes of the experiments; it is the custom at St. Thomas's to surround the principal mansion by the dwellings of the slaves, each of which has its separate though small plantation, principally of cocoa nuts, and other lofty palms; great part of the town is thus entirely overshadowed, and the view of the heavens is generally limited to a small space around the zenith. The temporary circumstances of the family who inhabited the other stone house placed it also out of the question; so that there appeared no other choice, than to apply for apartments in a stone fort situated on a sandy beach at the entrance of the harbour, and which must have proved in many respects a very inconvenient residence: the application was accordingly made.

It was late the following evening before the result of the consideration of the Junta was made known, being not only a refusal of the accommodation in the fort, but a denial of permission to land the instruments at all upon the island, grounded on the positive instructions of the Portuguese Government to its Colonies generally, not to permit foreigners to make any observations whatsoever in them, except by a special order from the Court itself; instructions which, it was further observed, were of such general notoriety, and so positively enjoined, that if the Marquess de Souza had not viewed the application from the British Secretary of State as made solely on behalf of an individual, and not as one in the object of which the Government itself was interested, the Marquess could not have failed to have referred to his Court for the only proper and sufficient authority.

As this communication was made verbally by Colonel André, and was not therefore necessarily conclusive, I endeavoured to see the President on the same evening, and again on the following morning, but in both instances without success, his secretary acquainting me that he had been obliged to decline public business for a few days, in consequence of an attack of fever. Having reason to suspect, that the unfavourable disposition of the Junta might have been in some degree influenced by the communications having hitherto passed through Mr. Fernandez, whose acknowledgment as temporary vice consul, I found, had been refused at the same time, I addressed a letter directly to the Junta; in which, after recapitulating the circumstances, and referring to the presence of a ship of war as sufficiently indicating the interest of the British Government, I requested, in the event of the Junta persisting in a refusal, its communication in writing; as Captain Clavering would not feel justified in quitting the island without an official document, which should enable the affair to be brought in due course under the consideration of the Court of Portugal, with which it would rest to judge between the Government of St. Thomas and the Marquess de Souza, and to decide by which of these authorities the request of the British Government, communicated with all due formality, had been frustrated. Whilst awaiting a reply to this letter, the Pheasant proceeded to Man of War Bay, a few miles to the westward of Santa Anna, to wood and water.

Close to the landing place at Man of War Bay, is a large stone mansion belonging to the extensive plantation of Fernandilla, which had been uninhabited for some months before our arrival; and at a short distance in-land, on the summit of an eminence of no great elevation, is a well-built brick church, at this time also unemployed; we had remarked these buildings in our passage along the coast a few days before, as likely to answer the purpose of the observations, if nothing better should offer at Santa Anna; we now landed for their further examination, and were not a little surprised to find them occupied by a guard of sixty soldiers, which had been despatched from the town, at the same time that the Pheasant sailed from the harbour, for the purpose of watching her proceedings. The officer commanding, who spoke English well, acquainted us, that he was ordered to oppose if possible, and otherwise to remonstrate, in the sole case of our attempting to land instruments, but that we were at liberty to cut wood, or to obtain any supplies of which we stood in need, and that he should be happy to render us assistance, or aid in making our stay agreeable; we readily assured him that we should not attempt to land instruments on the island without permission, and on this assurance he accompanied us to examine the buildings. I found the mansion-house the best suited for my purpose of any that I had seen, being too substantial for decay to have made much progress; it was also very convenient to the ship and well under her protection, as she was moored immediately abreast of it, and near the shore. The church would have been more healthy, but it was at a greater elevation than I had supposed, and very difficult of access. The following morning I accompanied Captain Clavering to examine a small rocky island about two miles off the shore, and uninhabited; and finding that the perpendicular face of the rock would admit of being bored to receive the screws of the supports of the clock and pendulums, it was arranged, that in the event of the refusal to land on St. Thomas's being persisted in, we should take possession of this island, which is called in the charts the Isla das Cabras, and build a protection from the weather with materials from the ship; the principal difficulty which we anticipated, and with which we should have had to contend, would have been the regulation of the temperature, which in spite of every precaution must have undergone great fluctuations; a much longer period than usual would therefore have been required, to have obtained results equally satisfactory; but the delay would have been preferable to quitting the neighbourhood of the equator, without at least an attempt to accomplish experiments, which we had gone so far to make.

On returning to the Pheasant we received the reply of the Junta, in which the instructions of their Court, and the insufficiency of any other than a direct authority to set them aside, were formally stated; but the general effect of the communication was not that of a decided refusal; as it concluded by expressing regret that the British Government had not obtained the proper authority for a purpose, which it was much the inclination of the government of the island to forward; and that in consideration of its importance, and that St. Thomas's was the only station in the immediate vicinity of the equator, on the African side of the Atlantic, at which the experiments could conveniently be made, the Provisional Government was still disposed to forward it, so far as their responsibility could be extended.

I received at the same time a private message from Colonel André requesting me to return to Santa Anna to confer personally with him;

and on compliance, I had the satisfaction of finding that every difficulty could be got over, except that of my admission into the fort: but it was suggested that a convent, situated at a short distance in the country, might contain suitable apartments, the occupancy of which, in such case, was offered, and that they should be fitted up in any way that I should direct. I found the convent a large stone building, with a handsome suite of reception rooms in tolerable repair; these would have answered well, had not the convent been entirely embosomed in wood, with no advantage over the house at Fernandilla to compensate for the greater distance from the ship; I therefore applied for, and obtained permission to occupy the latter; the officer with whom we were already acquainted, was ordered to remain with half his guard in the rooms which were not required, and was made responsible that no interruption should be offered, and nothing stolen. From this gentleman, Senor Manuel Gomez, a native of Portugal, Captain Clavering and myself received the most obliging and disinterested attention during our stay; having married the heiress of extensive plantations on the Island, he possesses considerable influence; and being partial to England, of which he speaks the language well, his good offices, which may be fully depended on, may prove of much avail to the ships of war on the African station, when they may have occasion to visit St. Thomas's.

In consequence of the delay caused by these preliminary measures, it was not until the 23rd of May that the instruments were disembarked at Fernandilla. The house was built of the compact basaltic stone of which the island is composed, and which proved not less impracticable to chisels than the granite of Sierra Leone; fortunately the stones were individually of less size, so that they could be removed without cutting, whenever it was necessary to drive a wooden picket for the screws, and without weakening the general strength of the wall, which exceeded a foot and a half in thickness. A great advantage was derived from the substantial nature of the walls, in the preservation of uniformity of temperature, to which, also, the foliage of the surrounding and lofty trees contributed in no small degree, so that although the range of the exterior thermometer was greater than at Sierra Leone, in consequence of the descent at night of the cold air from the very high land in the interior of the island, the extremes in twenty-four hours of a register thermometer, in the clock-room, were never so much as three degrees apart.

It not being possible to supply the deficiency of glass here as at Sierra Leone, the windows were closed with boards and matting, and light admitted through a small opening when actually required in observation; the clock and pendulum had also an additional protection from currents, by a screen of African matting, which enclosed them above and on either side.

The transit instrument was placed on a very substantial pillar of masonry, which had been designed to support one of the corners of the roof of a detached store, the building of which had not been proceeded in; being distant a few feet only from the house, I connected the upper part of it by a stage of communication with one of the windows of the clock room, and as the pillar was above 12 feet high, the instrument was thus inaccessible except from the house, and was consequently secure from disturbance; by cutting down the wood, in the direction of the meridian, through a screen of about 200 feet which interposed between the house and sea, an uninterrupted view was obtained of the north horizon; the transit was placed in the meridian on the 26th of May, and the going of the clock having been ascertained to be sufficiently steady, the observation of coincidences was commenced with No. 3 pendulum on the 28th of May, and closed on the 3rd of June; and with No. 4, on the 4th of June, and terminated on the 10th. The weather during the whole of this period was very unfavourable for celestial observations, being continually clouded, especially towards the horizon, but without rain; of twenty-five stars with which the transit list commenced on the 28th and 29th of May, I was obliged to be content with intervals of time obtained by eight, (exclusive of the sun,) on the 8th and 10th of June, their accordance being such as did not justify the delay of another day, for the chance of unnecessarily multiplying observations.

The subjoined tables, containing the details, are arranged in a similar manner to those of the preceding station, and do not appear to require any particular explanation.

The height of the pendulums above half tide was ascertained by levelling to be 21 feet.

The Island of St. Thomas is about 30 miles in length from North to South, and half as much in breadth; the equator passes five or six miles to the north of its Southern extremity; it is composed of a very compact and heavy Basaltic Rock, covered by a rich soil principally of vegetable decomposition, and is thickly wooded in every part. The interior of the Island is of considerable elevation; when first seen by the Pheasant, on the 13th of May, the principal peak subtended an angle of 13 minutes with the horizon, when by careful chronometric observation it was not less than 88 miles distant, whence its height may be inferred to exceed 7000 feet; the ascent is practicable, the principal difficulty being the absence of frequented paths.

The general result of the experiments at this Island indicated, as at Sierra Leone, a greater compression than the prevailing expectation; the retardation of the pendulum was even comparatively less here than at Sierra Leone; which circumstance, however, I had been prepared to expect, from the greater specific gravity of the Basalt of St. Thomas than of the Granite of Sierra Leone.

It is with great concern that I have now to notice the distressing consequences which attended on the experiments at this station. Captain Clavering had been again kind enough at my particular request, and from the obvious exigency, to land a guard of three marines, for the more effectual protection of the instruments, and to render me such personal attendance as was usually performed by my servant, who had been trained in the Northern Expeditions to be a very useful assistant in an observatory, but who had been taken ill at Sierra Leone, and having suffered severely from the fever of the country; was not yet sufficiently recovered to resume his duties; the marines were stationed in the outer room of the principal suite of apartments, the inmost of which was occupied by the clock and pendulums; it was a large and airy room in the first-story, the Portuguese guard occupying the ground-floor; the marines had no duties whatsoever to perform which required an exposure to the climate; nevertheless, on the 9th of June, being the day before the experiments were completed, one of the men shewed symptoms of fever, a second on the following morning, and the third in the afternoon, and unhappily all the cases proved fatal; of the two marines who had been landed to attend on me at Sierra Leone, one had died on the passage to St. Thomas's, having been taken ill the day after his embarkation; and the other was one of the present sufferers. It was thus my misfortune to witness the death of every individual landed for my assistance in Africa, with the exception of my servant, whose recovery from a relapse which occurred at St. Thomas's, was long very doubtful; it will readily be imagined, that we rejoiced in departure from a climate, which has shewn itself so generally fatal to European life.

The instruments were re-embarked in the afternoon of the 10th of June, and the Pheasant sailed on the same evening for the Island of Ascension.

TRANSITS OBSERVED AT THE ISLAND OF ST. THOMAS.

1822.	STARS.	TIMES OF TRANSIT BY THE CHRONOMETER 423.					Mean by the Chronometer.
		1st Wire.	2nd Wire.	Meridian Wire.	4th Wire.	5th Wire.	
		M. S.	M. S.	M. S.	M. S.	M. S.	H. M. S.
May 28.	δ Leonis	11 01.2	11 28	11 55.6	12 23.2	12 50.4	6 11 55.67
,,	ι Leonis	20 45.6	21 12	21 38.8	22 05.6	22 32	6 21 38.8
,,	β Leonis	46 00	46 27.6	46 54.8	47 22	47 49.2	6 46 54.73
,,	δ Corvi	...	...	27 27.2	27 54.4	28 21.6	7 27 27.2
,,	Spica	21 36	22 02.4	22 28.8	22 55	23 22	8 22 28.83
29.	Sun's 1st Limb..	24 46.4	25 14.8	25 42.8	26 11.2	26 39.2	23 26 51.15
	Sun's 2d Limb..	27 03.2	27 31.2	27 59.2	28 27.6	28 56	
,,	ι Leonis	16 51.2	17 17.2	17 44	18 10.8	18 37.6	6 17 44.13
,,	υ Leonis	30 00.8	30 27.2	30 53.2	31 19.2	31 45.2	6 30 53.13
,,	ε Corvi	02 58.8	03 26.8	03 54.8	04 22.8	04 50.8	7 03 54.8
,,	α Crucis	17 43.2	18 38.8	19 34.4	20 29.6	21 25.2	7 19 34.27
June 2.	β Leonis	26 26.8	26 54	27 21.2	27 48.4	28 15.2	6 27 21.13
3.	Sun's 1st Limb..	25 34.4	26 02.8	26 30.8	26 59.2	27 27.6	23 27 39.33
	Sun's 2d Limb..	27 50.8	28 19.2	28 47.6	29 16.4	29 44.8	
7.	δ Leonis	31 55.2	32 12	32 49.2	33 16.8	33 44	5 32 49.4
,,	ι Leonis	41 39.2	42 06	42 32.4	42 59.6	43 26	5 42 32.6
,,	β Leonis	6 53.2	7 20	7 47.2	8 15.2	8 42	6 07 47.67
8.	ι Leonis	37 44.8	38 11.2	38 38	39 04.8	39 32	5 38 38.13
,,	υ Leonis	50 54.4	51 20.4	51 46.8	52 13.2	52 39.6	5 51 46.87
,,	ε Corvi	...	...	24 49.2	...	...	6 24 49.2
9.	Sun's 2d Limb	29 02.4	29 31.2	29 59.6	30 28	30 56.8	11 28 51.2
10.	δ Leonis	20 12	20 39.2	21 06.4	21 34	22 01.2	5 21 06.53
,,	ε Corvi	16 04.8	...	...	...	...	6 17 00.8
,,	α Crucis	...	31 45.6	32 41.2	33 36.8	...	6 32 41.2
,,	δ Corvi	...	...	36 38.4	37 05.6	37 32.8	6 36 38.4
,,	Spica	30 48	31 14	31 40	32 06.8	32 33.6	7 31 40.4

ISLAND OF ST. THOMAS.

DEDUCTION of the RATE of the Chronometer No. 423 from TRANSITS; between the 28th of May and the 10th of June, 1822.

STARS.	28 to 29	29 to 30	30 to 31	31 to Ju. 1	1 to 2	2 to 3	3 to 4	4 to 5	5 to 6	6 to 7	7 to 8	8 to 9	9 to 10
	s.	s.	s.	s.	s.	s.	s.	s.	s.	s.	s.	s.	s.
δ Leonis . .	1.28	1.28	1.28	1.28	1.28	1.28	1.28	1.28	1.28	1.28	1.62	1.62	1.62
ι Leonis . .	1.24	1.3	1.3	1.3	1.3	1.3	1.3	1.3	1.3	1.3	1.44	..	..
υ Leonis. .	..	1.28	1.28	1.28	1.28	1.28	1.28	1.28	1.28	1.28	1.28	..	..
β Leonis. .	1.19	1.19	1.19	1.19	1.19	1.22	1.22	1.22	1.22	1.22	..	..	..
ε Corvi. . .	..	1.35	1.35	1.35	1.35	1.35	1.35	1.35	1.35	1.35	1.35	1.71	1.71
α Crucis . .	..	1.5	1.5	1.5	1.5	1.5	1.5	1.5	1.5	1.5	1.5	1.5	1.5
δ Corvi. . .	1.4	1.4	1.4	1.4	1.4	1.4	1.4	1.4	1.4	1.4	1.4	1.4	1.4
Spica. . .	1.4	1.4	1.4	1.4	1.4	1.4	1.4	1.4	1.4	1.4	1.4	1.4	1.4
The Sun (solar.)	..	1.25	1.25	1.25	1.25	1.25	1.51	1.51	1.51	1.51	1.51	1.51	..
MEANS— Gaining per Diem......	1.3	1.32	1.32	1.32	1.32	1.33	1.36	1.36	1.36	1.36	1.44	1.52	1.53
	1.318 (Sidereal)=1.322 in a Solar Day.						1.42 (Sidereal)=1.424 in a Solar Day.						

ISLAND of ST. THOMAS.—Comparisons of the Astronomical Clock with the Chronometer, No. 423, from the 27th of May to the 10th of June, 1822; with the Clock's Rate on Mean Solar Time deduced.

1822.		Chronometer.	Clock.	Clock's loss on 423.		Daily Rates. Chron.	Daily Rates. Clock.
		H. M. S.	H. M. S.	S.		Gaining.	Losing.
May 27	P.M.	8 00 00	8 31 13	124.1			
" 28	A.M.		8 30 11				
	P.M.		8 29 08.9	124.2			
" 29	A.M.		8 28 07.5				
	P.M.		8 27 04.7	123.5			
" 30	A.M.		8 26 03				
	P.M.		8 25 01.2	124.4	s. 124.1	s. 1.32	s. 122.78
" 31	A.M.		8 23 59.2				
	P.M.		8 22 56.8	124.2			
June 1	A.M.		8 21 55				
	P.M.		8 20 52.6	124.			
" 2	A.M.		8 19 50.8				
	P.M.		8 18 48.6	124.3			
" 3	A.M.		8 17 46.8				
	P.M.		8 16 44.3	124.3			
" 4	A.M.		8 15 42.4				
	P.M.		8 14 40	124.2			
" 5	A.M.		8 13 38.2				
	P.M.		8 12 35.8	124.3			
" 6	A.M.		8 11 34				
	P.M.		8 10 31.5	123.9	124.186	1.424	122.76
" 7	A.M.		8 9 29.5				
	P.M.		8 8 27.6	124.2			
" 8	A.M.		8 7 25.5				
	P.M.		8 6 23.4	124.			
" 9	A.M.		8 5 21 6				
	P.M.		8 4 19.4	124.4			
" 10	A.M.		8 3 17.2				
	P.M.		8 2 15				

ISLAND OF ST. THOMAS.

COINCIDENCES OBSERVED with PENDULUM No. 3; the Clock making 86277.22 Vibrations in a Mean Solar Day.

DATE.	Baro-meter.	No. of Coinci dence.	Tempe-rature.	Time of Disap-pearance	Time of Re-ap-pearance.	True Time of Coincidence.	Arc of Vibra-tion.	Mean Tempe-rature.	Mean Interval.	Correc-tion for the Arc.	Vibrations in 24 hours.	Reduction to a mean Tempera-ture.	Reduced Vibrations at 82°.1 Faht.
1822.	IN.		°	M. S.	M. S.	H. M. S.	°	°	S.	S.			
May 28 P.M.	30.068	1	81.2	29 46	29 47	3 29 46.5	1.2	81.45	644.8	+ 1.38	86010.98	−0.27	86010.71
		11	81.7	17 10	17 19	5 17 14.5	0.66						
,, 29 A.M.	30.060	1	82.2	43 05	43 07	9 43 06	1.16	82.3	644.5	1.26	86010.73	+0.08	86010.81
		11	82.4	30 26	30 36	11 30 31	0.62						
,, 29 P.M.	30.020	1	85.4	35 28	35 33	1 35 30.5	1.16	85.6	641.4	1.26	86009.45	+1.47	86010.92
		11	85.8	22 20	22 29	3 22 24.5	0.62						
,, 30. A.M.	30.080	1	80.6	25 49	25 51	8 25 50	1.2	81.45	644.8	1.38	86010.98	−0.27	86010.71
		11	82.3	13 14	13 22	10 13 18	0.66						
June 1 A.M.	30.020	1	79.7	6 03	6 07	8 06 05	1.16	80	646.9	1.26	86011.73	−0.88	86010.85
		11	80.3	53 48	54 00	9 53 54	0.62						
,, 1 P.M.	30.110	1	84	00 20	00 22	1 00 21	1.24	83.4	642.35	1.47	86010.06	+0.55	86010.61
		11	82.8	47 19	47 30	2 47 24 5	0.68						
,, 2 A.M.	30.138	1	81.1	01 49	01 51	9 01 50	1.18	81.4	644.75	1.32	86010.89	−0.29	86010.60
		11	81.7	49 12	49 23	10 49 17.5	0.63						
,, 3 P.M.	30.110	1	00.0	2 41	2 46	0 2 48	1.2	81.1	644.85	1.38	86010.99	−0.42	86010.57
		11	81.6	50 07	50 16	10 50 11.5	0.66						
Means	30.076							82.1			86010.726		86010.726

ISLAND OF ST. THOMAS.

COINCIDENCES OBSERVED with PENDULUM No. 4; the Clock making 86277.24 Vibrations in a Mean Solar Day.

DATE.	Barometer.	No. of Coincidence.	Temperature.	Time of Disappearance	Time of Re-appearance	True Time of Coincidence.	Arc of Vibration.	Mean Temperature.	Mean Interval.	Correction for the Arc.	Vibrations in 24 hours.	Reduction to a mean Temperature.	Reduced Vibrations at 83.°1 Faht.
1822.	IN.		°	M. S.	M. S.	H. M. S.	°	°	S.	S. +			
June 4 A.M.	30.095	1	81	50 58	51 03	9 51 00.5	1.2	81.4	669.6	1.38	86020.90	−0.71	86020.19
		11	81.8	42 30	42 43	11 42 36.5	0.66						
" 5 A.M.	30.085	1	80.6	25 19	25 23	9 25 21	1.18	81.8	669.05	1.32	86020.64	−0.55	86020.09
		11	83	16 45	16 58	11 16 51.5	0.63						
" 5 P.M.	30.026	1	86.4	02 06	02 09	2 02 07.5	1.22	85.6	664.9	1.44	86019.14	+1.05	86020.19
		11	84.8	52 49	53 04	3 52 56.5	0.68						
" 6 P.M.	30.020	1	85.2	08 05	08 09	3 08 07	1.18	84.1	666.6	1.32	86019.70	+0.42	86020.12
		11	83	59 06	59 20	4 59 13	0.63						
" 7 A.M.	30.120	1	81	07 56	08 01	10 07 58.5	1.18	82.4	668.05	1.32	86020.24	−0.29	86019.95
		11	83.8	59 13	59 25	11 59 19	0.63						
" 8 A.M.	30.120	1	83.8	45 15	45 18	10 45 16.5	1.26	83.9	665.85	1.53	86019.63	+0.34	86019.97
		11	84	36 09	36 21	12 36 15	0.7						
" 8 P.M.	30.080	1	84.4	40 59	41 05	3 41 02	1.2	84.4	665.9	1.38	86019.28	+0.55	86019.83
		11	84.4	31 53	32 09	5 32 01	0.66						
" 9 A.M.	30.100	1	81.3	02 25	02 29	9 02 27	1.22	81.9	668.25	1.44	86020.44	−0.50	86019.94
		11	82.5	53 43	53 56	10 53 49.5	0.68						
" 9 P.M.	30.100	1	83.2	49 22	49 29	2 49 25.5	1.18	83.2	667.35	1.32	86019.98	+0.04	86020.02
		11	83.2	40 31	40 47	4 40 39	0.63						
" 10 A.M.	30.100	1	82.1	15 16	15 17	9 15 16.5	1.22	82.2	667.85	1.44	86020.30	−0.38	86019.92
		11	82.3	06 30	06 40	11 06 35	0.68						
Means	30.084							83.1			86020.02		86020.02

ISLAND OF ASCENSION.

The Island of Ascension, which was previously uninhabited, was taken possession of by the British Government, in the year in which the late Emperor of France was sent in captivity to St. Helena, and has since remained in the occupation of a small detachment of seamen or marines, who are its only inhabitants; the Garrison, at the period of the Pheasant's visit, consisted of a party of the Royal Marines, commanded by Major John Campbell, whom we found in expectation of our arrival, in consequence of a letter which had preceded us from Commodore Sir Robert Mends, on whose command Ascension was considered a dependency; it is scarcely necessary to add that we were received by Major Campbell and the officers under his command with the utmost kindness and hospitality, and with a disposition to render every assistance in their power.

The only buildings on the island were those which had been erected for the accommodation of the garrison; they consisted of an officer's and soldiers' barrack, and a store-house, forming three sides of a barrack square, situated near the landing-place on the Northwest side of the island; the barracks had been described to me as being constructed of stone and mortar, with walls exceeding a foot in thickness, and I had relied on this information in selecting Ascension as one of the stations of experiment; I was therefore greatly disappointed on examining the walls, to find that the mortar of which they were principally composed, (the stones being comparatively few,) had been made without a due proportion of lime, and that it was to be feared in consequence that they would not prove sufficiently substantial to support such heavy weights as the clock and pendulum-frame; the walls of the store-house were indeed of an opposite character, being built of large masses of the heavy volcanic rock of which the Island chiefly consists, roughly squared so as to rest on each other, and forming a very compact wall of unusual thickness and great stability; from its appearance however, it was judged to present fewer facilities in the operations of boring or driving pickets, than had been experienced on any former occasion; the house itself was also very ill adapted in other respects for the experiments; it contained the whole of the provisions and stores of the garrison, including those of daily consumption; and the process of issuing and distributing the latter to the several messes took place within its walls, and could not be removed elsewhere, without such excessive inconvenience as amounted to impracticability; the only light, when the door was closed, was admitted through a small and grated window at the end of the building, several feet from the ground, and so inconveniently placed as to be quite unavailing in the observations, which would have therefore to depend on the opening of the door-way for the admission of light; the store-house was however the only alternative if the barracks should fail, rendering the prospect altogether so unfavourable, as appeared scarcely to justify the double risk, of injury in landing and putting up the instruments, and of the time which would have been consumed in what might have proved an unsuccessful attempt, or at least an unsatisfactory experiment; but I had learnt by experience to confide in the resources of a ship of war in surmounting difficulties almost of every kind, and I was well assured of Captain Clavering's ready disposition to spare no exertions which might assist me; no time was therefore lost in disembarking the instruments, which was accomplished on the evening of our arrival, through a surf which frequently interrupts all communication between the anchorage and shore for days together, but which was fortunately very moderate on this occasion.

On the following morning a trial was made of the barrack wall, which was found, as had been apprehended, too incohesive and unsubstantial to answer the purpose; a part of the wall of the store-house was therefore selected for an attempt, and after considerable labour, by the joint operation of chisels and wedges, three oaken pickets were established in an horizontal interstice between two of the largest masses of stone; the pickets were of sufficient size, and at the proper distances, to receive the three upper screws of the pendulum support; one of the two remaining screws at the ends of the brackets, happening to coincide with another interstice, was secured to a picket the same evening; and the other screw on the following morning, by splitting off and removing a part of the stone opposed to it, and substituting a junk of wood tightly wedged; the pendulum support, which was most important, was thus attached to the wall as firmly as could be desired; the clock was then fixed at the proper distance beneath it, by removing the stones which were opposed to its screws, (having previously wedged up the superincumbent stones,) and supplying their places with junks of wood cut for the occasion and bound in by wedges; the clock was less securely fastened by this method than the pendulum frame had been, but its immobility was of less consequence.

Aware of the inconveniences attendant on an unsteady and therefore uncertain temperature, and that greater errors might be apprehended from that source, than from any other whatsoever, every precaution was taken which might contribute to impede its fluctuations; the clock and pendulum were enclosed above, and at the sides, by a double skreen of African matting, stretched on a wooden frame, which projected about six feet from the wall and was continued to the ground; the light which was admitted by a very small opening in the door-way, was reflected upon the disk of the pendulum by mirrors properly disposed; the store was closed and the key remained in my possession, excepting for the short interval which was required in the daily issue of provisions, which took place in a distant part of the room separated by a walled partition from the instruments, and always at a certain hour of the forenoon, previously to which I had completed the morning series of coincidences; it was not possible, however, to prevent greater changes of temperature in the course of the twenty-four hours, than had occurred at either of the preceding stations; the surface of the soil, or rather the rock, in the neighbourhood of the barracks, being unprotected by foliage, and situated on the leeward side of the island, became extremely heated during the day by the power of the sun; and although from the great thickness of the walls his direct influence was little felt withinside, the occasional entrance of the heated air from without could not be altogether prevented.

In the embarrassment which a range of 8 or 9 degrees of the thermometer in the twenty-four hours produced, I felt the propriety of the determination which I had formed in its anticipation, of confining myself, whilst in the tropics, to those stations where the instruments could have the protection of the roof and walls of a substantial house; the store-room was certainly far superior in this respect to any temporary covering which could have been made with materials from the ship; but the variations of temperature were sufficiently perplexing, and required much watchfulness and attention in selecting the most favourable periods of the day for the observation of coincidences; by these means, however, it may be seen that the changes of temperature, whilst the pendulum was actually in vibration, rarely exceeded 2°, and only in a single instance 3°; and as the changes were progressive, and in opposite senses, being gaining in the forenoon, and losing in the afternoon series, the errors which might be apprehended from the pendulum being more slowly affected than the thermometer, would in great measure balance each other.

An unfinished wall near the Barrack-Square afforded a suitable and convenient situation for the transit instrument; the agreement in the results with stars differing so widely in declination, as those in the table of "observed transits," is a sufficient indication that the plane of the vertical motion of the Instrument was preserved throughout.

It is but justice to the chronometers of Messrs. Parkinson and Frodsham, that the attention of the reader should be directed to the opportunities, which incidentally occurred in the course of these experiments, and are exemplified in the Tables, of proving the steadiness of their going; such is the table shewing the rate of No. 423, at Ascension, deduced from the transits; the going of this chronometer is the more worthy of notice, as it was almost incessantly employed in observations, and exposed in consequence to continual changes of temperature and position. It would be impossible indeed to express the advantage which these chronometers proved to me on all occasions; or how much the thorough reliance which I could place on their time facilitated, and which is more important, how much it conduced to the accuracy of, the variety of observations which successively occupied my attention, and which I was usually pressed to complete within the least possible time. I may take the present occasion also to mention, as a circumstance well worthy of notice, that of twelve chronometers, which Messrs. Parkinson and Frodsham have at different times intrusted to my care in voyages of long duration and unusual exposure, not only has there not been a single failure, but I should find it difficult to say that any one chronometer has been decidedly inferior to the others.

It has been already remarked that from the mode in which the clock was attached to the wall and supported, it was not considered as perfectly secure from motion; and it is probable that its weight acting on the blocks of wood to which it was fastened, caused them to yield a little for the first three or four days, until effectually stopped by the resistance of the wedges ; as the level which marks the horizontality of the hollow cylinder in which the pendulum works, was observed to undergo occasional slight derangements during that period ; whenever these were noticed, they were immediately corrected by the screws in the lower plate designed for that purpose, until the adjustment was no longer disturbed ; but their effect on the rate of the clock in the first days of its going may be perceived by its comparisons with the chronometer ; and may be further traced, with remarkable correspondence, by a close examination of the table exhibiting the coincidences of Pendulum No. 3 ; in which table as a mean rate is taken for the clock, its daily irregularities are transferred in appearance to the Invariable Pendulum. Had the rate of the clock, as indicated from day to day by the chronometer, been introduced into the table, instead of the mean rate, the results with the detached pendulum on the several days would have been shewn to be not less in accord, than those of pendulum 4 appear in the succeeding table, when the irregularities in the clocks going had ceased ; and I may remark, that this mode of constructing the table would have been the more correct on this occasion, but as it is obvious that in either case, the alternate result must have been the same, the form adopted at the preceding station has been adhered to.

The height of the pendulums above the mean level of the sea was ascertained by direct measurement to be 17 feet.

The Pheasant arrived at Ascension on the 26th of June ; the observation of coincidences with No. 3, was commenced on the morning of the 30th, and ended on the evening of the 3rd of July ; and with No. 4, on the morning of the 5th, terminating on the evening of the 8th ; the instruments were re-embarked on the 9th, when we were again fortunate in the state of the surf ; and on the following day the Pheasant sailed for South America.

TRANSITS OBSERVED AT ASCENSION.

1822.	STARS.	TIMES OF TRANSIT BY THE CHRONOMETER 423. 1st Wire.	2nd Wire.	Meridian Wire.	4th Wire.	5th Wire.	Mean by the Chronometer.
		M. S.	M. S.	H. M. S.	M. S.	M. S.	H. M. S.
Jun. 29	Sun's 1st Limb.	56 04.8	56 33.2	12 57 01.6	57 30	57 58.4	12 57 01.6
	Sun's 2d Limb.	58 22.4	58 50.4	12 59 18.8	59 47.2	00 15.2	12 59 18 8
,,	β Crucis. . .	01 51.2	02 41.6	7 03 32	04 22.4	05 12.8	7 03 32
,,	ε Ursæ . . .	10 38	11 25.6	7 12 13.6	13 01.6	13 50	7 12 13.73
,,	γ Hydræ . .	34 20.4	34 49.2	7 35 17 6	35 46	36 14.4	7 35 17.53
,,	Spica . . .	40 56.8	41 23.6	7 41 50.4	42 17.2	42 44	7 41 50.4
,,	ξ Virginis . .	. . .	51 10	7 51 36.8	52 02.8	. . .	7 51 36.6
,,	ι Centauri. .	00 32.4	01 03.2	8 01 34	02 04.8	02 35.6	8 01 34
,,	η Ursæ Maj. .	05 04.8	05 45.6	8 06 26.4	07 07.2	07 48	8 06 26.4
,,	η Bootis. . .	. . .	11 40.4	8 12 08.4	12 35.6	. . .	8 12 08.2
,,	β Centauri. .	15 36	16 27.2	8 17 18.8	18 09.6	19 01.2	8 17 18 6
,,	π Hydræ . .	21 12	21 41.2	8 22 10.4	22 39.6	23 08.4	8 22 10.33
,,	κ Virginis . .	28 24.4	28 51.2	8 29 18	29 44.4	30 11.2	8 29 17.87
,,	Arcturus. .	32 29.2	32 56.8	8 33 24.8	33 52.8	34 20.8	8 33 24.87
,,	γ Bootis. . .	49 36	50 09.6	8 50 43.2	51 17.2	51 50.8	8 50 43.33
,,	α Centauri 2.	. .	. . .	8 54 27.2	. . .	. . .	8 54 27.2
,,	μ Virginis . .	58 36	59 04	8 59 30	59 56.4	00 22.8	8 59 30.2
,,	ε Bootis. . .	02 00.8	02 30.4	9 02 59.6	03 29.2	03 59.2	9 02 59.8
,,	α Libræ 2 . .	05 56	06 23.2	9 06 50.4	07 17.6	07 54.8	9 06 50.4
July 4	β Crucis. . .	42 25.2	43 15.6	6 44 06	44 56.4	45 46.8	6 44 06
,,	ε Ursæ . . .	51 11.2	51 59.2	6 52 47.2	53 35.6	54 23.6	6 52 47.33
,,	γ Hydræ . .	14 54.0	15 22.0	7 15 51.2	16 10.6	16 48	7 15 51.27
,,	Spica . . .	21 30.4	21 56.8	7 22 23.6	22 50	23 16.4	7 22 23.47
,,	ξ Virginis . .	31 18	31 44	7 32 10	32 36	33 02	7 32 10
,,	ι Centauri. .	41 06	41 36.8	7 42 07.6	42 38.8	43 09.6	7 42 07.73
,,	η Ursæ . . .	45 37.2	46 18	7 46 58.8	47 39.6	48 20.8	7 46 58.93
,,	η Bootis. . .	51 46	52 13.2	7 52 41.2	53 08.8	53 36.4	7 52 41.13
,,	γ Bootis. . .	30 08.4	30 42	8 31 16	31 50	32 24	8 31 16.07
,,	ε Bootis. . .	42 33.6	43 03.2	8 43 32.8	44 02.4	44 32	8 43 32.8

TRANSITS OBSERVED AT ASCENSION, *continued.*

1822.	STARS.	TIMES OF TRANSIT BY THE CHRONOMETER 423.					Mean by the Chronometer.
		1st Wire.	2nd Wire.	Meridian Wire.	4th Wire.	5th Wire.	
		M. S.	M. S.	H. M. S.	M. S.	M. S.	H. M. S.
July 4	α Libræ 2. .	46 29.2	46 56.4	8 47 23.6	47 50.8	48 17.6	8 47 23.53
5	Sun's 1st Limb.	57 27.6	57 56	12 58 24.4	58 52.8	59 21.6	12 58 24.47
	Sun's 2d Limb.	59 45.2	00 13.6	1 00 42	01 10.4	01 38.8	1 00 42
8	Sun's 1st Limb.	58 06	58 34.4	12 59 02.8	59 31.2	59 59.6	12 59 02.8
	Sun's 2d Limb.	00 23.2	00 51.6	1 01 20	01 48.4	02 16.8	1 01 20
,,	β Crucis. . .	26 51.8	27 42.2	6 28 32.6	29 23	30 13.4	6 28 32.6
,,	ε Ursæ . . .	35 37.6	36 25.6	6 37 13.6	38 01.6	39 50	6 37 13.66
,,	γ Hydræ . .	. .	. . .	7 00 17.6	. . .	. . .	7 00 17.6
,,	Spica . . .	05 56.8	06 23.6	7 06 50	07 16.8	07 43.6	7 06 50.13
,,	ι Centauri. .	. . .	26 03.6	7 26 34.4	27 05.2	. . .	7 26 34.4
,,	η Ursæ . . .	30 03.6	30 44.4	7 31 25.2	32 06.4	32 47.2	7 31 25.33
,,	β Centauri. .	. . .	. . .	7 42 18.8	. . .	. . .	7 42 18 8
,,	π Hydræ . .	46 12	46 40.8	7 47 10	47 38.8	48 08	7 47 09.93
,,	κ Virginis. .	53 24	53 50.8	7 54 17.6	54 44.4	55 10.8	7 54 17.6
,,	Arcturus. .	57 28.4	57 56	7 58 23.6	58 51.6	59 19.2	7 58 23.73
,,	γ Bootis. . .	14 35.2	15 08 8	8 15 42.4	16 16.4	16 50	8 15 42.53
,,	α Centauri 2 .	. . .	. . .	8 18 27.7	. . .	. . .	8 18 27.7
,,	μ Virginis. .	23 36.8	24 03.6	8 24 30	24 56.8	25 23.2	8 24 30.07
,,	ε Bootis. . .	27 00.4	27 30	8 27 59.6	28 29.2	28 58.8	8 27 59.6
,,	α Libræ 2. .	30 56.4	31 23.6	8 31 50.8	32 18	32 45.2	8 31 50.8

ASCENSION.—DEDUCTION of the RATE of the Chronometer, No. 423, from TRANSITS, between the 29th of June, and the 8th of July, 1822.

STARS.	29 to 30	30 to July 1	1 to 2	2 to 3	3 to 4	4 to 5	5 to 6	6 to 7	7 to 8
	s.	s.	s.	s.	s.	s.	s.	s.	s.
The sun (solar) . . .	2.6	2.6	2.6	2.6	2.6	2.6	2.72	2.72	2.72
β Crucis	2.71	2.71	2.71	2.71	2.71	2.56	2.56	2.56	2.56
ι Ursæ	2.63	2.63	2.63	2.63	2.63	2.49	2.49	2.49	2.49
γ Hydræ	2.66	2.66	2.66	2.66	2.66	2.5	2.5	2.5	2.5
Spica	2.52	2.52	2.52	2.52	2.52	2.57	2.57	2.57	2.57
ξ Virginis	2.59	2.59	2.59	2.59	2.59	. .	. .	. .	. .
ι Centauri	2.66	2.66	2.66	2.66	2.66	2.58	2.58	2.58	2.58
η Ursæ Maj. . . .	2.41	2.41	2.41	2.41	2.41	2.52	2.52	2.52	2.52
η Bootis	2.49	2.49	2.49	2.49	2.49	. .	. .	. .	. .
β Centauri	2.59	2.59	2.59	2.59	2.59	2.59	2.59	2.59	2.59
π Hydræ	2.54	2.54	2.54	2.54	2.54	2.54	2.54	2.54	2.54
κ Virginis	2.54	2.54	2.54	2.54	2.54	2.54	2.54	2.54	2.54
Arcturus	2.46	2.46	2.46	2.46	2.46	2.46	2.46	2.46	2.46
γ Bootis	2.46	2.46	2.46	2.46	2.46	2.54	2.54	2.54	2.54
α Centauri . . .	2.63	2.63	2.63	2.63	2.63	2.63	2.63	2.63	2.63
μ Virginis	2.56	2.56	2.56	2.56	2.56	2.56	2.56	2.56	2.56
ι Bootis	2.51	2.51	2.51	2.51	2.51	2.61	2.61	2.61	2.61
α Libræ	2.54	2.54	2.54	2.54	2.54	2.73	2.73	2.73	2.73
MEANS—Gaining per Diem	2.56	2.56	2.56	2.56	2.56	2.56	2.57	2.57	2.57
	2.56 = 2.57 Solar.					2.569 = 2.576 Solar.			

ASCENSION.—Comparisons of the Astronomical Clock with the Chronometer No. 423, from the 30th of June to the 9th of July, 1822; with the Clock's Rate on Mean Solar Time deduced.

1822.	Chronometer.	Clock.	Clock's Loss on 423.		Daily Rates. Chron.	Daily Rates. Clock.
	H. M. S.	M. S.	s.	s.	Gaining. s.	Losing. s.
June 30 A. M.		41 12.6				
" 30 P. M.		40 12.8	118.3			
July 1 A. M.		39 14.3				
" 1 P. M.		38 15.1	117.8			
" 2 A. M.		37 16.5		117.72	2.57	115.15
" 2 P. M.		36 17.4	117.5			
" 3 A. M.		35 19				
" 3 P. M.		34 20.6	117.3			
" 4 A. M.	10 00 00	33 21.7				
" 4 P. M.		32 23.5	117.3			
" 5 A. M.		31 24.4				
" 5 P. M.		30 26	116.8			
" 6 A. M.		29 27.6				
" 6 P. M.		28 29	116.8			
" 7 A. M.		27 30.8		116.8	2.58	114.22
" 7 P. M.		26 32.6	116.8			
" 8 A. M.		25 34				
" 8 P. M.		24 35.7	116.8			
" 9 A. M.		23 37.2				

Ascension.—COINCIDENCES OBSERVED with PENDULUM No. 3; the Clock making 86284.85 Vibrations in a Mean Solar Day.

DATE.	Barometer.	No. of Coincidence.	Temperature.	Time of Disappearance.	Time of Re-appearance.	True Time of Coincidence.	Arc of Vibration.	Mean Temperature.	Mean Interval.	Correction for the Arc.	Vibrations in 24 hours.	Reduction to a mean Temperature.	Reduced Vibrations at 81°.47 Faht.
1822.	IN.		°	M. S.	M. S.	H. M. S.	°	°	S.	S.			
June 30 A.M.	30.140	1	79.5	25 32	25 34	8 25 33	1.18	80.25	638	+ 1.29	86015.65	−0.51	86015.14
		11	81	11 48	11 58	10 11 53	0.6						
„ 30 P.M.	30.130	1	84	23 32	23 33	2 23 32.5	1.22	83.75	634.4	1.35	86014.17	+0.96	86015.13
		11	83.5	09 12	09 21	4 09 16.5	0.64						
July 1 A.M.	30.180	1	79	30 08	30 09	8 30 08.5	1.22	79.75	637.6	1.35	86015.55	−0.72	86014.83
		11	80.5	16 19	16 30	10 16 24.5	0.64						
1 P.M.	30.120	1	83.6	21 39	21 43	2 21 41	1.2	83.7	633.15	1.32	86013.62	+0.94	86014.56
		11	83.8	07 09	07 16	4 07 12.5	0.62						
„ 2 A.M.	30.140	1	77	40 46	40 48	8 40 47	1 2	78.1	639.55	1.32	86016.34	−1.39	86014.95
		11	79.2	27 17	27 28	10 27 22.5	0.62						
„ 2 P.M.	30.080	1	83.7	24 35	24 37	2 24 36	1.2	83.4	634	1.32	86013.96	+0.81	86014.77
		11	83.1	10 11	10 21	4 10 16	0.62						
„ 3 A.M.	30.170	1	76.4	17 18	17 20	8 17 19	1.2	77	640	1.32	86016.52	−1.88	86014.64
		11	77.6	03 53	04 05	10 03 59	0.62						
„ 3 P.M.	30.120	1	86.2	03 49	03 50	2 03 49.5	1.22	85.8	630.25	1.35	86012.37	+1.82	86014.19
		11	85.4	48 47	48 57	3 48 52	0.64						
Means	30.135							81.47			86014.77		86014.77

ASCENSION.—COINCIDENCES OBSERVED with PENDULUM No. 4; the Clock making 86285.78 Vibrations in a Mean Solar Day.

DATE.	Barometer.	No. of Coincidence.	Temperature.	Time of Disappearance.	Time of Re-appearance.	True Time of Coincidence.	Arc of Vibration.	Mean Temperature.	Mean Interval.	Correction for the Arc.	Vibrations in 24 hours.	Reduction to a mean Temperature.	Reduced Vibrations at 82.87 Faht.
1822.	IN.		°	M. S.	M. S.	H. M. S.	°	°	S.	S.			
July 5 A.M.	30.180	1	77.9	14 34	14 39	7 14 36.5	1.2	79.85	658.35	+ 1.32	86024.96	− 1.27	86023.69
		11	81.8	04 14	04 26	9 04 20	0.62						
„ 5 P.M.	30.150	1	85.6	34 42	34 43	12 34 42.5	1.22	85.65	651.95	1.35	86022.41	+ 1.17	86023.58
		11	85.7	23 16	23 28	2 23 22	0.64						
„ 5 P.M. (by lamp light.)	30.130	1	80	57 00	57 02	8 57 01	1.26	81	656.5	1.54	86024.46	− 0.79	86023.67
		11	82	46 19	46 33	10 46 26	0.72						
„ 6 A.M.	30.170	1	78.4	40 08	40 09	7 40 08.5	1.26	79.4	658.4	1.54	86025.20	− 1.44	86023.76
		11	80.4	29 45	30 00	9 29 52.5	0.72						
„ 6 P.M.	30.140	1	86	58 20	58 22	12 58 21	1.26	86.6	651	1.54	86022.22	+ 1.57	86023.79
		11	87.2	46 44	46 58	2 46 51	0.72						
„ 7 A.M.	30.170	1	79.8	55 17	55 20	7 55 18.5	1.18	81	657.9	1.29	86024.73	− 0.79	86023.94
		11	82.2	44 50	45 05	9 44 57.5	0.6						
„ 7 P.M.	30.140	1	85.9	21 18	21 23	2 21 20.5	1.2	85.45	652.05	1.32	86022.44	+ 1.08	86023.52
		11	85	09 55	10 07	4 10 01	0.62						
„ 8 P.M.	30.140	1	83.8	7 43	7 48	12 7 45.5	1.18	84.05	654.25	1.29	86028.31	+ 0.49	86023.80
		11	84.3	56 40	56 56	2 56 48	0.6						
Means . .	30.155							82.87			86023.72		86023.72

BAHIA.

Agreeably to the original design which I had given Sir Robert Mends, and on which his instructions to Captain Clavering were founded, the Pheasant should have proceeded from Ascension direct to Maranham as her next station; but, whilst at Ascension, Captain Clavering had been induced to land all the provisions which could be spared from the Pheasant, in consequence of a representation from Major Campbell that the provisions of the Garrison were much reduced, and that he had reason to apprehend that the vessel containing an expected supply must have missed the island and gone to Leeward, which we afterwards learnt to have been actually the case; it became necessary therefore that the Pheasant should stop at Bahia on the passage to Maranham to obtain a fresh supply.

On our arrival at Bahia on the 19th of July, we were apprized of the revolution which had commenced in the Brazils in the preceding February, and had already become so general, that the city of Bahia was the only possession retained by the Portuguese; we found them in daily expectation of an attack by sea and land, as the Independant troops were in force in the adjoining villages, awaiting the arrival of a squadron from Rio to commence their operations in concert; the city was in great measure deserted by its principal inhabitants, whose slaves being left to provide for themselves, added much to the causes of alarm; the British merchants were anxiously looking for the arrival of the Blossom sloop of war, which Sir Thomas Hardy, commanding at Rio, had promised for the protection of themselves and their property in case of exigency; she had not yet however appeared, and it was feared might not do so before the attack should take place. In this state of general insecurity and apprehension, the arrival of the Pheasant occasioned great joy, and Captain Clavering was met with a most pressing solicitation from the merchants, conveyed through Mr. Pennell, His Majesty's Consul, to remain at Bahia until tranquillity should be restored, or at least until he should be relieved by the Blossom: on Mr. Pennell's being made acquainted with the particular service on which we were employed, he readily undertook that the necessary accommodation and convenience for obtaining the rates of the pendulums at Bahia should not be wanting, if Captain Clavering would determine on remaining for a sufficient time: the situation of the merchants being such as fully justified a compliance with their request, and being desirous that the delay should not be altogether unproductive of advantage to ourselves, Captain Clavering acceded to this arrangement.

Having waited on General Madera, commanding the Portuguese troops, and on the civil authorities, to obtain permission, the instruments were landed and conveyed to Mr. Pennell's residence at Vittoria, where Captain Clavering and myself were kindly invited to remain as inmates during our stay. Vittoria is a suburb, a mile and a half from the city, situated on a sandstone cliff which descends abruptly about 200 feet to the sea; the great road by the coast to Pernambuco and generally towards the northward passes through the village, which at the time of our landing was occupied as an advanced post by the Portuguese troops, who were throwing up field works for its defence. The houses being generally abandoned by the inhabitants, a suitable one for the reception of the instruments was easily obtained in the vicinity of Mr. Pennell's, from which it was only separated by the road; the walls being of brick, the apparatus was quickly and very firmly put up in the manner which I most approved, namely, by separate pickets for each screw; the house was altogether extremely well adapted for the purpose, except that it possessed no convenient situation for a transit instrument, as was also the case in the grounds belonging to Mr. Pennell's house, which were on the rapid slope of the cliff; and as it would not have been prudent, at such an unsettled period, to have stationed an instrument in any public exposure, I was obliged to change the mode in which I had hitherto proceeded, in comparing the time-pieces with the heavens, and to substitute the Repeating Circle for the Transit.

As this was the first instance, I believe, of the application of the principle of repetition to a circle of this description, of so small a diameter as six inches, it may be proper to mention that, in consequence of its size, both the level and telescope could be attended by the same person; and with so much ease that I was in the habit also of noting the times of the observations myself, and of thus dispensing altogether with an assistant; which is not only an advantage in convenience, but also in accuracy, as the instant of contact can be marked with more exactness to parts of a second, by the observer himself by means of the beats of a chronometer, than by an assistant to whom it must be notified by voice or signal. In estimating the practical merits of this little instrument in comparison with those of larger size, its portability is the most obvious, and possibly may be the principal consideration; but the advantage which it possesses in not requiring two observers, will be acknowledged by those who have had much experience of both, to be scarcely of less value.

The mode of comparing the chronometer with Astronomical Time pursued at Bahia, furnished a much severer test of the uniformity of its rate in short intervals, than the observation of Transits as at Ascension at the commencement and at the close of the whole interval occupied by the experiments, with few intermediate comparisons; the Table which exhibits the results must be regarded as highly creditable.

The height of the pendulums above the sea, being 218 feet, was ascertained by several barometrical measurements, the particulars of which are arranged in a table, and are subjoined.

The observations were continued from the morning of the 24th of July to the morning of the 2nd of August; the period being divided as usual between the two pendulums.

General Madera having been re-inforced by troops from Lisbon, who had arrived off the harbour nearly at the same time as the Independent squadron from Rio, and had slipped past them, the intended attack was converted into a strict blockade, which though slow in operation, was ultimately successful. The Blossom having arrived, and the apprehension of immediate danger having subsided, Captain Clavering felt himself at liberty to pursue his voyage to Maranham, for which he accordingly sailed on the 7th of August.

BAHIA.—OBSERVATIONS to DETERMINE the RATE of the Chronometer No. 423, by ZENITH DISTANCES of the Sun, with a Repeating Circle; from the 23d of July to the 2d of August, 1822.

Latitude of the Place of Observation 12° 59′ 22″ S.; Longitude 38° 32′ W.

July, 23d, A.M.; Barometer 29.98; Thermometer 73°; ☉'s L.L.

Chronometer.	Level.		Readings, &c.		Chronometer.	Level.		Readings, &c.	
H. M. S.				° ′ ″	H. M. S.				° ′ ″
10 45 07.2	+2	0	First Vernier	263 06 40	10 59 04.5	+1	0	First Vernier	154 48 10
10 47 53	+4	+2	Second „	06 10	11 00 57.5	−1	−2	Second „	48 05
10 50 22	+4	+2	Third „	07 00	11 03 34	+3	+1	Third „	48 45
10 52 07.5	−5	−7	Fourth „	06 10	11 06 20	−4	−3	Fourth „	48 20
Mean...... 10 48 52.42	+5	−3	Mean........	263 06 30	Mean...... 11 02 29	−1	−4	Mean........	154 48 20
True time .. 8 16 06.87	+1		Level........	+1	True time .. 8 29 42.27	−2.5		Index.......	+96 53 30
Chron. fast . 2 32 45.55			Index........	+8.5*	Chron. fast . 2 32 46.73			Level........	−2.5
				263 06 39.5					251 41 48.5
			Observed Z.D.	65 46 40				Observed Z.D.	62 55 27
			Ref. and Paral.	+ 1 54.5				Ref. and Paral.	+ 1 40.5
			Semidiam	− 15 46.5	360° − 263° 06′ 30″ = 96° 53′ 30″			Semidiam	− 15 46.5
			True Z.D.....	65 32 48				True Z.D.....	62 41 21

Chronometer, Fast { H. M. S. 2 32 45.55, 2 32 46.73 } H. M. S. 2 32 46.14

July 23d, P.M.; Barometer 29.96; Thermometer 77°; ☉'s U.L.

Chronometer.	Level.		Readings, &c.		Chronometer.	Level.		Readings, &c.	
H. M. S.				° ′ ″	H. M. S.				° ′ ″
6 09 50	0	0	First Vernier	13 25 45	6 24 19	+2	+1	First Vernier	44 12 10
6 11 23.5	+4	+3	Second „	25 28	6 26 09	+4	+4	Second „	12 00
6 14 10	0	0	Third „	26 00	6 27 33	+2	+2	Third „	12 30
6 15 55	−2	−3	Fourth „	25 12	6 29 05	−2	−2	Fourth „	11 20
6 17 35	+5	+5	Mean........	13 25 36.25	6 30 43	−2	−2	Mean........	44 12 00
6 18 51.5	+5	+4	Level........	+10.5	6 32 33.5	−2	−3	Level........	+1
Mean...... 6 14 37.5	+12	+9	Index......	+360 00 08.5	Mean...... 6 28 23.75	+2	0	Index......	+346 34 24
True time .. 3 41 50.13	+10.5			373 25 55	True time .. 3 55 36.67	+1			390 46 25
Chron. fast.. 2 32 47.37			Observed Z.D.	62 14 19	Chron. fast . 2 32 47.08			Observed Z.D.	65 07 44
			Ref. and Paral.	+ 1 37.5				Ref. and Paral.	+ 1 51.5
360° + 08″.5 = 360° 00′ 08″.5			Semidiam	+15 46.5	360° − 13° 25′ 36″ = 346° 34′ 24″			Semidiam	+ 15 46.5
			True Z.D.....	62 31 43				True Z.D.....	65 25 22

Chronometer, Fast { H. M. S. 2 32 47.37, 2 32 47.08 } H. M. S. 2 32 47.22

* When the First Vernier of the Repeating Circle was set at Zero, the Index Correction, obtained by reading the other Verniers also, was +08″.5

Bahia.—Determination of the Rate of the Chronometer by Zenith Distances, *continued.*

July 24th A.M.; Barometer 29.97; Thermometer 71°; ⊙'s L.L.

Chronometer.	Level.		Readings, &c.	
H. M. S.				° ′ ″
10 57 06	−1	−4	First Vernier	16 20 55
10 59 26.5	+9	+7	Second „	35
11 02 13	+7	+4	Third „	50
11 04 34.5	+7	+4	Fourth „	20
11 06 49.5	+3	0	Mean........	16 20 40
11 08 09.5	+8	+6	Level........	+25
Mean......11 03 03.17	+33	+17	Index......	+360 00 08.5
True time.. 8 30 14.6	+25			376 21 13.5
Chron.fast.. 2 32 48.57			Observed Z.D.	62 43 32
			Ref. and Paral.	+ 1 39.2
363° + 08.5″ = 360° 00′ 08.5″			Semidiam....	− 15 46.5
			True Z.D.....	62 29 25

Chronometer.	Level.		Readings, &c.	
H. M. S.				° ′ ″
11 16 34	−9	−7	First Vernier	10 54 55
11 18 18	+10	+8	Second „	54 47
11 20 17	−7	−9	Third „	55 10
11 21 24.5	+5	+3	Fourth „	54 25
11 23 03	+8	+6	Mean........	10 54 49
11 24 26	−3	−5	Level........	0
Mean......11 20 40.4	+4	−4	Index......	+343 39 20
True time.. 8 47 51.6	0			354 34 09
Chron.fast.. 2 32 48.8			Observed Z.D.	59 05 41.5
			Ref. and Paral.	+ 1 24.5
360° − 16° 20′ 40″ = 343° 39′ 20″			Semidiam....	− 15 46.5
			True Z.D.....	58 51 19.5

Chronometer, Fast {H. M. S. 2 32 48.57; 2 32 48.8} H. M. S. 2 32 48.68

July 24th P.M.; Barometer 29.96; Thermometer 76°; ⊙'s U. L.

Chronometer.	Level.		Readings, &c.	
H. M. S.				° ′ ″
6 12 19	−5	−5	First Vernier	15 13 40
6 13 53.5	+3	+2	Second „	13 28
6 16 06.5	0	0	Third „	14 00
6 17 28.5	+5	+5	Fourth „	13 10
6 19 07	−2	−2	Mean........	15 13 34.5
6 20 37	−2	−2	Level........	−1.5
Mean...... 6 16 35.25	−1	−2	Index......	+360 00 08.5
True time.. 3 43 45.13	−1.5			375 13 41.5
Chron.fast.. 2 32 50.12			Observed Z.D.	62 32 17
			Ref. and Paral.	+ 1 39
360° + 08.5″ = 360° 00′ 08.5″			Semidiam....	+ 15 46.5
			True Z.D.....	62 49 42.5

Chronometer.	Level.		Readings, &c.	
H. M. S.				° ′ ″
6 27 14.5	+1	+1	First Vernier	48 48 25
6 28 25.2	+2	+2	Second „	00
6 30 07.5	−1	−1	Third „	30
6 32 01.5	+5	+5	Fourth „	05
6 33 47.2	+4	+3	Mean........	48 48 15
6 35 00	+4	+3	Level........	+ 14
Mean...... 6 31 06	+15	+13	Index......	+344 46 25
True time.. 3 58 15.73	+14			393 34 54
Chron.fast.. 2 32 50.27			Observed Z.D.	65 35 49
			Ref. and Paral.	+ 1 54
360° − 15° 13′ 35″ = 344° 46′ 25″			Semidiam....	+ 15 46
			True Z.D.....	65 53 30

Chronometer, Fast {H. M. S. 2 32 50.12; 2 32 50.27} H. M. S. 2 32 50.2

BAHIA.—Determination of the Rate of the Chronometer by Zenith Distances, *continued.*

July 25th A.M.; Barometer 30.05; Thermometer 71°; ☉'s L.L.

Chronometer.	Level.		Readings, &c.	Chronometer.	Level.		Readings, &c.
H. M. S.			° ′ ″	H. M. S.			° ′ ″
10 58 09.2	0	−1	First Vernier 14 43 00	11 14 39.6	−4	−5	First Vernier 11 30 15
11 01 13	−1	−3	Second „ 42 50	11 15 40	−4	−5	Second „ 05
11 02 56.8	−8	−6	Third „ 43 20	11 17 51.2	−4	−5	Third „ 30
11 05 03.6	+5	+8	Fourth „ 42 40	11 18 45.4	−2	−2	Fourth „ 00
11 07 31.7	−2	−4	Mean 14 42 57.5	11 20 06.4	+2	+1	Mean 11 30 12.5
11 08 56.4	+2	+1	Level . . . . − 7	11 23 30	+10	+8	Level . . . . −5
Mean . . 11 03 58.42	−4	−10	Index . . . +360 00 08.5	Mean . . . 11 18 25.45	−2	−8	Index . . +345 17 02.5
True time . 8 31 06.48	−7		374 42 59	True time . 8 45 33.48	−5		356 47 10
Chron. fast . 2 32 51.94			Observed Z.D. 62 27 10	Chron. fast . 2 32 51.97			Observed Z.D. 59 27 51.7
			Ref. and Paral. + 1 38				Ref. and Paral. +1 26
360° + 08″.5 = 360° 00′ 08″.5			Semidiam . . − 15 47	360° − 14° 42′ 57″.5 = 345° 17′ 02″.7			Semidiam . . −15 47
			True Z.D. . . 62 13 01				True Z.D. . . 59 13 31

Chronometer, Fast {H. M. S. 2 32 51.91; 2 32 51.97} H. M. S. 2 32 51.95

July 25th P.M.; Barometer 30.08; Thermometer 73°; ☉'s U.L.

Chronometer.	Level.		Readings, &c.	Chronometer.	Level.		Readings, &c.
H. M. S.			° ′ ″	H. M. S.			° ′ ″
6 11 57.5	+3	+3	First Vernier 13 42 35	6 25 10	−1	−2	First Vernier 44 37 12
6 13 42.5	+8	+8	Second „ 20	6 26 30	+1	+1	Second „ 30 40
6 15 16	+2	+1	Third „ 45	6 28 27.8	−1	−1	Third „ 37 12
6 16 32.4	−3	−3	Fourth „ 10	6 30 22.3	0	0	Fourth „ 36 50
6 18 24.7	−1	−1	Mean 13 42 27.5	6 32 29	−3	−2	Mean 44 36 58.5
6 19 36	−4	−1	Level . . . . +4.5	6 33 51.3	−1	−1	Level . . . . −5
Mean . . . 6 15 54.85	+5	+4	Index . . . +360 00 08.5	Mean . . . 6 29 28.9	−5	−5	Index 346 17 32.5
True time . 3 43 01.5	+4.5		373 42 40.5	True time . 3 56 36.63	−5		390 54 25
Chron. fast . 2 32 53.35			Observed Z.D. 62 17 07	Chron. fast . 2 32 52.27			Observed Z.D. 65 09 04
			Ref. and Paral. +1 37				Ref. and Paral. +1 51
360° + 08″.5 = 360° 00′ 08″.5			Semidiam . +15 47	360° − 13° 42′ 27″.5 = 346° 17′ 32″.5			Semidiam . . +15 47
			True Z.D. . . 62 34 31				True Z.D. . . 65 26 42

Chronometer, Fast {H. M. S. 2 32 53.35; 2 32 52.27} H. M. S. 2 32 52.81

Bahia.——Determination of the Rate of the Chronometer by Zenith Distances, *continued.*

July 26 A.M.; Barometer 29.98; Thermometer 72°; ☉'s L.L.

Chronometer.		Level.		Readings, &c.	
	H. M. S.				° ′ ″
	10 56 08	−4	−6	First Vernier .	19 58 30
	10 57 12.8	−3	−5	Second „	58 20
	10 58 52.6	+3	+2	Third „	58 55
	10 59 51.5	+2	+1	Fourth „	58 30
	11 01 23	−0	−1	Mean . . .	19 58 34
	11 02 45.8	0	0	Index . . .	+360 00 08.5
Mean . . .	10 59 22.3	−2	−9	Level . . .	−5.5
True Time.	8 26 28.13	−5.5			379 58 37
Chron. fast.	2 32 54.17			Observed Z.D.	63 19 46
				Ref. and Paral.	+1 42.5
360°+08″.5 = 360° 00′ 08″.5				Semidiam . .	−15 46.7
				True Z.D. . .	63 05 42

Chronometer.		Level.		Readings, &c.	
	H. M. S.				° ′ ″
	11 09 57.8	−2	−2	First Vernier .	21 24 50
	11 11 33	+4	+5	Second „	24 40
	11 13 52.9	−3	−5	Third „	25 15
	11 14 58	0	0	Fourth „	24 40
	11 16 54.8	0	0	Mean . . .	21 24 49
	11 17 52.6	−1	−1	Index . . .	+340 01 26
Mean . . .	11 14 11.52	−2	−3	Level . . .	−2.5
True time .	8 41 17.27	−2.5			361 26 13.5
Chron. fast.	2 32 54.25			Observed Z.D.	60 14 22
				Ref. and Paral.	+1 29
360° − 19° 58′ 34″ = 340° 01′ 26″				Semidiam . .	−15 47
				True Z.D. .	60 00 04

Chronometer Fast { H. M. S. 2 32 54.17 ; 2 32 54.25 } H. M. S. 2 32 54.21

July 26 P.M.; Barometer 29.98; Thermometer 75°; ☉'s U.L.

Chronometer.		Level.		Readings, &c.	
	H. M. S.				° ′ ″
	6 09 39	−4	−4	First Vernier .	9 58 10
	6 11 05.3	0	0	Second „	58 10
	6 12 49.7	−3	−4	Third „	58 25
	6 14 16.8	−2	−2	Fourth „	58 15
	6 15 46.5	−4	−4	Mean . . .	9 58 15
	6 16 58.7	0	0	Index . . .	+360 00 08.5
Mean . .	6 13 26	−13	−14	Level . . .	−13.5
True time .	3 40 31.73	−13.5			369 58 10
Chron fast.	2 32 54.27			Observed Z.D.	61 39 42
				Ref. and Paral.	+1 35
360°+08″.5 = 360° 00′ 08″.5				Semidiam . .	+15 47
				True Z.D. . .	61 57 04

Chronometer.		Level.		Readings, &c.	
	H. M. S.				° ′ ″
	6 23 36	+2	+3	First Vernier .	37 26 55
	6 25 03.2	+4	+4	Second „	26 35
	6 26 26	−1	−2	Third „	27 00
	6 27 42.4	+8	+7	Fourth „	26 50
	6 29 46	+4	+4	Mean . .	37 26 50
	6 31 12.7	0	−1	Index . . .	+350 01 45
Mean . . .	6 27 17.7	+17	+15	Level . . .	+16
True time .	3 51 22.2	+16			387 28 51
Chron. fast.	2 32 55.5			Observed Z.D.	64 34 48.5
				Ref. and Paral.	+1 48.5
360° − 9° 58′ 15″ = 350° 01′ 45″				Semidiam . .	+15 47
				True Z.D. .	64 52 24

Chronometer Fast { H. M. S. 2 32 54.27 ; 2 32 55.5 } H. M. S. 2 32 54.9

BAHIA.——Determination of the Rate of the Chronometer by Zenith Distances, *continued.*

July 27th A.M.; Barometer 30.03; Thermometer 75°; ⊙'s L.L.

Chronometer.	Level.		Readings, &c.	
H. M. S.				° ′ ″
11 04 53	−4	−5	First Vernier	8 54 30
11 05 47.4	+8	+6	Second „	54 15
11 07 11.2	−2	−4	Third „	54 55
11 08 02	+6	+5	Fourth „	54 10
11 09 47.6	−1	−3	Mean	8 54 27.5
11 10 39.8	−2	−4	Index . . .	+360 00 08.5
Mean . . . 11 07 43.5	+5	−5	Level	0
True time . 8 34 47.2	0			368 54 36
Chron. fast . 2 32 56.3			Observed Z.D.	61 29 06
			Ref. and Paral.	+ 1 34
360° + 08″.5 = 360° 00′ 08″.5			Semidiam .	−15 47
			True Z.D. . .	61 14 53

Chronometer.	Level.		Readings, &c.	
H. M. S.				° ′ ″
11 16 17.9	−6	−8	First Vernier	3 06 00
11 18 09.9	+3	+2	Second „	05 55
11 19 20.3	−7	−9	Third „	06 30
11 20 07	−2	−2	Fourth „	06 00
11 21 17.8	+5	+3	Mean	3 06 06
11 22 13.9	+2	+1	Index . . .	+351 05 32.5
Mean . . . 11 19 34.47	−5	−13	Level	−9
True time . 8 46 37.9	−9			354 11 30
Chron. fast . 2 32 56.57			Observed Z.D.	59 01 55
			Ref. and Paral.	+ 1 24
360° − 8° 54′ 27″.5 = 351° 05′ 32″.5			Semidiam . .	−15 47
			True Z.D. . .	58 47 32

Chronometer, Fast {H. M. S. 2 32 56.3, 2 32 56.57} H. M. S. 2 32 56.43

July 27th P.M.; Barometer 30.02; Thermometer 76°; ⊙'s U.L.

Chronometer.	Level.		Readings, &c.	
H. M. S.				° ′ ″
6 13 29.7	−2	−1	First Vernier	13 03 35
6 11 33	+8	+7	Second „	03 10
6 15 54.5	+7	+6	Third „	03 45
6 16 59.5	−1	0	Fourth „	03 00
6 18 15.3	+1	0	Mean	13 03 22.5
6 19 24	−3	−4	Index . . .	+360 00 08.5
Mean . . . 6 16 26	+10	+8	Level	+9
True time . 3 43 28.13	+9			373 03 40
Chron. fast . 2 32 57.87			Observed Z.D.	62 10 36.7
			Ref. and Paral.	+ 1 37
360° + 08″.5 = 360° 00′ 08″.5			Semidiam . .	+15 47
			True Z.D. . .	62 28 01

Chronometer.	Level.		Readings, &c.	
H. M. S.				° ′ ″
6 29 02.4	+5	+6	First Vernier	45 30 10
6 30 55.8	+10	+8	Second „	30 15
6 31 16.9	0	0	Third „	30 10
6 32 02.8	+2	+1	Fourth „	29 45
6 33 21.8	−5	−6	Mean	45 29 57.5
6 34 16	−2	−2	Index . . .	+346 56 37.5
Mean . . . 6 31 39.3	+10	+7	Level . . .	+8.5
True time . 3 58 41.73	+8.5			392 26 43.5
Chron. fast . 2 32 57.57			Observed Z.D.	65 24 27.5
			Ref. and Paral.	+ 1 53
360° − 13° 03′ 22″.5 = 346° 56′ 37″.5			Semidiam . .	+15 47
			True Z.D. . .	65 42 07

Chronometer, Fast {H. M. S. 2 32 57.87, 2 32 57.57} H. M. S. 2 32 57.72

Bahia.——Determination of the Rate of the Chronometer by Zenith Distances, *continued.*

July 28th A.M.; Barometer 30.05; Thermometer 71°; ⊙'s L.L. (*strong breezes.*)

Chronometer.	Level.		Readings, &c.		Chronometer.	Level.		Readings, &c.	
H. M. S.				° ′ ″	H. M. S.				° ′ ″
11 07 16.2	+7	+5	First Vernier	4 29 20	11 20 29	−2	−5	First Vernier	353 24 15
11 08 31.6	−2	−5	Second „	29 05	11 21 28.7	+2	+1	Second „	24 05
11 10 20.8	−0	−2	Third „	29 50	11 22 48.8	+5	+3	Third „	24 35
11 11 20	−4	−6	Fourth „	29 25	11 23 46	+6	+4	Fourth „	24 00
11 13 07.6	−0	−2	Mean	4 29 25	11 25 10.5	+2	+1	Mean . . .	353 24 14
11 14 10	+6	+4	Index . . .	+360 00 08.5	11 26 16.8	+8	+7	Index . . .	−4 29 25
Mean . . 11 10 47.7	+7	−6	Level . . .	+0.5	Mean . . . 11 23 19.97	+21	+11	Level . . .	+16
True time . 8 37 48.07	+0.5			364 29 34	True time. . 8 50 21.57	+16			348 55 05
Chron. fast. 2 32 59.63			Observed Z.D.	60 44 56	Chron. fast.. 2 32 58.4			Observed Z.D.	58 09 11
			Ref. and Paral.	+1 30				Ref. and Paral.	+1 21
$360^{\circ} + 08''.5 = 360^{\circ}\ 00'\ 08''.5$			Semidiam . .	−15 47	Index − $4^{\circ}\ 29'\ 25''$			Semidiam . .	−15 47
			True Z.D. .	60 30 39				True Z.D.	. 57 54 45

Chronometer Fast { H. M. S. 2 32 59.63 / 2 32 58.4 } H. M. S. 2 32 59.01

July 28th P.M.; Barometer 30.00; Thermometer 71°; ⊙'s U.L.

Chronometer.	Level.		Readings, &c.		Chronometer.	Level.	Readings, &c.
H. M. S.				° ′ ″			
6 06 28	−6	−4	First Vernier	5 03 35			
6 08 02.4	−3	−4	Second „	03 10			
6 09 44.8	−3	−5	Third „	03 55			
6 11 18.8	+3	+2	Fourth „	03 05			
6 13 16.8	−2	−3	Mean	5 03 26			
6 14 46.8	+2	+1	Index	360 00 08.5			
Mean . . . 6 10 36.27	−9	−13	Level	−11			
True time . 3 37 35.68	−11 ...			365 03 23			
Chron. fast . 2 33 00.59			Observed Z.D.	60 50 34			
			Ref. and Paral.	+1 31			
$360^{\circ} + 08''.5 = 360^{\circ}\ 00'\ 08''.5$			Semidiam . .	+15 47			
			True Z.D. . .	61 07 52			

Chronometer Fast H. M. S. 2 33 00.59

BAHIA.—Determination of the Rate of the Chronometer by Zenith Distances, *continued.*

July 29 A.M.; Barometer 30.05; Thermometer 71°; ☉'s U.L.

Chronometer.	H. M. S.	Level.		Readings, &c.	° ′ ″
	11 09 03.6	+2	0	First Vernier .	00 31 00
	11 11 00	+5	+3	Second „	30 50
	11 13 08.8	−2	−4	Third „	31 30
	11 14 27.2	+6	+4	Fourth „	30 40
	11 15 59.6	−3	−6	Mean . . .	00 31 00
	11 17 08.8	0	0	Level . .	+2.5
Mean. . .	11 13 28	+8	−3	Index . . .	+360 00 08.5
True time .	8 40 26.38	+2.5			360 31 11
Chron. fast.	2 33 01.62			Observed Z.D.	60 05 12
				Ref. and Paral.	+1 28
360° + 08″.5 = 360° 00′ 08″.5				Semidiam . .	−15 47
				True Z.D. . .	59 50 53

Chronometer.	H. M. S.	Level.		Readings, &c.	° ′ ″
	11 22 09.6	0	−1	First Vernier.	345 50 10
	11 23 28.8	+7	+9	Second „	05
	11 25 21.2	+5	+7	Third „	30
	11 26 31.6	−2	0	Fourth „	05
	11 27 50	−2	−4	Mean . .	345 50 12.5
	11 29 06.8	−2	0	Level . . .	+08.5
Mean. . .	11 25 44.67	+6	+11	Index . . .	−0 31 00
True time .	8 52 42.32	+8.5			345 19 21
Chron. fast.	2 33 02.35			Observed Z.D.	57 33 13.5
				Ref. and Paral.	+1 18.5
Index 00° 31′ 00″				Semidiam . .	−15 47
				True Z.D. . .	57 18 45

Chronometer Fast { H. M. S. 2 33 01.62 / 2 32 02.35 } H. M. S. 2 33 01.98

July 29 P.M.; Barometer 30.02; Thermometer 72°; ☉'s U.L.

Chronometer.	H. M. S.	Level.		Readings, &c.	° ′ ″
	5 51 32	0	0	First Vernier.	346 57 50
	5 53 23.6	+8	+7	Second „	57 40
	5 55 34 4	+3	+2	Third „	58 05
	5 57 54.8	−9	−10	Fourth „	57 30
	6 00 00	−2	−1	Mean	346 57 46
	6 01 26.8	0	0	Level . .	−1
Mean . . .	5 56 38.6	0	−2	Index . . .	+ 08.5
True time .	3 23 35.2	−1			346 57 54
Chron. fast.	2 33 03.4			Observed Z.D.	57 49 39
				Ref. and Paral.	+1 18
Index 00° 00′ 08″.5				Semidiam . .	+15 47
				True Z.D. . .	58 06 44

Chronometer.	H. M. S.	Level.		Readings, &c.	° ′ ″
	6 09 12.4	+2	+1	First Vernier.	354 57 55
	6 10 49.4	+5	+4	Second „	58 00
	6 12 35.6	0	−1	Third „	58 30
	6 14 15.6	−7	−8	Fourth „	58 05
	6 16 22	0	0	Mean . . .	351 58 07.5
	6 17 42.4	+2	0	Level . . .	−1
Mean. . .	6 13 28.4	+2	−4	Index . . .	+13 02 11
True time.	3 30 25.47	−1			368 00 22.5
Chron. fast.	2 33 02.93			Observed Z.D	61 20 02
				Ref. and Paral.	+1 33
360° − 346° 57′ 46″ = 13° 02′ 14″				Semidiam . .	+15 47
				True Z.D. . .	61 37 22

Chronometer Fast { H. M. S. 2 33 03.4 / 2 33 02.93 } H. M. S. 2 33 03.16

BAHIA.—Determination of the Rate of the Chronometer by Zenith Distances, *continued*.

July 30th P.M.; Barometer 30.10; Thermometer 72°; ☉'s U.L.

Chronometer.		Level.		Readings, &c.	
	H. M. S.				° ′ ″
	6 25 33.6	0	−1	First Vernier	28 22 30
	6 27 33.2	+6	+4	Second „	22 05
	6 29 13.6	−4	−6	Third „	22 30
	6 30 45.6	−2	−3	Fourth „	21 48
	6 32 29.2	−6	−8	Mean	28 22 13
	6 33 56.8	0	0	Level	−10
Mean . . .	6 29 55.33	−6	−14	Index . . .	+360 00 08.5
True time .	3 56 49.63	−10			388 22 11.5
Chron. fast .	2 33 05.7			Observed Z.D.	64 43 42
				Ref. and Paral.	+ 1 49
360° + 08″.5 = 360° 00′ 08″.5				Semidiam . .	+15 47
				True Z.D. . . .	65 01 18

Chronometer.		Level.		Readings, &c.	
	H. M. S.				° ′ ″
	6 45 19.6	0	0	First Vernier	82 37 50
	6 46 42	+10	+8	Second „	37 30
	6 48 32	−1	−3	Third „	38 10
	6 50 28	−1	−3	Fourth „	37 40
	6 52 15.6	0	0	Mean	82 37 47.
	6 55 38.8	−2	0	Level	+3
Mean . . .	6 49 49.33	+4	2	Index . . .	+331 37 47
True time .	4 16 43.23	+3			414 15 37.5
Chron. fast .	2 33 06.1			Observed Z.D.	69 02 36
				Ref. and Paral.	+ 2 15
360° − 28° 22′ 13″ = 331° 37′ 47″				Semidiam . .	+15 47
				True Z.D. . . .	69 20 38

Chronometer, Fast $\left\{\begin{array}{l}\text{2 33 05.7}\\ \text{2 33 06.1}\end{array}\right\}$ 2 33 05.9 (H. M. S.)

July 31st P.M.; Barometer 30.05; Thermometer 71°; ☉'s U.L.

Chronometer.		Level.		Readings, &c.	
	H. M. S.				° ′ ″
	6 11 43.2	0	−1	First Vernier	95 36 50
	6 16 07.2	+4	+3	Second „	36 20
	6 17 37.6	−4	−3	Third „	37 00
	6 19 06.4	−1	−2	Fourth „	36 25
	6 20 50	−1	+2	Mean	95 36 39
	6 22 10.8	0	0	Level	−1.5
Mean . . .	6 18 25.87	−2	−1	Index	277 22 14
True time .	3 45 16.97	−1.5			372 58 53
Chron. fast .	2 33 08.9			Observed Z.D.	62 09 49
				Ref. and Paral.	+ 1 37
360° − 82° 37′ 46″ = 277° 22′ 14″				Semidiam . .	+15 47
				True Z.D. . . .	62 27 13

Chronometer.		Level.		Readings, &c.	
	H. M. S.				° ′ ″
	6 27 16	0	−1	First Vernier	124 45 15
	6 28 31.6	+2	0	Second „	44 50
	6 30 18.8	−2	−3	Third „	45 45
	6 31 42	+1	+3	Fourth „	45 10
	6 33 29.2	0	−2	Mean . . .	124 45 15
	6 34 42.8	−1	−2	Level . . .	−2.5
Mean . . .	6 31 00.07	0	−5	Index	264 23 21
True time .	3 57 51.33	−2.5			389 08 34
Chron. fast .	2 33 08.84			Observed Z.D.	64 51 26
				Ref. and Paral.	+ 1 50
360° − 95° 36′ 39″ = 264° 23′ 21″				Semidiam . .	+15 47
				True Z.D. . . .	65 09 03

Chronometer, Fast $\left\{\begin{array}{l}\text{2 33 08.9}\\ \text{2 33 08.84}\end{array}\right\}$ 2 33 08.87 (H. M. S.)

Bahia.——Determination of the Rate of the Chronometer by Zenith Distances, *continued.*

August 1 A.M.; Barometer 30.10; Thermometer 71°; ☉'s L.L.

	Chronometer.	Level.		Readings, &c.	
	H. M. S.				° ′ ″
	11 03 28.4	−2	−3	First Vernier .	87 03 10
	11 04 56	−2	−3	Second „	2 40
	11 06 49.2	+2	+1	Third „	3 20
	11 08 36.8	+4	+3	Fourth „	2 50
	11 14 59.6	−2	−3	Mean . . .	87 03 00
	11 16 47.6	−5	−6	Index . . .	276 45 09
Mean. . .	11 09 16.27	−5	−11	Level . . .	−8
True time .	8 36 05.77	−8			363 48 01
Chron. fast.	2 33 10.5			Observed Z.D	60 38 00
				Ref. and Paral.	+1 30
360°−83° 14′ 51″=276 45′ 09″				Semidiam . .	−15 47
				True Z.D. .	60 23 43

	Chronometer.	Level.		Readings, &c.	
	H. M. S.				° ′ ″
	11 21 13.6	−5	−4	First Vernier .	69 51 35
	11 22 58	−0	−1	Second „	51 10
	11 25 22	+3	+2	Third „	51 40
	11 26 43.6	+5	+4	Fourth „	50 50
	11 28 26.4	−1	−2	Mean . . .	69 51 19
	11 31 28.8	+3	+4	Index . .	272 57 00
Mean. . .	11 26 02.07	+5	+3	Level . .	+4
True time.	8 52 51.77	+4			342 48 23
Chron. fast.	2 33 10.3			Observed Z.D.	57 08 04
				Ref. and Paral.	+1 18
360°−87° 03′ 00″=272° 57′ 00″				Semidiam . .	−15 47
				True Z.D. . .	56 53 35

Chronometer Fast { H. M. S. 2 33 10.5, 2 33 10.3 } H. M. S. 2 33 10.4

August 1st P.M.; Barometer 30.05; Thermometer 73; ☉'s U.L.

	Chronometer.	Level.		Readings, &c.	
	H. M. S.				° ′ ″
	6 08 39	−3	−1	First Vernier	75 01 25
	6 10 28	+0	+1	Second „	01 05
	6 12 11.2	−3	−1	Third „	01 40
	6 13 35.2	−4	0	Fourth „	00 50
	6 15 21.2	−3	−5	Mean . . .	75 01 15
	6 16 35.6	−2	−4	Index	290 08 41
Mean	6 12 48.37	−12	−10	Level	−11
True time	3 39 37	−11			365 09 45
Chron. fast .	2 33 11.37			Observed Z.D.	60 51 37.5
				Ref. and Paral.	+1 31.5
360° − 69° 51′ 19″ = 290° 08′ 41″				Semidiam . .	+15 47
				True Z.D. . .	61 08 56

Chronometer Fast H. M. S. 2 33 11.37

BAHIA.—Determination of the Rate of the Chronometer by Zenith Distances, *continued.*

August 2d, A.M.; Barometer 30.10; Thermometer 70°; ☉'s L.L.

Chronometer.	Level.		Readings, &c.	
H. M. S.				° ′ ″
11 15 30	+4	+2	First Vernier	349 47 20
11 17 04	+4	+2	Second „	47 20
11 19 07.4	+6	+4	Third „	47 40
11 20 46	+6	+4	Fourth „	47 20
11 22 22.8	−7	−9	Mean........	349 47 25
11 24 04.4	−3	−5	Level........	+4
Mean...... 11 19 49.1	+10	−2	Index........	+8.5
True time.. 8 46 35.6	+4			349 47 37
Chron. fast. 2 33 13.5			Observed Z.D.	58 17 56
			Ref. and Paral.	+ 1 23
			Semidiam....	− 15 47
			True Z.D.....	58 03 32

Chronometer.	Level.		Readings, &c.	
H. M. S.				° ′ ″
11 28 32	0	0	First Vernier	323 59 20
11 29 46.4	+4	+2	Second „	59 30
11 31 46	0	−2	Third „	59 50
11 32 54.4	−6	−7	Fourth „	59 20
11 34 40	−3	−5	Mean........	323 59 30
11 36 40.4	+2	0	Index	+10 12 35
Mean...... 11 32 23.2	−3	−12	Level.......	− 7.5
True time.. 8 59 09.85	−7.5			334 11 58
Chron. fast. 2 33 13.35			Observed Z.D.	55 41 59.7
			Ref. and Paral.	+ 1 14.3
360° − 349° 47′ 25″ = 10° 12′ 35″			Semidiam....	− 15 47
			True Z.D.....	55 27 27

Chronometer, Fast { H. M. S. 2 33 13.5 ; 2 33 13.35 } H. M. S. 2 33 13.42

August 2d, P.M.; Barometer 30.06; Thermometer 70°; ☉'s U.L.

Chronometer.	Level.		Readings, &c.	
H. M. S.				° ′ ″
5 54 02.2	+1	0	First Vernier	312 56 00
5 55 41.8	−5	−5	Second „	55 50
6 01 02	+1	0	Third „	56 20
6 02 28.2	−4	−5	Fourth „	55 40
6 04 13.2	−6	−7	Mean........	312 55 57.5
6 05 34.8	+2	+1	Index.......	+36 00 30
Mean...... 6 00 34.37	−11	−16	Level........	+13.5
True time.. 3 27 16.27	−13.5			348 56 14
Chron. fast.. 2 33 14.1			Observed Z.D.	58 09 23
			Ref. and Paral.	+ 1 21
360° − 323° 59′ 30″ = 36° 00′ 30″			Semidiam....	+15 47
			True Z.D....	58 26 31

Chronometer.	Level.	Readings, &c.

Chronometer Fast H. M. S. 2 33 14.1

RATE DEDUCED from the PRECEDING OBSERVATIONS.

A.M. to A.M.				P.M. to P.M.					
Date.	S.	Date.	S.	Date.	S.	Date.	S.	Date.	S.
July 23 to 24	2.55	July 25 to 29	2.51	July 23 to 24	2.98	July 25 to 26	2.09	July 27 to 31	2.79
„ 25	2.9	„ Aug. 1	2.63	„ 25	2.79	„ 27	2.46	„ Aug. 1	2.73
„ 26	2.69	„ 2	2.69	„ 26	2.56	„ 28	2.59	„ 2	2.73
„ 27	2.57	July 26 to 27	2.22	„ 27	2.62	„ 29	2.59	July 28 to 29	2.57
„ 28	2.57	„ 28	2.4	„ 28	2.67	„ 30	2.62	„ 30	2.65
„ 29	2.64	„ 29	2.59	„ 29	2.66	„ 31	2.68	„ 31	2.76
„ Aug. 1	2.7	„ Aug. 1	2.7	„ 30	2.67	„ Aug. 1	2.65	„ Aug. 1	2.69
„ 2	2.73	„ 2	2.74	„ 31	2.71	„ 2	2.66	„ 2	2.66
July 24 to 25	3.26	July 27 to 28	2.58	„ Aug. 1	2.68	July 26 to 27	2.82	July 29 to 30	2.74
„ 26	2.76	„ 29	2.77	„ 2	2.69	„ 28	2.84	„ 31	2.85
„ 27	2.58	„ Aug. 1	2.79	July 24 to 25	2.61	„ 29	2.75	„ Aug. 1	2.74
„ 28	2.58	„ 2	2.83	„ 26	2.35	„ 30	2.75	„ 2	2.73
„ 29	2.66	July 28 to 29	2.97	„ 27	2.51	„ 31	2.8	July 30 to 31	2.97
„ Aug. 1	2.71	„ Aug. 1	2.85	„ 28	2.67	„ Aug. 1	2.74	„ Aug. 1	2.73
„ 2	2.75	„ 2	2.88	„ 29	2.59	„ 2	2.74	„ 2	2.73
July 25 to 26	2.26	Ju.29 to Au.1	2.81	„ 30	2.62	July 27 to 28	2.87	Ju.31 to Au.1	2.5
„ 27	2.24	„ 2	2.86	„ 31	2.67	„ 29	2.72	„ 2	2.62
„ 28	2.35	Aug. 1 to 2	3.02	„ Aug. 1	2.65	„ 30	2.74	Aug. 1 to 2	2.73
				„ 2	2.65				
Means . . .	2.64		2.71		2.65		2.67		2.72
2.675				2.68					

Gaining 2.68 Seconds per Diem.

Bahia.——Comparisons of the Astronomical Clock with the Chronometer No. 423, from the 23d of July to the 2d of August, 1822, inclusive; with the Clock's Rate on Mean Solar Time deduced.

1822.	Chronometer.	Clock.	Clock's Loss on 423.		Daily Rates. Chron.	Daily Rates. Clock.
	H. M. S.	H. M. S.	S.	S.	Gaining.	Losing.
July 23 P. M.		8 35 49				
„ 24 A. M.		8 34 50.3	58.7	118.4		
„ 24 P. M.		8 33 50.6	59.7			
„ 25 A. M.		8 32 51.4	59.2	118.4	S.	S.
„ 25 P. M.		8 31 52.2	59.2		2.68	115.9
„ 26 A. M.		8 30 53.2	59	118.5		
„ 26 P. M.		8 29 53.7	59.5			
„ 27 A. M.		8 28 54.4	59.3	119		
„ 27 P. M.		8 27 54.7	59.7			
„ 28 A. M.		8 26 55.7	59	118.9		
„ 28 P. M.	9 55 00	8 25 55.8	59.9			
„ 29 A. M.		8 24 56	59.8	118.8		
„ 29 P. M.		8 23 57	59			
„ 30 A. M.		8 22 57.4	59.6	119.4		
„ 30 P. M.		8 21 57.6	59.8		2.68	116.56
„ 31 A. M.		8 20 58	59.6	119.5		
„ 31 P. M.		8 19 58.1	59.9			
Aug. 1 A. M.		8 18 58.6	59.5	119.3		
„ 1 P. M.		8 17 58.8	59.8			
„ 2 A. M.		8 16 59	59.8	119.5		
„ 2 P. M.		8 15 59.3	59.7			

BAHIA.——COINCIDENCES OBSERVED with PENDULUM No. 3; the Clock making 86284.1 Vibrations in a Mean Solar Day.

DATE.	Barometer.	No. of Coincidence.	Temperature.	Time of Disappearance.	Time of Re-appearance.	True Time of Coincidence.	Arc of Vibration.	Mean Temperature.	Mean Interval.	Correction for the Arc.	Vibrations in 24 hours.	Reduction to a mean Temperature.	Reduced Vibrations at 75.2 Faht.
1822.	IN.		°	M. S.	M. S.	H. M. S.	°	°	S.	S.			
July 24 A.M.	29.970	1	74.2	58 46	58 48	9 58 47	1.18	74.95	642.5	+ 1.33	86016.83	−0.10	86016.73
		11	75.7	45 48	45 56	11 45 52	0.62						
„ 24 P.M.	29.960	1	75.7	06 51	06 55	3 06 53	1.2	75.8	641.55	1.36	86016.48	+0.25	86016.73
		11	75.9	53 41	53 56	4 53 48.5	0.64						
„ 25 A.M.	29.980	1	75.2	57 31	57 35	9 57 33	1.2	75.2	642.25	1.36	86016.78	. . .	86016.78
		11	75.2	44 29	44 42	11 44 35.5	0.64						
25 P.M.	29.990	1	76.8	22 01	22 03	3 22 02	1.2	76.8	640.2	1.36	86015.9	+0.67	86016.57
		11	76.8	8 39	8 49	5 8 44	0.64						
„ 26 A.M.	29.980	1	73.3	8 48	8 53	8 08 50.5	1.16	74.5	643	1.22	86016.92	−0.29	86016.63
		11	75.7	55 55	56 06	9 56 00.5	0.6						
„ 26 P.M.	29.980	1	74.2	14 10	14 13	1 14 11.5	1.18	74.7	642.9	1.29	86016.99	−0.21	86016.78
		11	75.2	01 16	01 25	3 01 20.5	0.6						
„ 27 A.M.	30.020	1	73.5	9 22	9 25	8 9 23.5	1.18	73.9	644.75	1.26	86017 72	−0.55	86017.17
		11	74.3	56 45	56 57	9 56 51	0.58						
„ 27 P.M.	30.020	1	76.2	30 41	30 45	3 30 43	1.16	75.95	642.95	1.19	86016.89	+0.31	86017.20
		11	75.7	17 47	17 58	5 17 52.5	0.58						
Means	29.990							75.2			86016.82		86016.82

BAHIA.——COINCIDENCES OBSERVED with PENDULUM No. 4; the Clock making 86283.44 Vibrations in a Mean Solar Day.

DATE.	Barometer.	No. of Coincidence.	Temperature.	Time of Disappearance.	Time of Re-appearance.	True Time of Coincidence.	Arc of Vibration.	Mean Temperature.	Mean Interval.	Correction for the Arc.	Vibrations in 24 hours.	Reduction to a mean Temperature.	Reduced Vibrations at 72.9 Faht.
1822.	IN.		°	M. S.	M. S.	H. M. S.	°	°	S.	S. +			
July 28 P.M.	30.050	1	73.3	55 52	55 57	1 55 54.5	1.2	73.3	670.35	1.28	86027.28	+0.17	86027.45
		11	73.3	47 32	47 44	3 47 38	0.6						
„ 29 A.M.	30.050	1	72.8	47 06	47 07	9 47 06.5	1.22	73	670.9	1.35	86027.57	+0.04	86027.61
		11	73.2	38 49	39 02	11 38 55.5	0.64						
„ 29 P.M.	30.020	1	72.8	20 16	20 17	8 20 16.5	1.24	73.05	670.25	1.41	86027.41	+0.06	86027.47
		11	73.3	11 53	12 05	10 11 59	0.64						
„ 30 A.M.	30.100	1	71.4	59 14	59 21	9 59 17.5	1.18	72.05	671.95	1.29	86027.89	−0.36	86027.53
		11	72.7	51 08	51 26	11 51 17	0.6						
„ 30 P.M.	30.050	1	73.8	37 41	37 45	2 37 43	1.18	73.7	669.95	1.29	86027.13	+0.34	86027.47
		11	73.6	29 16	29 29	4 29 22.5	0.6						
„ 31 A.M.	30.100	1	71	57 11	57 16	7 57 13.5	1.2	71.4	673	1.32	86028.32	−0.63	86027.69
		11	71.8	49.17	49 30	9 49 23.5	0.62						
August 1 A.M.	30.100	1	72.8	29 24	29 26	10 29 25	1.24	73	671.15	1.41	86027.73	+0.04	86027.77
		11	73.2	21 10	21 23	12 21 16.5	0.64						
„ 1 P.M.	30.100	1	74	58 52	58 56	2 58 54	1.2	73.9	670.35	1.32	86027.32	+0.42	86027.74
		11	73.8	50 26	50 49	4 50 37.5	0.62						
„ 2 A.M.	30.000	1	72.1	21 30	21 34	9 21 32	1.2	72.6	671.75	1.32	86027.84	−0.13	86027.71
		11	73.1	13 22	13 37	11 13 29.5	0.62						
Means . .	30.063							72.9			86027.61		86027.61

BAHIA.—BAROMETRICAL OBSERVATIONS to DETERMINE the HEIGHT of the PENDULUM STATION.

DATE.	PENDULUM STATION.			AT THE SEA.				Height above the Level of Half Tide.		REMARKS.
	Barometer.	Thermometer. Att.	Thermometer. Det.	Barometer.	Thermometer. Att.	Thermometer. Det.				
1822.	IN.	°	°	IN.	°	°		Fathoms.		
July 24, 6 A.M.	29.964	73	70.5							
„ 24, 7 A.M.	...	...	...	30.176	72.4	72.4	12 ft. above ½ tide	35.4		
„ 24, 8 A.M	29.976	73	71							
„ 25, 6½ A.M.	29.951	73	70							
„ 25, 7 A.M.	....	...	...	30.177	72.5	72.5	do. do.	35.88		
„ 25, 7½ A.M.	29.987	74	71							
„ 26, 6½ A.M.	29.976	69	69							$\frac{1}{68}$th of the difference in the Mercurial Columns has been added, on account of the relative capacities of the tube and cistern.
„ 26, 7¼ A.M.	....	...	...	30.196	70	70	at the level of ½ tide	35.2	Fms. 35.53 or Feet. 213	
„ 26, 8 A.M.	29.980	72	71							
„ 31, 6½ A.M.	30.068	70	69							
„ 31, 7 A.M.	....	...	...	30.302	71	71	do. do.	35.87		
„ 31, 7½ A.M.	30.086	71.5	69							
Aug. 1, 6½ A.M.	30.106	71.75	70.75							
„ 1, 7 A.M.	....	...	...	30.329	71.5	71.5	do. do.	35.3		
„ 1, 7½ A.M.	30.114	72.75	71.75							

MARANHAM.

The City of Maranham is built on a low and chiefly alluvial island, situated within the entrance of a large river of the same name. Being in latitude 2° 12′ S., it is only a few miles more distant from the equator than the Pendulum Station at the Island of St. Thomas; but the character of the two stations, in respect to the density of the materials near the surface, could scarcely have been more dissimilar, if they had been purposely selected; I felt, therefore, a more than ordinary interest in the experiments at Maranham, because I considered that, conjointly with those at St. Thomas's, they were calculated to furnish a very notable practical exemplification of the influence which the superficial density has on the general attraction of the mass; and of the extent to which the rate of a clock or pendulum may be made liable to differ, in the same latitude, by the circumstances of the locality alone.

Even more than our usual good fortune attended us, in the exceeding kindness with which we were received by Mr. Hesketh, His Majesty's Consul; whose anxious desire to forward the inquiry in which I was engaged, and to render our stay in every respect agreeable, cannot be sufficiently acknowledged, but will ever be most gratefully remembered. The credentials with which I had been furnished by the Portuguese Ambassador, were addressed to the authorities of Para, a city a few miles nearer the equator, to which I had designed to have gone, before I was aware that Maranham had advantages in many respects which made it preferable. Through the good offices of Mr. Hesketh, the members of the Provisional Government at Maranham, were induced to overlook the informality, and to receive the introduction with the same consideration as if it had been addressed directly to themselves.

A room on the ground-floor of Mr. Hesketh's house was better adapted for the pendulums than any in which they had hitherto been accommodated; it was an inner room, with brick partitions, and borrowed light, in which the temperature did not vary two degrees in the twenty-four hours; Mr. Hesketh was also kind enough to relinquish for my use an apartment in the upper story, with windows opening in the four principal directions, the sills of which were sufficiently stable to support the Repeating Circle, and enabled me to employ it with great convenience in the determination of the latitude, as well as in the comparison of the chronometers with time: the house being in the middle of the city and without grounds, had no situation in which a transit instrument might have been placed; but it adjoined a meadow belonging to the Cathedral, of which the Bishop, who was also the President of the Provisional Government, offered me the use; the times of sunrise and sunset, however, when a transit can be employed to most advantage in tropical climates, are also the most healthy and agreeable in taking exercise; and as I had had experience of the sufficiency of the repeating circle for the purposes for which a transit is used, I determined to employ the former in preference, whilst in the latitudes in which the heavenly bodies move most rapidly in altitude.

The tables, on this occasion, appear to require no particular comment, as every part of the operation was very satisfactorily accomplished.

By repeated barometrical measurements, the particulars of which are given in a table, the height of the pendulums above half-tide was shewn to be 77 feet.

The discussion of the apparent irregularities in the action of gravity, produced by the different quality of the superficial materials of the globe, will be best pursued when the whole of the stations shall have been gone through; but without entering further into the discussion at present, it may be proper to observe, that the result of the experiments at Maranham confirmed both the inferences which I had drawn at Sierra Leone; *viz.*, that the effects of differences in the geological character of the surface on the sum of the attractive forces are greater than had been anticipated, or at least greater than any expressed anticipation; and that the Ellipticity of the earth, as deducible by the pendulum, agreeably to the present modes of operation and deduction, appears greater than the extreme limit within which previous expectation had been bounded. I had also the satisfaction of perceiving by them, that although the results, at the several stations which I had hitherto visited, were not strictly correspondent with each other, their deviations were systematic, and such as I should have assigned to each respectively, from a knowledge of its geological character, agreeably to the scale of apportionment, with which the experiments at St. Thomas's and Maranham had furnished.

In the increased confidence with which I now looked forward to a deduction of the figure of the Earth, of a far more satisfactory and decisive character than had yet been obtained, as the ultimate result of a sufficiently extended multiplication of the experiments, I ventured to write from Maranham to Sir Humphry Davy, to propose the extension of the series to the higher latitudes in the summer of the following year.

We arrived at Maranham on the 21st of August, and quitted it with much regret on the 7th of September.

MARANHAM.——OBSERVATIONS to DETERMINE the RATE of the Chronometer No. 423, by ZENITH DISTANCES of the Sun, with a Repeating Circle; from the 24th of August to the 4th of September, 1822.

Latitude of the Place of Observation 2° 31′ 41″ S.; Longitude 44° 21′ W.

August 24th A.M.: Barometer 30.05; Thermometer 80°; ☉'s L.L.

	Chronometer. H. M. S.	Level.		Readings, &c.	° ′ ″
	11 11 08.4	+3	+2	First Vernier	108 56 05
	11 12 50	−2	−3	Second „	56 00
	11 14 40.4	0	0	Third „	56 30
	11 15 41.8	−2	−2	Fourth „	55 50
	11 17 44	−3	−3	Mean . . .	108 56 06
	11 19 46.8	0	0	Level	− 5
Mean . . .	11 15 19.07	−1	−6	Index . .	236 19 31
True time .	8 18 31.5				345 15 32
Chron. fast .	2 56 47.57	−5		Observed Z.D.	57 32 35
				Ref. and Paral.	+ 1 17
360° − 123° 40′ 29″ = 236° 19′ 31″				Semidiam . .	− 15 51
				True Z.D. . .	57 18 01

	Chronometer. H. M. S.	Level.		Readings, &c.	° ′ ″
	11 28 05.2	+2	+1	First Vernier	68 38 30
	11 30 32.4	+2	+1	Second „	20
	11 32 05.2	+3	+2	Third „	40
	11 33 42.8	−1	−2	Fourth „	15
	11 35 42.4	−4	−4	Mean	68 38 26
	11 38 03.6	+1	+1	Level	+1
Mean . . .	11 33 01.93	+3	−1	Index . .	251 03 54
True time .	8 36 14.23				319 42 21
Chron. fast .	2 56 47.7	+1		Observed Z.D.	53 17 03.5
				Ref. and Paral.	+1 05.5
360° − 108° 56′ 06″ = 251° 03′ 54″				Semidiam . .	−15 51
				True Z.D. . .	53 02 18

Chronometer, Fast $\left\{\begin{array}{l}\text{H. M. S.}\\ 2\ 56\ 47.57\\ 2\ 56\ 47.7\end{array}\right\}$ H. M. S. 2 56 47.63

August 24th P.M.; Barometer 29.95; Thermometer 80°; ☉'s U.L.

	Chronometer. H. M. S.	Level.		Readings, &c.	° ′ ″
	6 42 04	−6	−7	First Vernier	57 22 20
	6 43 46.8	0	0	Second „	21 50
	6 45 42	+2	+2	Third ..	22 20
	6 47 19.6	+5	+5	Fourth ..	21 30
	6 49 14	−6	−6	Mean	57 22 00
	6 50 40.4	−1	−1	Index . . .	+290 00 34
Mean . . .	6 46 28.13	−6	−7	Level	−6.5
True time .	3 49 39.66				347 22 28
Chron. fast .	2 56 48.47	−6.5		Observed Z.D.	57 53 45
				Ref. and Paral.	+ 1 20
360° − 69° 59′ 26″ = 290° 00′ 34″				Semidiam . .	+15 51
				True Z.D. . .	58 10 56

	Chronometer. H. M. S.	Level.		Readings, &c.	° ′ ″
	6 55 30	−5	−4	First Vernier	63 48 10
	6 56 34	+2	+3	Second „	47 50
	6 58 22.4	−4	−3	Third „	46 10
	7 00 29.2	+4	+5	Fourth „	47 40
	7 02 23.2	−6	−5	Mean	63 47 58
	7 04 26	+7	+7	Index . . .	+302 38 00
Mean . . .	6 59 37.47	−2	+3	Level	0
True time .	4 02 48.33				366 25 58
Chron. fast .	2 56 49.14	+0		Observed Z.D.	61 04 20
				Ref. and Paral.	+1 31
360° − 57° 22′ 00″ = 302° 38′ 00″				Semidiam . .	+15 51
				True Z.D. . .	61 21 42

Chronometer, Fast $\left\{\begin{array}{l}\text{H. M. S.}\\ 2\ 56\ 48.17\\ 2\ 56\ 49.14\end{array}\right\}$ H. M. S. 2 56 48.8

MARANHAM.——Determination of the Rate of the Chronometer by Zenith Distances, *continued.*

August 25th A.M.; Barometer 30.05; Thermometer 80°; ☉'s L.L.

Chronometer.	Level.		Readings, &c.	
H. M. S.				° ′ ″
10 54 38.4	−2	−3	First Vernier	7 35 10
10 56 36.8	+2	0	Second ,,	35 00
10 58 34	0	0	Third ,,	35 20
11 00 20	−8	−9	Fourth ,,	34 40
11 02 01.2	0	0	Mean........	7 35 02.5
11 04 40.8	−5	−6	Level........	−15
Mean......10 59 28.53	−13	−17	Index......+360 00 08.5	
True time.. 8 02 37.44	−15			367 34 55
Chron. fast.. 2 56 51.09			Observed Z.D.	61 15 49
			Ref. and Paral. +	1 30
360° + 08″.5 = 360° 00′ 08″.5			Semidiam .... −	15 52
			True Z.D.....	61 01 27

Chronometer.	Level.		Readings, &c.	
H. M. S.				° ′ ″
11 13 16	+1	+1	First Vernier	318 09 10
11 15 07.6	+5	+4	Second ,,	09 10
11 17 09.2	−3	−3	Third ,,	09 40
11 18 57.6	−6	−7	Fourth ,,	09 20
11 20 49.2	+1	0	Mean........	318 09 20
11 23 17.2	+1	0	Level........	−3
Mean......11 18 06.13	−1	−5	Index........−7 35 02	
True time.. 8 21 14.5	−3			340 34 15
Chron. fast.. 2 56 51.63			Observed Z.D.	56 45 43
			Ref. and Paral. +	1 15
Index − 7° 35′ 02″			Semidiam .... −	15 52
			True Z.D.....	56 31 06

Chronometer, Fast {H. M. S. 2 56 51.09; 2 56 51.63} H. M. S. 2 56 51.36

August 26th A.M.; Barometer 30.05; Thermometer 80°; ☉'s L. L.

Chronometer.	Level.		Readings, &c.	
H. M. S.				° ′ ″
10 43 37.2	+2	+1	First Vernier	22 42 30
10 45 29.2	−4	−4	Second ,,	42 20
10 47 17.2	+1	+1	Third ,,	42 40
10 49 34.8	−5	−5	Fourth ,,	41 50
10 51 50.8	0	0	Mean........	22 42 20
10 54 09.2	0	0	Level........	−6.5
Mean......10 48 39.73	−6	−7	Index......+360 00 08.5	
True time.. 7 51 44.9	−6.5			382 42 22
Chron. fast.. 2 56 54.83			Observed Z.D.	63 47 04
			Ref. and Paral. +	1 42
360° + 08″.5 = 360° 00′ 08″.5			Semidiam	15 52
			True Z.D.....	63 32 54

Chronometer.	Level.		Readings, &c.	
H. M. S.				° ′ ″
11 05 51.2	0	0	First Vernier	14 20 30
11 07 06.8	+2	+1	Second ,,	20 20
11 09 04.8	−3	−4	Third ,,	20 30
11 10 49.6	+6	+6	Fourth ,,	20 00
11 12 41.2	−9	−9	Mean........	14 20 20
11 14 24	0	0	Level........	−5
Mean......11 09 59.6	−4	−6	Index......+337 17 40	
True time.. 8 13 05.9	5			351 37 55
Chron. fast.. 2 56 53.7			Observed Z.D.	58 36 19
			Ref. and Paral.	+1 22
360° − 22° 42′ 20″ = 337° 17′ 40″			Semidiam	−15 52
			True Z.D.....	58 21 49

Chronometer, Fast {H. M. S. 2 56 54.83; 2 56 53.7} H. M. S. 2 56 54.26

MARANHAM.——Determination of the Rate of the Chronometer by Zenith Distances, *continued.*

August 27 A.M.; Barometer 30.04; Thermometer 80°; ☉'s L.L.

Chronometer.		Level.		Readings, &c.	
	H. M. S.				° ′ ″
	10 40 59.2	+3	+2	First Vernier	40 37 45
	10 42 52	0	0	Second „	37 30
	10 44 56	+5	+3	Third „	38 00
	10 46 45.2	0	−1	Fourth „	37 20
	10 48 41.6	0	−1	Mean . . .	40 37 39
	10 50 22	0	−1	Index . . .	345 39 40
Mean. . .	10 45 46	+8	+2	Level . . .	+5
True time .	7 48 49	+5			386 17 24
Chron. fast.	2 56 57			Observed Z.D.	64 22 54
				Ref. and Paral.	+1 45
360° − 14° 20′ 20″ = 345° 39′ 40″				Semidiam. .	−15 52
				True Z.D. . .	64 08 47

Chronometer.		Level.		Readings, &c.	
	H. M. S.				° ′ ″
	10 55 45.6	0	−1	First Vernier	285 01 10
	10 57 44	0	−2	Second „	01 00
	11 00 43.6	0	0	Third „	01 40
	11 02 49.2	0	0	Fourth „	00 40
Mean . . .	10 59 15.6	0	−3	Mean. . . .	285 01 07.5
True time .	8 02 18.9			Index . . .	−40 37 39
Chron. fast .	2 56 56.7	−1.5		Level . .	−1.5
					244 23 27
				Observed Z.D.	61 05 52
				Ref. and Paral.	+ 1 30
Index − 40° 37′ 39″				Semidiam . .	−15 52
				True Z.D. . .	60 51 30

Chronometer, Fast {H. M. S. 2 56 57, 2 56 56.7} H. M. S. 2 56 56.85

August 28 P.M.; Barometer 29.95; Thermometer 81°; ☉'s U.L.

Chronometer.		Level.		Readings, &c.	
	H. M. S.				° ′ ″
	6 01 55.6	−3	−2	First Vernier	132 59 40
	6 03 15.2	0	0	Second „	132 59 30
	6 05 00.8	0	0	Third „	133 00 20
	6 06 21.2	−4	−3	Fourth „	132 59 50
	6 08 00	+4	+3	Mean . . .	132 59 50
	6 09 44.4	−4	−2	Level . . .	−5.5
Mean. . .	6 05 42.87	−7	−4	Index . . .	+242 26 37.5
True time .	4 08 44.2				375 26 22
Chron. fast.	2 56 58.67	−5.5		Observed Z.D.	62 34 23.7
				Ref. and Paral.	+1 37.3
360° − 117° 33′ 22.5″ = 242° 26′ 37.5″				Semidiam. .	+15 52
				True Z.D. . .	62 51 53

Chronometer.		Level.		Readings, &c.	
	H. M. S.				° ′ ″
	6 14 08.8	−4	−2	First Vernier	167 11 50
	6 15 40	−15	−12	Second „	11 30
	6 17 33.6	+2	+4	Third „	12 10
	6 19 19.2	−6	−4	Fourth „	11 40
	6 21 10.4	+1	+3	Mean . . .	167 11 47.5
	6 23 20.6	+6	+7	Level . . .	−10
Mean. . .	6 18 32.1	−16	−4	Index . . .	+227 00 10
True time.	4 21 33.07				394 11 47.5
Chron. fast.	2 56 59.03	−10		Observed Z.D	65 41 58
				Ref. and Paral.	+1 52
360° − 132° 59′ 50″ = 227° 00′ 10″				Semidiam. .	+15 52
				True Z.D. .	65 59 42

Chronometer, Fast {H. M. S. 2 56 58.67, 2 56 59.03} H. M. S. 2 56 58.85

MARANHAM.——Determination of the Rate of the Chronometer by Zenith Distances, *continued.*

August 29th A.M.; Barometer 30.04; Thermometer 80°; ☉'s L.L.

Chronometer.	Level.		Readings, &c.		Chronometer.	Level.	Readings, &c.
H. M. S.				° ′ ″			
10 54 36.4	0	0	First Vernier	141 40 20			
10 57 02.8	−1	−2	Second „	00			
10 58 50.8	0	0	Third „	20			
11 00 12	0	0	Fourth „	10			
11 02 00	0	0	Mean	141 40 12.5			
11 03 25.6	0	0	Level	−1.5			
Mean . . . 10 59 21.27	−1	−2	Index	223 25 00			
True time . 8 02 20.07	−1.5			365 05 11			
Chron. fast. 2 57 01.2			Observed Z.D.	60 50 52			
			Ref. and Paral.	+ 1 30			
360° − 136° 35′ 00″ = 223° 25′ 00″			Semidiam . .	−15 52			
			True Z.D. . .	60 36 30			

H. M. S.
Chronometer Fast 2 57 01.2

August 29th P.M.; Barometer 29.95; Thermometer 81°; ☉'s U.L.

Chronometer.	Level.		Readings, &c.		Chronometer.	Level.		Readings, &c.	
H. M. S.				° ′ ″	H. M. S.				° ′ ″
6 24 19.6	−6	−6	First Vernier	90 28 40	7 08 19.2	−2	−2	First Vernier	116 26 10
6 26 28.8	+4	+5	Second „	28 30	7 09 57.6	0	0	Second „	25 50
6 28 32.8	0	0	Third „	29 20	7 11 44.8	+5	+7	Third „	26 20
6 29 51.2	+3	+4	Fourth „	28 30	7 13 48.4	0	0	Fourth „	25 55
6 32 16	+3	+4	Mean	90 28 45	7 15 38.4	−10	−8	Mean	116 26 04
6 33 42	−10	−8	Index . . .	+231 42 45	7 17 36.4	+3	+4	Index . . .	+269 31 15
Mean . . . 6 29 11.73	−6	−1	Level	−3	Mean . . . 7 12 50.8	−4	+1	Level	−1.5
True time . 3 32 10.78	−3			322 11 27	True time. . 4 15 48.57	−1.5			385 57 17
Chron. fast. 2 57 00.95			Observed Z.D.	53 41 54	Chron. fast. . 2 57 02.23			Observed Z.D.	64 19 33
			Ref. and Paral.	+ 1 07				Ref. and Paral.	+ 1 45
360° − 128° 17′ 15″ = 231° 42′ 45″			Semidiam . .	+15 52	360° − 90° 28′ 45″ = 269° 31′ 15″			Semidiam . .	−15 52
			True Z.D. . .	53 58 54				True Z.D. . .	64 37 10

Chronometer Fast {H. M. S. 2 57 00.95; 2 57 02.23} H. M. S. 2 57 01.59

MARANHAM.—Determination of the Rate of the Chronometer by Zenith Distances, *continued.*

August 30th A.M.; Barometer 30.03; Thermometer 80°; ⊙'s L.L.

Chronometer.		Level.		Readings, &c.		Chronometer.	Level.	Readings, &c.
	H. M. S.				° ′ ″			
	10 52 21.4	+3	+2	First Vernier .	83 41 40			
	10 54 54.8	+3	+3	Second ,,	41 15			
	10 58 28.8	−8	−9	Third ,,	42 00			
	11 00 32.8	−2	−3	Fourth ,,	41 20			
	11 03 07.2	+2	+3	Mean	83 41 34			
	11 05 12.4	−1	−1	Index . . .	+281 05 37.5			
Mean. . .	10 59 06.23	−3	−5	Level . . .	−4			
True time .	8 02 02.42				364 47 07			
Chron. fast.	2 57 03.81	−4		Observed Z.D.	60 47 51			
				Ref. and Paral.	+ 1 30			
360° − 78° 54′ 22″.5 = 281° 05′ 37″.5				Semidiam .	−15 52			
				True Z.D. . .	60 33 29			

Chronometer, Fast 2 H. 57 M. 03.81 S.

August 30th P.M.; Barometer 30.00; Thermometer 81°; ⊙'s U.L.

Chronometer.		Level.		Readings, &c.	
	H. M. S.				° ′ ″
	6 28 27.6	+7	+8	First Vernier	327 32 00
	0 20 52	9	9	Second ,,	01 50
	6 32 06	+3	+4	Third ,,	32 20
	6 33 34.4	+3	+4	Fourth ..	32 00
	6 35 50	−4	−2	Mean	327 32 02.5
	6 37 25.2	−5	−4	Index	+08.5
Mean . .	6 32 52.53	+2	+8	Level	+05
True time	3 35 48.07				327 32 16
Chron. fast .	2 57 04.46	+5		Observed Z.D.	54 35 23
				Ref. and Paral.	+ 1 08
				Semidiam .	+15 52
				True Z.D. . .	54 52 23

Chronometer.		Level.		Readings, &c.	
	H. M. S.				° ′ ″
	6 42 41.8	0	0	First Vernier	315 11 35
	0 44 21.2	0	0	Second ,,	11 20
	6 45 56.8	+5	+7	Third ,,	11 40
	6 47 16	−9	−7	Fourth ,,	11 00
	6 49 04.4	+9	+11	Mean	315 11 24
	6 50 30.4	−8	−6	Index . . .	+ 32 27 57.5
Mean. . .	6 46 38.93	−3	+5	Level	+1
True time.	3 49 34.27				347 39 22.5
Chron. fast.	2 57 04.66	+1		Observed Z.D.	57 56 34
				Ref. and Paral.	+ 1 22
360° − 327° 32′ 02″.5 = 32° 27′ 57″.5				Semidiam . .	+15 52
				True Z.D. . .	58 13 48

Chronometer, Fast { 2 H. 57 M. 04.46 S. / 2 57 04.66 } 2 H. 57 M. 04.56 S.

MARANHAM.——Determination of the Rate of the Chronometer by Zenith Distances, *continued*.

August 31st A.M.; Barometer 30.05; Thermometer 80°; ☉'s L.L.

Chronometer.		Level.		Readings, &c.		Chronometer.		Level.		Readings, &c.	
	H. M. S.				° ′ ″		H. M. S.				° ′ ″
	10 54 30	−4	−4	First Vernier	10 02 50		11 09 37.2	−2	−4	First Vernier	350 45 00
	10 56 18.4	+2	+2	Second „	02 30		11 11 38.8	+4	+5	Second „	45 00
	10 59 13.2	0	0	Third „	03 00		11 14 27.2	0	0	Third „	45 40
	11 00 55.2	+4	+5	Fourth „	02 10		11 16 06	0	0	Fourth „	45 00
	11 02 58	0	0	Mean . . .	10 02 37		11 17 59.6	+1	+1	Mean . . .	350 45 10
	11 04 54.8	−3	−3	Index . . .	+353 00 49		11 20 31.2	+5	+5	Index . . .	−10 02 37
Mean. . .	10 59 48.27	−1	0	Level . . .	−0.5	Mean. . .	11 15 03.33	+8	+7	Level . . .	+07.5
True time .	8 02 42.17	−0.5			363 03 25	True time .	8 17 57.4	+7.5			340 42 51
Chron. fast	2 57 06.1			Observed Z.D.	60 30 34	Chron. fast	2 57 05.93			Observed Z.D.	56 47 09
				Ref. and Paral.	+1 30					Ref. and Paral.	+1 15
360−6° 59′ 11″=353° 00′ 49″				Semidiam . .	−15 53	Index −10° 02′ 37″				Semidiam . .	−15 53
				True Z.D. . .	60 16 11					True Z.D. . .	56 32 31

Chronometer, Fast {H. M. S. 2 57 06.1, 2 57 05.93} H. M. S. 2 57 06.01

August 31st P.M.; Barometer 39.95; Thermometer 81°; ☉'s U.L.

Chronometer.		Level.		Readings, &c.		Chronometer.		Level.		Readings, &c.	
	H. M. S.				° ′ ″		H. M. S.				° ′ ″
	6 18 08.4	0	0	First Vernier	302 59 20		6 35 08.4	+7	+8	First Vernier	280 39 10
	6 19 49.2	−2	−1	Second „	59 20		6 36 56	0	0	Second „	39 20
	6 21 46	+2	+1	Third „	59 50		6 38 45.6	+1	+2	Third „	39 40
	6 23 24	0	0	Fourth „	59 20		6 40 12.8	0	0	Fourth „	39 00
	6 24 59.2	0	0	Mean . . .	302 59 27.5		6 43 06.4	+2	0	Mean . .	280 39 17.5
	6 26 10	−2	−1	Index . . .	+9 14 50		6 44 35.2	0	0	Index . .	+57 00 32.5
Mean . . .	6 22 22.8	−2	−1	Level . . .	−1.5	Mean . . .	6 39 47.4	+10	+10	Level . . .	+10
True time .	3 25 15.9	−1.5			312 14 16	True time .	3 42 39.86	+10			337 40 00
Chron fast .	2 57 06.9			Observed Z.D.	52 02 23	Chron. fast .	2 57 07.54			Observed Z.D.	56 16 40
				Ref. and Paral.	+1 03					Ref. and Paral.	+1 15
360° − 350° 45′ 10″ = 9° 14′ 50″				Semidiam . .	+15 53	360° − 302° 59′ 27″.5 = 57° 00′ 32″.5				Semidiam . .	+15 58
				True Z.D. . .	52 19 19					True Z.D. . .	56 33 48

Chronometer, Fast {H. M. S. 2 57 06.9, 2 57 07.54} H. M. S. 2 57 07.2

MARANHAM.——Determination of the Rate of the Chronometer by Zenith Distances, *continued.*

September 1st P.M.; Barometer 29.95; Thermometer 81°. ☉'s U.L.

	Chronometer.	Level.		Readings, &c.	
	H. M. S.				° ′ ″
	6 24 15.6	+4	+3	First Vernier	322 29 20
	6 26 50.4	+4	+3	Second "	29 30
	6 28 41.2	−5	−3	Third "	30 00
	6 30 22.4	0	0	Fourth "	29 30
	6 32 21.6	−4	−2	Mean . . .	322 29 35
	6 33 51.6	+5	+7	Index . . .	+08.5
Mean. . .	6 29 23.8	+4	+8	Level . . .	+06
True time .	3 32 13.2	+6			322 29 49
Chron. fast	2 57 10.6			Observed Z.D.	53 44 58
				Ref. and Paral.	+ 1 06
				Semidiam .	+15 52
				True Z.D. . . .	54 01 56

	Chronometer.	Level.		Readings, &c.	
	H. M. S.				° ′ ″
	6 40 40.4	+6	+4	First Vernier	308 07 20
	6 42 37.6	−2	−1	Second "	07 10
	6 44 26	+2	+1	Third "	07 50
	6 45 59.6	+4	+2	Fourth "	07 10
	6 47 56	0	0	Mean . . .	308 07 22
	6 49 29.2	+4	+5	Index . . .	+37 30 25
Mean. . .	6 45 11.47	+14	+11	Level . . .	+12.5
True time .	3 48 00.37	+12.5			345 37 59
Chron. fast	2 57 11.1			Observed Z.D.	57 36 17.5
				Ref. and Paral.	+ 1 21
				Semidiam . .	+15 52
				True Z.D. . . .	57 53 30

360° − 322° 29′ 35″ = 37° 30′ 25″

Chronometer, Fast { H. M. S. 2 57 10.6 / 2 57 11.1 } H. M. S. 2 57 10.85

September 2d P.M.; Barometer 29.94; Thermometer 81°; ☉'s U.L.

	Chronometer.	Level.		Readings, &c.	
	H. M. S.				° ′ ″
	6 39 07.2	−5	−5	First Vernier	290 33 00
	6 40 25.2	+3	+4	Second "	32 50
	6 41 55.6	−4	−3	Third "	33 20
	6 43 24.6	−7	−5	Fourth "	32 20
	6 45 29.6	−7	−7	Mean . . .	290 32 52.5
	6 47 21.2	+6	+7	Index . . .	51 52 38
Mean. . .	6 42 57.23	−14	−9	Level . . .	−11.5
True time .	3 45 43.11	−11.5			342 25 19
Chron. fast	2 57 14.12			Observed Z.D.	57 04 13
				Ref. and Paral.	+1 17
				Semidiam . .	+15 53
				True Z.D. . . .	57 21 23

360° − 308° 17′ 22″ = 51° 52′ 38″

	Chronometer.	Level.		Readings, &c.	
	H. M. S.				° ′ ″
	6 52 31.6	0	0	First Vernier	292 33 15
	6 53 58	0	0	Second "	33 05
	6 55 38.8	−4	−2	Third "	33 30
	6 57 04.8	0	0	Fourth "	32 45
	6 58 32	+3	+2	Mean . . .	292 33 09
	6 59 52.4	+2	+4	Index . . .	69 27 07.5
Mean. . .	6 56 16.27	+1	+4	Level . . .	+2.5
True time .	3 59 02.37	+2.5			362 00 19
Chron. fast	2 57 13.9			Observed Z.D.	60 20 03
				Ref. and Paral.	+1 30
				Semidiam . .	+15 53
				True Z.D. . . .	60 37 26

360° − 290° 32′ 52″.5 = 69° 27′ 07″.5

Chronometer, Fast { H. M. S. 2 57 14.12 / 2 57 13.9 } H. M. S. 2 57 14.01

MARANHAM.—Determination of the Rate of the Chronometer by Zenith Distances, *continued*.

September 3d A.M.; Barometer 30.05; Thermometer 80°; ☉'s L.L.

	Chronometer.	Level.			
	H. M. S.				° ′ ″
	10 49 28.4	−4	−4	First Vernier	192 51 15
	10 51 32.8	+8	+7	Second „	51 00
	10 53 18.4	+3	+2	Third „	51 40
	10 54 57.6	0	0	Fourth „	51 00
	10 57 32.4	+2	+2	Mean	192 51 14
	10 59 07.6	−1	−1	Index . . .	+176 14 09
Mean . . .	10 54 19.53	+8	+6	Level . . .	+7
True Time.	7 57 04.6	+7			369 05 30
Chron. fast.	2 57 14.93			Observed Z.D.	61 30 55
				Ref. and Paral.	+ 1 30
360° − 183° 45′ 51″ = 176° 14′ 09″				Semidiam. . .	−15 54
				True Z.D. . .	61 16 32

	Chronometer.	Level.			
	H. M. S.				° ′ ″
	11 03 46.8	0	0	First Vernier	182 33 00
	11 05 15.2	0	0	Second „	33 00
	11 06 48.4	−5	−6	Third „	33 30
	11 08 17.2	−2	−1	Fourth „	32 30
	11 09 51.2	+3	+3	Mean	182 33 00
	11 11 06	0	0	Index . . .	+167 08 46
Mean . . .	11 07 30.8	−4	−4	Level	−4
True time.	8 10 14.54	−4			349 41 42
Chron. fast.	2 57 16.26			Observed Z.D.	58 16 57
				Ref. and Paral.	+ 1 23
360° − 192° 51′ 14″ = 167° 08′ 46″				Semidiam. . .	−15 51
				True Z.D. . .	58 02 26

Chronometer Fast {H. M. S. 2 57 14.93; 2 57 16.26} H. M. S. 2 57 15.59

September 3d P.M.; Barometer 29.95; Thermometer 81°; ☉'s U.L.

	Chronometer.	Level.		Readings, &c.	
	H. M. S.				° ′ ″
	6 23 07.6	+1	+0	First Vernier	141 39 00
	6 24 35.4	+7	+8	Second „	38 40
	6 26 26.8	−4	−3	Third „	39 30
	6 27 44.4	+10	+11	Fourth „	39 00
	6 29 20.4	−5	−5	Mean	141 39 02.5
	6 30 52	−3	−2	Index . . .	+177 27 00
Mean . . .	6 27 01.1	+6	+9	Level	+7.5
True time.	3 29 44.77	+7.5			319 06 10
Chron. fast.	2 57 16.33			Observed Z.D.	53 11 02
				Ref. and Paral.	+ 1 04
360° − 182° 33′ 00″ = 177° 27′ 00″				Semidiam. . .	+15 54
				True Z.D. . .	53 28 00

	Chronometer.	Level.		Readings, &c.	
	H. M. S.				° ′ ″
	6 36 03.6	+2	+4	First Vernier	124 51 10
	6 38 22.8	0	0	Second „	51 00
	6 43 15.4	−5	−5	Third „	51 20
	6 44 40.4	−5	−4	Fourth „	50 40
	6 46 07.2	−1	−2	Mean . .	124 51 02.5
	6 51 54.4	+6	+7	Index . . .	+218 20 58
Mean . . .	6 43 23.97	−3	0	Level	−1.5
True time.	3 46 08.37	−1.5			343 11 59
Chron. fast.	2 57 15.6			Observed Z.D.	57 11 59
				Ref. and Paral.	+ 1 17
				Semidiam. . .	+15 54
				True Z.D. .	57 29 10

Chronometer Fast {H. M. S. 2 57 16.33; 2 57 15.6} H. M. S. 2 57 15.96

MARANHAM.—Determination of the Rate of the Chronometer by Zenith Distances, *continued.*

September 4th A.M.; Barometer 30.04; Thermometer 80°; ☉'s L.L. (Flying Clouds.)

	Chronometer. H. M. S.	Level.		Readings, &c.	° ′ ″
	10 51 40	−2	−1	First Vernier	128 18 00
	10 53 27.2	+4	+2	Second ,,	17 40
	10 56 06.8	−13	−13	Third ,,	18 20
	11 00 06	0	0	Fourth ,,	17 50
	11 01 40	+2	+2	Mean . . .	128 17 57.5
	11 03 04.4	+4	+4	Index . . .	+235 08 57.5
Mean . . .	10 57 40.73	−5	−6	Level . . .	−5.5
True time .	8 00 23.67	−5.5			363 26 50
Chron. fast	2 57 17.06			Observed Z.D.	60 34 28
				Ref. and Paral.	+1 28
360° − 124° 51′ 02″.5 = 235° 08′ 57″.5				Semidiam . .	−15 54
				True Z.D. . .	60 20 02

	Chronometer. H. M. S.	Level.		Readings, &c.	° ′ ″
	11 06 44	0	0	First Vernier	109 41 30
	11 07 57.6	+5	+4	Second ,,	41 10
	11 09 58	−3	−3	Third ,,	41 50
	11 15 26.4	+8	+7	Fourth ,,	41 00
	11 17 02.4	−3	−4	Mean . . .	109 41 22.5
	11 18 44.4	0	0	Index . . .	+231 42 02
Mean. . .	11 12 38.8	+7	+4	Level . . .	+5.5
True time .	8 15 21.2	+5.5			341 23 30
Chron. fast	2 57 17.6			Observed Z.D.	56 53 55
				Ref. and Paral.	+1 17
360° − 128° 17′ 58″ = 231° 42′ 02″				Semidiam . .	−15 54
				True Z.D. . .	56 39 18

Chronometer Fast { H. M. S. 2 57 17.06 / 2 57 17.6 } H. M. S. 2 57 17.33

September 4th P.M.; Barometer 29.95; Thermometer 81°; ☉'s U.L.

	Chronometer. H. M. S.	Level.		Readings, &c.	° ′ ″
	7 01 07.2	0	0	First Vernier	124 45 40
	7 02 33.2	+1	+4	Second ,,	45 30
	7 04 17.2	+3	+4	Third ,,	46 10
	7 05 32.4	−14	−14	Fourth ,,	45 30
	7 07 22.8	+2	+3	Mean . . .	124 45 42.5
	7 08 39.2	0	0	Index . . .	+250 18 37.5
Mean. . .	7 04 55.33	−8	−3	Level . . .	−5.5
True time .	4 07 37.77	−5.5			375 04 15
Chron. fast	2 57 17.56			Observed Z.D.	62 30 42.5
				Ref. and Paral.	+1 41
360° − 109° 41′ 22″.5 = 250° 18′ 37″.5				Semidiam . .	+15 53.5
				True Z.D. . .	62 48 17

	Chronometer. H. M. S.	Level.		Readings, &c.	° ′ ″
	7 15 10.4	+7	+8	First Vernier	160 19 40
	7 16 58	+1	+2	Second ,,	19 40
	7 18 08	+1	+2	Third ,,	20 00
	7 19 19.6	−7	−6	Fourth ,,	19 40
	7 21 04.4	−5	−4	Mean . . .	160 19 45
	7 22 31.6	+5	+7	Index . . .	+235 14 17.5
Mean. . .	7 18 48.67	+2	+9	Level . . .	+5.5
True time .	4 21 30.37	+5.5			395 34 08
Chron. fast	2 57 18.3			Observed Z.D	65 55 41
				Ref. and Paral.	+1 59
360° − 124° 45′ 42″.5 = 235° 14′ 17″.5				Semidiam . .	+15 54
				True Z.D. . .	66 13 34

Chronometer Fast { H. M. S. 2 57 17.56 / 2 57 18 3 } H. M. S. 2 57 17.93

RATE DEDUCED from the PRECEDING OBSERVATIONS.

A.M. to A.M.				P.M. to P.M.			
Date.	S.	Date.	S.	Date.	S.	Date.	S.
Aug. 24 to Au. 25	3.73	Aug. 26 to Au. 31	2 35	Aug. 24 to 28	2.51	Aug. 29 to Sept. 2	3.1
„ 26	3.31	„ Sept. 3	2.67	„ 29	2.56	„ 3	2.87
„ 27	3.07	„ 4	2.56	„ 30	2.63	„ 4	2.72
„ 29	2.71	Aug. 27 to Au. 29	2.17	„ 31	2.63	Aug. 30 to 31	2.64
„ 30	2.7	„ 30	2.32	„ Sept. 1	2.75	„ Sept. 1	3.15
„ 31	2.63	„ 31	2.29	„ 2	2.8	„ 2	3.15
„ Sept. 3	2.8	„ Sept. 3	2.69	„ 3	2.72	„ 3	2.85
„ 4	2.7	„ 4	2.59	„ 4	2.65	„ 4	2.67
Aug. 25 to Au. 26	2.9	Aug. 29 to Au. 30	2.61	Aug. 28 to 29	2.74	Aug. 31 to Sept. 1	3.65
„ 27	2.74	„ 31	2.4	„ 30	2 85	„ 2	3.4
„ 29	2.46	„ Sept. 3	2.88	„ 31	2.78	„ 3	2.92
„ 30	2.49	„ 4	2.69	„ Sept. 1	3.00	„ 4	2.68
„ 31	2.44	Aug. 30 to Au. 31	2.2	„ 2	3.03	Sept. 1 to Sept. 2	3.16
„ Sept. 3	2.69	„ Sept. 3	2.95	„ 3	2.85	„ 3	2.55
„ 4	2.60	„ 4	2.7	„ 4	2.73	„ 4	2.53
Aug. 26 to Au. 27	2.59	Aug. 31 to Sept. 3	3.19	Aug. 29 to 30	2.57	Sept. 2 to Sept. 3	1.95
„ 29	2.31	„ 4	2.83	„ 31	2.8	„ 4	1.96
„ 30	2.39	Sept. 3 to Sept. 4	1.74	„ Sept. 1	3.09	Sept. 3 to Sept. 4	1.97
Means	2.645				2.765		

Gaining 2.7 Seconds per Diem.

MARANHAM.——Comparisons of the Astronomical Clock with the Chronometer, No. 423, from the 24th of August to the 4th of September, 1822; with the Clock's Rate on Mean Solar Time deduced.

1822.	Chronometer.	Clock.	Clock's loss on 423.			DAILY RATES. Chron.	DAILY RATES. Clock.
	H. M. S.	H. M. S.	S.	S.		Gaining.	Losing.
Aug. 24 A. M.	12 05 00	2 22 03					
„ 24 P. M.		2 20 54.6	68.4	136.6			
„ 25 A. M.		2 19 46.4	68.2				
„ 25 P. M.		2 18 38.4	68	135.9			
„ 26 A. M.		2 17 30.5	67.9				
„ 26 P. M.		2 16 22.6	67.9	135.8			
„ 27 A. M.		2 15 14.7	67.9		s. 136.05	s. 2.7	s. 133.35
„ 27 P. M.		2 14 06.9	67.8	135.7			
„ 28 A. M.		2 12 59	67.9				
„ 28 P. M.		2 11 50.9	68.1	136			
„ 29 A. M.		2 10 43	67.9				
„ 29 P. M.		2 09 34.7	68.3	136.3			
„ 30 A. M.		2 08 26.7	68				
„ 30 P. M.		2 07 18.6	68.1	136.1			
„ 31 A. M.		2 06 10.6	68				
„ 31 P. M.		2 05 02.4	68.2	136.4			
Sept. 1 A. M.		2 03 54.2	68.2				
„ 1 P. M.		2 02 45.9	68.3	136.2			
„ 2 A. M.		2 01 38	67.9		135.7	2.7	133
„ 2 P. M.		2 00 30	68	135.6			
„ 3 A. M.		1 59 22.4	67.6				
„ 3 P. M.		1 58 15.1	67.3	134.7			
„ 4 A. M.		1 57 07.7	67.4				
„ 4 P. M.		1 56 00.1	67.6	135.2			

MARANHAM.——COINCIDENCES OBSERVED with PENDULUM 3; the Clock making 86266.65 Vibrations in a Mean Solar Day.													
DATE.	Barometer.	No. of Coincidence.	Temperature.	Time of Disappearance.	Time of Re-appearance.	True Time of Coincidence.	Arc of Vibration.	Mean Temperature.	Mean Interval.	Correction for the Arc.	Vibrations in 24 hours.	Reduction to a Mean Temperature.	Reduced Vibrations at 38°.1 Faht.
1822.			°	M. S.	M. S.	H. M. S.	°	°	S.	° +			
Aug. 24 A.M.	30.050	1	83	48 25	48 26	10 48 25.5	1.22	83.25	644.95	1.36	86000.50	+0.6	86001.10
		11	83.5	35 50	36 00	12 35 55	0.64						
„ 24 P.M.	29.960	1	83.3	29 14	29 17	2 29 15.5	1.18	83.55	644.9	1 32	86000.42	+0.72	86001.14
		11	83.8	16 40	16 49	4 16 44 5	0.62						
„ 25 A.M.	30.050	1	82.8	02 59	03 00	8 02 59.5	1.2	82.85	645.6	1.32	86000.72	+0.42	86001.14
		11	82.9	50 31	50 40	9 50 35.5	0.62						
„ 25 P.M.	29.950	1	81.9	34 42	34 47	2 34 44.5	1.16	81.8	648	1.19	86001.57	....	86001.57
		11	81.7	22 40	22 49	4 22 44.5	0.58						
„ 26 A.M.	30.050	1	81.2	11 27	11 28	8 11 27.5	1.18	81.35	648.3	1.22	86001.72	−0.20	86001.52
		11	81.5	59 25	59 36	9 59 30.5	0.58						
„ 26 P.M.	29.970	1	82	30 48	30 51	1 30 49.5	1.18	82	647.65	1.22	86001.46	+0.07	86001.53
		11	82	18 41	18 51	3 18 46	0.58						
„ 27 A.M.	30.050	1	81	8 08	8 09	8 8 08.5	1.18	81.1	648.65	1.22	86001.86	−0.31	86001.55
		11	81.2	56 10	56 20	9 56 15	0.58						
„ 27 P.M.	29.950	1	81.6	9 16	9 18	2 9 17	1.2	81.25	648.25	1.28	86001.78	−0.24	86001.54
		11	80.9	57 14	57 25	3 57 19.5	0.6						
„ 28 P.M.	30.000	1	80.2	00 14	00 17	1 00 15.5	1.18	80.3	649.25	1.22	86002.12	−0.64	86001.48
		11	80.4	48 24	48 32	2 48 28	0.58						
„ 29 A.M.	30.050	1	80.8	21 42	21 45	8 21 43.5	1.18	80.85	648.3	1.22	86001.72	−0.42	86001.3
		11	80.9	9 42	9 51	10 9 46.5	0.58						
Means . . .	30.008							81.83			86001.39		86001.39

MARANHAM.—COINCIDENCES OBSERVED with PENDULUM 4; the Clock making 86267 Vibrations in a Mean Solar Day.

DATE.	Barometer.	No. of Coincidence.	Temperature.	Time of Disappearance.	Time of Re-appearance.	True Time of Coincidence.	Arc of Vibration.	Mean Temperature.	Mean Interval.	Correction for the Arc.	Vibrations in 24 hours.	Reduction to a Mean Temperature.	Reduced Vibrations at 82°.1 Faht.
1822.			°	M. S.	M. S.	H. M. S.	°	°	°	S. +			
Aug. 30 P.M.	29.950	1	82.6	32 54	32 55	2 32 54.5	1.24	82.5	668.45	1.43	86010.33	+0.21	86010.56
		11	82.4	24 12	24 26	4 24 19	0.64						
„ 31 A.M.	30.050	1	81.1	16 28	16 30	8 16 29	1.16	81.55	670.15	1.19	86010.73	−0.20	86010.53
		11	82	8 04	8 17	10 8 10.5	0.58						
„ 31 P.M.	29.950	1	82.2	28 52	28 55	2 28 53.5	1.18	82.1	670.85	1.22	86011.02	+0.04	86011.06
		11	82	20 36	20 48	4 20 42	0.58						
Sept. 1 A.M.	30.050	1	80.9	59 59	00 03	9 00 01	1.18	81.3	670.7	1.22	86010.96	−0.30	86010.66
		11	81.7	51 42	51 54	10 51 48	0.58						
„ 1 P.M.	29.940	1	82.4	50 34	50 37	2 50 35.5	1.2	82.3	669.25	1.28	86010.48	+0.12	86010.60
		11	82.2	42 01	42 15	4 42 08	0.6						
„ 2 P.M.	30.000	1	82.7	41 19	41 25	1 41 22	1.18	82.7	669.5	1.22	86010.52	+0.29	86010.81
		11	82.7	32 51	33 03	3 32 57	0.58						
„ 3 A.M.	00.030	1	81.5	05 55	05 57	9 05 56	1.2	81.4	670.7	1.28	86011.02	−0.26	86010.76
		11	81.3	57 37	57 49	10 57 43	1.6						
„ 3 P.M.	29.950	1	83.2	55 04	55 09	2 55 06.5	1.16	82.8	669.05	1.19	86010.31	+0.34	86010.65
		11	82.4	46 32	46 42	4 46 37	0 58						
„ 4 A.M.	30.050	1	81.1	01 39	01 45	9 01 42	1.18	81.55	670.7	1.22	86010.96	−0.20	86010.76
		11	82	53 24	53 36	10 53 30	0.58						
Means . . .	30.000							82.02			86010.7		86010.7

MARANHAM——BAROMETRICAL OBSERVATIONS to DETERMINE the HEIGHT of the PENDULUM STATION.

DATE.	At the Pendulum Station.			The Cistern one foot above High Water Mark.			Height above Mean Tide.	REMARKS.
	Barometer.	Thermometer.		Barometer.	Thermometer.			
		Att.	Det.		Att.	Det.		
1822	IN.	°	°	IN.	°	°	Feet.	
H. M. Sept. 5, 8 00 A. M.	30.059	85	82	....	..	..	81	The difference in the Mercurial Column equal 67ft. to which 12ft. are added for half the Fall of Tide, and 1ft. for the Height of the Cistern, above High Water Mark.
„ 5, 8 20 A. M.	....	..	..	30 125	85	82		
„ 5, 8 50 A. M.	30.059	85	82	....	..	..		
„ 5, 9 15 A. M.	....	..	..	30.124	85	82		
„ 5, 9 45 A. M.	30.061	85	82	....	..	..		
MEANS . . .	30.060	85	82	30.125	85	82		

TRINIDAD.

HAVING had the good fortune to meet Sir Ralph Woodford, Governor of Trinidad, in London, soon after I had determined on undertaking the present inquiry, he was kind enough to offer me letters of introduction to Port Spain, the seat of Government in Trinidad, which should ensure me a favourable reception, and the means of accomplishing the objects which I had in view. Port Spain being very desirably situated near the tenth parallel of latitude, I did not hesitate to avail myself of so advantageous an offer, and thus early anticipated the very agreeable and satisfactory visit, which I had now the opportunity of making.

In the absence of Sir Ralph Woodford in Europe, the administration was carried on by the Lieutenant Governor, Colonel Aretas William Young of the 3rd West India Regiment, with whom I had had a former though slight acquaintance, whilst serving together at Gibraltar in 1805. The Pheasant arrived at Port Spain on the 18th of September, which was a month later than Sir Ralph Woodford's letters had caused her to be expected, being the consequence of her detention at Bahia; Colonel Young had been so kind as to provide me an apartment at his house, and invited Captain Clavering and myself to be his guests during our residence.

The pendulums were admirably accommodated in the Vestry-Room of the new and very beautiful Protestant Church, which does so much credit to the architectural taste and skill of Mr. Reinagle by whom it was designed and built; and is one of the many improvements and decorations for which Port Spain is indebted to its present Governor, and which have rendered it one of the handsomest towns in the British Colonies. The walls of the vestry being of an extraordinary thickness, the temperature was more than ordinarily uniform; the going of the pendulums, of the clock, and of the chronometer, at this station, deserve to be particularly noticed.

Port Spain is built on a bed of gravel, between 30 and 40 feet deep, resting on a substratum of clay; it furnished therefore a second station in the low latitudes, Maranham being the other, in the opposite extreme in respect of local density, to the stations of St. Thomas and Ascension; the number of stations in each extreme being thus the same, the undue influence of either is counteracted, on the deductions which may be derived from a general summary.

The height of the pendulums was ascertained by direct measurement to be twenty-one feet above half tide.

It was with great regret that Captain Clavering and myself felt the propriety of pursuing the voyage, as soon as the immediate object which had occasioned our visit to Port Spain was completed; as the very agreeable society, and the many natural beauties and curiosities of Trinidad, were strong incitements to delay: but our departure was pressed by an anxiety to arrive in England as early as possible in the ensuing winter, in order to make the necessary preparations for proceeding to the high latitudes, as soon as the northern navigation should open in the spring, so that we might have the whole of the following season at our disposal in the Arctic Circle. The Pheasant therefore sailed for Jamaica as soon as the instruments were embarked, for the purpose of undergoing in the Dock Yard the repairs which she required, to enable her to encounter the gales which might be expected in crossing the Atlantic at so late a period of the year.

TRINIDAD.——OBSERVATIONS to DETERMINE the RATE of the Chronometer No. 423, by ZENITH DISTANCES of the Sun, with a Repeating Circle; from the 23d of September to the 4th of October, 1822.

Latitude of the Place of Observation 10° 38′ 43″ N.; Longitude 61° 36′ W.

September 23d A.M.; Barometer 29.97; Thermometer 81°; ☉'s L.L.

Chronometer.	Level.		Readings, &c.	Chronometer.	Level.		Readings, &c.
H. M. S.			° ′ ″	H. M. S.			° ′ ″
11 45 46	−4	−4	First Vernier 221 34 40	12 01 35.6	0	0	First Vernier 215 51 40
11 47 36.8	+5	+5	Second „ 34 40	12 03 39.6	+2	+2	Second „ 51 15
11 49 40.4	+1	+1	Third „ 35 20	12 05 34	+5	+5	Third „ 52 00
11 51 23.2	+1	+2	Fourth „ 34 30	12 07 44	+6	+6	Fourth „ 51 10
11 53 16.4	−5	−5	Mean . . 221 34 48	12 09 48.4	−2	−2	Mean . . . 215 51 31
11 54 44.4	+1	+1	Level . . . 0	12 11 26.4	−4	−4	Level . . . +7
Mean . . . 11 50 24.53	−1	0	Index . . . +156 29 58	Mean . . . 12 06 38	+7	+7	Index . . +138 25 12
True time . 7 48 24	0		378 04 46	True time . 7 59 39.1	+7		354 16 50
Chron. fast . 4 07 00.53			Observed Z.D. 63 00 48	Chron. fast . 4 06 58.9			Observed Z.D. 59 02 48
			Ref. and Paral. + 1 41				Ref. and Paral. + 1 24
360° − 203° 30′ 02″ = 156° 29′ 58″			Semidiam . . − 15 58	360° − 221° 34′ 48″ = 138° 25′ 12″			Semidiam . . −15 58
			True Z.D. . . 62 46 31				True Z.D. . 58 48 14

Chronometer, Fast { H. M. S. 4 07 00.53 ; 4 06 58.9 } H. M. S. 4 06 59.7

September 23d P.M.; Barometer 29.97; Thermometer 86°; ☉'s U.L.

Chronometer.	Level.		Readings, &c.	Chronometer.	Level.		Readings, &c.
H. M. S.			° ′ ″	H. M. S.			° ′ ″
8 23 04.4	−2	0	First Vernier 258 00 15	8 36 08	+2	+5	First Vernier 319 31 35
8 24 35.6	+2	0	Second „ 257 59 40	8 37 47.2	+2	+4	Second „ 31 40
8 26 18	−3	0	Third „ 258 00 20	8 39 38	−5	−7	Third „ 31 50
8 27 37.2	+3	0	Fourth „ 257 59 40	8 41 00.4	0	0	Fourth „ 31 15
8 29 04.8	0	0	Mean . . . 257 59 59	8 42 25.2	−4	−2	Mean . . . 319 31 35
8 30 40	−5	−3	Index . . . +144 08 29	8 43 39.6	0	0	Index . . . +102 00 01
Mean . . . 8 26 53.33	−5	−3	Level . . . −4	Mean . . 8 40 06.4	−5	0	Level . . . −02.5
True time . 4 19 53.53	−4		402 08 24	True time . 4 33 06	−2.5		421 31 34
Chron. fast . 4 06 59.8			Observed Z.D. 67 01 24	Chron. fast . 4 07 00.4			Observed Z.D. 70 15 16
			Ref. and Paral. + 2 02				Ref. and Paral. + 2 26
360° − 215° 51′ 31″ = 144° 08′ 29″			Semidiam . . +15 58	360° − 257° 59′ 59″ = 102° 00′ 01″			Semidiam . . +15 58
			True Z.D. . 67 19 24				True Z.D. . . 70 33 40

Chronometer, Fast { H. M. S. 4 06 59.8 ; 4 07 00.4 } H. M. S. 4 07 00.1

TRINIDAD.—Determination of the Rate of the Chronometer by Zenith Distances, *continued.*

September 24th A.M.; Barometer 30.01; Thermometer 81°; ☉'s L.L.

	Chronometer. H. M. S.	Level.		Readings, &c.	° ′ ″
	11 25 40.8	+3	+3	First Vernier	7 56 10
	11 27 01.6	−3	−3	Second ,,	56 00
	11 28 41.6	0	0	Third ,,	56 10
	11 30 31.2	−12	−11	Fourth ,,	56 00
	11 32 34	+2	+3	Mean . . .	7 56 05
	11 33 57.2	+4	+5	Index . . +	40 28 25 360 00 00
Mean . . .	11 29 44.4	−6	−3	Level	− 4.5
True time .	7 22 41.83	− 4.5			408 24 25
Chron. fast	4 07 02.57			Observed Z.D.	68 04 01
				Ref. and Paral.	+ 2 08
360° − 319° 31′ 35″ = 40° 28′ 25″				Semidiam. . .	−15 59
				True Z.D. . .	67 50 13

	Chronometer. H. M. S.	Level.		Readings, &c.	° ′ ″
	11 39 00	−3	−2	First Vernier	37 05 00
	11 40 16.8	0	0	Second ,,	05 00
	11 42 17.6	+7	+7	Third ,,	05 30
	11 43 40.8	−2	−1	Fourth ,,	05 00
	11 45 16	−4	−2	Mean . . .	37 05 07.5
	11 46 46.8	+1	+2	Index . . .	+352 03 55
Mean . . .	11 42 53	−1	+4	Level . . .	+ 1.5
True time .	7 35 50	+1.5			389 09 04
Chron. fast	4 07 03			Observed Z.D.	64 51 31
				Ref. and Paral.	+ 1 49
360° − 7° 56′ 05″ = 352° 03′ 55″				Semidiam. . .	−15 59
				True Z.D. . .	64 37 21

Chronometer, Fast {H. M. S. 4 07 02.57 / 4 07 03} H. M. S. 4 07 02.78

September 25th A.M.; Barometer 30.01; Thermometer 80°; ☉'s L.L.

	Chronometer. H. M. S.	Level.		Readings, &c.	° ′ ″
	11 22 18.8	0	0	First Vernier	91 01 45
	11 23 46	0	0	Second ,,	01 20
	11 25 13.2	0	0	Third ,,	01 45
	11 26 17.2	+4	+3	Fourth ,,	01 10
	11 28 36.8	0	0	Mean . . .	91 01 30
	11 29 46	0	0	Index . . .	+322 54 53
Mean . . .	11 25 59.67	+4	+3	Level . . .	+3.5
True time .	7 18 54	+3.5			413 56 26
Chron. fast	4 07 05.67			Observed Z.D.	68 59 25
				Ref. and Paral.	+ 2 15
360° − 37° 05′ 07″ = 322° 54′ 53″				Semidiam. . .	−15 59
				True Z.D. . .	68 45 41

	Chronometer. H. M. S.	Level.		Readings, &c.	° ′ ″
	11 35 12	−6	−6	First Vernier	126 46 00
	11 36 31.2	−5	−4	Second ,,	45 40
	11 37 50	+6	+7	Third ,,	46 20
	11 39 04	+2	+3	Fourth ,,	45 20
	11 40 21.2	0	0	Mean . . .	126 45 50
	11 41 34.8	0	0	Index . . .	+268 58 30
Mean . . .	11 38 25.53	−3	−0	Level . . .	−1.5
True time .	7 31 19.4	−1.5			395 44 19
Chron. fast	4 07 06.13			Observed Z.D.	65 57 23
				Ref. and Paral.	+1 55
360° − 91° 01′ 30″ = 268° 58′ 30″				Semidiam. . .	−15 59
				True Z.D. . .	65 43 19

Chronometer, Fast {H. M. S. 4 07 05.67 / 4 07 06.13} H. M. S. 4 07 05.9

TRINIDAD.—Determination of the Rate of the Chronometer by Zenith Distances, *continued*.

September 26th A.M.; Barometer 30.00; Thermometer 81°; ☉'s L.L.

	Chronometer.	Level.		Readings, &c.	
	H. M. S.				° ′ ″
	11 28 48	0	0	First Vernier	344 68 00
	11 30 07.6	−7	−7	Second „	07 50
	11 31 52.8	+2	+2	Third „	08 20
	11 33 19.2	−3	−2	Fourth „	08 00
	11 35 24	+7	+8	Mean . . .	344 08 02.5
	11 36 52.8	+1	+2	Index . . .	+59 59 58
Mean. . .	11 32 44.07	0	+3	Level . . .	+1.5
True time .	7 25 34.97	+1.5			404 08 01
Chron. fast	4 07 09.1			Observed Z.D.	67 21 20
				Ref. and Paral.	+2 04
360°−300° 00′ 02″=59° 59′ 58″				Semidiam . .	−16 00
				True Z.D. . .	67 07 24

	Chronometer.	Level.		Readings, &c.	
	H. M. S.				° ′ ″
	11 43 08	+3	+4	First Vernier	367 50 10
	11 44 35.6	0	0	Second „	50 00
	11 46 00	0	0	Third „	50 10
	11 47 18.4	0	0	Fourth „	50 00
	11 48 53.2	+1	+2	Mean . . .	367 50 05
	11 50 13.2	+1	+2	Index . . .	+15 51 57.5
Mean. . .	11 46 41.4	+5	+8	Level . . .	+6.5
True time .	7 39 32.8	+6.5			383 42 09
Chron. fast	4 07 08.6			Observed Z.D.	63 57 01
				Ref. and Paral.	+1 46
360°−344° 08′ 02″.5=15° 51′ 57″.5				Semidiam . .	−16 00
				True Z.D. . .	63 42 47

Chronometer, Fast { H. M. S. 4 07 09.1 ; 4 07 08.6 } H. M. S. 4 07 08.85

September 27th A.M.; Barometer 30.01; Thermometer 80°; ☉'s L.L.

	Chronometer.	Level.		Readings, &c.	
	H. M. S.				° ′ ″
	11 31 18	0	0	First Vernier	196 37 20
	11 32 53.6	+3	+2	Second „	37 20
	11 34 25.6	+2	+1	Third „	37 50
	11 35 34.4	+5	+4	Fourth „	37 20
	11 37 10.8	−2	−1	Mean . . .	196 37 27.5
	11 38 12.4	−3	−3	Index . . .	+204 22 29
Mean. . .	11 34 55.8	+5	+3	Level . . .	+4
True time .	7 27 43.6	+4			401 00 00
Chron. fast	4 07 12.2			Observed Z.D.	66 50 00
				Ref. and Paral.	+2 00
360°−155° 37′ 31″=204° 22′ 29″				Semidiam . .	−16 00
				True Z.D. . .	66 36 00

	Chronometer.	Level.		Readings, &c.	
	H. M. S.				° ′ ″
	11 42 44	0	0	First Vernier	221 52 40
	11 43 40.8	+8	+8	Second „	52 40
	11 45 04.8	0	0	Third „	53 00
	11 46 20	−3	−3	Fourth „	52 10
	11 47 40	0	0	Mean . .	221 52 37.5
	11 48 42	−2	−2	Index . . .	+163 22 32.5
Mean. . .	11 45 41.93	+3	+3	Level . . .	+3
True time .	7 38 29.8	+3			385 15 13
Chron. fast	4 07 12.13			Observed Z.D.	64 12 32
				Ref. and Paral.	+1 46
360°−196° 37′ 27″.5=163° 22′ 32″.5				Semidiam . .	−16 00
				True Z.D. . .	63 58 18

Chronometer, Fast { H. M. S. 4 07 12.2 ; 4 07 12.13 } H. M. S. 4 07 12.16

TRINIDAD.——Determination of the Rate of the Chronometer by Zenith Distances, *continued.*

September 28th A.M.; Barometer 29.97; Thermometer 81°, ⊙'s L.L.

	Chronometer.	Level.		Readings, &c.	
	H. M. S.				° ′ ″
	11 58 50.8	0	0	First Vernier	114 27 00
	12 00 04	+2	+2	Second ,,	26 45
	12 01 53.2	+2	+2	Third ,,	27 10
	12 03 31.6	0	0	Fourth ,,	26 50
	12 05 00	0	0	Mean . . .	114 26 59
	12 06 08.8	−4	−4	Index . . .	+246 18 27.5
Mean. . .	12 02 34.73	0	0	Level . . .	0
True time .	7 55 19.2	0			360 45 26
Chron. fast	4 07 15.53			Observed Z.D.	60 07 34
				Ref. and Paral.	+1 27
360°−113° 41′ 32″.5=246° 18′ 27″.5				Semidiam . .	−16 00
				True Z.D. . .	59 53 01

	Chronometer.	Level.		Readings, &c.	
	H. M. S.				° ′ ″
	12 10 34.8	0	0	First Vernier	98 40 40
	12 11 46.8	−4	−2	Second ,,	40 20
	12 13 15.2	−7	−6	Third ,,	40 50
	12 14 36	+6	+7	Fourth ,,	40 20
	12 16 10	−7	−6	Mean . . .	98 40 32.5
	12 17 20	+2	+1	Index . . .	+245 33 01
Mean . . .	12 13 57.13	−9	−7	Level . . .	−8
True time .	8 06 41.17	−8			344 13 25
Chron. fast	4 07 15.96			Observed Z.D.	57 22 14
				Ref. and Paral.	+1 17
360°−114° 26′ 59″=245° 33′ 01″				Semidiam . .	−16 00
				True Z.D. . .	57 07 31

Chronometer, Fast { H. M. S. 4 07 15.53 ; 4 07 15.96 } H. M. S. 4 07 15.75

September 29th A.M.; Barometer 30.03; Thermometer 81°; ⊙'s L.L.

	Chronometer.	Level.		Readings, &c.	
	H. M. S.				° ′ ″
	11 45 35.2	+3	+3	First Vernier	119 13 00
	11 47 16	0	0	Second ,,	12 30
	11 48 29.2	−3	−3	Third ,,	13 20
	11 49 49.2	−2	−2	Fourth ,,	12 30
	11 51 12	+2	+2	Mean . . .	119 12 55
	11 52 13.2	−2	−2	Index . . .	+261 19 27
Mean. . .	11 49 05.8	−2	−2	Level . . .	−2
True time .	7 41 46.73	−2			380 32 20
Chron. fast	4 07 19.13			Observed Z.D.	63 25 23
				Ref. and Paral.	+1 41
360°−98° 40′ 33″=261° 19′ 27″				Semidiam . .	+16 00
				True Z.D. . .	63 11 04

	Chronometer.	Level.		Readings, &c.	
	H. M. S.				° ′ ″
	11 59 30	−3	−2	First Vernier	119 43 20
	12 00 47.2	0	0	Second ,,	43 00
	12 02 20	−5	−5	Third ,,	43 40
	12 03 33.2	−3	−2	Fourth ,,	43 10
	12 04 59.2	−2	−1	Mean . . .	119 43 17.5
	12 05 59.2	+6	+7	Index . . .	+240 47 05
Mean. . .	12 02 51.47	−7	−3	Level . . .	−5
True time .	7 55 32.1	−5			360 30 17
Chron. fast	4 07 19.37			Observed Z.D.	60 05 03
				Ref. and Paral.	+1 27
360°−119° 12′ 55″=240° 47′ 05″				Semidiam . .	+16 00
				True Z.D. . .	59 50 30

Chronometer, Fast { H. M. S. 4 07 19.13 ; 4 07 19.37 } H. M. S. 4 07 19.25

Trinidad.—Determination of the Rate of the Chronometer by Zenith Distances, *continued.*

September 29th P.M.; Barometer 29.97; Thermometer 85°; ☉'s U.L.

	Chronometer.	Level.		Readings, &c.	
	H. M. S.				° ′ ″
	7 34 54.8	+3	+2	First Vernier	96 53 30
	7 36 30	0	0	Second ,,	53 10
	7 37 51.6	0	0	Third ,,	53 50
	7 39 07.6	+2	+1	Fourth ,,	53 10
	7 41 04.8	+3	+4	Mean . . .	96 53 25
	7 42 04.4	+1	+3	Index . . .	+240 16 50
Mean. . .	7 38 35.53	+5	+8	Level . . .	+6.5
True time .	3 31 15.2	+6.5			337 10 21
Chron. fast.	4 07 20.33			Observed Z.D.	56 11 43
				Ref. and Paral.	+1 13
360° − 119° 43′ 10″ = 240° 16′ 50″				Semidiam . .	+16 00
				True Z.D. . .	56 28 56

	Chronometer.	Level.		Readings, &c.	
	H. M. S.				° ′ ″
	7 50 10	0	0	First Vernier	95 12 50
	7 51 06.8	+2	0	Second ,,	12 20
	7 52 30	+5	+7	Third ,,	13 00
	7 53 43.6	−2	0	Fourth ,,	12 10
	7 55 03.2	−8	−6	Mean. . . .	95 12 35
	7 56 16	−4	−2	Index . . .	+263 06 35
Mean. . .	7 53 08.27	−7	−1	Level . . .	−4
True time .	3 45 48.67	−4			358 19 06
Chron. fast	4 07 19.6			Observed Z.D.	59 43 11
				Ref. and Paral.	+1 27
360° − 96° 53′ 25″ = 263° 06′ 35″				Semidiam . .	+16 00
				True Z.D. . .	60 00 38

Chronometer, Fast { H. M. S. 4 07 20.33 / 4 07 19.6 } H. M. S. 4 07 19.96

September 30th A.M.; Barometer 30.01; Thermometer 81°; ☉'s L.L.

	Chronometer.	Level.		Readings, &c.	
	H. M. S.				° ′ ″
	11 20 52	+7	+6	First Vernier	152 07 20
	11 22 02	+3	+2	Second ,,	7 20
	11 23 38.4	0	0	Third ,,	8 10
	11 24 41.2	+5	+5	Fourth ,,	7 40
	11 26 12.8	−3	−3	Mean . . .	152 7 37.5
	11 27 56.4	+2	+2	Level . . .	+13
Mean. . .	11 24 13.8	+14	+12	Index′. . .	+264 47 25
True time .	7 16 52	+13			416 55 14
Chron. fast.	4 07 21.8			Observed Z.D.	69 29 12
				Ref. and Paral.	+2 18
360° − 95° 12′ 35″ = 264° 47′ 25″				Semidiam . .	−16 00
				True Z.D. . .	69 15 30

	Chronometer.	Level.		Readings, &c.	
	H. M. S.				° ′ ″
	11 32 18.4	0	0	First Vernier	192 17 40
	11 33 33.6	+2	+3	Second ,,	17 30
	11 35 08.8	−3	−2	Third ,,	18 00
	11 36 23.6	+7	+7	Fourth ,,	17 20
	11 37 54.8	−4	−4	Mean . . .	192 17 37.5
	11 39 00.8	0	0	Level . . .	+3
Mean. . .	11 35 43.33	+2	+4	Index . . .	+207 52 22.5
True time.	7 28 20.8	+3			400 10 03
Chron. fast.	4 07 22.53			Observed Z.D	66 41 40.5
				Ref. and Paral.	+1 59.5
360° − 152° 07′ 37″.5 = 207° 52′ 22″.5				Semidiam . .	−16 00
				True Z.D. . .	66 27 40

Chronometer, Fast { H. M. S. 4 07 21.8 / 4 07 22.53 } H. M. S. 4 07 22.16

TRINIDAD.——Determination of the Rate of the Chronometer by Zenith Distances, *continued*.

October 1st A.M.; Barometer 30 03; Thermometer 81°; ☉'s L.L.

Chronometer.	Level.		Readings, &c.	
H. M. S.				° ′ ″
11 46 58	0	0	First Vernier	316 15 50
11 48 09.6	+1	+5	Second „	15 50
11 50 24.8	−3	−2	Third „	16 10
11 51 29.2	+3	+2	Fourth „	15 40
11 52 48	−3	−1	Mean	316 15 52.5
11 53 54	+2	+3	Level . . .	+0.5
Mean . . . 11 50 37.27	+3	+7	Index . . .	+62 20 00
True time . 7 43 12.1				378 35 57.5
Chron. fast. 4 07 25.17			Observed Z.D.	63 05 59.6
			Ref. and Paral.	+ 1 40.4
360° − 297° 40′ 00″ = 62° 20′ 00″			Semidiam . .	−16 01
			True Z.D. . . .	62 51 39

Chronometer.	Level.		Readings, &c.	
H. M. S.				′ ″
11 57 19.2	−3	−2	First Vernier.	320 09 10
11 58 32	+4	+5	Second „	09 20
12 00 03.6	−4	−3	Third „	09 50
12 01 20	+2	+3	Fourth „	09 20
12 02 53.6	−2	−0	Mean . .	320 09 25
12 04 18	+6	+8	Level . .	+ 7
Mean . 12 00 44.4	+3	+11	Index . . .	+43 44 08
True time . 7 53 19.6	+7			363 53 40
Chron. fast. 4 07 24.8			Observed Z.D.	60 38 57
			Ref. and Paral.	+1 28
360° − 316° 15′ 52″ = 43° 44′ 08″			Semidiam . .	−16 01
			True Z.D. . .	60 24 24

Chronometer, Fast { H. M. S. 4 07 25.17 / 4 07 24.8 } H. M. S. 4 07 24.98

October 3d A.M.; Barometer 30.02; Thermometer 81°; ☉'s L.L.

Chronometer.	Level.		Readings, &c.	
H. M. S.				° ′ ″
11 52 25.6	0	0	First Vernier	330 44 30
11 53 31.2	0	0	Second „	44 20
11 55 04	0	0	Third „	45 00
11 56 01.4	+3	+5	Fourth „	44 10
11 57 46.4	0	0	Mean . . .	330 44 30
11 58 51.2	0	0	Index . . .	+40 58 55
Mean . . . 11 55 36.63	+3	+5	Level . . .	+ 4
True time . 7 48 04 87	+4			371 43 29
Chron. fast . 4 07 31.76			Observed Z.D.	61 57 15
			Ref. and Paral.	+ 1 35
360° − 319° 01′ 05″ = 40° 58′ 55″			Semidiam . .	−16 01
			True Z.D. . . .	61 42 49

Chronometer.	Level.		Readings, &c.	
H. M. S.				° ′ ″
12 02 35.2	−4	−4	First Vernier	327 30 20
12 03 53.2	0	0	Second „	30 10
12 05 26.4	−5	−4	Third „	30 40
12 06 40.2	+7	+7	Fourth „	30 10
12 07 56.4	−3	−2	Mean . . .	327 30 20
12 09 10.4	0	0	Index . . .	+29 15 30
Mean . . . 12 05 56.97	−5	+3	Level . . .	− 4
True time. . 7 58 25.3	−4			356 45 46
Chron. fast. . 4 07 31.67			Observed Z.D.	59 27 38
			Ref. and Paral.	+ 1 26
360° − 330° 44′ 30″ = 29° 15′ 30″			Semidiam . .	−16 02
			True Z.D. . . .	59 13 02

Chronometer Fast { H. M. S. 4 07 31.76 / 4 07 31.67 } H. M. S. 4 07 31.71

TRINIDAD.—Determination of the Rate of the Chronometer by Zenith Distances, *continued.*

October 4th A.M.; Barometer 30.03; Thermometer 81°; ☉'s L.L.

	Chronometer.	Level.		Readings, &c.	
	H. M. S.				° ′ ″
	11 43 03.2	+3	+2	First Vernier	91 17 30
	11 44 43.2	0	0	Second „	17 20
	11 46 21.6	−4	−4	Third „	18 00
	11 47 38	−2	−2	Fourth „	17 30
	11 49 26.4	+4	+4	Mean . . .	91 17 35
	11 51 14.8	+6	+6	Index . .	+292 59 01
Mean. . .	11 47 04.53	+7	+6	Level . . .	+6.5
True time .	7 39 30.07	+6.5			384 16 42
Chron. fast.	4 07 34.46			Observed Z.D	64 02 47
				Ref. and Paral.	+1 45
				Semidiam . .	−16 02
				True Z.D. . .	63 48 30

360° − 67° 00′ 59″ = 292° 59′ 01″

	Chronometer.	Level.		Readings, &c.	
	H. M. S.				° ′ ″
	11 56 06.4	0	0	First Vernier	97 14 50
	11 57 24.4	0	0	Second „	14 30
	11 59 06.6	0	0	Third „	14 50
	12 00 25.6	0	0	Fourth „	14 40
	12 02 03.2	0	0	Mean	97 14 42.5
	12 03 21.2	0	0	Index . . .	+268 42 25
Mean . .	12 59 44.57	0	0	Level	0
True time .	7 52 09.97	0			365 57 07.5
Chron. fast..	4 07 34.6			Observed Z.D.	60 59 31
				Ref. and Paral.	+ 1 31
				Semidiam. .	−16 02
				True Z.D. . .	60 45 00

360° − 91° 17′ 35″ = 268° 42′ 25″

Chronometer, Fast { H. M. S. 4 07 34.46 / 4 07 34.6 } H. M. S. 4 07 34.53

October 4th P.M.; Barometer 29.96; Thermometer 84°; ☉'s U.L.

	Chronometer.	Level.		Readings, &c.	
	H. M. S.				° ′ ″
	8 19 48	−2	0	First Vernier	144 32 50
	8 21 19.2	−3	−1	Second „	32 40
	8 22 55.2	+2	0	Third „	33 00
	8 24 16	0	0	Fourth „	32 30
	8 25 58.8	−4	−2	Mean . .	144 32 45
	8 27 27.2	−1	0	Index . . .	+262 45 18
Mean . .	8 23 37.4	−8	−3	Level . . .	−5.5
True time .	4 16 02.6	−5.5			407 17 57
Chron. fast	4 07 34.8			Observed Z.D.	67 53 00
				Ref. and Paral.	+2 06
				Semidiam . .	+16 01
				True Z.D. . .	68 11 07

360° − 97° 14′ 42″ = 262° 45′ 18″

	Chronometer.	Level.		Readings, &c.	
	H. M. S.				° ′ ″
	8 33 20	+1	+4	First Vernier	211 41 25
	8 34 37.6	+2	0	Second „	15
	8 36 23.2	−5	−3	Third „	55
	8 38 15.6	−6	−4	Fourth „	05
	8 39 49.6	−5	−3	Mean . . .	211 41 25
	8 41 04.8	+4	+1	Index . . .	+215 27 15
Mean. . .	8 37 15.13	−13	−5	Level . . .	−9
True time.	4 29 40.8	−9			427 08 31
Chron. fast.	4 07 34.33			Observed Z.D.	71 11 25
				Ref. and Paral.	+2 30
				Semidiam . .	+16 02
				True Z.D. . .	71 29 57

360° − 144° 32′ 45″ = 215° 27′ 15″

Chronometer, Fast { H. M. S. 4 07 34.8 / 4 07 34.33 } H. M. S. 4 07 34.56

RATE DEDUCED from the PRECEDING OBSERVATIONS.

Date.	S.	Date.	S.	Date.	S.	Date.	S.	Date.	S.
A.M. to A.M.		Sept. 24 to 26	3.03	Sept. 25 to 30	3.25	Sept. 27 to 29	3.54	Sep.29 to Oc.3	3.11
Sept. 23 to 24	3.08	„ 27	3.13	„ Oct. 1	3.18	„ 30	3.33	„ 4	3.05
„ 25	3.10	„ 28	3.24	„ 3	3.23	„ Oct. 1	3.20	„ 30 to Oc. 1	2.82
„ 26	3.05	„ 29	3.29	„ 4	3.18	„ 3	3.26	„ 3	3.18
„ 27	3.11	„ 30	3.23	Sept. 26 to 27	3.31	„ 4	3.20	„ 4	3.09
„ 28	3.21	„ Oct. 1	3.17	„ 28	3.23	Sept. 28 to 29	3 50	Oct. 1 to 3	3.36
„ 29	3.26	„ 3	3.21	„ 29	3.47	„ 30	3.20	„ 4	3.18
„ 30	3.21	„ 4	3.17	„ 30	3.33	„ Oct. 1	3.07	Oct. 3 to 4	2.82
„ Oct. 1	3.16	Sept. 25 to 26	2.95	„ Oct. 1	3.23	„ 3	3.19	P.M. to P.M.	
„ 3	3.20	„ 27	3.13	„ 3	3.27	„ 4	3.13	Sept. 23 to 29	3.31
„ 4	3.16	„ 28	3.28	„ 4	3.21	Sept. 29 to 30	2.91	„ Oct. 4	3.13
Sept. 24 to 25	3.12	„ 29	3.34	Sept. 27 to 28	3.59	„ Oct. 1	2.87	Sep.29 to Oc.4	2.92
Means . . .	3.15		3.18		3.29		3.20		3.09

Gaining 3.19 Seconds per Diem.

TRINIDAD.——Comparisons of the Astronomical Clock with the Chronometer No. 423, from the 23d of September to the 4th of October, 1822; with the Clock's Rate on Mean Solar Time deduced.

1822.	Chronometer.	Clock.	Clock's Loss on 423.		DAILY RATES.	
					Chron.	Clock.
					Gaining.	Losing.
	H. M. S.	H. M. S.	s	s.	s.	s
Sept. 23 P. M.		8 33 15.2				
			127			
„ 24 P. M.		8 31 08.2				
			127.2			
„ 25 P. M.		8 29 01				
			127	127.06	3.19	123.87
„ 26 P. M.		8 26 54				
			127.1			
„ 27 P. M.		8 24 46.9				
			127			
„ 28 P. M.		8 22 39.9				
	9 10 00		127.1			
„ 29 P. M.		8 20 32.8				
			127			
„ 30 P. M.		8 18 25.8				
			127.1			
Oct. 1 P. M.		8 16 18.7		127.12	3.19	123.93
			127.1			
„ 2 P. M.		8 14 11.6				
			127.2			
„ 3 P. M.		8 12 04.4				
			127.2			
„ 4 P. M.		8 09 57.2				

TRINIDAD.—COINCIDENCES OBSERVED with PENDULUM No. 3; the Clock making 86276.13 Vibrations in a Mean Solar Day.

DATE.	Barometer.	No. of Coincidence.	Temperature.	Time of Disappearance.	Time of Re-appearance.	True Time of Coincidence.	Arc of Vibration.	Mean Temperature.	Mean Interval.	Correction for the Arc.	Vibrations in 24 hours.	Reduction to a mean Temperature.	Reduced Vibrations at 84.45 Faht.
1822.	IN.		°	M. S.	M. S.	H. M. S.	°	°	S.	S.			
Sept. 23 P.M.	29.980	1	85.5	39 49	39 53	2 39 51	1.18	85.5	638.8	+ 1.22	86007.22	+0.44	86007.66
		11	85.5	26 12	26 26	4 26 19	0.58						
„ 24 A.M.	30.030	1	83.8	35 11	35 13	7 35 12	1.26	84.8	638.3	1.48	86007.28	+0.15	86007.43
		11	85.8	21 31	21 39	9 21 35	0.68						
„ 24 P.M.	29.990	1	83.9	11 04	11 07	12 11 05.5	1.2	84.75	638.75	1.28	86007.28	+0.13	86007.41
		11	85.6	57 27	57 39	1 57 33	0.6						
„ 24 P.M.	29.970	1	85	51 08	51 13	2 51 10.5	1.26	84.9	638.55	1.48	86007.38	+0.19	86007.57
		11	84.8	37 26	37 46	4 37 36	0.68						
„ 25 A.M.	30.010	1	84.5	11 58	12 01	7 11 59.5	1.2	84.5	639.05	1.35	86007.47	. . .	86007.47
		11	84 5	58 25	58 35	8 58 30	0.64						
„ 25 P.M.	29.980	1	84.3	9 36	9 42	2 9 39	1.14	84.55	639.65	1.14	86007.52	. .	86007.52
		11	84.8	56 08	56 23	3 56 15.5	0.56						
„ 26 A.M.	30.150	1	83.2	38 00	38 03	8 38 01.5	1.2	83.5	640.3	1.35	86007.97	−0.40	86007.57
		11	83.8	24 37	24 52	10 24 44.5	0.64						
„ 27 P.M.	29.970	1	84.2	23 30	23 33	1 23 31.5	1.16	83.8	640.4	1.21	86007.91	−0.27	86007.64
		11	83.4	10 08	10 23	3 10 15.5	0.58						
„ 28 P.M.	29.980	1	84	51 42	51 47	12 51 44.5	1.2	83.8	640.15	1.35	86007.93	−0.27	86007.66
		11	83.6	38 19	38 33	2 38 26	0.64						
Means . .	30.007							84.45			86007.55		86007.55

TRINIDAD.—COINCIDENCES OBSERVED with PENDULUM 4; the Clock making 86276.13 Vibrations in a Mean Solar Day.

DATE.	Barometer.	No. of Coincidence.	Temperature.	Time of Disappearance.	Time of Re-appearance.	True Time of Coincidence.	Arc of Vibration.	Mean Temperature.	Mean Interval.	Correction for the Arc.	Vibrations in 24 hours.	Reduction to a Mean Temperature.	Reduced Vibrations at 83.35 Faht.
1822.	IN.		°	M. S.	M. S.	H. M. S.	°	°	°	S. +			
Sept. 29 A.M.	30.010	1	83	29 37	29 42	10 29 39.5	1.2	83.1	665.7	1.26	86018.16	−0.10	86018.06
		11	83.2	20 29	20 44	12 20 36.5	0.58						
,, 29 P.M.	29.990	1	83	42 36	42 41	12 42 38.5	1.18	83.3	665.25	1.22	86017.96	−0.02	86017.94
		11	83.6	33 24	33 38	2 33 31	0.58						
,, 30 A.M.	30.150	1	83.3	19 49	19 53	8 19 51	1.22	83.3	664.8	1.31	86017.87	−0.02	86017.85
		11	83.3	10 31	10 47	10 10 39	0.62						
,, 30 P.M.	29.970	1	83 8	44 18	44 22	2 44 20	1.18	83.7	665.1	1.21	86017.89	+0.15	86018.04
		11	83.6	35 04	35 18	4 35 11	0.58						
Oct. 1 A.M.	30.050	1	83	36 34	36 38	8 36 36	1.18	83.25	665.5	1.22	86018 04	−0.04	86018.00
		11	83.5	27 25	27 37	10 27 31	0.58						
,, 1 P.M.	29.980	1	83.3	47 22	47 27	2 47 24.5	1.22	83.15	665.55	1.34	86018.16	−0.08	86018.08
		11	83	38 11	38 29	4 38 20	0.62						
,, 2 A.M.	30.000	1	83	20 41	20 45	10 20 43	1.2	83.5	665.1	1.28	86017.96	+0.06	86018.02
		11	84	01 25	01 43	12 01 34	0.6						
,, 3 A.M.	30.020	1	83	12 28	12 34	10 12 31	1.18	83.5	665.35	1.22	86017.98	+0.06	86018.04
		11	84	3 17	3 32	12 3 24.5	0.58						
,, 3 P.M.	30.020	1	83 2	14 01	14 07	1 14 04	1.18	83.3	665.45	1 22	86018.02	−0 02	86018.00
		11	83.4	4 49	5 08	3 4 58 5	0.58						
Means .	30.023							83.35			86018.00		86018.00

JAMAICA.

THE Pheasant arrived at Jamaica on the 17th of October, when her repairs were immediately proceeded in at the Dock Yard at Port Royal, and ordered to be completed with all despatch, by Admiral Sir Charles Rowley, commanding on the station; it appearing, however, on examination, that her refittal would require a detention of three weeks, as she had been between three and four years within the tropics, I availed myself of the opportunity to determine the rates of the pendulums.

I was so fortunate as to obtain a house at Port Royal exceedingly well adapted for the purpose, which was lent me by my friend Major William Nicolls of the Royal Artillery, to whom it belonged in quality of acting Governor of Fort Charles, but was at that period unoccupied, as Major Nicolls was also in the temporary charge of the Quarter Master General's Department, and resided at the Head Quarters of the forces at Kingston.

The proceedings at Port Royal differed in no respect from those at the three preceding Stations: the Pendulums were in a room on the ground floor, which was kept carefully closed and darkened, and in which the temperature was consequently very uniform. October is accounted one of the rainy months at Jamaica; but the rains at that part of the season are not continuous, and proved no serious interruption to the observations with the Repeating Circle. The yellow fever was prevalent during our stay amongst the troops in Port Royal, and the daily deaths were deemed considerable, even in Jamaica, in proportion to the strength of the Garrison; they certainly appeared very considerable to persons unhabituated to the great and almost unceasing mortality of the West

India Islands; happily the fever did not communicate itself to the Pheasant, and she quitted Jamaica without a single instance of its appearance amongst her crew.

Fort Charles is built on a calcareous rock, nearly on a level with the surface of the sea, and at the extremity of the tongue of sand which forms the harbour of Kingston, and on which Port Royal is situated. The height of the pendulums above half tide, was nine feet.

The Pheasant's repairs being completed, she sailed under Sir Charles Rowley's orders for Havanna, on the 6th of November, in convoy of several British and American merchant vessels, as a protection against the Pirates who infested the shores of Cuba; and from Havanna she proceeded to New York.

JAMAICA.——OBSERVATIONS to DETERMINE the RATE of the Chronometer No. 423. by ZENITH DISTANCES of the Sun, with a Repeating Circle, from the 22d to the 30th of October, 1822.

Latitude of the Place of Observation 17° 55′ 55″ N.; Longitude 76° 54′ W.

October 22d, A.M.; Barometer 30.05; Thermometer 80°; ☉'s L.L.

	Chronometer.	Level.		Readings, &c.	
	H. M. S.				° ′ ″
	1 14 14	0	0	First Vernier	126 40 30
	1 15 14.8	0	0	Second ,,	40 10
	1 16 40.4	0	0	Third ,,	40 30
	1 18 02	−5	−4	Fourth ,,	40 10
	1 19 34.4	+10	+10	Mean . . .	126 40 20
	1 20 49.2	+2	+1	Index . . .	+360 00 08.5
	1 21 56.4	+2	+2	Level . . .	+11.5
	1 23 13.6	+3	+2		486 40 40
Mean. . .	1 18 43.1	+12	+11		
True time .	8 08 52.57	+11.5		Observed Z.D.	60 50 04
Chron. fast.	5 09 50.53			Ref. and Paral.	+ 1 29
				Semidiam. . .	−16 07
				True Z.D. . .	60 35 26

	Chronometer.	Level.		Readings, &c.	
	H. M. S.				° ′ ″
	1 31 13.6	0	0	First Vernier	335 15 50
	1 32 27.6	−2	−2	Second ,,	15 30
	1 33 37.2	+2	+2	Third ,,	15 50
	1 34 57.6	−2	−3	Fourth ,,	15 40
	1 36 31.6	−4	−5		
	1 38 04.8	−2	−1	Mean. . . .	335 15 42.5
	1 39 12	+2	+2	Index. . . .	233 19 40
	1 40 44	+5	+4	Level. . . .	0
	1 42 24.4	+5	+4		
	1 43 38.4	−2	−3		568 35 22.5
Mean. . .	1 37 17.1	+2	−2		
True time .	8 27 26.37	0		Observed Z.D.	56 51 32
Chron. fast.	5 09 50.74			Ref. and Paral.	+ 1 16
				Semidiam. . .	−16 07
360° − 126° 40′ 20″ = 233° 19′ 40″				True Z.D. . .	56 36 41

Chronometer, Fast { H. M. S. 5 09 50.53 / 5 09 50.74 } H. M. S. 5 09 50.63

October 22d, P.M.; Barometer 30.01; Thermometer 83°; ☉'s U.L.

	Chronometer.	Level.		Readings, &c.	
	H. M. S.				° ′ ″
	9 21 46.4	+4	+3	First Vernier	74 11 30
	9 23 06	−2	−3	Second ,,	11 15
	9 24 41.2	+2	+1	Third ,,	11 40
	9 26 09.2	0	0	Fourth ,,	11 00
	9 27 27.6	+3	+2	Mean. . . .	74 11 21
	9 28 28.8	−3	−4	Index. . +	{ 1 00 00 / 360 00 00 }
Mean. . .	9 25 16.53	+4	−1	Level. . . .	+1.5
True time .	4 15 24.2	+1.5			435 11 23
Chron. fast .	5 09 52.33			Observed Z.D.	72 31 54
				Ref. and Paral.	+ 2 42
				Semidiam. . .	+16 07
				True Z.D. . .	72 50 43

	Chronometer.	Level.		Readings, &c.	
	H. M. S.				° ′ ″
	9 32 10	0	0	First Vernier	164 44 45
	9 34 24.8	0	0	Second ,,	44 30
	9 35 55.2	0	0	Third ,,	45 00
	9 37 27.2	0	0	Fourth ,,	44 30
	9 39 12.8	0	0	Mean. . . .	164 44 41
	9 40 26.8	+2	+2	Index. . .	+285 48 39
Mean. . .	9 36 36.13	+2	+2	Level. . . .	−2
True time .	4 26 44.1	+2			450 33 22
Chron. fast .	5 09 52.03			Observed Z.D.	75 05 34
				Ref. and Paral.	+ 3 14
360° − 74° 11′ 21″ = 285° 48′ 39″				Semidiam. . .	+16 07
				True Z.D. . .	75 24 55

Chronometer, Fast { H. M. S. 5 09 52.33 / 5 09 52.03 } H. M. S. 5 09 52.18

JAMAICA.—Determination of the Rate of the Chronometer by Zenith Distances, *continued.*

October 23d A.M.; Barometer 30.06; Thermometer 80°; ☉'s L.L.

	Chronometer. H. M. S.	Level.		Readings, &c.	° ′ ″
	00 39 34.4	0	0	First Vernier	215 15 50
	00 41 32	0	0	Second ,,	15 40
	00 43 38.8	0	0	Third ,,	16 00
	00 45 04.4	+3	+2	Fourth ,,	15 40
	00 48 42.8	0	0	Mean . . .	215 15 48
	00 50 09.6	0	0	Index . . .	+195 15 20
Mean . . .	00 44 47	+3	+2	Level . . .	+2
True time .	7 34 52.57	+2.5			410 31 10
Chron. fast	5 09 54.43			Observed Z.D.	68 25 12
				Ref. and Paral.	+2 10
360°−164° 44′ 40″=195° 15′ 20″				Semidiam . .	−16 07
				True Z.D .	68 11 15

	Chronometer. H. M. S.	Level.		Readings, &c.	° ′ ″
	00 55 21.6	−4	−4	First Vernier	245 33 40
	00 57 36.4	+4	+4	Second ,,	33 30
	00 59 22.8	0	0	Third ,,	34 00
	01 00 53.6	0	0	Fourth ,,	33 20
	01 02 49.6	+2	+1	Mean . . .	245 33 38
	01 04 12.4	0	0	Index . . .	+144 44 12
Mean. . .	01 00 02.73	+2	+1	Level . . .	+1.5
True time .	7 50 08.2	+1.5			390 17 51
Chron. fast	5 09 54.53			Observed Z.D.	65 02 58
				Ref. and Paral.	+ 1 49
360°−215° 15′ 48″=144° 44′ 12″				Semidiam . .	−16 07
				True Z.D.. .	64 48 40

Chronometer Fast {H. M. S. 5 09 54.43 / 5 09 54.53} H. M. S. 5 09 54.48

October 23d P.M.; Barometer 30.02; Thermometer 83°; ☉'s U.L.

	Chronometer. H. M. S.	Level.		Readings, &c.	° ′ ″
	8 45 28	0	0	First Vernier	288 23 10
	8 47 29.4	−6	−6	Second ,,	23 10
	8 49 32	0	0	Third ,,	23 40
	8 51 11.6	0	0	Fourth ,,	23 00
	8 52 58	0	0	Mean . . .	288 23 15
	8 55 08	0	0	Index . . .	+100 59 35
Mean. . .	8 50 17.83	−6	−6	Level . . .	−6
True time .	3 40 21.33	−6			389 22 44
Chron. fast	5 09 56.5			Observed Z.D.	64 53 47
				Ref. and Paral.	+ 1 48
360°−259° 00′ 25″=100° 59′ 35″				Semidiam . .	+16 07
				True Z.D. . .	65 11 42

	Chronometer. H. M. S.	Level.		Readings, &c.	° ′ ″
	9 06 21.2	0	0	First Vernier	345 33 50
	9 08 30.8	+4	+4	Second ,,	33 40
	9 10 28.8	0	0	Third ,,	34 00
	9 11 56.4	+2	+2	Fourth ,,	33 30
	9 13 53.6	+5	+5	Mean . . .	345 33 45
	9 16 03.2	0	0	Index . . .	+71 36 45
Mean. . .	9 11 12.33	+11	+11	Level . . .	+11
True time .	4 01 15.7	+11			417 10 41
Chron. fast	5 09 56.63			Observed Z.D	69 31 47
				Ref. and Paral.	+ 2 18
360°−288° 23′ 15″=71° 36′ 45″				Semidiam . .	+16 07
				True Z.D.. .	69 50 12

Chronometer Fast {H. M. S. 5 09 56.5 / 5 09 56.63} H. M. S. 5 09 56.56

JAMAICA.—Determination of the Rate of the Chronometer by Zenith Distances, *continued.*

October 24th P.M.; Barometer 30.01; Thermometer 84°; ☉'s U.L.

	Chronometer.	Level.		Readings, &c.	
	H. M. S.				° ′ ″
	8 22 20.8	−3	−3	First Vernier	345 21 50
	8 23 48.4	0	0	Second „	22 00
	8 26 12.4	+5	+5	Third „	22 10
	8 27 37.6	+2	+3	Fourth „	21 40
	8 29 52.8	0	0	Mean	345 21 55
	8 31 02	+6	+5	Index . . .	+14 26 15
Mean . . .	8 26 49	+8	+10	Level	+9
True Time.	3 16 48.1	+9			359 48 19
Chron. fast.	5 10 00.9			Observed Z.D.	59 58 03
				Ref. and Paral.	+ 1 26
				Semidiam . .	+16 07
				True Z.D. . .	60 15 36

360° − 345° 33′ 45″ = 14° 26′ 15″

	Chronometer.	Level.		Readings, &c.	
	H. M. S.				° ′ ″
	8 42 08.8	0	0	First Vernier	25 07 35
	8 43 44	−2	−2	Second „	07 20
	8 45 36	+5	+7	Third „	07 50
	8 47 03.2	0	0	Fourth „	07 15
	8 48 39.6	+10	+10	Mean	25 07 30
	8 50 37.6	0	0	Index . . +	{ 08.5 360 00 00
Mean . . .	8 46 18.2	+13	+15	Level	+14
True time .	3 36 16.8	+14			385 07 52
Chron. fast.	5 10 01.4			Observed Z.D.	64 11 17
				Ref. and Paral.	+ 1 35
				Semidiam . .	+16 07
				True Z.D. . .	64 28 59

Chronometer, Fast { H. M. S. 5 10 00.9 / 5 10 01.4 } H. M. S. 5 10 01.15

October 25th P.M.; Barometer 30.02; Thermometer 83°; ☉'s U.L.

	Chronometer.	Level.		Readings, &c.	
	H. M. S.				° ′ ″
	8 28 14	0	0	First Vernier	34 24 40
	8 30 46	+3	+4	Second „	24 20
	8 32 37.2	+5	+5	Third „	24 50
	8 34 14	0	0	Fourth „	24 20
	8 36 14.8	+5	+5	Mean	34 24 32
	8 37 56.4	−3	−2	Index . . .	+334 52 30
Mean . . .	8 33 20.4	+11	+12	Level	+12.5
True time .	3 23 16.17	+11.5			369 17 14
Chron. fast.	5 10 04.23			Observed Z.D.	61 32 52
				Ref. and Paral.	+ 1 29
				Semidiam . .	+16 07
				True Z.D. . .	61 50 28

360° − 25° 07′ 30″ = 334° 52′ 30″

	Chronometer.	Level.		Readings, &c.	
	H. M. S.				° ′ ″
	8 42 25.6	0	0	First Vernier	60 52 40
	8 43 58.8	−3	−3	Second „	52 00
	8 45 43.2	−3	−3	Third „	52 45
	8 47 25.2	0	0	Fourth „	52 10
	8 49 13.6	−2	−3	Mean	60 52 24.5
	8 50 31.6	−6	6	Index . . .	+325 35 28
Mean . . .	8 46 33	−8	−9	Level	−8.5
True time .	3 36 28.73	−8.5			386 27 43
Chron. fast.	5 10 04.27			Observed Z.D.	64 24 37
				Ref. and Paral.	+ 1 43
				Semidiam . .	+16 08
				True Z.D. . .	64 42 28

360° − 34° 24′ 32″ = 325° 35′ 28″

Chronometer, Fast { H. M. S. 5 10 04.23 / 5 10 04.27 } H. M. S. 5 10 04.25

JAMAICA.——Determination of the Rate of the Chronometer by Zenith Distances, *continued.*

October 26th A.M.; Barometer 30.06; Thermometer 80°; ☉'s L.L.

Chronometer.		Level.		Readings, &c.	
	H. M. S.				° ′ ″
	1 23 46.8	0	0	First Vernier	65 57 15
	1 25 19	−8	−9	Second „	57 10
	1 27 28	0	0	Third „	57 50
	1 29 04.8	0	0	Fourth „	57 15
	1 30 56	−2	−2	Mean . . .	65 57 22.5
	1 32 33.6	0	0	Index . . .	+290 39 15
Mean. . .	1 28 11.37	−10	−11	Level . . .	−10
True time .	8 18 04.3	−10.5			356 36 27
Chron. fast	5 10 07.07			Observed Z.D.	59 26 04
				Ref. and Paral.	+1 24
$360° - 69° 20′ 45″ = 290° 39′ 15″$				Semidiam .	−16 08
				True Z.D. . .	59 11 20

Chronometer.		Level.		Readings, &c.	
	H. M. S.				° ′ ″
	1 36 52.8	+0	0	First Vernier	46 08 00
	1 38 18.8	−3	−2	Second „	07 40
	1 40 02	+5	+6	Third „	08 10
	1 41 51.6	+5	+5	Fourth „	07 30
	1 44 20.8	+2	+1	Mean . . .	46 07 55
	1 46 02.4	+5	7	Index . . .	+294 02 37.5
Mean. . .	1 41 14.73	+14	+17	Level . . .	+15.5
True time .	8 31 07.9	+15.5			340 10 48
Chron. fast	5 10 06.83			Observed Z.D.	56 41 48
				Ref. and Paral.	+1 16
$360° - 65° 57′ 22.5″ = 294° 02′ 37.5″$				Semidiam . .	−16 08
				True Z.D. . .	56 26 56

Chronometer, Fast $\left\{\begin{array}{l} 5^{H.}\ 10^{M.}\ 07.07^{S.} \\ 5\ 10\ 06.83 \end{array}\right\}$ 5 H. 10 M. 06.95 S.

October 26th P.M.; Barometer 30.02; Thermometer 83°; ☉'s U.L.

Chronometer.		Level.		Readings, &c.	
	H. M. S.				° ′ ″
	8 11 06	0	0	First Vernier	33 24 15
	8 12 34.8	0	0	Second „	23 50
	8 14 20.8	+11	+12	Third „	24 20
	8 16 05.6	0	0	Fourth „	23 50
	8 18 00.8	0	0	Mean . . .	33 24 04
	8 19 33.6	+4	+5	Index . . .	+313 52 05
Mean. . .	8 15 16.93	+15	+17	Level . . .	+16
True time .	3 05 08	+16			347 16 25
Chron. fast	5 10 08.93			Observed Z.D.	57 52 45
				Ref. and Paral.	+ 1 16
$360° - 46° 07′ 55″ = 313° 52′ 05″$				Semidiam . .	+16 08
				True Z.D. . .	58 10 09

Chronometer.		Level.		Readings, &c.	
	H. M. S.				° ′ ″
	8 23 26	0	0	First Vernier	36 06 45
	8 25 02.4	+4	+5	Second „	06 30
	8 26 53.6	0	0	Third „	07 10
	8 28 12.8	+6	+7	Fourth „	06 50
	8 29 39.2	0	0	Mean . .	36 06 49
	8 31 20.8	−2	−1	Index . . .	+326 35 56
Mean. . .	8 27 25.8	+8	+11	Level . . .	+9
True time .	3 17 17.43	+9.5			362 42 54
Chron. fast	5 10 08.37			Observed Z.D.	60 27 09
				Ref. and Paral.	+ 1 28
$360° - 33° 24′ 04″ = 326° 35′ 56″$				Semidiam . .	+16 08
				True Z.D. . .	60 44 45

Chronometer, Fast $\left\{\begin{array}{l} 5^{H.}\ 10^{M.}\ 08.93^{S.} \\ 5\ 10\ 08.37 \end{array}\right\}$ 5 H. 10 M. 08.65 S.

JAMAICA.——Determination of the Rate of the Chronometer by Zenith Distances, *continued.*

October 27th P.M.; Barometer 30.02; Thermometer 83°, ☉'s U.L.

	Chronometer.	Level.		Readings, &c.	
	H. M. S.				° ′ ″
	8 15 25.2	+2	+3	First Vernier	29 34 30
	8 16 54.8	−2	−2	Second „	34 20
	8 18 23.6	+2	+1	Third „	34 40
	8 19 44	+6	+7	Fourth „	34 10
	8 21 10.4	+5	+6	Mean . .	29 34 25
	8 24 27.6	−2	−1	Index . . .	+323 53 12
Mean. . .	8 19 20.93	+11	+15	Level . . .	+13
True time .	3 09 08.1	+13			353 27 50
Chron. fast	5 10 12.83			Observed Z.D.	58 54 38
				Ref. and Paral.	+1 23
$360^\circ - 36^\circ\ 06'\ 48'' = 323^\circ\ 53'\ 12''$				Semidiam . .	+16 08
				True Z.D. . .	59 12 09

	Chronometer.	Level.		Readings, &c.	
	H. M. S.				° ′ ″
	8 30 15.2	+2	+3	First Vernier	42 04 10
	8 31 53.2	+5	+6	Second „	04 00
	8 33 30.4	0	0	Third „	04 15
	8 34 59.6	+3	+4	Fourth „	03 50
	8 36 41.6	−4	−2	Mean . . .	42 04 04
	8 38 21.2	−2	−1	Index . . .	+330 25 35
Mean . .	8 34 16.87	+4	+10	Level . . .	+7
True time .	3 24 03.43	+7			372 29 46
Chron. fast	5 10 13.44			Observed Z.D.	62 04 58
				Ref. and Paral.	+1 33
$360^\circ - 29^\circ\ 34'\ 25'' = 330^\circ\ 25'\ 35''$				Semidiam . .	+16 08
				True Z.D. . .	62 22 38

Chronometer, Fast { H. M. S. 5 10 12.83 / 5 10 13.44 } H. M. S. 5 10 13.13

October 28th P.M.; Barometer 30.01; Thermometer 83°; ☉'s U.L.

	Chronometer.	Level.		Readings, &c.	
	H. M. S.				° ′ ″
	8 39 00	+2	+1	First Vernier	307 55 25
	8 40 27.6	0	0	Second „	55 10
	8 41 38.4	+2	+1	Third „	55 45
	8 42 58	+6	+7	Fourth „	55 05
	8 44 25.2	−2	−1	Mean . . .	307 55 25
	8 45 51.6	−2	−1	Index . . .	+75 59 35
Mean. . .	8 42 23.47	−5	−7	Level . . .	+6
True time .	3 32 06.07	−6.5			383 55 06
Chron. fast	5 10 17.4			Observed Z.D.	63 59 11
				Ref. and Paral.	+ 1 45
$360^\circ - 284^\circ\ 00'\ 25'' = 75^\circ\ 59'\ 35''$				Semidiam . .	+16 08
				True Z.D. . .	64 17 04

Chronometer.	Level.	Readings, &c.

Chronometer, Fast H. M. S. 5 10 17.4

JAMAICA.——Determination of the Rate of the Chronometer by Zenith Distances, *continued.*

October 29th P.M.; Bar. 30.01; Therm. 83°; ☉'s U.L.

Chronometer.	H. M. S.	Level.		Readings, &c.	° ′ ″
	9 16 57.6	0	0	First Vernier	335 00 15
	9 18 16.8	0	0	Second „	00 10
	9 20 01.6	0	0	Third „	00 20
	9 21 13.2	+3	+4	Fourth „	00 00
	9 23 05.6	0	0	Mean . . .	335 00 11
	9 24 35.6	−2	−0	Index . . .	+100 03 22.5
Mean. . .	9 20 41.73	+1	+4	Level . . .	+2.5
True time .	4 10 20	+2.5			435 03 36
Chron. fast.	5 10 21.73			Observed Z.D.	72 30 36
				Ref. and Paral.	+2 45
360° − 259° 03′ 22″.5 = 100° 03′ 22″.5				Semidiam . .	+16 09
				True Z.D. . .	72 49 30

Chronometer, Fast (H. M. S.) 5 10 21.73

October 30th A.M.; Bar. 30.05; Ther 81°, ☉'s L.L.

Chronometer.	H. M. S.	Level.		Readings, &c.	° ′ ″
	1 06 08.4	0	0	First Vernier	168 26 40
	1 11 37.2	0	0	Second „	26 20
	1 15 00	−5	−3	Third „	26 40
	1 16 26.4	−4	−3	Fourth „	26 20
Mean. . .	1 12 18	−9	−6	Mean. . . .	168 26 30
True time .	8 01 54.17	−7.5		Index. . .	+ 85 23 05
Chron. fast.	5 10 23.83			Level. . . .	−07
					253 49 28
(Cloudy.)				Observed Z.D.	63 27 22
				Ref. and Paral.	+ 1 40
360° − 274° 36′ 55″ = 85° 23′ 55″				Semidiam. . .	−16 09
				True Z.D. . .	63 12 53

Chronometer, Fast (H. M. S.) 5 10 23.83

October 30th P.M.; Barometer 30.02; Thermometer 83°; ☉'s U.L.

Chronometer.	H. M. S.	Level.		Readings, &c.	° ′ ″
	8 18 27.6	+2	+1	First Vernier	169 23 20
	8 20 06	+4	+4	Second „	23 10
	8 21 50	+5	+5	Third „	23 40
	8 23 34.2	−1	−2	Fourth „	23 00
	8 25 56.4	+2	−2	Mean . . .	169 23 17.5
	8 27 24.8	+2	+1	Index . . .	+191 33 30
Mean. . .	8 22 53	+14	+11	Level . . .	+12.5
True time .	3 12 28.13	+12.5			360 57 00
Chron. fast.	5 10 24.87			Observed Z.D.	60 09 30
				Ref. and Paral.	+1 27
360° − 168° 26′ 30″ = 191° 33′ 30″				Semidiam . .	+16 09
				True Z.D. . .	60 27 06

Chronometer.	H. M. S.	Level.		Readings, &c.	° ′ ″
	9 06 12.4	0	0	First Vernier	230 12 50
	0 07 08.6	0	0	Second „	12 30
	9 08 42.8	0	0	Third „	13 00
	9 09 54	+15	+15	Fourth „	12 10
	9 11 11.2	+3	+3	Mean . . .	230 12 37.5
	9 12 31.6	0	0	Index . . .	190 36 42.5
Mean. . .	9 09 19.27	+18	+18	Level . . .	+ 18
True time.	3 58 54.03	+18			420 49 38
Chron. fast.	5 10 25.24			Observed Z.D.	70 08 16
				Ref. and Paral.	+2 23
360° − 169° 23′ 17″ 5 = 190° 36′ 42″.5				Semidiam . .	+16 09
				True Z.D. . .	70 26 48

Chronometer, Fast (H. M. S.) {5 10 24.87, 5 10 25.24} (H. M. S.) 5 10 25.05

RATE DEDUCED from the PRECEDING OBSERVATIONS.							
Date.	S.	Date.	S.	Date.	S.	Date.	S.
P.M. to P.M.		October 23 to 26	4.03	October 25 to 26	4.40	October 27 to 30	3.97
October 22 to 23	4.38	„ 27	4.14	„ 27	4.44	October 28 to 29	4.33
„ 24	4.48	„ 28	4.17	„ 28	4.38	„ 30	3.82
„ 25	4.02	„ 29	4.19	„ 29	4.37	October 29 to 30	3.32
„ 26	4.12	„ 30	4.07	„ 30	4.16	A.M. to A.M.	
„ 27	4.19	October 24 to 25	3.10	October 26 to 27	4.48	October 22 to 23	3.88
„ 28	4.20	„ 26	3.75	„ 28	4.37	„ 26	4.09
„ 29	4.22	„ 27	4.00	„ 29	4.36	„ 30	4.15
„ 30	4.11	„ 28	4.06	„ 30	4.10	October 23 to 26	4.16
October 23 to 24	4.59	„ 29	4.12	October 27 to 28	4.27	„ 30	4.19
„ 25	3.84	„ 30	3.99	„ 29	4.30	October 26 to 30	4.22
MEANS.—Gaining per Diem	4.22		3.96		4.33		4.01
	4.14 Seconds.						

JAMAICA.——Comparisons of the Astronomical Clock with the Chronometer, No. 423, from the 22d to the 30th of October, 1822; with the Clock's Rate on Mean Solar Time deduced.

1822.	Chronometer.	Clock.	Clock's loss on 423.		Daily Rates. Chron.	Daily Rates. Clock.
	H. M. S.	M. S.	S.	S.	S. Gaining.	S. Losing.
Oct. 22 A.M.	1 55 00	55 19.8				
			109.9			
" 23 A.M.		53 29.9				
			109.9			
" 24 A.M.		51 40		109.8	4.14	105.66
			109.7			
" 25 A.M.		49 50.3				
			109.7			
" 26 A.M.		48 00.6				
			109.9			
" 27 A.M.		46 10.7				
			109.8			
" 28 A.M.		44 20.9		109.45	4.14	105.31
			109.5			
" 29 A.M.		42 31.4				
			108.6			
" 30 A.M.		40 42.8				

JAMAICA.——COINCIDENCES OBSERVED with PENDULUM 3; the Clock making 86294.34 Vibrations in a Mean Solar Day.

DATE.	Barometer.	No. of Coincidence.	Temperature.	Time of Disappearance.	Time of Reappearance.	True Time of Coincidence.	Arc of Vibration.	Mean Temperature.	Mean Interval.	Correction for the Arc.	Vibrations in 24 hours.	Reduction to a Mean Temperature.	Reduced Vibrations at 81.77.
1822			°	M. S.	M. S.	H. M. S.	°	°	S.	°			
Oct. 22 A.M.	30.060	1	81.1	13 31	13 33	10 13 32	1.22	81.55	642.05	+ 1.38	86026.90	−0.09	86026.81
		11	82	00 26	00 39	12 00 32.5	0.66						
„ 22 P.M.	30.010	1	81.8	19 08	19 12	1 19 10	1.2	81.7	642.5	1.34	86027.06	−0.03	86027.03
		11	81.6	06 09	06 21	3 06 15	0.64						
„ 23 A.M.	30.026	1	80	25 29	25 33	7 25 31	1.18	80.5	643	1.28	86027.20	−0.53	86026.67
		11	81	12 33	12 49	9 12 41	0.62						
„ 23 P.M.	30.020	1	81.6	11 06	11 11	2 11 08.5	1.18	81.6	642.4	1.22	86026.9	−0.07	86026.83
		11	81.6	58 07	58 18	3 58 12.5	0.58						
„ 24 A.M.	30.070	1	80.8	47 30	47 35	8 47 32.5	1.18	81	642.9	1.22	86027.12	−0.32	86026.80
		11	81.2	34 34	34 49	10 34 41.5	0.58						
„ 24 P.M.	30.030	1	81.8	18 35	18 39	1 18 37	1.14	81.85	642.25	1.14	86026.76	+0.03	86026.79
		11	81.9	5 32	5 47	3 05 39.5	0.56						
„ 25 A.M.	30.070	1	82.6	12 41	12 42	11 12 41.5	1.22	82.9	640.85	1.35	86026.37	+0.47	86026.84
		11	83.2	59 24	59 36	12 59 30	0.64						
„ 25 P.M.	30.030	1	83.2	18 38	18 40	3 18 39	1.22	83.1	640.35	1.35	86026.19	+0.55	86026.74
		11	83	5 14	5 31	5 05 22.5	0.64						
„ 26 A.M.	30.070	1	81.2	57 49	57 51	9 57 51.5	1.18	81.75	641.45	1.25	86026.53	....	86026.53
		11	82.3	44 38	44 54	11 44 46	0.6						
Means . . .	30.043							81.77			86026.78		86026.78

JAMAICA.——COINCIDENCES OBSERVED with PENDULUM No. 4; the Clock making 86294.69 Vibrations in a Mean Solar Day.

DATE.	Barometer.	No. of Coincidence.	Temperature.	Time of Disappearance.	Time of Re-appearance.	True Time of Coincidence.	Arc of Vibration.	Mean Temperature.	Mean Interval.	Correction for the Arc.	Vibrations in 24 hours.	Reduction to a mean Temperature.	Reduced Vibrations at 83.6.
1822.	IN.		°	M. S.	M. S.	H. M. S.	°	°	S.	S.			
Oct. 27 A.M.	30.030	1	83.5	31 39	31 45	11 31 42	1.2	83.5	662.9	+ 1.34	86035.69	−0.04	86035.65
		11	83.5	22 01	22 21	1 22 11	0.64						
„ 27 P.M.	30.010	1	83.5	44 19	44 24	3 44 21.5	1.2	83.35	663.15	1.34	86035.77	−0.10	86035.67
		11	83.2	34 44	35 02	5 34 53	0.64						
„ 28 P.M.	30.020	1	83.8	59 08	59 15	12 59 11.5	1.2	84	662.6	1.38	86035.61	+0.17	86035.78
		11	84.2	49 28	49 47	2 49 37.5	0.66						
28 P.M.	30.000	1	84.2	00 10	00 13	3 00 11.5	1.2	84	662.4	1.38	86035.51	+0.17	86035.68
		11	83.8	50 27	50 44	4 50 35.5	0.66						
„ 29 A.M.	30.060	1	82	35 27	35 29	11 35 28	1.3	83	662.9	1.52	86035.87	−0.25	86035.62
		11	84	25 50	26 04	1 25 57	0.66						
„ 29 P.M.	30.020	1	84.1	58 47	58 50	2 58 48.5	1.28	84.45	661.65	1.48	86035.33	+0.36	86035.69
		11	84.8	48 53	49 17	4 49 05	0.66						
„ 30 A.M.	30.100	1	81.8	36 22	36 26	9 36 24	1.28	82.3	663.4	1.48	86036.03	−0.55	86035.48
		11	82.8	26 50	27 06	11 26 58	0.66						
„ 30 P.M.	30.080	1	83.8	8 44	8 49	12 8 46.5	1.28	84	662.05	1.48	86035.44	+0.17	86035.61
		11	84.2	58 57	59 17	1 59 07	0.66						
Means	30.040							83.6			86035.65		86035.65

NEW YORK.

PREVIOUSLY to entering into a detail of the proceedings at New York, it may be proper to notice the considerations which induced me to attach a more than ordinary interest to the experiments at that station; and to entertain a hope that the rates of the pendulums, obtained in one of the principal cities of the United States, might have a value beyond that of adding another station towards the more precise determination of the figure of the earth.

The Government of the United States, excited by the assiduity with which the principal governments of Europe were occupied in the regulation of the weights and measures of their respective dominions, and in devising methods of ensuring their perpetuity, had recently directed its attention to the procurement of a national Scale of linear measure, and to an inquiry into the modes of determining the value of its divisions, so as to enable its verification at any subsequent period, or its replacement in case of loss or accident. An official report on these subjects had been drawn up in considerable detail, by one of the leading members of the administration, and was published in 1821; the Scale, which that report recommended to be obtained and adopted as a standard, to which the several present measures of the States should be referred, and by which they should be perpetuated, was proposed to be itself a duplicate, so far as instrumental and executive accuracy would admit, of the national Scale of France; having consequently its foundation, nominally, in a certain aliquot part of the terrestrial meridian, but its real and practical verification in the length of the pendulum vibrating seconds at the observatory at Paris; the report contained no specific recommendation of measures for determining the value of the scale by a reference proper to the United States, nor were indeed any proceedings for that purpose, apparently, contemplated in its provisions; but it could scarcely be deemed probable, that a Nation, characteristically jealous of independence, and in which a disposition to scientific discussion and inquiry existed, and was rapidly progressive, would long rest satisfied with a means of verifying its scale, which would require the operations, on any future occasion of reference, to be conducted in a foreign Capital, and which would therefore be at the will of a foreign Nation; it was more reasonable to expect, that in the eventual prosecution of the purposes of the government, of which the attainment of a scale was necessarily the first step, the value of its divisions would be ultimately determined by a reference to an invariable length in nature which should exist within the territory of the United States; and that the length of the pendulum vibrating some definite portion of time at some selected station would be adopted for that purpose, because the pendulum exclusively possesses an essential quality in a natural standard, that of being easily accessible. In the event of so probable an undertaking being carried into effect, it would become highly desirable to compare the measurement thus made in the United States, with the results of the similar operations in Great Britain and France; as by the comparison, their accuracy would receive reciprocal confirmation, and a decisive practical demonstration would be afforded, of the identity with which the pendulum can be measured by different experimentors, and of its consequent effective value, in its application as a standard of reference.

The experiments made at the principal stations of the European Arc, have manifested that the difference in the length of the pendulum at two places on the globe, cannot be inferred from a knowledge of their respective latitudes, even were the general Ellipticity of the meridian correctly known; because the strict relation of the force of gravity to the square of the sine of the latitude, does not exist in nature, being interfered with by the variable density of the materials near the surface; the experiments contained in the present volume, afford still more decisive evidence of the same fact, and manifest the great extent of the irregularity which is induced thereby. A comparison, therefore, between the measurements of the pendulum made at two places on the globe, requires, and can only be accomplished by, a direct experiment. By employing at New York the pendulums with which I was furnished, I had it in my power to convey the measurement of the seconds' pendulum in London, so carefully made by Captain Kater, and adopted by the British Parliament to fix in perpetuity the divisions of its national Scale, from the spot in which the measurement was actually accomplished, to a station within the United States; and I should consequently place on record the length of the seconds' pendulum at that station measured on the British Scale, with a precision only inferior to the original determination in London by the very limited errors which might be introduced in the operations with the intermediate pendulums; and as the relative proportion of the British and French Scales has been very carefully ascertained, the same process would also determine the length of the pendulum at New York in parts of the French Scale, and also of the American, presuming the execution of the latter to have been strictly conformable to its design: whatsoever station, therefore, the government of the United States might ultimately select for the operations of an original measurement, the means would thus be presented of convenient access and always within its command, of comparing the result with that of the British measurement; if reasons of expediency, unconnected with those of science, should determine the selection elsewhere than at New York, a second proceeding with pendulums of comparison would be required, intermediately between New York and the station so chosen; but if New York were itself the station, no other proceedings than those of the original measurement would be required for the comparison; and as this consideration might possibly operate in determining the choice, I was particularly desirous of obtaining permission to make the experiments in some public edifice at New York, which might be equally accessible on future occasions of similar or connected operations*.

* No comparison has yet been accomplished between the measurements of the natural standards of different countries. The comparison of the measurements of the seconds' pendulum in London and at Paris, the one effected by the method of Borda, the other by that of Kater, was undertaken at the instance of the Bureau des Longitudes, by M. Arago, with whom were associated Messrs. Biot and Humboldt; the execution was attempted by means of two invariable pendulums, of which the rates were obtained in Paris in October 1817, at Greenwich, in November 1817, and again at Paris in March and August 1818: from the summary account of these experiments, published at the close of the third volume of the Base du Systeme Metrique, it is obvious that from some accidental cause or causes, the several results were not attained with the precision which the occasion required, or of which the mode of experiment is capable. The failure in a precise determination, is, however, the less to be regretted, since if the rates of the invariable pendulums had been obtained at Paris and at *Greenwich*, with the full accuracy which is practicable in such proceedings, the comparison of the measurements in Paris and in *London* would not have been accomplished thereby—it is true that the latitudes of London and Greenwich are so nearly the same, that the difference in the length of their respective pendulums due to the Ellipticity of the earth, may be computed by an assumed Ellipticity, without endangering a sensible error; but the assumption that the disposition and nature of the materials near the surface at the observatory on Greenwich Hill, and in Portland Place, London, are the same, and consequently that what may be termed their irregular influence on the general attraction, is precisely alike in both cases, is one which may involve error of too much consequence to be hazarded in an inquiry which ought to be so rigorously exact. It is not probable that this circumstance was overlooked, although it is not expressly adverted to, by the eminent persons who conducted or were concerned in the experiments, especially as M. Arago dwells on the importance of effecting the comparison by a direct observation which should not involve supposition; it may rather be attributed to a circumstance, which if it be not timely attended to, may produce far more serious inconveniences hereafter, namely, that the spot to which the Parliamentary

A trigonometrical survey of the United States had also been for some time in contemplation, to be conducted as a national undertaking upon the same extended scale as that proceeding in Great Britain; the preparations had so far advanced, that the instruments required for its execution, which had been ordered from Europe, had already arrived at Washington; it was not unreasonable to hope, therefore, that the interest which the experiments at New York might excite, in the inquiry concerning the figure of the Earth, amongst persons of science in the United States, might dispose the government to participate with those of Great Britain and France in the operations for its determination, by directing the variation in the length of the seconds' pendulum to be ascertained at the principal stations of the American survey, in the same manner that has been done in Great Britain. In that case New York would serve as a connecting link between the American series, and the equatorial stations which I had already visited, and those in the high latitudes which I had it in prospect to visit; it would also connect the operations in America, with those of the British survey, and with those of the French philosophers in France and Spain; and might thus become the means, not only of producing a considerable extension of the inquiry, but of combining the whole operations into one general determination, the value of which would far exceed the partial results of the several series, considered independently of each other.

These were the motives, which being stated to Vice-Admiral Sir Charles Rowley at Jamaica, prevailed with him to give his sanction that the Pheasant should stop at New York on her way to England; a

standard of Great Britain is referred, to which foreigners must resort to obtain a direct comparison with the measurement on which the standard is founded, and in which alone it can strictly be verified by posterity, is in the house of an individual, instead of being in a public edifice.

measure which not having been contemplated when Sir Robert Mends's instructions were drawn up, had not been authorized by him.

The Pheasant arrived at New York on the 10th of December; I had the advantage of being previously known to Dr. David Hosack, of that city, whose ardour in the pursuit, and liberality in the promotion of philosophical research, may justly entitle him to rank amongst its most distinguished patrons; I had soon, through his means, the satisfaction of finding myself placed in a situation and in circumstances, which I could not but deem as highly favourable to the purposes which I had in view: by the permission of the President and Council of Columbia College most kindly tendered, an apartment opposite the door leading into the gallery of the college chapel was assigned for the pendulums, being in all respects extremely well adapted, and having an additional and great recommendation in the assurance that it would be equally accessible on future occasions of a similar nature; the use of the Cupola was likewise permitted as a temporary observatory, for which it was well suited by having windows with firm and broad sills opening in the four principal directions.

In one of the consequences of the accommodation which the instruments thus received at Columbia College, I must ever deem myself to have been most highly fortunate, namely, in the association which it procured me of the Professor of Natural and Experimental Philosophy, and of Chemistry, Mr. James Renwick, whose interest in the experiments was so strongly excited as to induce him to give me his unremitting co-operation, a circumstance peculiarly desirable and satisfactory on an occasion in which the results may hereafter come in question in the comparison of the standard measurements of the two countries.

By favour of the gentlemen who superintend the administration of the customs in New York the instruments were permitted to be landed without undergoing the customary formality of inspection; they were disembarked on the 11th of December, and were ready to have commenced the observation of coincidences on the morning of the 15th, had not the weather proved an obstruction until the 22nd; the delay, however, may ultimately have been beneficial, in giving time to the clock to take up a more steady rate, than it might possibly have had in the earlier days.

The apparatus connected with the clock and pendulums, was in every respect most satisfactorily set up; the room appropriated to them was entered only for the purposes of observation or of comparing the clock, which operations were allowed to occupy no more time than they absolutely required; a precaution unnecessary within the tropics, where the general temperature is so nearly that of the human body, but which becomes highly deserving of attention, where the disparity is so great as in the severe winters of New York; the room was also kept constantly dark at other times by skreens of matting and baize suspended before the windows; by these means the variations of temperature of the apartment rarely equalled a degree and half in the twenty-four hours.

No very suitable situation for a transit instrument occuring within a convenient distance, the repeating circle was employed in the comparison of the chronometer with astronomical time; being desirous however of the fullest corroboration of accuracy, an eighteen inch astronomical telescope was firmly attached by a brass plate and screws to the side wall of the eastern window of the College Chapel, having the vertical side of the tower of the Presbyterian Church in its field of view, behind which the times of disappearance of three stars in the constellation of the Great Bear were observed by Mr. Renwick and myself on the 24th of December and on the 2nd of January, comprising the interval through which the observation of coincidences was continued.

New York is built on a bed of sand above one hundred feet in depth, and resting on primitive rock; the height of the pendulums above the sea was obtained as follows:

	Feet.
A station in Murray Street adjoining Columbia College appears by the survey of New York to be above high water-mark - - - - - -	20.8
Add, the ground line of Columbia College above the station in Murray Street, by estimation - - -	14
Add, the pendulums above the ground line, by estimation	30
Add, the half rise of the tides, immediate between the springs and neaps - - - - - -	2.5
Total height of the pendulums above half-tide	67 feet.

An account of the experiments with the pendulums at Columbia College, and of the corresponding series in London, was presented in the spring of 1823 to the Literary and Philosophical Society of New York, and will make a part of the second volume of their Transactions. I have taken that opportunity of noticing the results in their connexion with the incidental purposes, which I hoped they might be instrumental in promoting; and I am happy to be enabled to add, that Mr. Renwick is proceeding, under the cognizance of the government of the United States, and with its assistance, in the direct measurement of the seconds' pendulum at Columbia College, by means of the pendulum with convertible axes.

NEW YORK.——OBSERVATIONS to DETERMINE the RATE of the Chronometer No. 423, by ZENITH DISTANCES of the Sun and Stars; from the 22d of December, 1822, to the 2d of January, 1823.

Latitude of the Place of Observation 40° 42′ 43″ N.; Longitude 74° 03.5′ W.

By Zenith Distances of the Sun, West of the Meridian.

December 22d; Barometer 30.34; Thermometer 37°; ☉'s U.L.

Chronometer.		Level.		Readings, &c.		Chronometer.		Level.		Readings, &c.	
	H. M. S.				° ′ ″		H. M. S.				° ′ ″
	7 55 00	+5	−2	First Vernier	97 18 10		8 10 17.2	+2	−5	First Vernier	206 18 20
	7 56 41.2	−5	+2	Second „	18 10		8 11 58	−2	−8	Second „	18 10
	7 58 56.8	−8	0	Third „	18 30		8 14 07.6	+2	−5	Third „	18 30
	8 00 27.6	+5	−2	Fourth „	18 10		8 15 42.4	+9	+3	Fourth „	18 20
	8 02 44	+1	−6	Mean . .	97 18 15		8 17 43.2	+5	−2	Mean	206 18 20
	8 04 23 2	0	0	Index . . .	+360 00 08.5		8 19 08	+7	0	Index . . .	+262 41 45
Mean . . .	7 59 42.13	+13	−23	Level . .	−5	Mean . .	8 14 49.4	+23	−17	Level . .	+3
True time .	2 58 40.9	−5			457 18 18	True time .	3 13 47.33	+3			469 00 08
Chron. fast	5 01 01.23			Observed Z.D.	76 13 03	Chron. fast	5 01 02.07			Observed Z.D.	78 10 01
				Ref. and Paral.	+3 54					Ref. and Paral.	+4 32.5
				Semidiam . .	+16 18	360° − 97° 18′ 15″ = 262° 41′ 45″				Semidiam . .	+16 17.5
				True Z.D. . .	76 33 14					True Z.D. . .	78 30 51

Chronometer, Fast { H. M. S. 5 01 01.23; 5 01 02.07 } H. M. S. 5 01 01.65

December 23d; Barometer 30.26; Thermometer 20°.5; ☉'s U.L

Chronometer.		Level.		Readings, &c.		Chronometer.		Level.		Readings, &c.	
	H. M. S.				° ′ ″		H. M. S.				° ′ ″
	8 01 10	+3	−7	First Vernier	352 56 30		8 15 20.4	+10	0	First Vernier	105 20 45
	8 03 47.2	0	0	Second „	56 20		8 16 48	0	0	Second „	20 20
	8 05 18.8	0	−10	Third „	56 40		8 18 34	+8	−1	Third „	20 50
	8 06 54	0	0	Fourth „	56 30		8 20 12.8	+2	−8	Fourth „	20 10
	8 08 39.2	0	0	Mean . . .	352 56 30		8 22 29.6	0	0	Mean . . .	105 20 31.5
	8 10 35.2	−6	+4	Index . . .	+108 47 48		8 24 19.2	+12	+2	Index . . +	{ 7 03 30 360 00 00
Mean . . .	8 06 04.07	−23	+7	Level . . .	−8	Mean . . .	8 19 37.3	+34	−9	Level . . .	+12.5
True time	3 05 04.07	−8			461 44 10	True time .	3 18 36.23	+12.5			472 24 14
Chron. fast	5 01 00.00			Observed Z.D.	76 57 22	Chron. fast	5 01 01.07			Observed Z.D.	78 44 02
				Ref. and Paral.	+ 4 15					Ref. and Paral.	+ 4 56
360° − 251° 12′ 12″ = 108° 47′ 48″				Semidiam . .	+16 17.5	360° − 352° 56′ 30″ = 7° 03′ 30″				Semidiam . .	+16 18
				True Z.D. .	77 17 54.5					True Z.D. . .	79 05 16

Chronometer, Fast { H. M. S. 5 01 00; 5 01 01.07 } H. M. S. 5 01 00.53

NEW YORK.——Determination of the Rate of the Chronometer by Zenith Distances, *continued.*

By Zenith Distances of the Sun, West of the Meridian.

January 2d; Barometer 30.09; Thermometer 42°; ☉'s U.L.

Chronometer.	H. M. S.	Level.		Readings, &c.	° ′ ″
	8 02 24.4	+1	−8	First Vernier	331 42 30
	8 04 40.4	+6	−2	Second „	42 10
	8 06 22.4	+7	−1	Third „	42 50
	8 07 52.8	0	0	Fourth „	42 00
	8 09 29.2	+7	0	Mean . . .	331 42 22.5
	8 11 25.6	0	0	Index . .	+124 53 05
Mean. . .	8 07 02.47	+21	−11	Level . . .	+5
True time .	3 06 29.27	+5			456 35 32
Chron. fast.	5 00 33.2			Observed Z.D.	76 05 55.3
				Ref. and Paral.	+3 47.5
360° − 235° 06′ 55″ = 124° 53′ 05″				Semidiam . .	+16 18
				True Z.D. . .	76 26 01

Chronometer.	H. M. S.	Level.		Readings, &c.	° ′ ″
	8 14 51.6	+2	−5	First Vernier	78 17 50
	8 16 51.2	+8	+1	Second „	17 40
	8 18 55.2	−10	−3	Third „	18 20
	8 20 47.6	+2	−5	Fourth „	17 30
	8 22 41.2	0	−8	Mean. . .	78 17 50
	8 24 46	+6	−1	Index . . +	{ 28 17 37.5 { 360 00 00
Mean. . .	8 19 48.8	+19	−32	Level . . .	−6.5
True time .	3 19 16.18	−6.5			466 35 21
Chron. fast	5 00 32.62			Observed Z.D.	77 45 53
				Ref. and Paral.	+4 19
360° − 331° 42′ 22.5″ = 28° 17′ 37.5″				Semidiam .	+16 18
				True Z.D. . .	78 06 30

Chronometer, Fast { H. M. S. 5 00 33.2 / 5 00 32.62 } H. M. S. 5 00 32.91

January 3d; Barometer 30.20; Thermometer 41°; ☉'s U.L.

Chronometer.	H. M. S.	Level.		Readings, &c.	° ′ ″
	8 07 50	0	0	First Vernier	219 43 10
	8 09 16.8	−2	−10	Second „	43 00
	8 10 54.8	0	0	Third „	43 30
	8 12 03.2	+8	0	Fourth „	42 50
	8 13 32	+10	+2	Mean . . .	219 43 07
	8 18 31.6	+10	+2	Index . . .	+239 55 35
Mean. . .	8 12 01.4	+32	−12	Level . . .	+10
True time .	3 11 30.77	+10			459 38 52
Chron. fast.	5 00 30.63			Observed Z.D.	76 36 29
				Ref. and Paral.	+3 57
360° − 120° 04′ 25″ = 239° 55′ 35″				Semidiam . .	+16 18
				True Z.D. . .	76 58 44

Chronometer.	Level.	Readings, &c.

Chronometer, Fast H. M. S. 5 00 30.63

New York.——Determination of the Rate of the Chronometer by Zenith Distances, *continued.*

By Zenith Distances of the Sun, East of the Meridian.

December 24th; Barometer 30.40; Thermometer 18°; ☉'s L.L.

Chronometer.		Level.		Readings, &c.	
	H. M. S.				° ′ ″
	1 36 58	0	0	First Vernier	221 04 40
	1 38 40	0	0	Second ,,	04 30
	1 40 30.4	0	0	Third ,,	04 50
	1 41 45.6	+12	+1	Fourth ,,	04 40
	1 43 50.4	−6	+4	Mean . . .	221 04 40
	1 45 00.8	+3	−8	Index . . .	+254 54 45
Mean . . .	1 41 07.53	+20	−14	Level . . .	+ 3
True time .	8 40 09.32	+3			475 59 28
Chron. fast	5 00 58.21			Observed Z.D.	79 19 55
				Ref. and Paral.	+ 5 15
360° − 105° 05′ 15″ = 254° 54′ 45″				Semidiam. .	−16 18
				True Z.D. . .	79 08 52

Chronometer.		Level.		Readings, &c.	
	H. M. S.				° ′ ″
	1 47 48.4	+8	−4	First Vernier	328 26 15
	1 49 18	+8	−4	Second ,,	26 15
	1 51 00.4	+3	−8	Third ,,	26 35
	1 52 21.6	+10	0	Fourth ,,	26 10
	1 55 04.8	+9	−2	Mean . . .	328 26 19
	1 56 29.2	+7	−4	Index . . .	+138 55 20
Mean . . .	1 52 00.4	+45	−22	Level . . .	+11.5
True time .	8 51 02.6	+11.5			467 21 50
Chron. fast	5 00 57.8			Observed Z.D.	77 53 38
				Ref. and Paral.	+ 4 19
360° − 221° 04′ 40″ = 138° 55′ 20″				Semidiam. .	−16 18
				True Z.D. . .	77 41 59

Chronometer, Fast { H. M. S. 5 00 58.21; 5 00 57.8 } H. M. S. 5 00 58

December 29th; Barometer 30.30; Thermometer 26°; ☉'s L.L.

Chronometer.		Level.		Readings, &c.	
	H. M. S.				° ′ ″
	1 47 44	+8	−2	First Vernier	330 45 00
	1 50 13 2	+5	−6	Second ,,	45 00
	1 51 54	0	0	Third ,,	45 20
	1 53 12	0	0	Fourth ,,	44 40
	1 54 44	+9	−1	Mean . . .	330 45 00
	1 56 26.4	0	0	Index . . .	+137 10 47.5
Mean . . .	1 52 22.27	+22	−9	Level . . .	+6.5
True time .	8 51 38.03	+6.5			467 55 54
Chron. fast	5 00 44.24			Observed Z.D.	77 59 19
				Ref. and Paral.	+ 4 35
360° − 222° 49′ 12″.5 = 137° 10′ 47″.5				Semidiam. .	−16 18
				True Z.D. . .	77 47 36

Chronometer.		Level.		Readings, &c.	
	H. M. S.				° ′ ″
	1 59 52.8	−3	−12	First Vernier	70 53 35
	2 00 56	0	0	Second ,,	53 10
	2 01 50.8	0	0	Third ,,	53 55
	2 02 55.2	+10	+1	Fourth ,,	53 00
	2 04 02	+1	−8	Mean . . .	70 53 25
	2 05 22.4	+3	−6	Index . . +	{ 29 15 00; 360 00 00 }
Mean . . .	2 02 29.87	+15	−29	Level . .	−7
True time .	9 01 41.4	−7			460 08 18
Chron. fast	5 00 45.47			Observed Z.D.	76 41 23
				Ref. and Paral.	+4 07
360° − 330° 45′ 00″ = 29° 15′ 00″				Semidiam. .	−16 18
				True Z.D. . .	76 29 12

Chronometer, Fast { H. M. S. 5 00 44.24; 5 00 45.47 } H. M. S. 5 00 44.85

NEW YORK.——Determination of the Rate of the Chronometer by Zenith Distances, *continued.*

By Zenith Distances of the Sun, East of the Meridian.

January 2d; Barometer 29.98; Thermometer 32°; ⊙'s L.L.

Chronometer.		Level.		Readings, &c.		Chronometer.		Level.		Readings, &c.	
	H. M. S.				° ′ ″		H. M. S.				° ′ ″
	1 42 11.6	0	0	First Vernier	131 35 30		1 53 45.2	0	0	First Vernier	235 06 50
	1 43 31.2	+3	−6	Second "	5 20		1 55 22	0	0	Second "	07 00
	1 45 31.2	0	0	Third "	6 00		1 57 00.4	−2	−10	Third "	07 30
	1 47 02	+8	0	Fourth "	5 10		1 59 10.4	0	0	Fourth "	06 50
	1 49 46	−10	0	Mean . . .	131 35 30		2 00 33.6	0	0	Mean . .	235 07 02.5
	1 50 56	+7	−1	Index . . .	+341 01 46		1 02 02	+7	1	Index . . .	+228 24 30
Mean. . .	1 46 29.7	+18	+17	Level . . .	0	Mean. . .	1 57 58.93	+7	−13	Level . . .	−3
True time .	8 45 55.3	+0.5			472 37 16	True time .	8 57 24.1	−3			463 31 30
Chron. fast	5 00 34.4			Observed Z.D.	78 46 12.7	Chron. fast	5 00 34.83			Observed Z.D.	77 15 15
				Ref. and Paral.	+ 4 46.3					Ref. and Paral.	+ 4 12
360°−18° 58′ 14″=341° 01′ 46″				Semidiam . .	−16 18	360°−131° 35′ 30″=228° 24′ 30″				Semidiam .	−16 18
				True Z.D . .	78 34 41					True Z.D.. .	77 03 09

Chronometer Fast { H. M. S. 5 00 34.4, 5 00 34.83 } H. M. S. 5 00 34.61

By Zenith Distances of α Lyræ, West of the Meridian.

December 22d; Barometer 30.30; Thermometer 35°.

Chronometer.		Level.		Readings, &c.	
	H. M. S.				° ′ ″
	11 31 56	+6	−3	First Vernier	251 12 30
	11 34 06.4	+2	−7	Second "	12 00
	11 36 14	−1	−9	Third "	12 30
	11 38 26.8	+9	0	Fourth "	11 50
	11 41 48	+2	−7	Mean . . .	251 12 12.5
	11 44 08	+7	−2	Index . . .	+153 41 40
Mean. . .	11 37 46.53	+26	−29	Level . . .	−1.5
True time .	6 36 43.9	−1.5			404 53 51
Chron. fast	5 01 02.63			Observed Z.D.	67 28 58
				Refraction. .	+2 26
360°−206° 18′ 20″=153° 41′ 40″				True Z.D.. .	67 31 24

Chronometer, Fast H. M. S. 5 01 02.63

January 2d; Barometer 30.02; Thermometer 32.5°.

Chronometer.		Level.		Readings, &c.	
	H. M. S.				° ′ ″
	12 02 51	0	0	First Vernier	291 50 00
	12 04 21.8	0	0	Second "	50 00
	12 07 21	0	0	Third "	50 30
	12 08 35.2	+8	−1	Fourth "	49 30
	12 11 32	−9	−0	Mean . . .	291 50 00
	12 12 55.2	+5	−4	Index . . .	+180 00 00
Mean. . .	12 07 57.53	+13	−14	Level . . .	0
True time .	7 07 23.47	0			471 50 00
Chron. fast	5 00 34.06			Observed Z.D	78 38 20
				Refraction. .	+4 53
360°−180° 00′ 00″=180° 00′ 00″				True Z.D.. .	78 43 13

Chronometer, Fast H. M. S. 5 00 34.06

New York.——Determination of the Rate of the Chronometer by Zenith Distances, *continued.*

By Zenith Distances of Rigel East of the Meridian.

December 23d; Barometer 30.26; Thermometer 17°.

Chronometer.	Level.		Readings, &c.		Chronometer.	Level.		Readings, &c.	
H. M. S.				° ′ ″	H. M. S.				° ′ ″
12 21 39.2	+1	−9	First Vernier	53 20 30	12 41 22.8	+6	−4	First Vernier.	89 43 10
12 24 53.6	+8	−4	Second „	20 10	12 43 18	+11	0	Second „	42 50
12 27 21.2	0	0	Third „	20 30	12 45 28	+10	−1	Third „	43 20
12 30 15.2	+9	−2	Fourth „	20 00	12 47 13.6	+12	+2	Fourth „	42 40
12 32 24.8	0	0	Mean	53 20 17.5	12 49 58	+3	−7	Mean .	89 43 00
12 35 02	+1	−10	Level	−3	12 52 28.4	+7	−4	Level .	+ 17.5
Mean . . . 12 28 36	+19	−25	Index . . .	+360 00 08.5	Mean . . . 12 46 38.13	+51	−16	Index . . .	+306 39 42.5
True time . 7 27 33.9	−3			413 20 23	True time . 7 45 37.17	+17.5			396 23 00
Chron. fast. 5 01 02.1			Observed Z.D.	68 53 24	Chron. fast. 5 01 00.96			Observed Z.D.	66 03 50
			Refraction . .	+2 41				Refraction . .	+2 20
			True Z.D. . .	68 56 05	360° − 53° 20′ 17″.5 = 306° 39′ 42″.5			True Z.D. . .	66 06 10

Chronometer, Fast { H. M. S. 5 01 02.1 / 5 01 00.96 } H. M. S. 5 01 01.53

January 2d; Barometer 30.02; Thermometer 28°.

Chronometer.	Level.		Readings, &c.		Chronometer.	Level.		Readings, &c.	
H. M. S.				° ′ ″	H. M. S.				° ′ ″
12 25 43.2	+8	−1	First Vernier	308 01 45	12 38 39.2	+6	−3	First Vernier	313 26 10
12 27 15.2	0	0	Second „	01 35	12 40 01.6	0	0	Second „	26 00
12 28 39.6	0	0	Third „	02 15	12 41 31.2	+3	−5	Third „	26 30
12 30 04.4	0	0	Fourth „	01 25	12 42 43.2	+6	−2	Fourth „	25 50
12 31 57.2	0	0	Mean . .	308 01 45	12 44 54.4	+3	−5	Mean . . .	313 26 07.5
12 34 13.6	0	0	Index . . .	+68 10 00	12 47 40	0	0	Index . . .	+ 57 58 15
Mean . . . 12 29 38.87	+8	−1	Level	+ 3	Mean . . . 12 42 34.93	+18	−15	Level . . .	+ 1.5
True time. . 7 29 04.23	+3.5			376 11 48	True time . 7 41 59.94	+1.5			365 24 24
Chron. fast. . 5 00 34.64			Observed Z.D.	62 41 58	Chron. fast . 5 00 34.99			Observed Z.D.	60 54 04
			Refraction . .	1 57				Refraction . .	+1 49
360° − 291° 50′ 00″ = 68° 10′ 00″			True Z.D. . . .	62 43 55	360° − 308° 01′ 45″ = 51° 58′ 15″			True Z.D. . . .	60 55 53

Chronometer Fast { H. M. S. 5 00 34.64 / 5 00 34.99 } H. M. S. 5 00 34.81

New York.——Determination of the Rate of the Chronometer by Zenith Distances, *continued.*

By Zenith Distances of Sirius, East of the Meridian.

January 3d; Barometer 30.20; Thermometer 36.5°.

Chronometer.	Level.		Readings, &c.		Chronometer.	Level.		Readings, &c.	
H. M. S.				° ′ ″	H. M. S.				° ′ ″
13 33 42.8	0	0	First Vernier	343 24 00	13 46 24	+3	−6	First Vernier	43 02 10
13 35 08	0	0	Second „	24 00	13 48 26.4	+2	−8	Second „	02 00
13 36 36	0	0	Third „	24 30	13 50 17.6	+2	−7	Third „	02 35
13 39 06.8	0	0	Fourth „	23 30	13 53 46.8	0	0	Fourth „	02 00
13 41 54	+2	−6	Mean	343 24 00	13 57 05.6	0	0	Mean	43 02 11
13 43 15.6	+2	−7	Index . . .	+87 51 05	13 58 48	+9	0	Index . . +	16 36 00 360 00 00
Mean . . . 13 38 17.2	+4	−13	Level	−3.5	Mean . . . 13 52 28.07	+16	−21	Level	−2.5
True time . 8 37 44.4	−3.5			431 15 01.5	True time . 8 51 54.77	−2.5			419 38 09
Chron. fast . 5 00 32.8			Observed Z.D.	71 52 30	Chron. fast . 5 00 33.3			Observed Z.D.	69 56 21.5
			Refraction .	+3 02				Refraction . .	+2 43.5
360°−272° 08′ 55″=87° 51′ 05″			True Z.D. . . .	71 55 32	360°−343° 24′ 00″=16° 36′ 00″			True Z.D. . .	69 59 05

Chronometer, Fast { H. M. S. 5 00 32.8 ; 5 00 33.3 } H. M. S. 5 00 33.05

New York.——OBSERVATIONS to DETERMINE the RATE of the Chronometer No. 423, from the 24th of December to the 2d of January, by the DISAPPEARANCE of STARS behind the Steeple of the Presbyterian Church, viewed in a Telescope fixed to the Wall, in the Eastern Window of the Chapel of Columbia College.

	Time of Disapp. by the Chron.		Difference.	Interval of Sid. Days.	Difference between 9 Solar and Sid. Days.	Chronometer's Loss.		
	December 24.	January 2.				In the Interval.	per Sid. Day.	per Sol. Day.
	H. M. S.	H. M. S.	M. S.		M. S.	S.	S.	S.
1st.—Ursæ Majoris	12 11 35.2	11 35 49.6	35 45.6	9	35 23.19	22.41	2.49	2.5
2d.—Ursæ Majoris	12 15 16	11 39 29.6	35 46.4	9	35 23.19	23.21	2.58	2.59
3d.—Ursæ Majoris	12 23 56.8	11 48 10.4	35 46.4	9	35 23.19	23.21	2.58	2.59

RESULTS of the PRECEDING OBSERVATIONS.

	Interval.	No. of Days.	Chron.'s loss, on Mean Solar Time.
			s.
By the Sun, West of the Meridian	December 22d to January 2d	11	2.61
	„ 22d to „ 3d	12	2.59
	„ 23d to „ 2d	10	2.76
	„ 23d to „ 3d	11	2.72
By the Sun, East of the Meridian	„ 24th to „ 2d	9	2.59
By α Lyræ, West of the Meridian	„ 22d to „ 2d	11	2.59
By Rigel, East of the Meridian	„ 23d to „ 2d	10	2.67
By Rigel and Sirius, East of the Meridian	„ 23d to „ 3d	11	2.59
By the Diappearance of Stars No. 1.	„ 24th to „ 2d	9	2.50
By the Diappearance of Stars No. 2.	„ 24th to „ 2d	9	2.59
By the Diappearance of Stars No. 3.	„ 24th to „ 2d	9	2.59
Mean,—Chronometers' loss per Diem			2.62
Intermediate.			
By the Sun, East of the Meridian	December 24 to December 29.	5	2.68
	„ 29 to January 2.	4	2.56

New York.——Comparisons of the Astronomical Clock with the Chronometer No. 423, between the 22d of December, 1822, and the 3d of January, 1823; with the Clock's Rate on Mean Solar Time deduced.

1822.	Chronometer.	Clock.	Clock's Loss on 423.			Daily Rates. Chron.	Daily Rates. Clock.
	H. M. S.	H. M. S.	S.		S.	S. Losing.	S. Losing.
Dec. 22 A. M.		8 50 02.9					
			30.6				
,, 23 A. M.		8 49 32.3					
			30.3				
,, 24 A. M.		8 49 52					
			30.4	29.98			
,, 25 A. M.		8 48 31.6					
			29.6				
,, 26 A. M.		8 48 02					
			29				
,, 27 A. M.		8 47 33					
			30		30.01	2.62	32.63
,, 28 A. M.	7 00 00	8 47 03					
			29.9				
,, 29 A. M.		8 46 33.1					
			29.9				
,, 30 A. M.		8 46 03.2					
			29	30.04			
,, 31 A. M.		8 45 34.2					
			30.2				
1823 Jan. 1 A. M.		8 45 04					
			30				
,, 2 A. M.		8 44 34					
			31.3				
,, 3 A. M.		8 44 02.7					

NEW YORK.——COINCIDENCES OBSERVED with PENDULUM No. 3; the Clock making 86367.37 Vibrations in a Mean Solar Day.

DATE.	Observer.	Barometer.	No. of Coincidence.	Temperature.	Time of Disappearance.	Time of Re-appearance.	True Time of Coincidence.	Arc of Vibration.	Mean Temperature.	Mean Interval.	Correction for the Arc.	Vibrations in 24 hours.	Reduction to a mean Temperature.	Reduced Vibrations at 35.66°
1822.		IN.		°	M. S.	M. S.	H. M. S.	°	°	S.	S.			
Dec.											+			
23 A.M.	Capt. Sabine	30.140	1	40.3	33 49	33 53	10 33 51	1.18	40.2	684.15	1.22	86116.10	+1.91	86118.01
			11	40.1	27 46	27 59	12 27 52.5	0.58						
23 P.M.	Mr. Renwick	30.260	1	40.3	40 00	40 06	1 40 03	1.2	40.25	684.25	1.25	86116.15	+1.93	86118.08
			11	40.2	33 56	34 15	3 34 05.5	0.58						
24 A.M.	Capt. Sabine	30.400	1	33.7	11 02	11 05	9 11 03.5	1.2	34	691	1.25	86118.63	−0.70	86117.93
			11	34.3	06 06	06 21	11 06 13.5	0.58						
24 A.M.	Capt. Sabine	30.400	1	34.75	28 50	28 54	11 28 52	1.18	35.5	688.95	1.25	86117.87	−0.07	86117.80
			11	36.25	23 36	23 47	1 23 41.5	0.6						
24 P.M.	Mr. Renwick	30.400	1	36.8	35 41	35 45	1 35 43	1.26	36.55	686.8	1.46	86117.32	+0.37	86117.69
			11	36.3	30 01	30 21	3 30 11	0.66						
25 A.M.	Capt. Sabine	30.120	1	32.9	41 43	41 47	10 41 45	1.2	33.25	691.15	1.28	86118.72	−1.01	86117.71
			11	33.6	36 49	37 04	12 36 56.5	0.6						
25 P.M.	Mr. Renwick	30.000	1	33.5	59 30	59 37	12 59 33.5	1.21	34.55	690.4	1.26	86118.42	−0.47	86117.95
			11	35.6	54 27	54 48	2 54 37.5	0.58						
26 A.M.	Capt. Sabine	30.000	1	32.6	51 20	51 25	10 51 22.5	1.2	32.8	692.05	1.28	86119.04	−1.20	86117.84
			11	33	46 37	46 49	12 46 43	0.6						
26 P.M.	Mr. Renwick	30.180	1	33.1	58 26	58 30	12 58 28	1.24	33.85	691.35	1.31	86118.81	−0.76	86118.05
			11	34.6	53 31	53 52	2 53 41.5	0.59						
Means. .		30.210							35.66			86117.9		86117.9

New York.—COINCIDENCES OBSERVED with PENDULUM 4; the Clock making 86367.37 Vibrations in a Mean Solar Day.

DATE.	Observer.	Barometer.	No. of Coincidence.	Temperature.	Time of Disappearance.	Time of Re-appearance.	True Time of Coincidence.	Arc of Vibration.	Mean Temperature.	Mean Interval.	Correction for the Arc.	Vibrations in 24 hours.	Reduction to a Mean Temperature.	Reduced Vibrations at 33.81°.
1822.		IN.		°	M. S.	M. S.	H. M. S.	°	°	S.	S.			
Dec. 27 A.M.	Capt. Sabine	30.420	1	32.7	11 33	11 39	11 11 36	1.18	32.5	719.55	+1.22	86128.52	−0.55	86127.97
			11	32.3	11 23	11 40	1 11 31.5	0.58						
27 P.M.	Mr. Renwick	30.420	1	32.4	22 57	23 04	1 23 00.5	1.24	32.6	719.4	1.31	86128.57	−0.51	86128.06
			11	32.8	22 39	23 10	3 22 54.5	0.59						
28 A.M.	Mr. Renwick	30.290	1	32.6	13 22	13 24	10 13 23	1.3	32.8	718.95	1.44	86128.56	−0.42	86128.14
			11	33	12 59	13 26	12 13 12.5	0.62						
28 P.M.	Mr. Renwick	30.290	1	33.1	36 43	36 48	12 36 45.5	1.24	33	719.15	1.30	86128.46	−0.34	86128.12
			11	32.9	36 25	36 49	2 36 37	0.58						
29 A.M.	Capt. Sabine	30.290	1	32	7 22	7 30	9 07 26	1.11	32.85	720.1	1.17	86128.63	−0.40	86128.23
			11	33.7	7 20	7 34	11 07 27	0.58						
29 P.M.	Mr. Renwick	39.290	1	35	30 32	30 39	11 30 35.5	1.24	35.75	716.55	1.30	86127.58	+0.81	86128.39
			11	36.5	29 52	30 10	1 30 01	0.58						
30 A.M.	Mr. Renwick	30.460	1	33.4	3 35	3 40	10 03 37.5	1.28	34.25	717.85	1.38	86128.12	+0.18	86128.30
			11	35.1	3 03	3 29	12 03 16	0.6						
30 P.M.	Mr. Renwick	30.560	1	35.3	25 16	25 22	12 25 19	1.23	35.95	715.7	1.33	86127.33	+0.90	86128.23
			11	36.6	24 26	24 46	2 24 36	0.61						
31 P.M.	Mr. Renwick	30.790	1	33	47 02	47 09	12 47 05.5	1.2	33.6	718.05	1.27	86128.05	−0.09	86128.96
			11	34.2	46 35	46 57	2 46 46	0.59						
1823. Jan. 1 A.M.	Capt. Sabine	30.310	1	31.2	31 44	31 53	11 31 48.5	1.18	31.2	721.25	1.25	86129.11	−1.1	86128.01
			11	31.2	31 52	32 10	1 32 01	0.6						
2 A.M.	Capt. Sabine	29.980	1	33.3	15 33	15 39	9 15 36	1.2	34.5	718.2	1.28	86128.12	+0.29	86128.41
			11	35.7	15 10	15 26	11 15 18	0.6						
2 P.M.	Mr. Renwick	30.020	1	36.2	21 01	21 02	12 21 01.5	1.29	36.75	715.4	1.41	86127.31	+0.81	86128.12
			11	37.3	20 05	20 26	2 20 15.5	0.61						
Means		30.367							33.81			86128.2		86128.2

THE Pheasant anchored at Portsmouth on the 5th of February, 1823, after an unusually tedious passage of thirty-one days from New York; and proceeding to the river, landed the instruments at Deptford on the 18th of February.

On my arrival in London, I had the satisfaction of finding that the letter which I had written to Sir Humphry Davy from Maranham, proposing the extension of the experiments to the high latitudes, had met the approbation of the Commissioners of Longitude, and that Lord Melville's consent had been obtained for the employment of one of His Majesty's ships in its prosecution; accordingly the Griper sloop-of-war, which had been one of the vessels engaged in the Expedition of North-West Discovery in 1819-1820, on which occasion she had been strengthened for the encounter of ice, was commissioned by Captain Clavering on the 26th of February, and her equipment, for the particular navigation for which she was destined, proceeded in.

The plan of the voyage, approved by the Admiralty, and conveyed in their instructions to Captain Clavering, directed him to proceed in the first instance to Hammerfest, near the north cape of Norway, as a pendulum station adjoining the 70 degree of latitude; from thence to a second station in or near the 80th parallel, on the northern coast of Spitzbergen; afterwards to make the east coast of Greenland in as high a latitude as the barrier of ice, which renders that coast difficult of access, would permit; and having got within the barrier, and in the navigable channel which is usually found in the northern seas in the neighbourhood of land, to ascend the coast to the northward, as far as might be compatible with a return to England in the same year, and to make a third pendulum station at the highest latitude that might be thus attained; Captain Clavering was then directed to return to the southward in order to get off the coast of Greenland before the advance of the season might endanger his detention during the winter; after which he was at liberty to use his discretion in making a fourth station at Iceland, or elsewhere in or about the same parallel, and then to return to England. In case the Griper should be accidentally detained in the high latitudes during the winter, she was ordered to be provisioned for eighteen months; and the same liberal supply of preserved meats and warm clothing was furnished for the health and comfort of the seamen, and under the same regulations of issue, as had taken place in the Voyages of Discovery.

The interval of the Griper's equipment was occupied in repeating the trial of the pendulums in Portland Place, to ascertain that they had undergone no alteration in the course and by the events of the preceding voyage*; and in providing an apparatus for the support of the clock and pendulums, and for their protection against the weather, required at the stations of the north, where no other accommodation or convenience could be expected, than the rock which might serve as a foundation. It was desirable that the various parts of the apparatus for these purposes should be contrived with as much regard to portability and conveniency of stowage, as was compatible with the stability of the supports, and the sufficiency of the defence against the weather; the preparation for supporting the detached pendulums consisted of a cast-iron tripod stand, the legs of which screwed into the angles of a strong triangular frame which rested on the ground; the vertical front was an equilateral triangle, the sides of which were six feet and a half long; the three legs of the tripod screwed at the upper angle, opposite to the ground, to the vertical sides and back of a rectangular frame, the upper and horizontal side of which was fitted to receive the agate planes on which the pen-

* The particulars are reserved until the general account of the experiments with the deached pendulums in London.

dulum vibrated; by levelling the foundation so that the ground frame was nearly horizontal, the side of the rectangle supporting the planes was sufficiently so, to admit of their being brought into exact adjustment, by means of the screws for fixing and levelling them, attached to the brass box in which they were contained; the ground frame was furnished with screws at the angles, by which it could be raised from the ground on a sufficiently hard foundation, if it were preferred that the rest should be on three points of bearing, instead of on the frame generally.

The clock was provided with a similar support, but made of wood, and so contrived as to stand withinside the iron tripod, beneath the rectangular frame which carried the planes, and on the space comprehended by the ground frame; thus when the pendulum was suspended, and the clock placed in its proper position in relation to it, every part of the apparatus belonging to the pendulum was detached from and unconnected with the clock, and with its support; the wooden stand was furnished with a vertical and side adjustment, for placing the clock in beat: both the stands were so constructed as to be readily taken in pieces, for convenience in packing and carriage; the iron tripod stand weighed altogether above two hundred pounds; they were contrived and executed by Mr. Jones of Charing-Cross.

In providing a suitable cover for the instruments, I had the advantage of having had much experience in the Northern Expedition, and particularly at Melville Island; I was aware that in the Arctic Circle, during the summer months, a much less substantial protection would suffice, than in the temperate or torrid zones, (and especially than in the latter), because the constant presence of the sun above the horizon causes the range of the external thermometer in the twenty-four hours to be much less than in other parts of the globe; in fact, in the very high latitudes, where the difference in the sun's altitude on the northern and southern meridians does not exceed 20 or 30 degrees, and whilst his inferior altitude is still sufficiently high to cause the balance of radiation to be in favour of the earth, the temperature, independently of the variations occasioned by changes in the weather, is very nearly the same at all hours. It was thus fortunate that accommodation was least needed where it was most deficient, and that the absence of houses could be supplied by a temporary provision, which would not have been sufficient at the stations of the preceding voyage, in a point of so much importance as the maintenance of an uniform temperature.

I obtained from the Ordnance Department one of the large circular tents used by the artillery for the purposes of the laboratory, consisting of a canvass roof and walls without lining, and supported by a single central pole; the tent was of sufficient dimensions to include a wooden house of twelve feet square and ten feet high, having the tent pole in the middle; the house was constructed of very substantial frame work, with a boarded roof, floor. and walls, and was made to take in pieces and put together at pleasure, the pieces which fitted to each other being marked correspondingly; the roof was flat and divided into 16 compartments, each of which could be raised and folded back upon the adjoining one, with which it was connected by hinges, so that the light passing through the canvass of the tent might be admitted into the room in the direction and quantity which might be desired; the frame was every where strengthened with diagonal pieces so as to be extremely firm; the tent pole passed through a hole made to receive it in a cross beam of tho roof, and was stepped into a very strong cross beam of the floor; so that besides the usual security against the weather of guys and tent-cords, the tent had the additional support of the whole strength and weight of the house; it was intended that the room should be only partially floored, in order that the clock and pendulum stands might rest on an independent foundation on the ground or rock beneath, and be thus insulated with regard to the house, as they already were with regard to each other: the house was built in the dock-yard at Deptford, by an order from the Board of Admiralty, kindly obtained by Mr. Barrow; and was as firm and strong at the last station at which it was used, as when it was first completed.

A small but extremely portable observatory for the transit instrument was made under the direction of Mr. Dollond, the roof, sides, and floor, of which were framed in separate pieces and fastened together by copper screws; the pedestal for the support of the transit, passed through the floor, but was unconnected with it, so as to be entirely insulated; it exceeded four feet in length, of which a foot and a half was above the ground, and the lower end wedged firmly into a frame sunk to the proper depth for that purpose; the pedestal was octagonal in shape, and being hollow was filled in with earth and stones to increase its firmness; a flat slab of free stone was screwed to the upper end, and rendered horizontal by the insertion of wedges before the screws were tightened; on this stone rested a second slab of the same material to which the frame and pillars of the transit fitted and were attached by screws; the upper stone was moveable upon the surface of the lower, until the instrument was placed sufficiently near the meridian to be within reach of exact adjustment by its own means, when the stones were cemented to each other by Plaister of Paris, which united the qualities of becoming dry immediately, and of permitting them to be separated without injury, when the experiments at the station were concluded. When the stones were cemented, the transit and frame could be removed at pleasure, as the screws ensured its correct replacement, and the observatory was thus rendered disposable at such times for the use of other instruments; as all its fastenings were of copper, it was particularly adapted for experiments connected with Magnetism.

A marquee for myself and a tent for my servant, both of which had been supplied by the Ordnance Department at the commencement of the Arctic Expeditions in 1818, completed the preparation for the northern stations.

The compensation of the chronometers Nos. 423 and 493 requiring adjustment, they were returned to the makers for that purpose on my arrival in England, and Nos. 649 and 602 received in their stead; No. 649 was a pocket chronometer to be used in observation as No. 423 had been; I also received No. 423 again, a day or two before my departure for the north, having particularly requested that I might do so; its rate had not been examined by the makers since the adjustment of its compensation, and I found it somewhat wider than it had been previously; but as I used No. 649 from henceforth (with one accidental exception,) I did not take the pains to reduce the rate.

The equipment of the Griper being pressed with all the means which Captain Clavering could obtain, she was ready for sea by the second week in May, and sailed from the Nore on the 11th for Hammerfest, where she arrived on the 4th of June, having manifested in the passage that the heavy sailing by which she had been distinguished on her former voyage, was in no degree improved.

HAMMERFEST.

Hammerfest is a small trading and fishing town, built on one of the numerous islands which adjoin the coast of Finmarken, and is distant only a few miles from the northern extremity of Europe; the town is situated on the southern side of the harbour; and on the opposite side called Fugleness, Mr. John Crowe, an English merchant, has formed within a few years past, a very promising commercial establishment, occupying a long and narrow projection of rock, washed on each side by the sea, and raised but little above its level. I had had the good fortune to make Mr. Crowe's acquaintance in London during the spring, whilst the Griper was fitting, and he had sailed on his return to Hammerfest, some days before she was ready for sea: his arrival therefore, had preceded ours, and we found him prepared to receive us, with an hospitality characteristic of the country in which he has made his summer residence.

A spot was soon selected at Fugleness, in which the rock was sufficiently level for the few square feet required as a foundation for the house and instruments; the guys which stayed the tent pole, and the cords by which the roof and walls of the tent were fastened down, were secured by grapnels and ice anchors, and by heavy pieces of the rock brought for the purpose; the walls of the house were banked up with earth for a considerable part of their height, and folds of canvass were nailed over every crevice by which the air might gain admission into the room; these precautions were not unnecessary, as the weather proved most unfavourable during the greater part of our stay, being almost an incessant gale, with rain, sleet, and heavy fog; nor were they in vain, as notwithstanding the violence of the weather, the interior of the room remained perfectly dry, and free from drafts; even in the heaviest gale, floss silk suspended by the side of the pendulum shewed that the air within the house was in a tranquil state; the apparatus of the house and tent answered its purpose so effectually, that the going of the clock and pendulums sustained no inconvenience whatsoever. The comparison of the chronometer with astronomical time was, however, very much impeded by the weather; of twelve stars with which the transit list commenced on the 9th of June, three only could be observed on the 22nd, and none on any of the intermediate days, except η Ursæ, which was visible whilst passing two of the wires on the 14th; the 23d continuing obscure, the series of coincidences was closed on the preceding day, as the gain of the chronometer in the interval between the observations on the 9th and 22nd, appeared to have been satisfactorily determined by the transits of the sun and of the three stars, and as the daily rate of the chronometer and of the clock in the interim had been sufficiently uniform; the transit observations were also corroborated by the results deduced from those with the Repeating Circle, although the opportunities for the use of the latter instrument were by no means favourable, as in addition to the slow motion of the sun in altitude in the latitude of 70°, his limb became generally ill-defined when at the proper distance from the meridian; this interruption of continuity, or ragged appearance of the disk is a frequent impediment to exact observation in the Arctic Circle; it affects both limbs, commencing generally at about 30° or 35° of altitude, and increasing as the sun descends.

The clock room not being of sufficient size to receive the telescope for the observation of coincidences, when at the proper distance from the pendulum, the stand for its support was placed between the house and tent, and a window of a single pane was made in the wall which interposed between the telescope and its object; a wooden porch was built around the telescope, capable of containing a single observer, by whom, when seated at the telescope, the face of the clock was visible through the open window.

Qualoën or Whale Island on which Hammerfest is situated, as well as the neighbouring islands and adjacent continent, are composed of primitive rock; that of Qualoën is principally Gneis. The height of the pendulums above the sea was twenty-nine feet.

As the present occasion was the second in which Hammerfest had been visited from England for astronomical purposes, and as from the peculiarity of its situation in being the most northern town in the globe, future occasions may occur of a similar nature, it may be useful to notice, that the unfavourable weather, which we experienced during the month of June, is said to be very prevalent in the summer months, but to be confined to the islands; as we were informed, that at the same periods the weather on the adjacent continent would present the remarkable contrast of a serene and clear atmosphere; on this account, Alten would probably be a preferable station to Hammerfest for celestial observations; but in other respects, the harbour of Hammerfest is more easily accessible, and as being the residence of Mr. Crowe, and of the other members of his commercial establishment, Fugleness possesses an advantage of much consideration to English visitors.

The instruments being re-imbarked on the morning of the 23d of June, the Griper sailed in the afternoon, and anchored on the first of July in Fair Haven on the north of Spitzbergen; the only ice which had been seen on the passage, being a small stream, which frequently during the summer season is found to set with the current round the southern shore of Spitzbergen.

TRANSITS OBSERVED AT HAMMERFEST.

DATE.	STARS.	TIMES OF TRANSITS BY No. 649.					Mean by the Chronometer.	Chron. Slow on Mean Time.
		1st Wire.	2d Wire.	Meridian Wire.	4th Wire.	5th Wire.		
1823.		H. M.	H. M.	H. M. S.	H. M.	H. M.	H. M. S.	H. M. S.
June 9	Sun 1st Limb	.....	.....	......	.....	.....	22 24 51.43	1 33 47.82
	Sun 2d Limb	25 03.2	25 31.6	22 26 00.4	26 28.8	26 57.6		
„ „	η Ursæ	56 06.8	56 47.6	6 57 28	58 09.6	58 50	6 57 28.33	
„ „	Arcturus. . .	23 30	23 58	7 24 25.6	24 54	25 22	7 24 25.87	
„ „	α Lyræ	45 58	46 31.2	11 47 04.8	47 38	48 12	11 47 04.8	
„ 10	Sun 1st Limb	22 58.8	23 27.2	22 23 56	24 24.8	24 53.2	22 25 04.6	1 33 46.3
	Sun 2d Limb	25 15.6	25 44.4	22 26 13.2	26 42	27 10.8		
„ 12	Sun 1st Limb	23 25.2	23 54	22 24 22.4	24 51.2	25 19.2	22 25 31.6	1 33 43.25
	Sun 2d Limb	25 43.6	26 12	22 26 40.8	27 09.6	27 38		
„ 13	Sun 1st Limb	.....	.....	......	.....	.....	22 25 45.57	1 33 41 63
	Sun 2d Limb	25 57.6	26 25.6	22 26 54.4	27 23.2	27 51.6		
„ 14	Sun 1st Limb	23 53.2	24 21.6	22 24 50.4	25 18.8	25 47.2	22 25 59.17	1 33 40.38
	Sun 2d Limb	.....	.....	......	.....	.....		
„ „	η Ursæ	36 34.4	37 15.6	6	.....	.....	6 37 56 13	
„ 17	Sun 1st Limb	24 36	25 04.8	22 25 33.2	26 02	26 30.4	22 26 42	1 33 35.6
	Sun 2d Limb	26 53.2	27 22	22 27 50.8	28 19.6	28 48		
„ 22	Sun 1st Limb	25 46.4	26 15.2	22 26 43.6	27 12	27 41.2	22 27 52.07	1 33 29.33
	Sun 2d Limb	28 04.4	28 33.2	22 29 01.6	29 30	29 59.2		
„ „	η Ursæ	5 17.6	5 58.4	6 06 39.6	7 20	8 01.2	6 06 39.4	
„ „	Arcturus. . .	32 41.6	33 09.6	6 33 37.6	34 05.6	34 33.2	6 33 37.53	
„ „	α Lyræ	55 08.8	55 42.8	10 56 15.6	56 49.2	57 22.8	10 56 15.8	

HAMMERFEST.——OBSERVATIONS to DETERMINE the RATE of the Chronometer No. 649, on Mean Time, between the 9th and 22d of June, 1823, by ZENITH DISTANCES of the Sun, with a Repeating Circle.

Latitude of the Place of Observation 70° 40′ 04.5″ N.; Longitude 23° 45′ E.

June 9th P.M.; Barometer 30.10; Thermometer 57°; ☉'s U.L.

Chronometer.		Level.		Readings, &c.	
	H. M. S.				° ′ ″
	3 54 15.2	0	0	First Vernier	63 47 30
	3 56 02.4	+7	+3	Second „	47 10
	3 58 07.6	0	0	Third „	47 50
	4 59 55.2	+5	0	Fourth „	47 10
	4 01 28.4	0	0	Mean . . .	63 47 25
	4 02 54	+6	+1	Index . . .	+332 10 22
Mean . . .	3 58 47.13	+22	0	Level . . .	+11
True Time.	5 32 35	+11			395 57 58
Chron. slow	1 33 47.87			Observed Z.D.	65 59 40
				Ref. and Paral.	+ 2 00
360° − 27° 49′ 38″ = 332° 10′ 22″				Semidiam . .	+15 47
				True Z.D. . .	66 17 27

Chronometer.		Level.		Readings, &c.	
	H. M. S.				° ′ ″
	4 07 35.6	+5	0	First Vernier	106 13 00
	4 09 18.4	0	0	Second „	12 50
	4 11 02	0	0	Third „	13 10
	4 12 42.8	−2	0	Fourth „	12 20
	4 14 12.8	+4	0	Mean . . .	106 12 50
	4 16 06	+3	0	Index . . .	+296 12 35
Mean . . .	4 11 49.6	+12	−2	Level . . .	+5
True time .	5 45 38.1	+5			402 25 30
Chron. slow	1 33 48.5			Observed Z.D.	67 04 15
				Ref. and Paral.	+ 2 06
360° − 63° 47′ 25″ = 296° 12′ 35″				Semidiam . .	+15 47
				True Z.D. . .	67 22 08

Chronometer, Slow { H. M. S. 1 33 47.87 ; 1 33 48.5 } H. M. S. 1 33 48.19

June 22d P.M.; Barometer 29.85; Thermometer 47°; ☉'s U.L.

Chronometer.		Level.		Readings, &c.	
	H. M. S.				° ′ ″
	3 49 03.2	+3	−5	First Vernier	91 10 50
	3 52 37.2	+2	−5	Second „	10 00
	3 54 16.8	0	0	Third „	11 00
	3 56 06.4	0	−6	Fourth „	10 15
	3 57 37.6	+5	−3	Mean . . .	91 10 31
	3 59 08	0	0	Index . . .	+268 48 02.5
Mean . . .	3 54 48.2	+10	−19	Level . . .	+4.5
True time .	5 28 20.05	−4.5			359 58 38
Chron. slow	1 33 31.85			Observed Z.D.	59 59 46
				Ref. and Paral.	+1 34
360° − 91° 11′ 57.5″ = 268° 48′ 02.5″				Semidiam . .	+15 45
				True Z.D. . .	60 17 05

Chronometer.		Level.		Readings, &c.	
	H. M. S.				° ′ ″
	4 03 38	0	0	First Vernier	97 14 20
	4 05 08	+2	−5	Second „	14 00
	4 06 35.6	−1	+5	Third „	14 50
	4 08 10	+6	−1	Fourth „	14 15
	4 09 43.2	0	0	Mean . . .	97 14 21
	4 10 58.8	−2	−8	Index . . .	+268 49 29
Mean . . .	4 07 22.27	+13	−17	Level . . .	−2
True time	5 40 54.25	−2			366 03 48
Chron. fast.	1 33 31.98			Observed Z.D.	61 00 38
				Ref. and Paral.	+1 38
360° − 91° 10′ 31″ = 268° 49′ 29″				Semidiam . .	+15 45
				True Z.D. . .	61 18 01

Chronometer, Slow { H. M. S. 1 33 31.85 ; 1 33 31.98 ; 1 33 31.97* } H. M. S. 1 33 31.93

The particulars of this observation are in the next page.

HAMMERFEST.—Determination of the Rate of the Chronometer by Zenith Distances, *continued.*

June 22d P.M.; Bar. 29.85; Therm. 47°; ☉'s U.L.

	Chronometer.	Level.		Readings, &c.	
	H. M. S.				° ′ ″
	4 15 59.6	−7	0	First Vernier	109 26 00
	4 17 04.4	+6	−1	Second „	25 30
	4 19 16.4	−7	+1	Third „	26 20
	4 20 32	+5	−2	Fourth „	25 50
	4 22 36	0	0	Mean . . .	109 25 55
	4 23 56.4	+6	−1	Index . .	+262 45 39
Mean. . .	4 19 54.13	+18	−18	Level . . .	0
True time .	5 53 26.1				372 11 34
Chron. slow	1 33 31.97	0		Observed Z.D	62 01 56
				Ref. and Paral.	+1 42
360°−97° 14′ 21″ = 262° 45′ 39″				Semidiam. . .	+15 45
				True Z.D. . .	62 19 23

June 10th A.M.; Bar. 29.94; Ther 57°; ☉'s L.L.

	Chronometer.	Level.		Readings, &c.	
	H. M. S.				° ′ ″
	5 15 47.2	0	0	First Vernier	129 51 20
	5 18 04.4	+3	0	Second „	51 00
	5 20 42	0	0	Third „	51 40
	5 23 07.2	0	0	Fourth „	50 40
	5 25 21.6	−3	0	Mean . . .	129 51 10
	5 28 11.2	0	0	Index . . .	+253 46 42.5
Mean. . .	5 21 52.27	+3	−3	Level . . .	−0
True time .	6 55 41.1				283 37 52.5
Chron. slow	1 33 48.83	0		Observed Z.D.	63 56 19
				Ref. and Paral.	+ 1 50
360°−106° 13′ 17″.5 = 253° 46′ 42″.5				Semidiam. . .	−15 47
				True Z.D.	. 63 42 22

June 10th A.M.; Barometer 29.94; Thermometer 57°; ☉'s L.L.

	Chronometer.	Level.		Readings, &c.	
	H. M. S.				° ′ ″
	5 33 31.2	+3	0	First Vernier	145 00 20
	5 35 29.2	+5	0	Second „	00 00
	5 37 56	+2	0	Third „	00 30
	5 40 47.2	−3	0	Fourth „	00 00
	5 42 48	0	0	Mean . . .	145 00 12.5
	5 44 13.2	0	0	Index . . .	+230 08 50
Mean. . .	5 39 07.47	+10	−3	Level . . .	+3.5
True time .	7 12 55.7				375 09 06
Chron. slow	1 33 48.23	+3.5		Observed Z.D.	62 31 31
				Ref. and Paral.	+ 1 43
360°−129° 51′ 10″ = 230° 08′ 50″				Semidiam. . .	−15 47
				True Z.D. . .	62 17 27

	Chronometer.	Level.		Readings, &c.	
	H. M. S.				° ′ ″
	5 50 02.4	0	0	First Vernier	152 38 20
	5 51 39.6	0	0	Second „	38 10
	5 53 52.4	+5	0	Third „	38 40
	5 55 28.4	−5	0	Fourth „	38 10
	5 57 26	+8	0	Mean . . .	152 38 20
	5 59 00	0	0	Index . . .	+214 59 47.5
Mean. .	5 54 34.8	+8	−5	Level . . .	+1.5
True time .	7 28 23.6				367 38 09
Chron. slow	1 33 48.8	+1.5		Observed Z.D.	61 16 21.5
				Ref. and Paral.	+ 1 36.5
360°−145° 00′ 12″.5 = 214° 59′ 47″.5				Semidiam. . .	−15 47
				True Z.D. . .	61 02 11

	H. M. S.		H. M. S.
Chronometer, Slow	1 33 48.83	}	1 33 48.62
	1 33 48.23		
	1 33 48.8		

HAMMERFEST.——Determination of the Rate of the Chronometer by Zenith Distances, *continued.*

June 22d A.M.; Barometer 29.85; Thermometer 44°; ⊙'s L.L.

Chronometer.	Level.		Readings, &c.		Chronometer.	Level.		Readings, &c.	
H. M. S.				° ′ ″	H. M. S.				° ′ ″
6 09 44	+2	−5	First Vernier	118 23 20	6 24 40.8	0	0	First Vernier	108 11 50
6 11 13.2	−3	−11	Second „	23 05	6 26 13.2	+5	−2	Second „	11 20
6 12 39.6	+7	0	Third „	23 40	6 28 00.4	0	0	Third „	12 00
6 15 08	+6	−1	Fourth „	23 00	6 29 38	0	0	Fourth „	11 20
6 17 18	+2	−5	Mean . . .	118 23 16	6 31 56.8	−5	+2	Mean . . .	108 11 37.5
6 19 00.4	0	0	Index . . .	+238 24 34	6 33 55.2	+7	0	Index . . .	+241 36 44
Mean. . . 6 14 10.53	+17	−25	Level . . .	−4	Mean. . . 6 29 04.07	+14	−7	Level . . .	+3.5
True time . 7 47 29.4	−4			356 47 46	True time . 8 02 34.77	+3.5			349 48 25
Chron. slow 1 33 28.87			Observed Z.D.	59 27 58	Chron. slow 1 33 30.7			Observed Z.D.	58 18 04
			Ref. and Paral.	+1 33				Ref. and Paral.	+1 28
360° − 121° 35′ 26″ = 238° 24′ 34″			Semidiam . .	−15 46				Semidiam . .	−15 45
			True Z.D. . .	59 13 45				True Z.D. . .	58 03 47

Chronometer, Slow { H. M. S. 1 33 28.87 ; 1 33 30.7 } H. M. S. 1 33 29.79

RECAPITULATION.

		H. M. S.	H. M. S.
June 9th P.M. / June 10th A.M.	Chronometer Slow	1 33 48.19 / 1 33 48.62	1 33 48.4 at Midnight, June 9th.
June 22d A.M. / June 22d P.M.	Chronometer Slow	1 33 29.79 / 1 33 31.93	1 33 30.86 at Noon, June 22d.

Chronometer's Gain in 12½ Days . 0 00 17.54 = 1.4 Seconds per Diem.

HAMMERFEST.——DEDUCTION of the RATE of the Chronometer No. 649 from the 9th to 22d of June, 1823.

STARS.		9 to 10	10 to 11	11 to 12	12 to 13	13 to 14	14 to 15	15 to 16	16 to 17	17 to 18	18 to 19	19 to 20	20 to 21	21 to 22
		s.	s.	s.	s.	s.	s.	s.	s.	s.	s.	s.	s.	s.
By Transits of	The Sun . .	1.52	1.52	1.52	1.62	1.25	1.59	1.59	1.59	1.25	1.25	1.25	1.25	1.25
	η Ursæ . . .	1.47	1.47	1.47	1.47	1.47	1.29	1.29	1.29	1.29	1.29	1.29	1.29	1.29
	Arcturus . .	1.42	1.42	1.42	1.42	1.42	1.42	1.42	1.42	1.42	1.42	1.42	1.42	1.42
	α Lyræ . .	1.38	1.38	1.38	1.38	1.38	1.38	1.38	1.38	1.38	1.38	1.38	1.38	1.38
By Zenith Distances of the Sun out of the Meridian . .		1.4	1.4	1.4	1.4	1.4	1.4	1.4	1.4	1.4	1.4	1.4	1.4	1.4
MEANS—Gaining per Diem.		1.44	1.44	1.44	1.46	1.38	1.42	1.42	1.42	1.35	1.35	1.35	1.35	1.35
		1.43					1.38							

Hammerfest.—Comparisons of the Astronomical Clock with the Chronometer No. 649, from the 9th to the 22d of June, 1823; with the Clock's Rate on Mean Solar Time deduced.

1823.	Chronometer.	Clock.	Clock's Loss on 649.			Daily Rates. Chron.	Daily Rates. Clock.
	H. M. S.	H. M. S.	s.	s.		Gaining.	Gaining.
June 9 A.M.		9 05 18.7					
" 9 P.M.		9 05 52	33.3				
" 10 A.M.		9 06 25.4	33.4	66.7			
" 10 P.M.		9 06 58.6	33.2				
" 11 A.M.		9 07 32	33.4	66.6			
" 11 P.M.		9 08 05.4	33.4		s.	s.	s.
" 12 A.M.		9 08 38.7	33.3	66.7	66.77	1.43	68.2
" 12 P.M.		9 09 12.2	33.5				
" 13 A.M.		9 09 45.8	33.6	67.1			
" 13 P.M.		9 10 19 1	33.3				
" 14 A.M.		9 10 53	33.9	67.2			
" 14 P.M.		9 11 26	33				
" 15 A.M.		9 11 59.3	33.3				
" 15 P.M.	9 00 00	9 12 32.7	33.4	66.7			
" 16 A.M.		9 13 05.7	33				
" 16 P.M.		9 13 39.7	34	67			
" 17 A.M.		9 14 13.3	33.6				
" 17 P.M.		9 14 47	33.7	67.3			
" 18 A.M.		9 15 20.4	33.4				
" 18 P.M.		9 15 54.8	34.4	67.8	67.43	1.38	68.81
" 19 A.M.		9 16 20.6	33.8				
" 19 P.M.		9 17 03	34.4	68.2			
" 20 A.M.		9 17 36.8	33.8				
" 20 P.M.		9 18 10.5	33.7	67.5			
" 21 A.M.		9 18 44.2	33.7				
" 21 P.M.		9 19 17.8	33.6	67.3			
" 22 A.M.		9 19 51.4	33.6				
" 22 P.M.		9 20 25	33.6	67.2			

HAMMERFEST.—COINCIDENCES OBSERVED with PENDULUM 3; the Clock making 86468.2 Vibrations in a Mean Solar Day.

DATE.	Barometer.	No. of Coincidence.	Temperature.	Time of Disappearance.	Time of Re-appearance.	True Time of Coincidence.	Arc of Vibration.	Mean Temperature.	Mean Interval.	Correction for the Arc.	Vibrations in 24 hours.	Reduction to a Mean Temperature.	Reduced Vibrations at 55.05.
1823.	IN.		°	M. S.	M. S.	H. M. S.	°	°	S.	° +			
June 9 P.M.	30.18	1	48	40 47	40 50	1 40 48.5	1.22	49.5	680.25	1.41	86215.36	−2.33	86213.03
		11	51	34 05	34 17	3 34 11	0.66						
„ 9 P.M.	30.18	1	54.5	28 53	28 55	8 28 54	1.24	53.9	675.55	1.47	86213.56	−0.48	86213.08
		11	53.3	21 18	21 41	10 21 29.5	0.68						
„ 10 A.M.	30.00	1	52	56 56	57 00	6 56 58	1.2	53	677.35	1.34	86214.22	−0.86	86213.36
		11	54	49 45	49 58	8 49 51.5	0.64						
„ 10 P.M.	30.00	1	61.4	27 10	27 11	2 27 10.5	1.2	62.2	667.1	1.34	86210.29	+3.00	86213.29
		11	63	18 14	18 29	4 18 21.5	0.64						
„ 10 P.M.	30.04	1	62	19 50	19 52	6 19 51	1.24	61.8	666.65	1.47	86210.24	+2.84	86213.08
		11	61.6	10 50	11 05	8 10 57.5	0.68						
„ 11 A.M.	30.05	1	48	57 44	57 49	8 57 46.5	1.18	48.25	681.95	1.28	86215.89	−2.86	86213.03
		11	48.5	51 18	51 34	10 51 26	0.62						
„ 11 P.M.	30.09	1	51.5	11 52	11 53	2 11 52.5	1.2	53.75	675.85	1.34	86213.65	−0.55	86213.10
		11	56	4 22	4 40	4 04 31	0.64						
„ 11 P.M.	30.07	1	54	39 04	39 08	9 39 06	1.2	53.4	675.75	1.34	86213.61	−0.69	86212.92
		11	52.8	31 31	31 56	11 31 43.5	0.64						
„ 12 P.M.	29.86	1	66	11 12	11 17	2 11 14.5	1.16	65	664.15	1.23	86209.02	+4.18	86213.20
		11	64	1 49	2 03	4 01 56	0.6						
„ 13 A.M.	29.88	1	50.8	24 12	24 13	10 24 12.5	1.22	51.8	678.1	1.41	86214.56	−1.36	86213.20
		11	52.8	17 05	17 22	12 17 13.5	0.64						
„ 13 P.M.	29.84	1	53.7	23 31	23 33	2 23 32	1.22	53.75	676.35	1.41	86213.90	−0.55	86213.35
		11	53.8	16 07	16 24	4 16 15.5	0.66						
„ 13 P.M.	29.80	1	54	56 29	56 33	6 56 31	1.22	53.1	676.4	1.41	86213.98	−0.82	86213.16
		11	52.2	49 06	49 24	8 49 15	0.66						
„ 14 A.M.	29.56	1	52	12 46	12 51	10 12 48.5	1.16	54	676.2	1.23	86213.64	−0.44	86213.20
		11	56	5 23	5 38	12 05 30.5	0.6						
„ 14 P.M.	29.47	1	57.3	42 14	42 19	4 42 16.5	1.22	57.15	671.65	1.41	86212.12	+0.88	86213.00
		11	57	34 07	34 19	6 34 13	0.66						
Means . . .	29.93							55.05			86213.15		86213.15

HAMMERFEST.——COINCIDENCES OBSERVED with PENDULUM No. 4; the Clock making 86468.81 Vibrations in a Mean Solar Day.

DATE.	Barometer.	No. of Coincidence.	Temperature.	Time of Disappearance.	Time of Re-appearance.	True Time of Coincidence.	Arc of Vibration.	Mean Temperature.	Mean Interval.	Correction for the Arc.	Vibrations in 24 hours.	Reduction to a mean Temperature.	Reduced Vibrations at 44.34.
1823.	IN.		°	M. S.	M. S.	H. M. S.	°	°	S.	S.			
June 15 P.M.	29.62	1	48	53 37	53 42	5 53 39.5	1.2	47	709.05	+ 1.28	86226.16	+1.12	86227.28
		11	46	51 35	52 05	7 51 50	0.6						
„ 16 A.M.	29.59	1	50	19 21	19 24	9 19 22.5	1.2	49.4	706	1.28	86225.12	+2.12	86227.24
		11	48.8	16 53	17 12	11 17 02.5	0.6						
„ 16 P.M.	29.58	1	47	58 46	58 53	1 58 49.5	1.2	46.65	709.6	1.25	86226.35	+0 97	86227.10
		11	46.3	56 56	57 15	3 57 05.5	0.58						
„ 16 P.M.	29.58	1	46.2	38 18	38 23	5 38 20.5	1.2	46	710.05	1.28	86226.52	+0.70	86227.22
		11	45.8	36 30	36 52	7 36 41	0.6						
„ 17 A.M.	29.40	1	42.6	34 59	35 04	9 35 01.5	1.2	42.8	714.05	1.25	86227.87	−0.65	86227.22
		11	43	33 54	34 10	11 34 02	0.58						
„ 17 P.M.	29.35	1	44.2	18 33	18 36	2 18 34.5	1.2	44.9	711.1	1.28	36226.88	+0.23	86227.11
		11	45.6	16 57	17 14	4 17 05.5	0.6						
„ 19 A.M.	29.92	1	40	32 21	32 28	10 32 24.5	1.16	40.75	716.7	1.19	86228.70	−1.51	86227.19
		11	41.5	31 44	31 59	12 31 51.5	0.58						
„ 19 P.M.	29.85	1	41.8	27 51	27 58	1 27 54.5	1.16	41.9	715.15	1.19	86228.20	−1.02	86227.18
		11	42	26 54	27 18	3 27 06	0.58						
„ 19 P.M.	29.80	1	42	50 09	50 11	4 50 10	1.24	42	714.5	1.40	86228.10	−0.98	86227.18
		11	42	49 04	49 26	6 49 15	0.64						
„ 20 A.M.	29.66	1	41.5	54 37	54 44	9 54 40.5	1.18	41.75	715.05	1.25	86228.19	−1.09	86227.10
		11	42	53 44	53 58	11 53 51	0.6						
„ 20 P.M.	29.74	1	42.2	12 59	13 07	2 13 03	1.18	42	714.55	1.25	86228.01	−0.98	86227.03
		11	41.8	12 00	12 17	4 12 08.5	0.6						
„ 21 A.M.	29.93	1	38.8	15 51	15 57	10 15 54	1.2	39.4	717.85	1.28	86229.16	−2.07	86227.09
		11	40	15 24	15 41	12 15 32 5	0.6						
„ 22 P.M.	29.81	1	52	2 37	2 41	4 2 39	1.24	51.9	702.45	1.40	86224.00	+3.18	86227.18
		11	51.8	59 34	59 53	5 59 43.5	0.64						
Means	29.68							44.34			86227.18		86227.18

SPITZBERGEN.

The station selected for the pendulums at Spitzbergen was on one of the islands, called the "Norways," situated to the north-west of the main land of Spitzbergen, and forming with its coast the harbour of Fair-Haven: it is the south-eastermost of the group, and the inner of the two islands which are immediately to the eastward of Cloven Cliff, a head-land well known to the navigators of the high latitudes; as a further means of its recognition, it is the second when viewed from Fair-Haven, from Vogelsang, which is still, and probably will ever remain, as pre-eminently entitled to its appellation, as when the distinction was conferred nearly two centuries ago. The establishment of house, tents, and observatory, occupied a low, but dry and tolerably level and well-protected situation, at the south-western extremity of the island; the foundation on which the clock and pendulum supports rested was an extensive bed of quartz rock. The height of the pendulums was twenty-one feet above half-tide.

Captain Clavering, being desirous of employing himself during the experiments in examining the state of the ice to the northward of Spitzbergen, sailed for that purpose on the 4th of July, leaving Mr. Henry Foster, midshipman, and Mr. Rowland, assistant surgeon, of the Griper, with a boat and crew, and a sufficiency of fuel and provisions to have made good her passage to Norway in the course of the autumn, had any accident befallen the ship; we had, however, the satisfaction of witnessing her return on the 10th, having traced in the interval the continuity of the ice, from its abutment on the land eastward of the Norways, to the longitude of 11° west; the examination was not pursued further to the westward. because the line of the ice had taken the south-westerly direction, in which we should again fall in with it on the passage to Greenland. The experiments proceeded without interruption, and were concluded on the 19th of July.

In illustration of the small extent of the range of the external thermometer in the twenty-four hours in high latitudes in summer, I have subjoined a table shewing the extremes of a register thermometer in successive periods of six hours, from the 9th to the 18th of July; the thermometer was suspended in the open air about four feet from the ground, and protected from the sun's rays by a roof at a considerable distance above the thermometer: the exposure was free in every other direction.

TRANSITS OBSERVED AT SPITSBERGEN.

DATE.	STARS.	TIMES OF TRANSIT BY THE CHRONOMETER 649.					Mean by the Chronometer.	CLOCK.
		1st Wire.	2d Wire.	Meridian Wire.	4th Wire.	5th Wire.		
1823.		M. S.	M. S.	H. M. S.	M. S.	M. S.	H. M. S.	H. M. S.
July 5	Sun mid. 1st Limb	17 07.2	17 36	11 18 04.4	18 32.8	19 01.2	11 19 13.06	11 09 33.97
	Sun mid. 2d Limb	19 24 4	19 53.6	11 20 22	20 50	21 18.8		
„ 6	α Cygni. . . .	52 36.4	53 12.8	12 53 50	54 26.8	55 03 6	12 53 49.93	12 45 43.4
„ 7	Sun noon 1st Limb	17 24.8	17 53.2	23 18 20.8	18 49.6	19 17.6	23 19 29.63	23 12 00.51
	Sun noon 2d Limb	19 41.6	20 09.6	23 20 37.8	21 06.8	21 35 2		
„ „	η Ursæ	54 50.4	55 31.2	5 56 12	56 52.4	57 33.2	5 56 11.87	5 49 06.22
„ „	Arcturus. . .	22 13.6	22 41.2	6 23 09.6	23 37.2	24 05.2	6 23 09.4	6 16 05.36
„ „	γ Draconis . .	06 03.6	06 45.6	10 07 28	08 10	08 52 4	10 07 27.93	10 00 37.32
„ „	α Lyræ	44 41.2	45 14.8	10 45 48	46 21.6	46 55.6	10 45 48.2	10 38 59.9
„ 8	η Ursæ	50 55.2	51 36.4	5 52 16.8	52 57.6	53 38.8	5 52 16.93	5 46 37.14
„ 10	Sun noon 1st Limb	17 56	18 24	23 18 52.4	19 20.8	19 48.8	23 20 00.46	11 16 53.02
	Sun noon 2d Limb	20 12	20 40.4	23 21 08.4	21 36.8	22 05.2		
„ „	η Ursæ	43 04.8	43 45.6	5 44 26.4	45 07.2	45 48	5 44 26.4	5 41 41.96
„ „	Arcturus. . .	10 28.4	10 56.4	6 11 24	11 52	12 20	6 11 24.13	6 08 41.32
„ „	α Lyræ	. . . . .	. . . .	10 34 03.8	34 37.4	35 11	10 34 03.8	10 31 33.26
„ „	α Cygni. . . .	36 57.6	37 34	12 38 10.8	38 47.6	39 24.4	12 38 10.83	12 35 51.44
„ 11	η Ursæ	39 11.2	39 52	5 40 32.8	41 14	41 54.8	5 40 32.93	5 39 14.63
„ „	Arcturus. . .	06 34.4	07 02.4	6 07 30	07 58	08 26	6 07 30.13	6 06 13.49
„ 12	Arcturus. . .	. . . . .	. . . . .	6 03 34.4	04 02.4	04 30	6 03 34.2	6 03 45.74
„ 13	Sun noon 1st Limb	18 24.4	18 52.4	23 19 20 8	19 49.2	20 18	23 20 28.93	23 21 44.77
	Sun noon 2d Limb	20 40.4	21 08.4	23 21 36.8	22 05.6	22 33.6		
„ 14	Sun noon 1st Limb	18 32	19 00.4	23 19 28.8	19 57.2	20 25.2	23 20 36.65	23 23 21.22
	Sun noon 2d Limb	20 48.4	21 16.4	23 21 44.4	22 12.8	22 41.6		
„ 17	γ Draconis . .	26 46.4	27 28.4	9 28 10.4	28 52.6	29 31.4	9 28 10.43	9 36 00.8
„ „	α Lyræ	05 24	05 57.6	10 06 31.2	07 04.8	07 38.4	10 06 31.2	10 14 23.92
„ 19	Sun noon 1st Limb	18 55.6	19 23.6	23 19 51.6	20 20	20 48	23 20 59.3	23 31 12.42
	Sun noon 2d Limb	21 10.8	21 38.8	23 22 06.8	22 34.8	23 03.2		
„ „	η Ursæ	07 37.6	08 18.4	5 08 59.2	09 40	10 20.8	5 08 59.2	5 19 34.26
„ „	Arcturus. . .	35 01.2	35 29.2	5 35 57.2	36 24.8	36 52.4	5 35 57	5 46 33.78
„ „	γ Draconis . .	18 51.2	19 33.6	9 20 15.6	20 58	21 40.4	9 20 15.73	9 31 06.5
„ „	α Lyræ	57 29.6	58 02.8	9 58 36.4	59 10	59 43.6	9 58 36.47	9 09 29.63
„ „	Sun mid. 1st Limb	18 57.6	19 25.6	11 19 53.6	20 21.6	20 49.6	11 21 01.3	11 31 59.61
	Sun mid. 2d Limb	21 13.2	21 41.2	11 22 08.8	22 36.8	23 05.2		
„ „	α Cygni. . . .	01 29.6	02 06.4	12 02 43.6	03 20.4	03 57.2	12 02 43.47	12 22 07.02

SPITZBERGEN.—Comparisons of the Astronomical Clock with the Chronometer, No. 649, from the 6th to the 20th of July, 1823.

1823.		Chronometer.	Clock.	Clock slow or fast on 649.	
		H. M. S.	H. M. S.		M. S.
July 6	A. M.	3 00 00	2 50 33.6	Slow	9 26.4
„ 6	P. M.		2 51 17.6	„	8 42.4
„ 7	A. M.		2 52 01	„	7 59
„ 7	P. M.		2 52 43.8	„	7 16.2
„ 8	A. M.		2 53 26.9	„	6 33.1
„ 8	P. M.		2 54 09.8	„	5 50.2
„ 9	A. M.		2 54 53.8	„	5 06 2
„ 9	P. M.		2 55 38.5	„	4 21.5
„ 10	A. M.		2 56 22.7	„	3 37.3
„ 10	P. M.		2 57 05.6	„	2 54.4
„ 11	A. M.		2 57 49.3	„	2 10.7
„ 11	P. M.		2 58 31.8	„	1 28.2
„ 12	A. M.		2 59 16.2	„	0 43.8
„ 12	P. M.		3 00 00.2	Fast	0 00.2
„ 13	A. M.		3 00 44.7	„	0 44.7
„ 13	P. M.		3 01 29.6	„	1 29.6
„ 14	A. M.		3 02 13.7	„	2 13.7
„ 14	P. M.		3 02 58	„	2 58
„ 15	A. M.		3 03 42	„	3 42
„ 15	P. M.		3 04 26	„	4 26
„ 16	A. M.		3 05 11.6	„	5 11.6
„ 16	P. M.		3 05 56.9	„	5 56.9
„ 17	A. M.		3 06 41.7	„	6 41.7
„ 17	P. M.		3 07 26.6	„	7 26.6
„ 18	A. M.		3 08 10.7	„	8 10.7
„ 18	P. M.		3 08 56.7	„	8 56.7
„ 19	A. M.		3 09 41.7	„	9 41.7
„ 19	P. M.		3 10 27	„	10 27
„ 20	A. M.		3 11 12	„	11 12
„ 20	P. M.		3 11 56.8	„	11 56.8

SPITZBERGEN.——OBSERVATIONS to DETERMINE the RATE of the Chronometer 649, and of the Astronomical Clock, from 6th to 20th of July, 1823, by ZENITH DISTANCES of the Sun, with a Repeating Circle.

Latitude of the Place of Observation, 79° 49′ 58″ N.; Longitude 11° 40′ E.

July 6th A.M.; Barometer 29.96; Thermometer 41°, ☉'s L.L.

Chronometer.	H. M. S.	Level.		Readings, &c.	° ′ ″
	5 31 55.2	+5	−3	First Vernier	176 41 40
	5 34 00	+7	−1	Second ,,	41 30
	5 35 55.6	+6	−2	Third ,,	41 50
	5 37 58.8	+10	+2	Fourth ,,	41 10
	5 39 46.8	−13	−5	Mean . .	176 41 32.5
	5 41 50	0	0	Index . . .	+225 27 45
Mean. . .	5 36 54.4	+30	−24	Level . . .	+3
True time .	6 21 45.96	+3			402 09 20.5
Chron. slow	0 44 51.56			Observed Z.D.	67 01 33.5
				Ref. and Paral.	+2 11
360° − 134° 32′ 15″ = 225° 27′ 45″				Semidiam . .	−15 45.5
				True Z.D. . . .	66 47 59

Chronometer.	H. M. S.	Level.		Readings, &c.	° ′ ″
	5 48 37.2	0	0	First Vernier	214 35 45
	5 50 24	+12	+4	Second ,,	35 50
	5 52 11.2	−2	−10	Third ,,	36 20
	5 54 00.8	0	0	Fourth ,,	35 30
	5 55 37.6	+7	−2	Mean . . .	214 35 50
	5 57 26	0	−9	Index . . .	+183 18 27.5
Mean . .	5 53 02.8	+23	−23	Level . . .	0
True time .	6 37 52.94	0			397 54 17.5
Chron. slow	0 44 50.14			Observed Z.D.	66 19 03
				Ref. and Paral.	+2 06
360° − 176° 41′ 32″ 5 = 183° 18′ 27″.5				Semidiam . .	−15 45
				True Z.D. . . .	66 05 24

	H. M. S.	H. M. S.
Chronometer, Slow	0 44 51.56 0 44 50.14	0 44 50.85
Clock Slow of the Chronometer .		0 09 26.4
Clock Slow of Mean Time		0 54 17.25

July 6th P.M.; Barometer 29.93; Thermometer 44°; ☉'s U.L.

Chronometer.	H. M. S.	Level.		Readings, &c.	° ′ ″
	4 30 49.6	+8	0	First Vernier	247 22 50
	4 33 10	+10	+3	Second ,,	22 45
	4 35 15.6	+1	−7	Third ,,	23 30
	4 37 16.8	+7	0	Fourth ,,	22 30
	4 41 53.2	−1	−9	Mean . . .	247 22 54
	4 44 16.8	+3	−5	Index . . .	+145 24 10
Mean. . .	4 37 07	+32	−22	Level . . .	+5
True time .	5 21 59.9	+5			392 47 09
Chron. slow	0 44 52.9			Observed Z.D.	65 27 51.5
				Ref. and Paral.	+2 00.2
360° − 214° 35′ 50″ = 145° 24′ 10″				Semidiam . .	+15 45.5
				True Z.D. . . .	65 45 37

Chronometer.	H. M. S.	Level.		Readings, &c.	° ′ ″
	4 49 22.8	0	0	First Vernier	284 25 30
	4 50 52.8	0	0	Second ,,	25 30
	4 52 33.2	0	0	Third ,,	25 55
	4 54 01.6	0	0	Fourth ,,	25 10
	4 55 35.6	0	0	Mean . . .	284 25 31
	4 57 01.6	0	0	Index . . .	+112 37 06
Mean. . .	4 53 14.6	0	0	Level . . .	0
True time.	5 38 07.34	0			397 02 37
Chron. slow	0 44 52.74			Observed Z.D	66 10 26
				Ref. and Paral.	+2 04
360° − 247° 22′ 54″ = 112° 37′ 06″				Semidiam . .	+15 45
				True Z.D. . .	66 28 15

	H. M. S.	H. M. S.
Chronometer, Slow	0 44 52.9 0 44 52.74	0 44 52.82
Clock Slow of the Chronometer .		0 08 42.4
Clock Slow of Mean Time . . .		0 53 35.22

Spitzbergen.—Determination of the Rate of the Chronometer 649, and of the Astronomical Clock, by Zenith Distances, ***continued.***

July 7th P.M.; Barometer 29.80; Thermometer 40°; ☉'s U.L.

Chronometer.		Level.		Readings, &c.		Chronometer.		Level.		Readings, &c.	
	H. M. S.				° ′ ″		H. M. S.				° ′ ″
	4 29 53.6	+5	−3	First Vernier	162 23 00		5 00 08.4	+5	−3	First Vernier	203 00 00
	4 31 37.6	+5	−3	Second „	22 50		5 01 43.2	+7	−1	Second „	202 59 40
	4 34 12	+2	−6	Third „	23 30		5 03 20.4	0	−7	Third „	203 00 30
	4 36 04.4	+6	−2	Fourth „	22 45		5 05 35.2	0	0	Fourth „	202 59 50
	4 38 00.4	+8	0	Mean . . .	162 23 01		5 07 15.2	0	0	Mean . . .	203 00 00
	4 39 52.8	0	−7	Index . . .	+230 21 57.5		5 10 01.2	0	−6	Index . . .	+197 36 59
Mean . . .	4 34 56.8	+26	−21	Level . . .	+ 2.5	Mean . . .	5 04 40.6	+12	−17	Level . .	−2.5
True time.	5 19 46.7	+2.5			392 45 01	True time.	5 49 33.34	−2.5			400 36 56.5
Chron. slow	0 44 49.9			Observed Z.D.	65 27 30	Chron. slow	0 44 52.74			Observed Z.D.	66 46 09
				Ref. and Paral.	+ 2 00.5					Ref. and Paral.	+ 2 08.5
360° − 129° 38′ 02″.5 = 230° 21′ 57″.5				Semidiam. . .	+15 45.5	360° − 162° 23′ 01″ = 197° 36′ 59″				Semidiam. . .	+15 45.5
				True Z.D. . .	65 45 16					True Z.D. . .	67 04 03

	H. M. S.	H. M. S.
Chronometer, Slow	{0 44 49.9 0 44 52.74}	0 44 51.32
Clock Slow of the Chronometer .		0 07 16.2
Clock Slow of Mean Time . . .		0 52 07.52

July 8th A.M.; (Sun in Eclipse) Barometer 29.75; Thermometer 33.5°; ☉'s L.L.

Chronometer.		Level.		Readings, &c.		Chronometer.	Level.	Readings, &c.
	H. M. S.				° ′ ″			
	6 07 36.4	+6	−3	First Vernier	20 44 35			
	6 09 16.8	0	0	Second „	44 20			
	6 12 02.4	+4	−6	Third „	45 00			
	6 13 46.4	+7	−2	Fourth „	43 50			
	6 15 52.4	+2	−7	Mean . . .	20 44 26			
	6 17 37.6	+6	−4	Index . . +	{13 17 54 360 00 00}			
Mean . . .	6 12 42	+25	−22	Level . . .	+1.5			
True time.	6 57 31.14	+1.5			394 02 21.5			
Chron. slow	0 44 49.14			Observed Z.D.	65 40 23.6			
				Ref. and Paral.	+ 2 03			
360° − 346° 42′ 06″ = 13° 17′ 54″				Semidiam. . .	−15 45.5			
				True Z.D. . .	65 26 41			

	H. M. S.
Chronometer, Slow. . .	0 44 49.14
Clock Slow of Chronometer	0 06 33.1
Clock Slow of Mean Time	0 51 22.24

Spitzbergen.—Determination of the Rate of the Chronometer 649, and of the Astronomical Clock, by Zenith Distances, *continued.*

July 20th A.M.; Barometer 29.70; Thermometer 42°; ⊙'s L.L.

	Chronometer.	Level.		Readings, &c.	
	H. M. S.				° ′ ″
	5 44 44.4	+6	−2	First Vernier	50 30 30
	5 47 38	0	0	Second „	30 10
	5 49 34.4	0	0	Third „	30 10
	5 51 32	+10	+1	Fourth „	30 00
	5 53 52.4	+7	−2	Mean . .	50 30 12.5
	5 55 53.2	+6	−3	Index . . +	{ 00 00 02.5 360 00 00 }
Mean. . .	5 50 32.4	+30	−7	Level . . .	+11 5
True time .	6 35 23.23	+11.5			410 30 26.5
Chron. slow	0 44 50.83			Observed Z.D.	68 25 04.5
				Ref. and Paral.	+2 19
360°−359° 59′ 57″.5=00° 00′ 02″.5				Semidiam . .	−15 46.5
				True Z.D. . .	68 11 38

	Chronometer.	Level.		Readings, &c.	
	H. M. S.				° ′ ″
	6 00 34	+3	−5	First Vernier	97 17 40
	6 02 04.4	+3	−5	Second „	17 35
	6 04 06.4	0	0	Third „	18 10
	6 05 28.8	−1	−9	Fourth „	17 40
	6 06 57.2	+8	0	Mean.. . .	97 17 46
	6 09 05.6	+8	+1	Index . . .	+309 29 47.5
Mean. . .	6 04 42.73	+23	−20	Level . . .	+1.5
True time .	6 49 32.8	+1.5			406 47 35
Chron. slow	0 44 50.07			Observed Z.D.	67 47 56
				Ref. and Paral.	+2 14
360°−50° 30′ 12″.5=309° 29′ 47″.5				Semidiam . .	−15 46
				True Z.D. .	67 34 24

	H. M. S.	H. M. S.
Chronometer, Slow	{ 0 44 50.83 0 44 50.07 }	0 44 50.45
Clock Fast of the Chronometer .		0 11 12
Clock Slow of Mean Time . . .		0 33 38.45

July 20th P.M.; Barometer 29.70; Thermometer 42°; ⊙'s U.L.

	Chronometer.	Level.		Readings, &c.	
	H. M. S.				° ′ ″
	5 35 14.8	+5	−3	First Vernier	158 20 30
	5 37 32	+5	−3	Second „	20 20
	5 40 03.2	+5	−3	Third „	20 40
	5 42 48	0	0	Fourth „	20 10
	5 45 32.2	+10	+2	Mean . . .	158 20 25
	5 47 34	−11	−7	Index . . .	+262 42 14
Mean . .	5 41 27.37	+27	−27	Level . . .	0
True time .	6 26 20.3	0			421 02 39
Chron. slow	0 44 52.93			Observed Z.D.	70 10 26
				Ref. and Paral.	+2 33
360°−97° 17′ 46″=262° 42′ 14″				Semidiam . .	+15 46
				True Z.D. . .	70 28 45

Chronometer.	Level.	Readings, &c.

	H. M. S.
Chronometer, Slow	0 44 52.93
Clock Fast of the Chronometer	0 11 56.8
Clock Slow of Mean Time .	0 32 56.13

SPITZBERGEN.——RATE of the ASTRONOMICAL CLOCK from the 7th to the 19th of July, 1823, deduced from Transits.

STARS.	7 to 8	8 to 9	9 to 10	10 to 11	11 to 12	12 to 13	13 to 14	14 to 15	15 to 16	16 to 17	17 to 18	18 to 19
	s.	s.	s.	s.	s.	s.	s.	s.	s.	s.	s.	s
η Ursæ.	86 83	88.32	88.32	88.58	88.36	88.36	88.36	88.36	88.36	88.36	88.36	88.36
Arcturus	87.89	87.89	87.89	88.08	88.16	88.49	88.49	88.49	88.49	88.49	88.49	88.49
γ Draconis	88 26	88.26	88.26	88.26	88.26	88.26	88.26	88 26	88.26	88.26	88.76	88.76
α Lyræ	87.67	87.67	87.67	88.86	88.86	88.86	88.86	88.86	88.86	88.86	88 76	88.76
α Cygni	87.92	87.92	87.92	88.46	88.46	88.46	88.46	88.46	88.46	88.46	88.46	88.46
Mean by Stars on Sidereal Time.	87.71	88.01	88.01	88.45	88.42	88.49	88.48	88.49	88.48	88.49	88.56	88.57
Mean by Stars on Solar Time . .	87.95	88.25	88.25	88.69	88.66	88.73	88.72	88.73	88.72	88.73	88.80	88.81
Sun, Noon.	87.94	87.94	87.94	89.00	89.00	89.00	89.10	88.44	88.44	88.44	88.44	88.44
Sun, Midnight.	88.44	88.44	88.44	88 44	88.44	88.44	88.44	88.44	88.44	88.44	88.44	88.44
Mean by the Sun . . .	88.19	88.19	88.19	88.72	88.72	88.72	88.77	88.44	88.44	88.44	88.44	88.44
Mean by the Sun and Stars.—Gaining on Solar Time	88.02	88.23	88.23	88.70	88.68	88.73	88.73	88.65	88.64	88.65	88.70	88.70
	88.555 Seconds.											

SPITZBERGEN.—RATE of the ASTRONOMICAL CLOCK from the 6th to the 20th of July, deduced by ZENITH DISTANCES.

A.M. to A.M.		P.M. to P.M.	
	s.		s.
July 6th to July 20th; Clock gaining per diem . . .	88.48	July 6th to July 20th; Clock gaining per Diem . .	88 5
July 8th to July 20th; „ . . .	88.65	July 7th to July 20th; „ . . .	88.57
MEAN.—Clock gaining per Diem	88.565		88.535
	88.55 Seconds.		

SPITZBERGEN.—COINCIDENCES OBSERVED with PENDULUM No. 3.

DATE.	Barometer.	Clock gaining	No. of Coincidence.	Temperature.	Time of Disappearance.	Time of Re-appearance.	True Time of Coincidence.	Arc of Vibration.	Mean Temperature.	Mean Interval.	Correction for the Arc.	Vibrations in 24 hours.	Reduction to a Mean Temperature.	Reduced Vibrations at 45°.2.
1823.	IN.	S.		°	M. S.	M. S.	H. M. S.	°	°	S.	S. +			
July 8 A.M.	29.70	88.02	1	45	59 32	59 37	4 59 34.5	1.18	44.9	692.3	1.25	86239.43	−0.13	86239.30
			11	44.8	54 53	55 02	6 54 57.5	0.6						
„ 8 P.M.	29.73	88.02	1	48	26 54	26 59	2 26 56.5	1.2	47.9	688.65	1.31	86238.15	+1.13	86239.28
			11	47.8	21 37	21 49	4 21 43	0.62						
„ 9 A.M.	29.81	88.23	1	42.2	34 00	34 05	10 34 02.5	1.2	43.4	693.3	1.28	86240.04	−0.76	86239.28
			11	44.6	29 29	29 42	12 29 35.5	0.6						
„ 9 P.M.	29.85	88.23	1	45.4	59 11	59 16	2 59 13.5	1.22	47 2	688.1	1.35	86238.19	+0.84	86239.03
			11	49	53 48	54 01	4 53 54.5	0.64						
„ 9 P.M.	29.89	88.23	1	48.2	45 42	45 47	10 45 44.5	1.28	48.5	686.5	1.48	86237.76	+1.39	86239.15
			11	48.8	40 04	40 15	12 40 09.5	0.66						
„ 10 A.M.	29.90	88.23	1	42	32 10	32 14	10 32 12	1.2	44	691.65	1.31	86239.43	−0.50	86238.93
			11	46	27 23	27 34	12 27 28.5	0.62						
„ 10 P.M.	29.95	88.23	1	50.8	23 38	23 43	3 23 40.5	1.2	51.1	683.85	1.28	86236.54	+2.48	86239.02
			11	51.4	17 33	17 45	5 17 39	0.6						
„ 10 P.M.	29.95	88.23	1	50.8	52 50	52 53	9 52 51.5	1.28	50.4	683.9	1.48	86236.78	+2.18	86238.96
			11	50	46 45	46 56	11 46 50.5	0.66						
„ 11 A.M.	29.96	88.70	1	47.3	48 20	48 25	1 48 22.5	1.22	46.75	687.75	1 35	86238.55	+0.65	86239.20
			11	46.2	42 55	43 05	3 43 00	0.64						
„ 11 P.M.	29.97	88.70	1	51	48 50	48 53	2 48 51.5	1.28	51.2	682.35	1.48	86236.68	+2.52	86239.20
			11	51.4	42 29	42 41	4 42 35	0.66						
„ 11 P.M.	29.97	88.70	1	44.8	29 34	29 37	9 29 35.5	1.28	43.8	690.5	1.51	86239.71	−0.59	86239.12
			11	42.8	24 31	24 50	11 24 40.5	0.68						
„ 12 A.M.	29.95	88.68	1	44	10 22	10 27	11 10 24.5	1.2	44.25	691.6	1.31	86239.89	−0.40	86239.49
			11	44.5	5 35	5 46	1 05 40.5	0.62						
„ 12 P.M.	29.91	88.68	1	45.2	00 40	00 44	3 00 42	1.22	44.6	690.9	1.35	86239.67	−0.25	86239.42
			11	44	55 45	55 57	4 55 51	0.64						
„ 12 P.M.	29.90	88.68	1	41.8	9 06	9 10	8 09 08	1.28	40.8	694.45	1.48	86241.08	−1.85	86239.23
			11	39.8	4 46	4 59	10 04 52.5	0.66						
„ 13 P.M.	29.87	88 73	1	37	8 4	8 7	11 08 05.5	1.2	37.1	699.25	1.31	86242.65	−3.40	86239 25
			11	37.2	4 32	4 44	1 04 38	0.62						
„ 13 P.M.	29.88	88.73	1	37.2	46 33	46 34	5 46 33.5	1.3	37.15	698.9	1.55	86242.79	−3.36	86239.43
			11	37.1	42 56	43 09	7 43 02.5	0 68						
Means . .	29.89								45.2			86239.21		86239.21

SPITZBERGEN.——COINCIDENCES OBSERVED with PENDULUM No. 4.

DATE.	Barometer.	Clock gaining.	No. of Coincidence.	Temperature.	Time of Disappearance.	Time of Re-appearance.	True Time of Coincidence.	Arc of Vibration.	Mean Temperature.	Mean Interval.	Correction for the Arc.	Vibrations in 24 hours.	Reduction to a mean Temperature.	Reduced Vibrations at 37°.51.
1823.	IN.	S.		°	M. S.	M. S.	H. M. S.	°	°	S.	S. +			
July 14 P.M.	29.90	88.73	1	36.5	23 32	23 40	12 23 36	1.22	37	726.85	1.35	86252.07	−0.21	86251.86
			11	37.5	24 36	24 53	2 24 44.5	0.64						
" 14 P.M.	29.90	88.73	1	37	43 40	43 41	5 43 40.5	1.32	36.5	726.55	1.66	86252.32	−0.42	86251.90
			11	36	44 37	44 55	7 44 46	0.72						
" 15 A.M.	29.92	88.65	1	34	52 00	52 07	1 52 03.5	1.22	33.7	720	1.35	86253.05	−1.60	86251.45
			11	33.4	53 32	53 55	3 53 43.5	0.64						
" 15 P.M.	29.93	88.65	1	34.2	53 58	54 06	0 54 02	1.24	34.1	729.8	1.47	86253.11	−1.43	86251.68
			11	34	55 31	55 49	2 55 40	0.68						
" 15 P.M.	29.88	88.65	1	33.5	48 27	48 33	7 48 30	1.28	33.45	730.3	1.51	86253.31	−1.71	86251.60
			11	33.4	50 03	50 23	9 50 13	0.68						
" 16 A.M.	29.81	88.64	1	32.2	53 57	54 05	1 54 01	1.28	32.5	731.4	1.51	86253.65	−2.10	86251.55
			11	32.8	55 46	56 04	3 55 55	0.68						
" 16 P.M.	29.72	88.64	1	37.2	25 28	25 36	10 25 32	1.2	37.4	726.2	1.31	86251.77	−0.05	86251.72
			11	37.6	26 27	26 41	12 26 34	0.62						
" 17 A.M.	29.70	88.65	1	39.5	8 6	8 14	10 8 10	1.22	38.95	724.05	1.35	86251.09	+0 60	86251.69
			11	38.4	8 42	8 59	12 8 50.5	0.64						
" 17 P.M.	29.72	88.65	1	38.8	15 06	15 13	3 15 09.5	1.2	38.4	724.95	1.31	86251.35	+0.37	86251.72
			11	38	15 48	16 10	5 15 59	0.62						
" 17 P.M.	29.72	88.65	1	37.4	22 16	22 22	7 22 19	1.22	37.65	725.4	1.35	86251.45	+0.06	86251.51
			11	37.9	23 03	23 23	9 23 13	0.64						
" 18 A.M.	29.72	88.70	1	39.5	31 16	31 23	4 31 19.5	1.2	39.2	723.75	1.28	86250.98	+0.71	86251.69
			11	38.9	31 46	32 08	6 31 57	0.6						
" 18 P.M.	29.72	88.70	1	39.1	24 04	24 11	10 24 07.5	1.14	39.25	724.4	1.14	86251.06	+0.73	86251.79
			11	39.4	24 42	25 01	12 24 51.5	0.56						
" 19 A.M.	29.66	88.70	1	41.8	31 44	31 51	10 31 47.5	1.18	42.65	719.6	1.25	86249.57	+2.16	86251.73
			11	43.5	31 35	31 52	12 31 43.5	0.6						
" 19 P.M.	29.80	88.70	1	42 5	45 48	45 53	6 45 50.5	1.28	42.25	719.3	1.48	86249.70	+1.99	86251.69
			11	42	45 34	45 53	8 45 43.5	0.66						
" 19 P.M.	29.70	88.70	1	39.5	48 48	48 53	11 48 50.5	1.26	39.5	722.8	1.39	86250.79	+0.84	86251.63
			11	39.5	49 10	49 27	1 49 18.5	0.62						
Means. .	29.79								37.51			86251.68		86251.68

Spitzbergen.——TEMPERATURE of the AIR in the SHADE.

Date.		Thermometer. Max.	Thermometer. Min.	Mean	Mean—and Remarks.
July 8	9 P.M.	°	°		
		35.5	31.5	33.5	
„ 9	3 A.M.				
		36	31	33.5	
„ „	9 A.M.				
		38.5	34	36.25	36 Calm; with Fog.
„ „	3 P.M.				
		45	37	41	
„ „	9 P.M.				
		42	36.5	39.25	
„ 10	3 A.M.				
		41	35	38	
„ „	9 A.M.				
		48	37	42.5	40.7 Wind West; Clear Weather.
„ „	3 P.M.				
		47	41	43	
„ „	9 P.M.				
		42	40	41	
„ 11	3 A.M.				
		46	40	43	
„ „	9 A.M.				
		47	40	43.5	41.5 Wind West; Clear Weather.
„ „	3 P.M.				
		41	36.5	38.75	
„ „	9 P.M.				
		37	35	36	
„ 12	3 A.M.				
		38.5	35	36.75	
„ „	9 A.M.				
		37	34.5	35.75	35.8 Calm; Thick Fog.
„ „	3 P.M.				
		37	32	34.5	
„ „	9 P.M.				
		32.5	31	31.75	
„ 13	3 A.M.				
		35	31.5	33.25	
„ „	9 P.M.				
		34	33	33.5	33 Wind East; Fog.
„ „	3 P.M.				
		34.5	32	33.25	
„ „	9 P.M.				

Date.		Thermometer. Max.	Thermometer. Min.	Mean	Mean—and Remarks.
July 13	9 P.M.	°	°		
		33	31	32	
„ 14	3 A.M.				
		33	31	32	
„ „	9 A.M.				
		35	32.5	34.25	32.4 Wind East; Fog.
„ „	3 P.M.				
		32	31	31.5	
„ „	9 P.M.				
		31	30	30.5	
„ 15	3 A.M.				
		31.5	30	30.75	
„ „	9 A.M.				
		32	30	31	30.5 Wind N.E., Fresh.
„ „	3 P.M.				
		30.5	29.5	30	
„ „	9 P.M.				
		29.5	29.5	29.5	
„ 16	3 A.M.				
		34	29	31.5	
„ „	9 A.M.				
		35	32.5	33.75	32.5 Wind East; Fog.
„ „	3 P.M.				
		36	35	35.5	
„ „	9 P.M.				
		36.5	34	35.25	
„ 17	3 A.M.				
		37	32.5	34.75	
„ „	9 A.M.				
		37	35	36	35.3 Calm; Thick Fog.
„ „	3 P.M.				
		36	34.5	35.25	
„ „	9 P.M.				
		37	33	35	
„ 18	3 A.M.				
		37.5	35	36.25	
„ „	9 A.M.				
		37.5	36	36.75	35.8 Calm; Thick Fog.
„ „	3 P.M.				
		36	34	35	
„ „	9 P.M.				

GREENLAND.

Being desirous of preserving unbroken the continuity of the account of the pendulum experiments, I shall confine myself at present to such a brief notice of the outline of the Griper's voyage, from the time of her quitting Spitzbergen, until her departure from Greenland at the close of the season of navigation, as may be sufficient to explain the reasons which determined the choice of the pendulum station on that coast; and shall reserve until a subsequent occasion, the few remarks which it may be proper to make on the parts of Greenland, of which we were thus accidentally the first visitors.

Captain Clavering, having succeeded in forcing a passage through the barrier of ice, which impedes the access to the shores of East Greenland, in a higher latitude than it is recorded to have been previously traversed, arrived on the coast between the 74th and 75th degrees of latitude in the second week in August; he proceeded to ascend it in the open channel within the barrier, experiencing no obstruction until he had passed the 75th parallel, when his progress to the northward was checked by ice, closely packed against the land, and sustaining, apparently, a very heavy pressure from without, so as to be impassable, until an off-shore wind should relieve the pressure, and concur with re-action in producing a set of the ice in an opposite direction; how soon this event might take place was entirely conjectural, and there was an immediate necessity of seeking a situation, in which to await the change, of less exposure than the extreme advanced position, which from local circumstances, was one of considerable danger; Captain Clavering decided, therefore, to return to a harbour of safe anchorage, in the latitude of 74° 30′, which he had examined in passing, and to station the ship there for the period which the experiments should require; after which, if the weather and the state of the ice should authorize it, he might resume the attempt to examine the coast to the northward as far as the navigation might be open in the latter and more favourable part of the season. Captain Clavering also determined on employing the boats, during the ship's detention, in the examination of a very extensive opening in the coast, between the harbour in which the ship was stationed and the latitude of 74°; and on accompanying them himself.

The objects thus proposed being accomplished, the Griper sailed again to the northward on the 31st of August; but a continuance of the profoundly serene weather which had prevailed almost uninterruptedly during the preceding three weeks, and had been highly favourable for the operations of the boats, as well as for those on shore, proved eventually the source of disappointment, as it had previously been of satisfaction; the inactivity which it compelled was the more vexatious, as the state of the ice, when viewed from the hills which exceeded 3000 feet in height, appeared sufficiently open for navigation as far as the eye could reach, which was certainly beyond the 76th degree: what might have been undertaken with a favourable breeze and concurrent circumstances in the first week in September, could not be attempted a few days later without a most obvious risk of the detention during the winter, which Captain Clavering was strongly cautioned against in his instructions; he employed therefore the short remainder of the season, in which the sun's presence above the horizon for a sufficient number of hours of the day justified his continuance on the coast, in extending its examination and survey from the 74th to the 72d degrees of latitude. Having accomplished this purpose on the 15th of September, and finding the vicinity of the land no longer accessible, by reason of the ice which set in from seaward, and might be expected to do so with increased pressure in the approaching autumnal gales, the Griper quitted the further examination of the coast to the southward, and by favour of the first gale which had been experienced for several weeks, repassed the barrier of ice on the evening of the 17th of September.

The harbour, in which the Griper remained from the 15th to the 31st of August, was formed by the channel which separates the main land from an island in which the experiments were made, and was secured from the access of heavy ice from the ocean, by a smaller island, situated in the mid-channel at the entrance. The groupe, of which these islands form a part, consists of two, nearly of the same size, and of two others, much smaller, being rather rocks than islands; they extend from the latitude of of 74° 30′ to that of 74° 42′, and were distinguished by the officers and seamen of the Griper, by the appellation of the Pendulum Islands. They partake in the character and general appearance of the main land, which is that of the trap formation: although the principal islands are in no direction so much as ten miles across, the greater part of their surface exceeds 2000 feet in height, the elevated parts being remarkably tabled on the summits. The anchorage was abreast of a plain of considerable extent on the inner, or south westernmost of the two largest islands; the plain consisted of the debris of a sandstone rock, and was generally swampy; but a perfectly dry spot, which had been the site of an Esquimaux village, was found for the pendulums on the shore close to the anchorage.

As Captain Clavering was desirous of having a chronometer on which he could depend for the determination of longitudes, during his excursion in the boats, I supplied him with No. 649, which I had used in the comparison of the clock with astronomical time at Hammerfest and at Spitzbergen, reserving No. 423 for that purpose at Greenland; and as Messrs. Parkinson and Frodsham had led me to expect that No. 423 would keep a less uniform rate from day to day than in the former voyage, although an equally good mean rate in intervals of longer duration, in consequence of its not having been in their hands a sufficient time, when returned to have its compensation corrected, I compared it with astronomical time, by means of the transit instrument, from forenoon to forenoon, and from afternoon to afternoon, and transferred the rate thus daily obtained immediately to the clock.

Being desirous of ascertaining how near an approach could be made by a sextant and mercurial horizon in so high a latitude to the precision with which the daily rate is determined by transits, I made the observations, the particulars of which are subjoined in a table; they present a confirmation, if any were needed, of the high opinion which the most experienced observers have always expressed of the practical merits of the sextant; I have employed the method of absolute altitudes instead of that of equal altitudes, from having had much experience of both, which has induced me to give a decided practical preference to the method of absolute altitudes, especially in the high latitudes.

The height of the pendulums above the sea was ascertained by direct measurement, to be thirty-one feet and a half.

TRANSITS OBSERVED AT GREENLAND.

DATE.	STARS.	TIMES OF TRANSIT BY THE CHRONOMETER 423.					Mean by the Chronometer.
		1st Wire.	2d Wire.	Meridian Wire.	4th Wire.	5th Wire.	
1823.		M. S.	M. S.	H. M. S.	M. S.	M. S.	H. M. S.
Aug. 20	Capella (below the Pole.)	. . .	33 16.8	8 33 54.4	34 32	35 10	8 33 54.4
„ „	α Cygni	3 53.2	4 30	12 5 07.2	5 43.6	6 20.4	12 05 06.93
„ „	α Orionis	12 56.8	13 23.2	21 13 49.6	14 16	14 42	21 13 49.53
„ „	α Lyræ (below the Pole).	57 58.4	58 31.6	21 59 05.2	59 38.8	00 12	21 59 05.2
„ 21	Sun {1st Limb . .	24 49.2	25 16	1 25 42.4	26 09.2	26 36	1 25 42.53
	Sun {2d Limb . .	26 58.8	27 25.2	1 27 52	28 19.2	28 46	1 27 52.2
„ „	γ Draconis . . .	. . .	. . .	9 18 48	19 30	20 12	9 18 48
„ „	α Lyræ	56 02.4	56 36	9 57 09.6	57 43.2	58 16.8	9 57 09.6
„ „	α Aquilæ . . .	07 17.6	07 44	11 08 10.4	08 36.8	09 03.2	11 08 10.4
„ „	Capella	. . .	. . .	20 27 32.8	28 10.4	28 48	20 28 10.4
„ „	α Orionis	09 07.6	09 33.6	21 09 59.6	10 26	10 52.4	21 09 59.8
„ „	γ Draconis (below the Pole).	15 30.8	16 13.2	21 16 55.2	17 37.2	18 19.6	21 16 55.2
„ „	α Lyræ (below the Pole).	54 09.2	54 42.8	21 55 16.4	55 50	56 23.6	21 55 16.4
„ 22	Arcturus . . .	. . .	. . .	5 . .	. . .	31 35.2	5 30 39.6
„ „	Capella (below the Pole).	25 01.2	25 38.4	8 26 16	26 53.6	27 30.8	8 26 16
„ „	γ Draconis . . .	13 34	14 16	9 14 58.4	15 40.8	16 22.8	9 14 58.4
„ „	α Lyræ	52 12.8	52 46.4	9 53 20	53 53.2	54 26.8	9 53 19.87
„ „	α Cygni	56 14.8	56 51.2	11 57 28	58 04.8	58 41.6	11 57 28.07
„ 23	Sun {1st Limb . .	24 31.6	24 58.4	1 25 25.2	25 52	26 18.8	1 25 25.2
	Sun {2d Limb . .	26 41.6	27 08.4	1 27 34.8	28 01.6	28 28.4	1 27 34.93
„ „	Arcturus . . .	25 53.2	26 21.2	5 26 49.2	27 17.2	27 44.8	5 26 49.13
„ 24	Sun {1st Limb . .	24 22.4	24 49.2	1 25 16	25 42.8	26 09.6	1 25 16
	Sun {2d Limb . .	27 32.4	26 59.2	1 27 26	27 52.8	28 19.6	1 27 26
„ „	Arcturus . . .	22 03.6	22 31.2	5 22 59.6	23 27.2	23 54.8	5 22 59.33
„ „	Capella (below the Pole).	17 20.8	17 58.4	8 18 36	19 13.6	19 51.2	8 18 36
„ „	γ Draconis . . .	05 52.8	06 35.2	9 07 17.2	07 59.2	08 41.2	9 07 17.13
„ „	α Lyræ	44 32	45 05.6	9 45 39.2	46 12.4	46 46	9 45 39.07

TRANSITS OBSERVED AT GREENLAND, *continued.*

DATE.	STARS.	TIMES OF TRANSIT BY 423.					Mean by the Chronometer.
		1st Wire.	2d Wire.	Meridian Wire.	4th Wire.	5th Wire.	
1823.		M. S.	M. S.	H. M. S.	M. S.	M. S.	H. M. S.
Aug. 24	α Aquilæ	55 47.6	56 14	10 56 40.4	57 06.8	57 33.2	10 56 40.4
" "	α Cygni	48 33.6	49 10.4	11 49 47.2	50 23.6	51 00.8	11 49 47.13
" "	α Andromedæ . . .	12 07.2	12 36.8	15 13 06.4	13 36	14 05.6	15 13 06.4
" "	γ Pegasi	17 04.8	17 31.6	15 17 58.4	18 25.2	18 52.4	15 17 58.47
" "	Capella	15 25.6	16 03.2	20 16 40.8	17 18.4	17 56	20 16 40.8
" "	α Orionis	. . .	. . .	20 58 30	58 56.4	59 22.8	20 58 30
" "	γ Draconis (below the Pole).	04 00.4	04 42.4	21 05 24.4	06 06.4	06 48.4	21 05 24.4
" "	α Lyræ (below the Pole).	42 38.8	43 12.4	21 43 46	44 19.6	44 53.2	21 43 46
" 25	Sun { 1st Limb . .	24 12.4	24 39.2	1 25 06	25 32.8	25 59.6	1 25 06
	Sun { 2d Limb . .	26 22	26 48.8	1 27 15.6	27 42.4	28 08.8	1 27 15.53
" "	Arcturus	18 13.2	18 41.2	5 19 09.2	19 36.8	20 04.8	5 19 09.07
" "	Capella (below the Pole).	13 30.4	14 08	8 14 45.6	15 23.6	16 01.2	8 14 45.73
" "	γ Draconis	02 03.2	02 45.6	9 03 27.6	04 09.6	04 52	9 03 27.6
" "	α Lyræ	40 42	41 15.6	9 41 49.2	42 22.8	42 56 4	9 41 49.2
" "	α Aquilæ . . .	51 57.6	52 24	10 52 50.8	53 17.2	53 43.6	10 52 50.67
" "	α Cygni	44 43.2	45 19.6	11 45 56.4	46 33.2	47 10.4	11 45 56.53
" "	α Andromedæ . . .	08 17.2	08 46.8	15 09 16.8	09 46.4	10 16	15 09 16.67
" "	γ Pegasi	13 14.8	13 41.6	15 14 08.8	14 35.6	15 02.8	15 14 08.73
" "	α Arietis	05 59.6	06 28	17 06 56.4	07 24.4	07 52.8	17 06 56.27
" "	Capella	11 37.2	12 14.8	20 12 52.4	13 29.6	13 07.2	20 12 52.27
" "	α Orionis	53 48.4	54 15.2	20 54 41.6	55 08	55 34	20 54 41.47
" "	γ Dracon. (below the Pole).	00 11.6	00 53.6	21 01 35.6	02 17.6	02 59.6	21 01 35.6
" "	α Lyræ (below the Pole).	38 50	39 23.6	21 39 57.2	40 30.8	41 04.4	21 39 57.2
" 26	Sun { 1st Limb . .	24 04.4	24 30.8	1 24 57.6	25 24.4	25 50.8	1 24 57.6
	Sun { 2d Limb . .	26 13.6	26 40.4	1 27 06.8	27 33.6	28 00.4	1 27 06.93
" "	Capella (below the Pole).	09 42.8	10 20.4	8 10 58	11 35.6	12 13.2	8 10 58
" "	γ Draconis	58 15.6	58 57.6	8 59 40	00 22.4	01 04.4	8 59 40

TRANSITS OBSERVED AT GREENLAND, *continued.*

DATE.	STARS.	TIMES OF TRANSIT BY 423.					Mean by the Chronometer.
		1st Wire.	2d Wire.	Meridian Wire.	4th Wire.	5th Wire.	
1823.		M. S.	M. S.	H. M. S.	M. S.	M. S.	H. M. S.
Aug. 26	α Lyræ.	36 54.4	37 28	9 38 01.6	38 35.2	39 08.4	9 38 01.53
" "	Aldebaran . . .	30 23.2	30 50.4	19 31 17.6	31 44.8	32 12	19 31 17.6
" "	Capella	07 48.4	08 25.6	20 09 03.2	09 40.8	10 18	20 09 03.2
" "	α Orionis	50 00.8	50 27.2	20 50 53.6	51 20	51 46.4	20 50 53.6
" "	α Lyræ (below the Pole) .	35 02.4	35 35.6	21 36 09.2	36 42.8	37 16.4	21 36 09.27
" 27	Sun, 1st Limb . .	23 55.6	24 22	1 24 48.8	25 15.6	25 42	1 24 48.8
	Sun, 2d Limb . .	26 04.8	26 31.2	1 26 58	27 24.8	27 51.2	1 26 58
" "	Arcturus. . . .	10 37.2	11 04.8	5 11 33.2	12 00.8	12 28.4	5 11 33.27
" "	Capella (below the Pole)	05 55.2	06 32.8	8 07 10.4	07 48	08 25.6	8 07 10.4
" "	γ Draconis . . .	54 27.2	55 09.2	8 55 51.2	56 33.2	57 15.2	8 55 51.2
" "	α Lyræ.	33 05.6	33 39.2	9 34 12.8	34 46.4	35 19.6	9 34 12.73
" "	α Aquila	44 21.6	44 48	10 45 14.4	45 41.2	46 07.2	10 45 14.47
" "	α Cygni	37 06.8	37 43.6	11 38 20.4	38 57.2	39 34	11 38 20.4
" "	Aldebaran . . .	26 34.4	27 01.6	19 27 28.8	27 56	28 23.2	19 27 28.8
" "	Capella . ..	03 59.6	04 37.2	20 05 14.8	05 52.4	06 30	20 05 14.8
" "	α Lyræ (below the Pole) .	31 13.6	31 47.2	21 32 20.8	32 54	33 27.6	21 32 20.67
" 28	Sun, 1st Limb . .	23 46.8	24 13.2	1 24 39.6	25 06.4	26 32.8	1 24 39.73
	Sun, 2d Limb . .	25 55.6	26 22	1 26 48.8	27 15.6	27 42.4	1 26 48.87
" "	Capella (below the Pole)	02 06.8	02 44	8 03 21.6	03 59.2	04 36.8	8 03 21 67
" "	α Lyræ. . .	29 18	29 51.2	9 30 24.8	30 58.4	31 32	9 30 24.87
" "	Castor (below the Pole)	21 36.4	22 07.2	10 22 38	23 19.2	23 40.4	10 22 38.2
" "	Pollux (below the Pole).	32 47.6	33 17.2	10 33 47.2	34 16.8	34 46.4	10 33 47.07
" "	α Aquilæ . .	40 33.6	41 00	10 41 26.4	41 52.8	42 19.2	10 41 26.4
" "	Aldebaran . .	22 44.8	23 14	19 23 39.2	24 06.4	24 33.6	19 23 39.2
" "	Capella	00 09.6	00 47.2	20 01 24.8	02 02.4	02 39.6	20 01 24.73
" "	α Orionis . . .	42 22.4	42 48.8	20 43 15.2	43 41.6	44 08	20 43 15.2
" "	γ Dracon, (below the Pole)	48 45.2	49 27.2	20 50 09.2	50 51.2	51 33.2	20 50 09.2

TRANSITS OBSERVED AT GREENLAND, *continued.*

DATE.	STARS.	TIMES OF TRANSIT BY 423.					Mean by the Chronometer.
		1st Wire.	2d Wire.	Meridian Wire.	4th Wire.	5th Wire.	
1823.		M. S.	M. S.	H. M. S.	M. S.	M. S.	H. M. S.
Aug. 28	α Lyræ (below the Pole) .	27 23.2	27 56.8	21 28 30.4	29 04	29 37.2	21 28 30.33
„ 29	Sun 1st Limb . .	23 34.8	24 01.6	1 24 28.4	24 55.2	25 21.6	1 24 28.33
	Sun 2d Limb .	25 44.4	26 11.2	1 26 38	27 04.4	27 31.2	1 26 37.87
„ „	Arcturus. . . .	02 59.6	03 27.2	5 03 54.8	04 22.4	04 50.4	5 03 54.87
„ „	Capella (below the Pole)	58 17.2	58 54.4	7 59 32	00 09.6	00 47.2	7 59 32.07
„ „	γ Draconis . .	46 48.4	47 30.8	8 48 12.8	48 54 8	49 36.8	8 48 12.73
„ „	α Lyræ.	25 27.2	26 00.8	9 26 34.4	27 08	27 41.2	9 26 34.33
„ „	Castor (below the Pole).	17 46.4	18 17.2	10 18 48	19 18.8	19 49.6	10 18 48
„ „	Pollux (below the Pole).	28 57.6	29 27.6	10 29 57.2	30 27.2	30 56.8	10 29 57.27
„ „	α Aquilæ	36 43.2	37 09.6	10 37 36	38 02.8	38 29.2	10 37 36.13
„ „	Capella	56 20.4	56 58	19 57 35.6	58 13.2	58 50.4	19 57 35.53
„ „	α Orionis	38 32.8	38 59.2	20 39 25.6	39 52	40 18.4	20 39 25.6
„ „	γ Dracon. (below the Pole)	44 56.8	45 38.8	20 46 20.8	47 02.8	47 44.8	20 46 20.8
„ „	α Lyræ (below the Pole) .	23 34.8	24 08	21 24 41.6	25 15.2	25 48.4	21 24 41.6
„ 30	Sun 1st Limb . .	23 24	23 50.8	1 24 17.6	24 44.4	25 11.2	1 24 17.6
	Sun 2d Limb . .	25 34	26 00.4	1 26 27.2	26 54	27 20.4	1 26 27.2
„ „	Arcturus. . . .	. . .	. . .	4 59 05.6	00 34	01 01.6	4 59 05.6
„ „	Capella (below the Pole)	54 27.6	55 04.8	7 55 42.4	56 20	56 57.6	7 55 42.47
„ „	α Lyræ.	21 38	22 11.6	9 22 45.2	23 18.8	23 52.4	9 22 45.2
„ „	Castor (below the Pole).	13 56.8	14 27.6	10 14 58.8	15 30	16 01.2	10 14 58.87
„ „	Pollux (below the Pole).	25 08.8	25 38.4	10 26 08	26 38	27 07.6	10 26 08.13
„ „	α Aquilæ	32 54	33 20.4	10 33 46.8	34 13.2	34 39.6	10 33 46.8
„ „	α Cygni	25 39.2	26 16	11 26 52.8	27 29.6	28 06.4	11 26 52.8

GREENLAND.——DEDUCTION of the RATE of the CHRONOMETER No. 423, from TRANSITS, August, 1823.

STARS.	20 to 21		21 to 22		22 to 23		23 to 24		24 to 25		25 to 26		26 to 27		27 to 28		28 to 29		29 to 30	
	A.M. to A.M.	P.M. to P.M.	A.M. to A.M.	P.M. to P.M.	A.M. to A.M.	P.M. to P.M.	A.M. to A.M.	P.M. to P.M.	A.M. to A.M.	P.M. to P.M.	A.M. to A.M.	P.M. to P.M.	A.M. to A.M.	P.M. to P.M.	A.M. to A.M.	P.M. to P.M.	A.M. to A.M.	P.M. to P.M.	A.M. to A.M.	P.M. to P.M.
	s.	s.	s.	s.	s.	s.	s.	s.	s.	s.	s.	s.	s.	s.	s.	s.	s.	s.	s.	s.
Aldebaran	..	..	..	..	..	..	..	..	..	..	..	..	..	..	7.11	..	6.31	..	..	..
Capella	..	..	..	..	6.04	..	6.04	..	6.04	..	7.38	..	6.84	..	7.51	.	5.84	..	6.71	..
α Orionis	..	..	6.18	..	5.98	..	5.98	..	5.98	..	7.38	..	8.04	..	6.71	..	6.71	..	6.31	..
γ Dracon. (below the Pole)	..	..	..	..	5.64	..	5.64	..	5.64	..	7.11	..	7.11	..	7.11	..	7.11	..	7.51	..
α Lyræ (below the Pole.)	..	..	7.11	..	5.78	..	5.78	..	5.78	..	7.11	..	7.98	..	7.31	..	5.57	..	7.18	..
The Sun	..	..	6.12	..	6.12	..	6.46	..	5.6	..	7.8	..	7.76	..	7.92	..	6.42	..	6.98	..
Arcturus	..	..	..	..	..	5.44	..	6.11	..	5.65	..	8.01	..	8.01	..	6.71	..	6.71	..	6.64
Capella (below the Pole)	..	6.71	..	6.71	..	5.91	..	5.91	..	5.64	..	8.18	..	8.31	..	7.18	..	6.31	..	6.31
γ Draconis	..	..	..	6.31	..	5.28	..	5.26	..	6.38	..	8.31	..	7.11	..	6.67	..	6.67	..	.
α Lyræ	..	..	..	6.18	..	5.51	..	5.51	..	6.04	..	8.24	..	7.11	..	8.05	..	5.37	..	6.78
Castor (below the Pole)	..	..	..	..	..	..	..	.	..	..	..	..	..	..	..	..	..	5.71	..	6.78
Pollux (below the Pole)	..	..	..	..	..	..	..	..	..	..	..	..	..	..	.	..	..	6. 11	..	6.77
α Aquilæ	..	..	..	5.91	..	5.91	..	5.91	..	6.18	..	7.81	..	7.81	..	7.84	..	5.64	..	6.58
α Cygni	..	6.48	..	6.48	..	5.44	..	5.44	..	5.31	..	7.85	..	7.85	..	6.71	..	6.71	..	6.71
α Andromedæ . . .	..	..	..	..	..	..	..	..	..	6.18	..	..	..	..	..	..	..	..	..	..
γ Pegasi . . .	..	..	..	..	..	..	..	..	..	6.17	..	..	..	..	..	..	..	..	.	..
Means.— Gaining per Diem. Sidereal .	..	6.59	6.47	6.32	5.91	5.58	5.98	5.7	5.81	5.94	7.36	8.07	7.55	7.7	7.28	7.19	6.33	6.15	6.94	6.65
Solar . .	..	6.61	6.49	6.34	5.93	5.6	6.0	5.72	5.83	5.96	7.38	8.09	7.57	7.72	7.3	7.21	6.35	6.17	6.96	6.67

Mean 6″.61 in a Sidereal or 6″.63 in a Solar Day.

GREENLAND.——Observations to determine the Rate of the Chronometer 423, from the 20th to the 30th of August, 1823, by Altitudes of the Sun, taken with a Sextant and Mercurial Horizon.

Latitude of the Place of Observation, 74° 32′ 19″ N.; Longitude, 18° 49′ W.

DATE.	Chronometer.	App. Alt. Sun's centre.	Barometer.	Thermometer.	Chron. fast on Mean Time.	CHRON.'S RATE.	
						Altdes.	Trants.
1823.	H. M. S.	° ′ ″	IN.	°	H. M. S.	S.	S.
Aug. 20 P.M.	5 52 07.13	36 45 07	29.87	35	1 23 38.7		
,, 21 A.M.	8 54 27.8	35 27 09	29.87	32	1 23 45.7	6 53	6.61
,, 21 P.M.	5 44 25.53	37 03 32.5	29.90	35	1 23 45.23	6.7	6.49
,, 22 A.M.	8 46 41.8	33 50 45.5	29.90	36	1 23 52.4	5.44	6.34
,, 22 P.M.	5 51 50.27	35 28 12.5	29.95	39	1 23 50.67	5.91	5.93
,, 23 A.M.	....	....	...	...	....	5.5	5.6
,, 23 P.M.	5 37 17.27	36 36 13	29.90	42	1 23 56.27	5.91	6.0
,, 24 A.M.	....	....	...	...	....	6.8	5.72
,, 24 P.M.	5 32 28.87	36 30 51	29.90	40	1 24 03.07	5.91	5.83
,, 25 A.M.	9 23 08.53	36 26 52.5	29.71	33	1 24 10.13	5.26	5.96
,, 25 P.M.	5 29 35.73	36 10 48	27.73	36	1 24 08.23	6.7	7.38
,, 26 A.M.	9 14 06.73	34 42 43	29.68	36.5	1 24 16.83	8.84	8.09
,, 26 P.M.	5 36 19.67	34 40 39	29.73	45	1 24 17.17	7.45	7.57
,, 27 A.M.	....	....	...	...	....	7.71	7.72
,, 27 P.M.	....	....	...	...	....	7.45	7.3
,, 28 A.M.	0 18 11.03	33 57 20	30.03	39	1 24 31.73	7.71	7.21
,, 28 P.M.	5 50 21.8	31 33 37	30.00	36	1 24 32.6	6.44	6.35
,, 29 A.M.	9 18 12.07	33 14 10	30.15	34.5	1 24 38.17		
MEANS—Gaining per Diem						6.64	6.68

The "Apparent Altitudes" are each a mean of six observations, three of the Upper, and three of the Lower Limb: they are corrected for an Index Error of 1′ 20″. The Thermometer was suspended in the air and in the shade, near the place of observation. The times by the Chronometer were noted by the beats.

GREENLAND.——Comparisons of the Astronomical Clock with the Chronometer No. 423, from the 20th to the 30th of August, 1823; with the Clock's Rate on Mean Solar Time deduced.

1823.	Chronometer.	Clock.	Clock's Gain.		Daily Rates.	
			A.M. to A.M.	P.M. to P.M.	Chron.	Clock.
	H. M. S.	H. M. S.	s.	s.	Gaining. s.	Gaining. s.
Aug. 20 P. M.	9 00 00	9 44 24.2				
„ 21 A. M.		9 45 00.2	..	71.8	6.61	78.41
„ 21 P. M.		9 45 36	71.3	..	6.49	77.79
„ 22 A. M.		9 46 11.5	..	71.5	6.34	77.84
„ 22 P. M.		9 46 47.5	72	..	5.93	77.93
„ 23 A. M.		9 47 23.5	..	72.5	5.6	78.1
„ 23 P. M.		9 48 00	71.8	..	6.0	77.8
„ 24 A. M.		9 48 35.3	..	71.9	5.72	77.62
„ 24 P. M.		9 49 11.9	72.2	..	5.83	78.03
„ 25 A. M.		9 49 47.5	..	72.2	5.96	78.16
„ 25 P. M.		9 50 24.1	70.6	..	7.38	77.98
„ 26 A. M.		9 50 58.1	..	69.8	8.09	77.89
„ 26 P. M.		9 51 33.9	70.5	..	7.57	78.07
„ 27 A. M.		9 52 08.6	..	70.3	7.72	78.02
„ 27 P. M.		9 52 44.2	70.4	..	7.3	77.7
„ 28 A. M.		9 53 19	..	70	7.21	77.21
„ 28 P. M.		9 53 54.2	71.6	..	6.35	77.95
„ 29 A. M.		9 54 30.6	..	71.8	6.17	77.97
„ 29 P. M.		9 55 06	71	..	6.96	77.96
„ 30 A. M.		9 55 41.6	..	71.4	6.67	78.07
„ 30 P. M.		9 56 17.4				
		MEAN				77.94

GREENLAND.——COINCIDENCES OBSERVED with PENDULUM 3; the Clock making 86477.94 Vibrations in a Mean Solar Day.

DATE.	Barometer.	No. of Coincidence.	Temperature.	Time of Disappearance.	Time of Re-appearance.	True Time of Coincidence.	Arc of Vibration.	Mean Temperature.	Mean Interval.	Correction for the Arc.	Vibrations in 24 hours.	Reduction to a Mean Temperature.	Reduced Vibrations at 44°.
1823.	IN.		°	M. S.	M. S.	H. M. S.	°	°	S	° +			
Aug. 21 A.M.	29.885	1	39.5	41 22	41 26	8 41 24	1.2	41.5	688.45	1.25	86227.95	−1.05	86226.90
		11	43.5	36 04	36 13	10 36 08.5	0.58						
„ 21 P.M.	29.900	1	49.6	15 10	15 15	2 15 12.5	1.18	49.1	679.9	1.22	86224.76	+2.14	86226.90
		11	48.6	8 26	8 37	4 8 31.5	0.58						
„ 21 P.M.	29.900	1	49.1	00 08	00 15	6 00 11.5	1.18	48.45	680.65	1.22	86225.06	+1.87	86226.93
		11	47.8	53 30	53 46	7 53 38	0.58						
„ 22 A.M.	29.900	1	36.8	7 40	7 46	6 7 43	1.18	38.4	692.6	1.22	86229.44	−2.35	86227.09
		11	40	3 04	3 14	8 3 09	0.58						
„ 22 Noon.	29.935	1	43.8	46 04	46 09	11 46 06.5	1.2	44.4	685.5	1.25	86226.87	+0.17	86227.04
		11	45	40 16	40 27	1 40 21.5	0.58						
„ 22 P.M.	29.950	1	48	1 42	1 44	6 1 43	1.28	47.65	681.5	1.48	86225.62	+1.53	86227.15
		11	47.3	55 10	55 26	7 55 18	0.66						
„ 23 A.M.	29.880	1	41.3	15 28	15 30	10 15 29	1.28	43.15	686.55	1.48	86227.50	−0.36	86227.14
		11	45	09 48	10 01	12 9 54.5	0.66						
„ 23 P.M.	29.888	1	51.3	42 45	42 50	5 42 47.5	1.2	49.85	679.1	1.25	86224.51	+2.46	86226.97
		11	48.4	35 50	36 07	7 35 58.5	0.58						
„ 24 A.M.	29.965	1	40	27 23	27 25	10 27 24	1.28	42.2	687.9	1.48	86228.00	−0.76	86227.24
		11	44.4	21 55	22 11	12 22 03	0.66						
„ 24 P.M.	29.873	1	46.8	39 20	39 23	4 39 21.5	1.26	46.6	683.05	1.42	86226.14	+1.09	86227.23
		11	46.4	33 04	33 20	6 33 12	0.64						
„ 25 A.M.	29.715	1	34	38 01	38 08	6 38 04.5	1.18	36.1	695.2	1.22	86230.38	−3.32	86227.06
		11	38.2	33 49	34 04	8 33 56.5	0.58						
„ 25 P.M.	29.720	1	39.4	49 09	49 13	11 49 11	1.22	40.5	690.05	1.34	86228.62	−1.47	86227.15
		11	41.6	44 03	44 20	1 44 11.5	0.62						
Means . . .	29.876							44			86227.07		86227.07

GREENLAND.——**COINCIDENCES OBSERVED with PENDULUM No. 4; the Clock making 86477.94 Vibrations in a Mean Solar Day.**

DATE.	Barometer.	No. of Coincidence.	Temperature.	Time of Disappearance.	Time of Re-appearance.	True Time of Coincidence.	Arc of Vibration.	Mean Temperature.	Mean Interval.	Correction for the Arc.	Vibrations in 24 hours.	Reduction to a mean Temperature.	Reduced Vibrations at 41°.14.
1823.	IN.		°	M. S.	M. S.	H. M. S.	°	°	S.	S.			
Aug. 25 P.M.	29.710	1	45.8	33 26	33 34	4 33 30	1.18	45.3	711.65	+ 1.19	86236.07	+1.75	86237.82
		11	44.8	31 55	32 18	6 32 06.5	0.56						
„ 26 A.M.	29.675	1	34	58 47	58 55	5 58 51	1.2	36.1	722.4	1.25	86239.77	−2.12	86237.65
		11	38.2	59 08	59 22	7 59 15	0.58						
„ 26 P.M.	29.728	1	46.8	41 51	41 59	12 41 55	1 18	48	707.75	1.19	86234.75	+2.88	86237.63
		11	49.2	39 45	40 00	2 39 52.5	0 56						
„ 26 P.M.	29.728	1	50	29 49	29 54	4 29 51.5	1.22	49	706.35	1.31	86234.39	+ 3.3	86237.69
		11	48	27 24	27 46	6 27 35	0.6						
„ 27 A.M.	29.800	1	32.5	33 47	33 49	4 33 48	1.28	32.45	726.95	1.45	86241.49	−3.65	86237.84
		11	32.4	34 48	35 07	6 34 57.5	0.64						
„ 27 A.M.	29.900	1	39	10 58	11 05	9 11 01.5	1.18	40.5	717	1.19	36237.89	−0.27	86237.62
		11	42	10 23	10 40	11 10 31.5	0.56						
„ 27 P.M.	29.900	1	44	44 52	44 55	3 44 53.5	1.26	41.5	712.18	1.41	86236.69	+1.41	86238.10
		11	45	43 33	43 50	5 43 41.5	0.62						
„ 28 A.M.	30.000	1	33.1	36 42	36 46	5 36 44	1.28	34.45	724.45	1.45	86240.65	−2.81	86237.84
		11	35.8	37 18	37 39	7 37 28.5	0.64						
„ 28 A.M.	30.030	1	38	42 35	42 39	9 42 37	1.22	39	719.6	1.28	86238.84	−0.9	86237.94
		11	40	42 25	42 41	11 42 33	0.58						
„ 28 P.M.	30.000	1	46	8 21	8 23	4 8 22	1.3	45.4	711.05	1.48	86236.18	+1.79	86237.97
		11	44.8	6 43	6 02	6 6 52.5	0.64						
„ 29 A.M.	30.150	1	37.2	45 22	45 23	8 45 22.5	1.28	39.1	719.15	1.45	86238.89	−0.86	86238.03
		11	41	45 06	45 22	10 45 14	0.64						
„ 29 P.M.	30.160	1	44.3	17 17	17 22	4 17 19.5	1.22	43.55	713.8	1.28	86236.92	+1.01	86237.93
		11	42.8	16 09	16 26	6 16 17.5	0.58						
„ 30 A.M.	30.000	1	36	34 33	34 34	7 34 33.5	1.3	37.5	721.2	1.48	86239.60	−1.53	86238.07
		11	39	34 37	34 54	9 34 45.5	0.64						
Means . . .	29.906							41.14			86237.86		86237.86

DRONTHEIM.

It had been originally my intention to have made Reikivik in Iceland the concluding station of the pendulum experiments in the high latitudes; but the difficulties which had been experienced in getting a complement of seamen for the Griper, when fitting at Deptford, had delayed the commencement of the voyage until the early part of the season had passed, and had caused her to arrive later at every station than had been designed; consequently, when on the 17th of September, we found ourselves finally disengaged from the Greenland ice, the season of navigation was drawing towards a close; the autumnal gales had already commenced, and the nights were above sixteen hours long; under such circumstances it would not have been prudent to have risked the approach to the coast of Iceland, with which we were imperfectly acquainted, in a vessel which sailed so heavily as the Griper; and it was preferable to recross the northern ocean to seek a pendulum station on the coast of Norway, nearly in the same latitude as had been contemplated in Iceland; and as Drontheim, the ancient capital of Norway, appeared beyond comparison, the most eligible situation for the purpose, the Griper arrived there on the 8th of October.

As our visit to Drontheim was not premeditated, we were unfurnished with official introductions to the authorities; but our reception by his Excellency Count Trampe, Governor of the province of Drontheim, was not suffered to be the less cordial or unreserved, on account of the absence of that formality; the attentions which we experienced from that gentleman, and the facilities which we enjoyed in consequence of his sanction, were in every respect such as might have been expected from a friend of the late Sir Joseph Banks. Through the good offices of Mr. Schnitler,

His Britannic Majesty's Consul, I obtained the necessary accommodation and means of performing the experiments with great convenience and advantage, in a villa in the environs of the town, belonging to Mr. Hans Wensel, whose daughter Mr. Schnitler had married. Mr. Wensel was so kind as to permit Captain Clavering and myself to occupy his villa as our residence, and to allow me to take up the flooring of one of the rooms, to enable the clock and pendulum supports to rest on the ground beneath, and to be thus unconnected with the house or its foundation. Mr. Wensel's villa is situated about an English mile from Drontheim, on the right of the road which ascends the Steinberget hill, and is on a foundation of mica-slate, approaching very nearly to clay-slate. The observatory containing the transit instrument was established on a small eminence on the lawn belonging to the house.

It had been the good fortune of Captain Clavering and myself to have experienced at each of the inhabited stations which we had visited the most marked hospitality and kindness, but at none were our obligations in these respects greater than at Drontheim; to Mr. Schnitler especially we were indebted for the most assiduous and unremitting endeavours to render our residence agreeable; and if I may be permitted in a single instance to notice personal attentions, not directly conducing to the promotion of the experiments, but contributing materially to our comfort and pleasure whilst engaged in them, I would avail myself of the present occasion, to express for Captain Clavering and myself our very grateful remembrance of the exceeding kindness which we received from Mr. Knutzon, and from the younger branches of his amiable and excellent family.

Mr. Mandall, an officer of Norwegian Engineers, was kind enough at my request to undertake the trigonometrical measurement of the height of the pendulums at Mr. Wensel's house, which he found to be 118 Norwegian, or 121.5 British feet above the level of mean tide.

TRANSITS OBSERVED AT DRONTHEIM.

DATE.	STARS.	TIMES OF TRANSIT BY No. 649.					Mean by the Chronometer.	CLOCK.
		1st Wire.	2d Wire.	Meridian Wire.	4th Wire.	5th Wire.		
1823.		M. S.	M. S.	H. M. S.	M. S.	M. S.	H. M. S.	H. M. S.
Oct. 15	α Lyræ	16 40	17 13.6	4 17 46.8	18 20.4	18 54	4 17 46.93	4 24 54.27
„ „	β Lyræ	29 19.2	29 50.4	4 30 21.6	30 52.8	31 24	4 30 21.6	4 37 29.3
„ „	γ Lyræ	38 05.2	38 36	4 39 07.2	39 38	40 08.8	4 39 07.07	4 46 15.07
„ „	γ Aquilæ . . .	23 40.4	24 06.8	5 24 33.2	25 00	25 26.4	5 24 33.33	5 31 42.68
„ „	α Aquilæ . .	27 57.6	28 24	5 28 50.4	29 16.8	29 43.2	5 28 50.4	5 35 59.9
„ „	δ Aquilæ . . .	47 56.8	48 22.8	5 48 49.2	49 15.6	49 41.6	5 48 49.2	5 55 59.3
„ „	2 α Capri. . .	53 58.8	54 25.6	5 54 52	55 18.4	55 45.2	5 54 52	6 02 02.25
„ „	α Cygni. . .	20 40.8	21 17.6	6 21 54	22 30.4	23 07.2	6 21 54	6 29 05.06
„ „	α Aquarii . . .	42 09.4	42 36	7 43 02	43 28	43 54	7 43 02	7 50 15.5
„ 16	α Lyræ . . .	12 44.8	13 18.4	4 13 52	14 25.2	14 58.8	4 13 51.87	4 21 43.32
„ „	β Lyræ . . .	25 24.4	25 55.6	4 26 26.8	26 58	27 29.6	4 26 26.87	4 34 18.72
„ „	α Lyræ	34 09.8	34 41.2	4 35 12.4	35 43.2	36 14.4	4 35 12.23	4 43 04.36
„ „	α Pegasi . . .	37 18.8	37 45.6	8 38 12.8	38 39.6	39 06.8	8 38 12.73	8 46 12.44
„ 17	α Lyræ	08 49.6	09 22.8	4 09 56.4	10 30	11 03.6	4 09 56.47	4 18 32.95
„ „	β Lyræ	21 28.4	21 59.6	4 22 30.8	23 02	23 33.6	4 22 30.87	4 31 07.73
„ „	γ Lyræ	30 14.4	30 45.2	4 31 16.4	31 47.6	32 18.4	4 31 16.4	4 39 53.55
„ „	γ Aquilæ . .	15 49.6	16 16	5 16 42.4	17 08.8	17 35.6	5 16 42.47	5 25 21.04
„ „	α Aquilæ .	20 07.2	20 33.6	5 21 00	21 26.4	21 52.8	5 21 00	5 29 38.66
„ „	δ Aquilæ . . .	40 06.4	40 32.4	5 40 58.4	41 24.8	41 51.2	5 40 58.6	5 49 37.9
„ „	2 α Capri. .	46 07.6	46 34.8	5 47 01.6	47 28.4	47 55.2	5 47 01.33	5 55 41.07
„ „	α Cygni. . . .	12 50	13 26.8	6 14 03.6	14 40	15 16.8	6 14 03.47	6 22 43.86
„ „	α Aquarii . . .	34 18.8	34 45.2	7 35 11.6	35 37.6	36 03.6	7 35 11.4	7 43 54.33
„ „	α Pegasi . . .	33 22.4	33 49.6	8 34 16.4	34 43.6	35 10.4	8 34 16.47	8 43 01.28
„ 19	α Lyræ	00 58	01 31.6	4 02 05.2	02 38.4	03 12	4 02 05.07	4 12 11.53
„ „	β Lyræ	13 37.2	14 08.4	4 14 39.6	15 10.8	15 42	4 14 39.6	4 24 46.46
„ „	α Pegasi . . .	25 31.6	25 58.8	8 26 25.6	26 52.8	27 19.6	8 26 25.67	8 36 40.35
„ 23	α Arietis . . .	. . .		11 11 36.4	12 04.4	12 32.4	11 11 36.4	11 24 51.4

TRANSITS OBSERVED AT DRONTHEIM, *continued.*

DATE.	STARS.	TIMES OF TRANSIT BY No. 649.					Mean by the Chronometer.	CLOCK.
		1st Wire.	2d Wire.	Meridian Wire.	4th Wire.	5th Wire.		
1823.		M. S.	M. S.	H. M. S.	M. S.	M. S.	H. M. S.	H. M. S.
Oct. 23	- Tauri. . . .	. . .	. . .	13 28 39.6	29 07.2	29 34.8	13 28 39.6	13 41 58.74
,, ,,	- Tauri. . . .	. . .	. . .	13 32 18	32 45.6	33 13.2	13 22 18	13 45 37.25
,, ,,	Aldebaran. .	38 51.2	39 18.4	13 39 45.6	40 12.8	40 39.6	13 39 45.53	13 53 05.03
,, ,,	Capella. . .	16 15.6	16 52.8	14 17 30.4	18 08	18 45.2	14 17 30.4	14 30 51.13
,, ,,	β Tauri. . . .	27 58	28 27.6	14 28 57.2	29 26.8	29 56.8	14 28 57.27	14 42 18.37
,, 28	γ Draconis . .	47 13.2	47 55.2	2 48 37.2	49 19.2	50 01.2	2 48 37.2	3 05 12.01
,, ,,	α Lyræ	25 52.8	26 26	2 26 59.6	27 33.2	28 06.8	3 26 59.67	3 43 35.57
,, ,,	β Lyræ	38 32.4	39 03.6	3 39 34.8	40 06.4	40 37 6	3 39 34.93	3 56 11.21
,, ,,	γ Lyræ	. . .	. . .	3 48 20.4	48 51.6	49 22.4	3 48 20.4	4 04 56.96
,, ,,	γ Aquilæ . . .	. . .	. . .	4 33 46	34 12.4	34 38.8	4 33 46	4 50 23.94
Nov. 1	γ Draconis . .	31 38.4	32 20.4	2 33 02.8	33 44.8	34 26.8	2 33 02.67	2 52 30.71
,, ,,	α Lyræ	10 28.4	10 52	3 11 25.6	11 59.2	12 32.8	3 11 25.6	3 30 54.84
,, ,,	β Lyræ	22 57.6	23 28.8	3 24 00	24 31.6	25 02.8	3 24 00.13	3 43 29.79
,, ,,	γ Lyræ	31 43.6	32 14.4	3 32 45.6	33 16.8	33 47.6	3 32 45.6	3 52 15.54
,, ,,	α Aquilæ . .	21 36.4	22 02.8	4 22 29.2	22 55.6	23 22	4 22 29.2	4 42 00.69
,, ,,	α Pegasi. . . .	34 52	35 18.8	7 35 46	36 12.8	36 39.6	7 35 45.87	7 55 23.46
,, 3	γ Draconis . .	23 50.8	24 32.4	2 25 14.8	25 57.2	26 38.8	2 25 14.8	2 46 09.42
,, ,,	α Lyræ	02 30.4	03 04	3 03 37.6	04 11.2	04 44.8	3 03 37.6	3 24 33.30
,, ,,	α Arietis . . .	27 50.4	28 18.4	10 28 46.8	29 15.2	29 43.6	10 28 46.87	10 49 55.59
,, ,,	- Tauri. . . .	44 55.2	45 22.8	12 45 50.4	46 18	46 45.6	12 45 50.4	13 07 03.28
,, ,,	- Tauri. . . .	48 33.2	49 00.8	12 49 28.4	49 56	50 23.6	12 49 28.4	13 10 41.4
,, ,,	Aldebaran . .	56 01.2	56 28.8	12 56 56	57 23.6	57 50.8	12 56 56.07	13 18 09.57
,, ,,	Capella . . .	33 25.6	34 03.2	13 34 40.8	35 18	35 55.6	13 34 40.67	13 55 55.34
,, ,,	β Tauri. . . .	45 08.8	45 38.4	13 46 08	46 37.6	47 07.2	13 46 08	14 07 22.98

Drontheim.—Comparisons of the Astronomical Clock with the Chronometer No. 649, from the 15th of October to the 3d of November, 1823.

DATE.	Chronometer.	Clock.	Clock's gain in 12 hours.	DATE.	Chronometer.	Clock.	Clock's gain in 12 hours.
1823.	H. M. S.	H. M. S.	S.	1823.	H. M. S.	H. M. S.	S.
Oct. 15 P.M.	4 00 00	4 07 06.8		Oct. 25 A.M.		7 14 13	
,, 15 P.M.		7 07 12.2		,, 25 P.M.		7 14 34.8	21.8
,, 16 A.M.		7 07 34.2	22	,, 26 A.M.		7 14 56.9	22.1
,, 16 P.M.		7 07 56.6	22.4	,, 26 P.M.		7 15 18.4	21.5
,, 17 A.M.		7 08 19.1	22.5	,, 27 A.M.		7 15 39.9	21.5
,, 17 P.M.		7 08 41.8	22.7	,, 27 P.M.		7 16 00.7	20.8
,, 18 A.M.		7 09 04.1	22.3	,, 28 A.M.		7 16 21.1	20.4
,, 18 P.M.		7 09 26.9	22.8	,, 28 P.M.	7 00 00	7 16 42.6	21.5
,, 19 A.M.		7 09 49.6	22.7	,, 29 A.M.		7 17 04.5	21.9
,, 19 P.M.	7 00 00	7 10 12	22.4	,, 29 P.M.		7 17 25.9	21.4
,, 20 A.M.		7 10 34.5	22.5	,, 30 A.M.		7 17 47.6	21.7
,, 20 P.M.		7 10 56.7	22.2	,, 30 P.M.		7 18 09.2	21.6
,, 21 A.M.		7 11 18.9	22.2	,, 31 A.M.		7 18 30.8	21.6
,, 21 P.M.		7 11 40.7	21.8	,, 31 P.M.		7 18 52.4	21.6
,, 22 A.M.		7 12 02.8	22.1	Nov. 1 A.M.		7 19 14.5	22.1
,, 22 P.M.		7 12 24.2	21.4	,, 1 P.M.	2 00 00	2 19 27	22
,, 23 A.M.		7 12 45.7	21.5	,, 1 P.M.	7 00 00	7 19 36.5	
,, 23 P.M.		7 13 07.5	21.8	,, 1 P.M.	10 00 00	10 19 42	22.1
,, 23 P.M.	12 00 00	12 13 16.5		,, 2 A.M.		7 19 58.6	
,, 24 A.M.	3 00 00	3 13 22.4	22.1	,, 2 P.M.	7 00 00	7 20 20.7	22.1
,, 24 A.M.	7 00 00	7 13 29.6		,, 3 A.M.		7 20 41.7	21
,, 24 P.M.		7 13 51.8	22.2	,, 3 P.M.		7 21 02.6	20.9

The Clock was found by Transits to be gaining on Mean Solar time 45″.47 per diem, and by the above Comparisons 43″.7 per diem on the Chronometer; consequently the Chronometer was gaining 1″.77 on Mean Time.

DRONTHEIM.——DAILY RATE of the ASTRONOMICAL CLOCK on Mean Solar Time, deduced from TRANSITS.

STARS.	Oct. 15 to 16	16 to 17	17 to 18	18 to 19	19 to 20	20 to 21	21 to 22	22 to 23	23 to 24	24 to 25	25 to 26	26 to 27	27 to 28	28 to 29	29 to 30	30 to 31	31 to Nov. 1	1 to 2	2 to 3
	s.	s.	s.	s.	s.	s.	s.	s.	s.	s.	s.	s.	s.	s.	s.	s.	s.	s.	s.
γ Draconis. .	. .	. .	. .	. .	. .	. .	. .	. .	. .	. .	. .	. .	. .	45.58	45.58	45.58	45.58	45.26	45.26
α Lyræ . . .	44.96	45.54	45.2	45.2	45.25	45.25	45.25	45.25	45.25	45.25	45.25	45.25	45.25	45.73	45.73	45.73	45.73	45.17	45.17
β Lyræ . . .	45.33	44.92	45.27	45.27	45.33	45.33	45.33	45.33	45.33	45.33	45.33	45.33	45.33	45.55	45.55	45.55	45.55	. .	. .
γ Lyræ . . .	45.2	45.1	45.31	45.31	45.31	45.31	45.31	45.31	45.31	45.31	45.31	45.31	45.31	. .	. .	. .	. .	. .	. .
γ Aquilæ .	45.09	45.09	45.26	45.26	45.26	45.26	45.26	45.26	45.26	45.26	45.26	45.26	45.26	. .	. .	. .	. .	. .	. .
α Aquilæ . .	45.29	45.29	45.38	45.38	45.38	45.38	45.38	45.38	45.38	45.38	45.38	45.38	45.38	45.38	45.38	45.38	45.38	. .	. .
2 α Capri . .	45.32	45.32	. .	. .	. .	. .	. .	. .	. .	. .	. .	. .	. .	. .	. .	. .	. .	. .	. .
α Cygni . . .	45.31	45.31	. .	. .	. .	. .	. .	. .	. .	. .	. .	. .	. .	. .	. .	. .	. .	. .	. .
α Aquarii . .	45.32	45.32	. .	. .	. .	. .	. .	. .	. .	. .	. .	. .	. .	. .	. .	. .	. .	. .	. .
α Pegasi . .	. .	44.68	45.44	45.44	45.38	45.38	45.38	45.38	45.38	45.38	45.38	45.38	45.38	45.38	45.38	45.38	45.38	. .	.
α Arietis. . .	. .	. .	. .	.	. .	. .	. .	. .	45.38	45.38	45.38	45.38	45.38	45.38	45.38	45.38	45.38	45.38	45.38
– Tauri . . .	. .	. .	. .	. .	. .	. .	. .	. .	45.41	45.41	45.41	45.41	45.41	45.41	45.41	45.41	45.41	45.41	45.41
– Tauri . .	. .	. .	. .	. .	. .	. .	. .	. .	45.38	45.38	45.38	45.38	45.38	45.38	45.38	45.38	45.38	45.38	45.38
Aldebaran .	. .	. .	. .	. .	. .	. .	. .	. .	45.41	45.41	45.41	45.41	45.41	45.41	45.41	45.41	45.41	45.41	45.41
Capella . .	. .	. .	. .	. .	. .	. .	. .	. .	45.38	45.38	45.38	45.38	45.38	45.38	45.38	45.38	45.38	45.38	45.38
β Tauri . . .	. .	. .	. .	. .	. .	. .	. .	. .	45.42	45.42	45.42	45.42	45.42	45.42	45.42	45.42	45.42	45.42	45.42
MEANS, gaining on . . Sid. time	45.22	45.18	45.31	45.31	45.32	45.32	45.32	45.32	45.36	45.36	45.36	45.36	45.36	45.45	45.46	45.45	45.46	45.35	45.35
MEANS, gaining on . . Sol. time	45.35	45.31	45.44	45.44	45.45	45.45	45.45	45.45	45.49	45.49	45.49	45.49	45.49	45.58	45.59	45.58	45.59	45.48	45.48

Drontheim.—COINCIDENCES OBSERVED with PENDULUM No. 3.

DATE.	Barometer.	Clock gaining	No. of Coincidence.	Temperature.	Time of Disappearance.	Time of Re-appearance.	True Time of Coincidence.	Arc of Vibration.	Mean Temperature.	Mean Interval.	Correction for the Arc.	Vibrations in 24 hours.	Reduction to a Mean Temperature.	Reduced Vibrations at 47°.41.
1823.	IN.	S.		°	M. S.	M. S.	H. M. S.	°	°	S.	S. +			
Oct. 16 A.M.	29.47	45.35	1	47.4	59 24	59 29	7 59 26.5	1.18	47.7	684.35	1.22	86193.92	+0.12	86194.04
			11	48	53 22	53 38	9 53 30	0.58						
„ 16 P.M.	29.46	45.35	1	48.2	34 21	34 25	1 34 23	1.2	48.45	683.5	1.25	86193.65	+0.44	86194.09
			11	48.7	28 10	28 26	3 28 18	0.58						
„ 17 A.M.	29.53	45.31	1	46	47 40	47 41	7 47 40.5	1.26	46.05	685.75	1.39	86194.57	−0.57	86194.00
			11	46.1	41 50	42 06	9 41 58	0.62						
„ 17 P.M.	29.52	45.31	1	46.3	16 36	16 41	1 16 38.5	1.2	46.35	685.55	1.25	86194.37	−0.44	86193.93
			11	46.4	10 45	11 03	3 10 54	0.58						
„ 18 A.M.	29.42	45.44	1	45.1	32 30	32 35	9 32 32.5	1.2	45.5	686.45	1.28	86194.86	−0.80	86194.06
			11	45.9	26 50	27 04	11 26 57	0.6						
„ 18 P.M.	29.44	45.44	1	46.4	52 34	52 39	1 52 36.5	1.2	46.2	685.85	1.25	86194.63	−0.51	86194.12
			11	46	46 42	47 08	3 46 55	0.58						
„ 19 A.M.	29.55	45.44	1	45	19 08	19 13	9 19 10.5	1.2	45.65	686.35	1.28	86194.84	−0.74	86194.10
			11	46.3	13 26	13 42	11 13 34	0.6						
„ 19 P.M.	29.60	45.44	1	46.9	4 10	4 15	2 4 12.5	1.2	46.45	685.1	1.28	86194.36	−0.40	86193.96
			11	46	58 11	58 36	3 58 23.5	0.6						
„ 19 P.M. (by lamp light.)	29.67	45.44	1	47.5	35 52	35 55	7 35 53.5	1.2	47.5	684.05	1.28	86193.98	+0.04	86194.02
			11	47.5	29 46	30 02	9 29 54	0.6						
„ 20 A.M.	30.00	45.45	1	46.2	55 02	55 07	8 55 04.5	1.2	46.7	684.9	1.25	86194.27	−0.30	86193.97
			11	47.2	49 06	49 21	10 49 13.5	0.58						
„ 21 P.M.	30.06	45.45	1	46.9	09 40	09 45	1 09 42.5	1.18	46.9	684.55	1.22	86194.12	−0.21	86193.91
			11	46.9	03 40	03 56	3 03 48	0.58						
„ 22 A.M.	30.34	45.45	1	46.1	21 07	21 09	10 21 08	1.26	46.2	684.9	1.39	86194.41	−0.51	86193.90
			11	46.3	15 09	15 25	12 15 17	0.62						
„ 22 P.M.	30.30	45.45	1	46.2	4 00	4 05	2 4 02.5	1.18	46.2	685.3	1.22	86194.40	−0.51	86193.89
			11	46.2	58 00	58 31	3 58 15.5	0.58						
„ 22 P.M. (by lamp light.)	30.30	45.45	1	46.6	28 43	28 47	7 28 45	1.26	46.6	684.55	1.39	86194.29	−0.34	86193.95
			11	46.6	22 40	23 01	9 22 50.5	0.62						
„ 23 Noon.	29.92	45.45	1	50.1	41 00	41 07	11 41 03.5	1.18	50.1	681	1.22	86192.82	+1.13	86193.95
			11	50.1	34 23	34 44	1 34 33.5	0.58						
„ 23 P.M. (by lamp light.)	30.17	45.45	1	51.2	2 23	2 24	10 2 23.5	1.28	51.3	678.75	1.45	86192.21	+1.63	86193.84
			11	51.4	55 22	55 40	11 55 31	0.64						
„ 24 A.M.	30.17	45.49	1	52.2	17 56	17 59	11 17 57.5	1.24	52.1	678.6	1.37	86192.09	+1.97	86194.06
			11	52	10 54	11 13	1 11 03.5	0.62						
Means . .	29.82								47.41			86193.99		86193.99

DRONTHEIM.—COINCIDENCES OBSERVED with PENDULUM No. 4.

DATE.	Barometer.	Clock gaining.	No. of Coincidence.	Temperature.	Time of Disappearance.	Time of Reappearance.	True Time of Coincidence.	Arc of Vibration.	Mean Temperature.	Mean Interval.	Correction for the Arc.	Vibrations in 24 hours.	Reduction to a mean Temperature.	Reduced Vibrations at 46°.73.
1823.	IN.	S.		°	M. S.	M. S.	H. M. S.	°	°	S.	S.			
Oct. 25 A.M.	29.82	45.49	1	47.8	20 15	20 23	11 20 19	1.18	47.8	709.6	+ 1.25	86203.11	+0.45	86203.56
			11	47.8	18 23	18 47	1 18 35	0.60						
„ 25 P.M. (by lamp light.)	29.82	45.49	1	46.9	8 47	8 51	5 08 49	1.28	46.75	710	1.45	86203.47	. .	86203.47
			11	46.6	7 00	7 18	7 07 09	0.64						
„ 26 A.M.	29.78	45.49	1	50.3	15 11	15 19	11 15 15	1.22	50.55	706.25	1.31	86202.01	+1.60	86203.61
			11	50.8	12 45	13 10	1 12 57.5	0.6						
„ 27 A.M.	29.53	45.49	1	53	45 23	45 31	11 45 27	1.22	52.9	703.65	1.31	86201.11	+2.59	86203.70
			11	52.8	42 30	42 57	1 42 43.5	0.6						
„ 27 P.M. (by lamp light.)	29.35	45.49	1	52.9	14 08	14 15	10 14 11.5	1.22	52.15	704.4	1.33	86201.39	+2.28	86203.67
			11	51.4	11 24	11 47	12 11 35.5	0.62						
„ 28 A.M.	29.00	45.49	1	50.1	15 46	15 55	11 15 50.5	1.2	50.45	706.55	1.28	86202.08	+1.56	86203.64
			11	50.8	13 25	13 47	1 13 36	0.6						
„ 28 P.M. (by lamp light.)	29.00	45.49	1	51	37 58	38 03	8 38 00.5	1.28	51	705.15	1.45	86201.77	+1.79	86203.56
			11	51	35 23	35 41	10 35 32	0.64						
„ 29 A.M.	29.18	45.58	1	48	44 27	44 35	10 44 31	1.2	47.5	709.9	1.28	86203.32	+0.32	86203.64
			11	47	42 37	43 03	12 42 50	0.6						
„ 29 P.M. (by lamp light.)	29.13	45.58	1	47	01 02	01 07	9 01 04.5	1.22	47.1	710.45	1.31	86203.55	+0.15	86203.70
			11	47.2	59 17	59 41	10 59 29	0.6						
„ 30 A.M.	29.14	45.59	1	43.8	57 54	57 59	10 57 56.5	1.24	43.5	714.85	1.37	86205.11	−1.36	86203.75
			11	43.2	56 55	57 15	12 57 05	0.62						
„ 31 A.M.	29.58	45.58	1	42.7	47 15	47 21	4 47 18	1.2	43.25	715.05	1.28	86205.08	−1.46	86203.62
			11	43.8	46 17	46 40	1 46 28.5	0.6						
Nov. 1 A.M.	29.78	45.59	1	40.5	19 12	19 15	11 19 13.5	1.24	40.5	718.25	1.37	86206.25	−2.62	86203.63
			11	40.5	18 47	19 05	1 18 56	0.62						
„ 2 A.M.	29.87	45.48	1	39.8	07 05	07 12	11 07 08.5	1.18	40.5	718.7	1.22	86206.16	−2.62	86203.54
			11	41.2	06 42	07 09	1 06 55.5	0.58						
„ 3 A.M.	29.90	45.48	1	40.3	13 08	13 11	11 13 09.5	1.24	40.3	718.25	1.37	86206.13	−2.68	86203.45
			11	40.3	12 46	12 59	1 12 52.5	0.62						
Means. . .	29.49								46.73			86203.61		86203.61

RETURN TO ENGLAND.

THE port of Drontheim is situated at the head of a fiord, and is distant above ninety miles from the sea; it was considered desirable, therefore, to proceed to one of the outports, in order to await a favourable opportunity of putting to sea; as the few hours of daylight, which were reduced to seven in the twenty-four, the general prevalence of westerly winds towards the close of the year, and the very dull sailing of the Griper, rendered a sufficient offing an object of importance, and difficult to be obtained. With this intention therefore we weighed from Drontheim on the 13th of November with a head wind, and had succeeded on the 19th in beating down the fiord a distance of sixty miles, when the increasing badness of the weather obliged us to anchor in a small harbour between an island and the south shore of the fiord, and detained us there, against every exertion to proceed, until the 4th of December; on that day the wind sprung up from the E.N.E., and continuing thirty-six hours, carried the Griper about one hundred miles to the westward of the fiord, and thirty from the Stadtland, where the coast trending to the southward enabled her to lay along the land with the wind at west, to which quarter it had again shifted; and on its coming on soon after to blow with great violence, her situation became very critical upon a lee shore; by carrying a press of sail, Captain Clavering succeeded in making his way good along the land, although nearing it, from the 62d to the 58th degrees of latitude, when the line of coast opening to the Baltic relieved us from immediate danger.

This gale which lasted three days, during which period there was no intermission of its violence, was remarkable for the small amount of the effect produced on the barometer, either on its approach, during its continuance, or on its cessation; and by the indications which were afforded of its having originated in a disturbed state of electricity in the atmosphere; it was accompanied by very vivid lightning, which is particularly unusual in high latitudes in winter, and by the frequent appearance, and continuance for several minutes at a time, of balls of fire at the extremities of the yard-arms, and mast-heads; of these not less than eight were counted at one time.

Without further occurrence of note than a continuation of boisterous weather, the Griper made Flamborough-head on the 13th of December, and arrived in due course at Deptford on the 19th; from whence the instruments were landed in London, and deposited in Portland-place.

LONDON.

Before I proceed in the detail of the observations of coincidences in London, it is desirable to state the particulars of an examination which was made of the thermometer that had been used throughout the experiments in registering the temperature of the pendulums, with a view of ascertaining the accuracy or otherwise of its graduation, and of determining the value of the corrections, which might be required at different parts of the scale, to produce the corresponding indications of Fahrenheit's thermometer. In this examination, I had the advantage of the very valuable assistance of my friend Mr. John Frederic Daniell, Fellow of the Royal Society.

The thermometer was made by Mr. Jones of Charing-cross, in 1821, for the purpose of accompanying the pendulum belonging to the Board of Longitude; and as the occasion was one which required a more than usual accuracy, proportionate pains were understood to have been bestowed in its construction; with the second pendulum I ordered and received from Mr. Jones a second thermometer, which in appearance was a duplicate of the first. The scales comprised a range from zero to 150°, and were divided into half degrees, of sufficient size to admit a fair estimation to tenths. On my arrival at Sierra Leone, which was the first opportunity I had had of carefully comparing the thermometers with each other, I had the mortification to find that they differed more than a degree in their indications, at the temperatures which I might expect whilst within the tropics; occasioning an uncertainty in the deduction of the rate of the pendulum, amounting to not less than $\frac{4}{10}$ths of a vibration per diem; being greater, as I had reason to believe, than the sum of the uncertainties due to all other causes whatsover. As I had not the means at that time of referring any part of the scales to a natural standard, nor of comparing them with any other thermometer, in the accuracy of which I could confide, I took the precaution of registering the temperature of the pendulums, on all occasions, by both thermometers, suspending them for that purpose, one on each side of the pendulum, at equal distances from it, and at equal heights; so that if any accident should befall the one, which I could not but anticipate as an event of probable occurrence, I might still retain the means of assigning the true temperatures by the registry of the other, as soon as a favourable opportunity should present itself, of effecting a rigorous examination of its scale.

The particular attention which was paid to the safety of the thermometers, preserved them uninjured to the close of the experiments; as the registry of either, however, is sufficient for the record, I have selected for that purpose the thermometer of which the tube was most equable in its dimensions, and of which the ultimate correct graduation was in consequence attended with the least inconvenience, although it happened to have been the one in which the errors of greatest amount had prevailed. As the scale did not reach higher than 150°, the freezing point of water was the only point which could be verified by a direct reference to a natural standard; the reference was accordingly made, and the graduation at 32° was proved, on several trials, to be exact. In order to ascertain a second determinate point in the scale, it became necessary to compare the thermometer, under circumstances which might ensure a correct comparison, with one of which the scale should admit of verification in two points by a natural standard; for this purpose, Mr. Daniell was so kind as to allow me the opportunity of employing a thermometer in the construction and examination of which he had bestowed much pains; the points of boiling and freezing water had been determined experimentally, and proved by repeated subsequent trials; the accuracy of the intermediate division had been very carefully and minutely scrutinized, by ascertaining that detached portions of mercury occupied, in different parts of the tube, equal spaces as measured on the scale. The two thermometers were placed in boiling ether, with their bulbs on the same level and near the middle of the vessel which contained the fluid; the height of the mercury was then read on their respective scales, at intervals of ten minutes, by Mr. Daniell, Mr. Newman (by whom Mr. Daniell's thermometer had been made), and myself, as follows; P. being the pendulum thermometer, and D. the one belonging to Mr. Daniell; the barometer reduced to 32° stood at 30.368 inches.

	Mr. Daniell.		Mr. Newman.		Capt. Sabine.	
First Reading	P. 99°	D. 97°.6	P. 98°.9	D. 97°.7	P. 98°.8	D. 97°.5
Second ,,	P. 99.5	D. 98.4	P. 99.6	D. 98.4	P. 99.5	D. 98.2
Third ,,	P. 100	D. 98.7	P. 100	D. 98.7	P. 99.9	D. 98.6

Whence the difference between the thermometers, at the part of the scale which was under examination, appeared to be 1°.27; or 98°.27 of the pendulum thermometer corresponded with the 97th degree of Mr. Daniell's thermometer, and consequently with the 97th degree of Fahrenheit's scale.

The space between the points of 32° and 97°, which were thus determined on the scale of the pendulum thermometer, was then divided by an engine into sixty-five equal parts, on one side of the tube, the old division being suffered to remain on the other side; and the new graduation was extended throughout the whole length of the scale. The following table exhibits the comparative indications of the old and new divisions between the degrees of 32 and 97, including the extreme range of temperature during the pendulum experiments; the comparison was made by means of the micrometer screw of the dividing engine:

New Scale.	Old Scale.	New Scale.	Old Scale.	New Scale.	Old Scale.
32°	32°	51°.5	52°	80°	81°
35	35.1	55	55.6	85	86.1
40	40.2	60	60.7	90	90.2
43	43.25	64	64.75	94	95.22
45	45.3	70	70.8	97	98.27
48	48.4	76	76.9		

Now if the points of 32° and 97° were correctly assumed as according with Fahrenheit's scale, the first from experiment and the second from Mr. Daniell's thermometer, and if the tube were every where of equal capacity, then was the new graduation strictly that of Fahrenheit's thermometer; and the preceding table would furnish the corrections for the degrees of the old division, or of that in which the temperature of the pendulums had been recorded in the course of the experiments, into the true degrees of Fahrenheit. To prove, therefore, the equal capacity, a column of mercury was detached, and the tube gauged by Mr. Daniell and myself as follows; the degrees by which the length of the column in different parts of the tube was measured, being those of the new division:—

Mr. Daniell.			Capt. Sabine.		
Upper end of the Column.	Lower end of the Column.	Length of the Column.	Upper end of the Column.	Lower end of the Column.	Length of the Column.
°	°	°	°	°	°
99.23	64.25	34.98	101.15	66.23	34.92
94.28	59.3	34.98	96.5	61.28	35.02
89.3	54.4	34.9	91.3	56.4	34.9
84.4	49.55	34.85	86.37	51.5	34.87
79.6	44.7	34.9	81.6	46.65	34.95
74.8	39.8	35	76.7	41.75	34.95
69.9	34.9	35	71.85	36.85	35
65	30	35	66.98	32	34.98
Mean		34.93	Mean		34.95

Whence it appeared that the length of the detached column of mercury, thus measured on a scale of equal divisions, was so nearly the same in all parts of the tube, as to afford a satisfactory evidence, that the capacity of the tube was sufficiently equable in the space included between 32° and 97° to justify the intermediate graduation into 65 equal divisions.

The want of a standard thermometer, verified at all points of its scale by competent authority, having been felt in many other instances as well as in the present, for purposes of reference, the superintendence of the construction of such a thermometer has been undertaken by Mr. Daniell and Captain Kater, at the instance of a committee of the Royal Society: as soon as it shall have been completed, it is designed to obtain an additional proof of the correctness of the 97th degree of the new division of the scale of the pendulum thermometer (which rests at present upon the presumed accuracy of Mr. Daniell's thermometer), by comparing its indication with that of the standard during their immersion in boiling ether. As the construction of a standard thermometer was undertaken early in the present year (1824), it is hoped that its completion may be accomplished before the publication of these experiments, so that a notice of the result of the comparison may be appended at the close of the volume; but as it is confidently anticipated, from the habitual accuracy of Mr. Daniell, and from his justly high authority in the construction of meteorological instruments, that no difference, deserving of regard, will be found between the thermometer on which he has already bestowed much pains, and the one which he has undertaken to superintend, the comparisons in the table in page 184 have been employed, in reducing the temperature of the pendulum during the observation of coincidences at the several stations as registered by Mr. Jones's thermometer, to the corresponding degrees which would have been shewn by a correctly-graduated Fahrenheit's thermometer.

EXPERIMENTS IN 1821 AND 1822.

It has already been noticed that, previously to the employment of the pendulums at the stations adjoining the equator, their rates had been obtained in London. The time which intervened between the date of the order procured from the Board of Longitude for the construction of the instruments and their embarkation, being not more than was necessary for their preparation, it might have been very doubtful, whether so desirable a measure, as a preliminary trial of the pendulums in London, could have been accomplished, had I been obliged to await the completion of their own apparatus; fortunately, the provision which had been made for Captain Kater's experiments with the pendulum with convertible axes, had not been removed from Portland-place; as soon, therefore, as the pendulums themselves could be got ready, I availed myself of Capt. Kater's permission to employ the agate planes belonging to his pendulum, which still remained upon the support; and with the assistance of one of Mr. Browne's clocks, of which he was kind enough to supply the rate, I made the observations contained in the following Tables, I. and II. The thermometer used to register the temperature of the pendulums was one which had been employed by Captain Kater in his experiments, the degrees of which were true degrees of Fahrenheit's scale. The corrections for buoyancy, which express the value in vibrations per diem of the retardation of the pendulums from their oscillating in a medium of variable resistance instead of in a vacuum, have been computed on the data, that the specific gravity of the pendulums is 8.6; that water is 836 times heavier than air, when the thermometer is at 53°, and the barometer, of which the temperature of the mercury is also 53°, is at 29.27 inches; and that in observations which may be made in other states of the barometer and thermometer, the number 836 will vary inversely as the height of the barometer, and directly, its $\frac{1}{480}$th part, for each degree of Fahrenheit that the thermometer differs from 53°.

TABLE I. LONDON, 1821.——COINCIDENCES OBSERVED with PENDULUM No. 3.

DATE.	Barometer.	Clock making per Diem.	No of Coincidence.	Temperature.	Time of Disappearance.	Time of Re-appearance.	True Time of Coincidence.	Arc of Vibration.	Mean Temperature.	Mean Interval.	Correction for the Arc.	Vibrations in 24 hours.	Reduction to a mean Temperature.	Reduced Vibrations 65°.66.
	IN.	S.		°	M. S.	M. S.	H. M. S.	°	°	S.	S.			
July 6 P.M. (Observed by Mr. Browne.)	29.80	Arnold. 86399.44	1	66.2	27 52	27 59	2 27 55.5	1.22	66.35	676.2	+ 1.35	86145.23	+0.29	86145.5
			11	66.5	20 25	20 50	4 20 37.5	0.64						
„ 7 A.M.	29.86	86399.44	1	65	25 38	25 43	11 25 40.5	1.26	65.65	676.65	1.46	86145.52	. . .	86145.5
			11	66.3	18 22	18 32	1 18 27	0.66						
„ 8 A.M.	30.04	86399.44	1	64.9	7 23	7 27	11 7 25	1.37	65.25	677.1	1.71	86145.95	−0.17	86145.7
			11	65.6	00 11	00 21	1 00 16	0.71						
„ 13 A.M.	30.06	86399.44	1	65.2	00 15	00 22	11 00 18.5	1.3	65.4	677	1.55	86145.75	−0.10	86145.6
			11	65.6	53 00	53 17	12 53 08.5	0.68						
Means . . .	29.94								65.66			86145.61		86145.6

The correction for Buoyancy is + 5.96, making 86151.57 Vibrations in vacuo, at the temperature of 65.66 Fahrenhei

TABLE II. LONDON, 1821.——COINCIDENCES OBSERVED with PENDULUM No. 4.

DATE.	Barometer.	Clock making per Diem.	No. of Coincidence.	Temperature.	Time of Disappearance.	Time of Re-appearance.	True Time of Coincidence.	Arc of Vibration.	Mean Temperature.	Mean Interval.	Correction for the Arc.	Vibrations in 24 hours.	Reduction to a Mean Temperature.	Reduced Vibrations 68°.13.
	IN.	S.		°	M. S.	M. S.	H. M. S.	°	°	S.	S.			
Sep. 17 P.M.	30.10	Arnold. 86400.3	1	69	15 27	15 33	1 15 30	1.36	69	701.35	+ 1.64	86155.52	+0.37	86155.89
			11	69	12 17	12 30	3 12 23.5	0.68						
„ 18 P.M.	29.90	86400.4	1	69	28 45	28 50	2 28 47.5	1.38	69.4	699.8	1.74	86155.22	+0.53	86155.75
			11	69.8	25 18	25 33	4 25 25.5	0.71						
„ 19 A.M.	29.90	86400.6	1	66.7	01 34	01 40	10 01 37	1.34	67.25	702.55	1.57	86156.21	−0.37	86155.84
			11	67.8	57 35	57 50	11 57 42.5	0.66						
„ 20 P.M.	29.80	86400.7	1	67.3	37 27	37 33	2 37 30	1.36	67.5	701.6	1.64	86156.04	−0.26	86155.78
			11	67.7	34 18	34 34	4 34 26	0.68						
„ 21 A.M.	29.70	86400.7	1	67 2	7 39	7 45	9 07 42	1.36	67.5	702.05	1.64	86156.22	−0.26	86155.96
			11	67.8	4 35	4 49	11 04 42	0.68						
Means . . .	29.88								68.13			86155.84		86155.84

The correction for Buoyancy is +5.94, making 86161.78 Vibrations in vacuo, at the temperature of 68.13 Fahrenhei

As the course of the experiments proceeded at Sierra Leone and at the equatorial stations, I had occasion to remark, that whilst the difference in the rate of the two pendulums in corresponding circumstances, or the excess in the number of vibrations per diem of the one pendulum over that of the other, due to their actual difference in length, was constant at the several stations adjoining the equator, or as nearly so as the nature of the observations would authorize an expectation, its amount deducible from the experiments in London contained in the preceding Tables, appeared a much wider departure from the subsequent experience, than could be attributed to error in the observation of coincidences: this remark will be best illustrated by the following collected view of the respective differences at the stations visited in the voyage of 1822.

Stations.	Difference in the number of Vibrations, per diem, of the two Pendulums.	
St. Thomas . .	9.69	9.68 Mean.
Maranham . .	9.39	
Ascension . .	9.51	
Sierra Leone .	9.74	
Trinidad. . .	10.00	
Bahia. . . .	9.90	
Jamaica . . .	9.60	
New York . .	9.59	
London . . .	11.25	

Two modes suggested themselves, whilst the equatorial stations were in progress, of accounting plausibly for the difference which thus appeared between London and the other stations: the pendulums had been returned to Mr. Jones subsequently to the experiments in London, to be fitted with cases, and packed for embarkation; and it was possible that an accident might have befallen one of them in the course of those operations, and have occasioned a slight alteration in its length, equivalent to between one and two vibrations per diem; or I might have transcribed erroneously the sign prefixed to the rate of the clock during the coincidences with No. 3, in Mr. Browne's memorandum which I had left in England; as a gaining rate of 0.56 parts of a second, instead of a losing rate of the same amount, would have reduced the apparent difference within the reasonable limit of errors of observation. Deeming the latter supposition the more probable, I wrote to Mr. Browne to request him to refer to the rate of the clock at the period in question; but on my return to England in January, 1823, I received a fresh memorandum from him, by which I perceived that I had not been mistaken in the original transcription, either in the quality or in the amount of the rate. I then proceeded to repeat the trial of the pendulums in London, expecting to discover by the results compared with those of 1821, in which of the pendulums an alteration had taken place. Captain Kater's agate planes were still on the pendulum support, in Portland-place; the screws, by which the box containing the planes belonging to my own pendulums was fastened on its support, did not correspond with the holes which had been made to receive the screws of Captain Kater's planes in the mahogany plank described in the *Phil. Trans.* for 1819, Part III., p. 41; and Mr. Browne was unwilling that the strength of the plank should be impaired by fresh perforations. I was induced, therefore, to employ Captain Kater's planes a second time instead of my own, and to make no other difference from the proceedings of 1821, than by the substitution of my own thermometer for Captain Kater's. Mr. Browne was again kind enough to permit me to employ one of his clocks for the coincidences detailed in Tables III. and IV., and to supply its rate.

Table III. London, 1823.—COINCIDENCES OBSERVED with PENDULUM No. 3.

DATE.	Barometer.	Clock making per Diem.	No. of Coincidence.	Temperature.	Time of Disappearance.	Time of Re-appearance.	True Time of Coincidence.	Arc of Vibration.	Mean Temperature.	Mean Interval.	Correction for the Arc.	Vibrations in 24 hours.	Reduction to a mean Temperature.	Reduced Vibrations at 49°.1.
	IN.	S.		°	M. S.	M. S.	H. M. S.	°	°	S.	S. +			
Feb. 28 A.M.	29.45	Bolton. 86400.43	1	49.1	16 32	16 36	11 16 34	1.25	49.5	693.25	1.42	86152.56	+0.17	86152.73
			11	49.9	12 01	12 12	1 12 06.5	0.64						
„ 28 P.M.	29.45	86400.43	1	49	32 32	32 38	2 32 35	1.2	49.1	694	1.27	86152.73	. . .	86152.73
			11	49.2	28 09	28 21	4 28 15	0.6						
Mar. 1 A.M.	29.90	86400.74	1	48.4	34 52	34 57	11 34 54.5	1.2	49	693.75	1.27	86152.94	−0.04	86152.90
			11	49.6	30 25	30 39	1 30 32	0.6						
„ 1 P.M	29.93	86400.74	1	49.6	39 09	39 14	2 39 11.5	1.2	49.3	693.85	1.27	86152.97	+0.08	86153.05
			11	49	34 42	34 58	4 34 50	0.6						
„ 2 A.M.	30.00	86400.25	1	48.1	41 04	41 09	11 41 06.5	1.25	48.6	694.4	1.4	86152.80	−0.21	86152.59
			11	49.1	36 42	36 59	1 36 50.5	0.64						
„ 2 P.M.	29.95	86400.25	1	49.1	16 41	16 46	2 16 43.5	1.2	49.1	694.2	1.27	86152.60	. . .	86152.60
			11	49.1	12 20	12 31	4 12 25.5	0.6						
Means . .	29.78								49.1			86152.77		86152.77

The correction for Buoyancy is +6.15, making 86158.92 Vibrations in Vacuo; and 49.°1 of the scale of the registering Thermometer is equivalent to 48°.67 of Fahrenheit, being the temperature of the Pendulum.

Table IV. London, 1823.—COINCIDENCES OBSERVED with PENDULUM No. 4.

DATE.	Barometer.	Clock making per Diem.	No. of Coincidence.	Temperature.	Time of Disappearance.	Time of Re-appearance.	True Time of Coincidence.	Arc of Vibration.	Mean Temperature.	Mean Interval.	Correction of the Arc.	Vibrations in 24 hours.	Reduction to a mean Temperature.	Reduced Vibrations at 52°.39.
	IN.	S.		°	M. S.	M. S.	H. M. S.	°	°	S.	S. +			
Mar. 4 A.M.	29.30	Bolton. 86400.47	1	53	55 06	55 11	10 55 08.5	1.18	52.9	722.25	1.25	86162.51	+0.21	86162.72
			11	52.8	55 25	55 37	12 55 31	0.6						
„ 4 P.M.	29.35	86400.47	1	52.8	19 24	19 31	1 19 27.5	1.24	52.9	722.25	1.33	86162.59	+0.21	86162.80
			11	53	19 44	19 56	3 19 50	0.6						
„ 5 A.M.	29.45	86400.39	1	51.6	45 16	45 25	11 45 20.5	1.18	51.8	723.8	1.25	86162.89	−0.25	86162.64
			11	52	45 53	46 04	1 45 58.5	0.6						
„ 5 P.M.	29.50	86400.39	1	52	58 31	58 37	1 58 34	1.2	52.35	723.1	1.28	86162.69	. . .	86162.69
			11	52.7	58 56	59 14	3 59 05	0.6						
„ 6 A.M.	29.75	86400.29	1	52	52 11	52 17	11 52 14	1.24	52	723.5	1.36	86162.79	−0.16	86162.63
			11	52	52 40	52 58	1 52 49	0.62						
Means . .	29.47								52.39			86162.69		86162.69

The correction for Buoyancy is +6.02, making 86168.71 Vibrations in Vacuo; and 52°.39 of the scale of the registering Thermometer is equivalent to 51°.88 of Fahrenheit, being the temperature of the Pendulum.

LONDON.——COMPARISON of the VIBRATIONS of the PENDULUMS in 1821 and 1823.

	Date.	Vibrations per Diem.	Tempe-rature.	Reduction to a Mean Tempera-ture of 62°*	Reduced Vibrations at 62°.		Excess of Vibrations of Pendu-lum 4.
			°	s.			
Pendulum 3..	1821	86151.57	65.66	+1.54	86153.11	86153.21	11.20
	1823	86158.92	48.67	−5.61	86153.31		
Pendulum 4..	1821	86161.78	68.13	+2.58	86164.36	86164.41	
	1823	86168.71	51.88	−4.26	86164.45		

* The Reduction is in the proportion of 0,421 parts of a second per diem for each degree.

It is shewn by this comparison, that no alteration whatsoever had taken place, either in the absolute or in the relative length of the pendulums since their first construction; and in so far as the experiments just recorded afforded an evidence of this very important fact, their results were highly satisfactory. The reason still, however, remained to be inquired into, of the apparent difference in the length of the pendulums with relation to each other, in London and elsewhere, in which error of some kind was obviously involved, since it had been ascertained that no real difference had existed. As the employment of Captain Kater's agate planes was a departure from the strict correspondence of the proceedings in London with those at the other stations, it became the first object of suspicion; I caused, therefore, the box containing the planes belonging to the pendulum to be fitted with screws to suit the holes already existing in the mahogany plank, and having substituted them for Capt. Kater's on the support, the coincidences of the succeeding Tables V. and VI. were observed.

TABLE V. LONDON, 1823.——COINCIDENCES OBSERVED with PENDULUM No. 3.

DATE.	Barometer.	Clock making per Diem.	No. of Coincidence.	Temperature.	Time of Disappearance.	Time of Re-appearance.	True Time of Coincidence.	Arc of Vibration.	Mean Temperature.	Mean Interval.	Correction for the Arc.	Vibrations in 24 hours.	Reduction to a mean Temperature.	Reduced Vibrations at 53°.52.
	IN.	S.		°	M. S.	M. S.	H. M. S.	°	°	S.	S.			
March 19 A.M.	29.75	Bolton. 86400.6	1	53.5	24 57	25 02	11 24 59.5	1.28	53.6	691.6	+ 1.45	86152.21	+0.03	86152.24
			11	53.7	20 10	20 21	1 20 15.5	0.64						
„ 19 P.M.	29.75	86400.6	1	53.7	53 51	53 58	2 53 54.5	1.28	53.6	691.6	1.45	86152.21	+0.03	86152.24
			11	53.5	49 03	49 18	4 49 10.5	0.64						
„ 20 A.M.	29.60	86400.65	1	51.2	17 23	17 28	11 17 25.5	1 28	51.05	694.5	1.45	86153.30	−1.04	86152.26
			11	50.9	13 06	13 15	1 13 10.5	0.64						
„ 20 P.M.	29.60	86400.65	1	51	30 33	30 41	2 30 37	1.26	51	695.15	1.41	86153.47	−1.06	86152.41
			11	51	26 23	26 34	4 26 28.5	0.64						
„ 21 P.M.	29.25	86400.85	1	55	18 16	18 22	12 18 19	1.3	54.85	690.4	1.54	86152.10	+0.56	86152.66
			11	54.7	13 14	13 32	2 13 23	0.66						
„ 22 A.M.	29.15	86400.65	1	54.4	52 09	52 14	10 52 11.5	1.3	54.5	691.35	1.54	86152.24	+0.41	86152.65
			11	54.6	47 20	47 30	12 47 25	0.66						
„ 22 P.M.	29.15	86400.65	1	54.7	58 01	58 06	1 58 03.5	1.22	54.8	691.65	1.33	86152.13	+0.54	86152.67
			11	54.9	53 14	53 26	3 53 20	0.62						
„ 24 P.M.	30.20	86400.1	1	55	56 32	56 38	1 56 35	1.2	54.75	692.35	1.28	86151.78	+0.52	86152.30
			11	54.5	51 52	52 05	3 51 58.5	0.6						
„ 25 A.M.	30.30	86400.2	1	53	23 01	23 09	11 23 05	1.22	53.5	693.3	1.31	86152.27	. . .	86152.27
			11	54	18 31	18 45	1 18 38	0.6						
Means .	29.64								53.52			86152.41		86152.41

The correction for Buoyancy is + 6.06, making 86158.47 Vibrations in Vacuo; and 53°.52 of the scale of the registering Thermometer is equivalent to 52°.97 of Fahrenheit's scale, being the temperature of the Pendulum.

Table VI. London, 1823.——COINCIDENCES OBSERVED with PENDULUM No. 4.

DATE.	Barometer.	Clock making per Diem.	No. of Coincidence.	Temperature.	Time of Disappearance.	Time of Re-appearance.	True Time of Coincidence.	Arc of Vibration.	Mean Temperature.	Mean Interval.	Correction for the Arc.	Vibrations in 24 hours.	Reduction to a Mean Temperature.	Reduced Vibrations at 54°.98
	IN.	s.		°	M. S.	M. S.	H. M. S.	°	°	s.	°			
Mar. 23 A.M.	29.65	Bolton. 86400.3	1	54.5	50 32	50 39	11 50 35.5	1.24	54.75	719.4	+ 1.36	86161.46	−0.10	86161.36
			11	55	50 18	50 41	1 50 29.5	0.62						
„ 23 P.M.	29.65	86400.3	1	55	13 43	13 51	2 13 47	1.22	54.95	720.05	1.31	86161.61	. . .	86161.61
			11	54.9	13 40	13 55	4 13 47.5	0.6						
April 5 P.M.	29.07	86399.85	1	53.3	21 00	21 06	1 21 03	1.28	53.2	723	1.45	86162.31	−0.75	86161.56
			11	53.1	21 28	21 38	3 21 33	0.64						
„ 6 P.M.	29.25	86399.85	1	57	35 43	35 50	12 35 46.5	1.3	57	718.4	1.50	86160.84	+0.85	86161.69
			11	57	35 25	35 36	2 35 30.5	0.64						
Means . . .	29.40								54.98			86161.55		86161.55

The correction for Buoyancy is +6.02, making 86167.57 Vibrations in Vacuo; and 54°.98 of the scale of the registering Thermometer is equivalent to 54°.4 of Fahrenheit, being the temperature of the Pendulum.

DIFFERENCE in the VIBRATION of the PENDULUMS, on Captain Kater's Planes, and on their own.

	PLANES.	Vibrations per Diem.	Temperature.	Reduction to a Mean Temperature of 62° *.	Reduced Vibrations at 62°. Faht.	Difference in the Vibrations of each Pendulum, on Capt. Kater's Planes and on its own.
			°			
Pendulum 3.	Captain Kater's.	86153.21	62	. . .	86153.21	1.46
	Its own.	86158.47	52.97	−3.80	86154.67	
Pendulum 4.	Captain Kater's.	86164.41	62	. .	86164.41	0.04
	Its own.	86167.57	54.4	−3.20	86164.37	

* The Reduction is in the proportion of 0.421 s. for each degree of Fahrenheit.

Difference in the number of the Vibrations of the two Pendulums on their own Planes, in London, and at the Southern Stations.			
Station.	Pendulum 3.	Pendulum 4.	Excess of Vibrations of Pendulum 4.
London . . .	86154.68	86164.38	9.70
Southern Stations, Page 189.			9.68

The cause of the want of correspondence in the rate of the one pendulum compared with that of the other, in London and at the Southern Stations, had thus been traced to the accidental employment of Captain Kater's planes of suspension in the experiments in London of 1821, and in the first series in 1823. The most careful examination and comparison of the two sets of planes, made after the difference was known which their respective employment produced, failed in discovering its occasion; but it may well be conceived, that inequalities or irregularities of various kinds might exist, either in the planes, or in the knife-edges, or in both, which might become sensible in the application of so delicate a test as the vibration of a pendulum, though they might not be perceptible by other means. It is remarkable that one pendulum should have been thus affected by the change of planes, whilst the other was not so; that the rate of No. 4 should have been uninfluenced, whilst that of No. 3 varied so much as its 59-thousandth part: the knife-edges of the two pendulums were precisely of similar dimensions, and the Y's, by which they were lowered on the planes previously to oscillation, must have deposited them at all times as nearly as possible on the same points of bearing. It may not be superfluous to add, that the horizontal adjustment of the planes was in every instance most carefully attended to, and examined occasionally by Mr. Browne, as well as frequently by myself. Fortunately, it is of the effect only, and not of the cause, that it is important to the experiments to be assured; and whilst the effect is placed beyond question by repeated experiment, it may readily be conceded that the cause may be too minute to admit of a satisfactory investigation.

Were an illustration wanting of the importance in experiments of this nature, of maintaining the strictest correspondence in the proceedings at the different stations, even in the most minute and apparently inconsequential particulars, this instance of the effect of the change of planes upon the vibration of pendulum 3 affords a strong one. It is probable that such instances might be of rare occurrence; that the rates of pendulums generally, as that of No. 4, would be the same on different planes, supposing their construction to be similar, and the adjustments properly regarded; but in the evidence which is here presented of an alteration being produced in a single instance, that condition of the experiment, which requires the adoption of every precaution conducive to the utmost attainable accuracy, is not fulfilled, unless the same planes are used in all the experiments which are designed to be comparative.

I must not omit to notice, that the knowledge of the existence of error in the earlier experiments with No. 3 in London, is one of the incidental advantages derived from the employment of two pendulums; had I been furnished only with the pendulum belonging to the Board of Longitude, I might not have been led to suspect the inaccuracy arising from the use of Captain Kater's planes; and I should thus have assigned, from correct experiments elsewhere, an erroneous value to the length of the seconds' pendulum at every one of the other stations which I visited.

In the following Table, No. VII., the particulars are arranged which I received from Mr. Browne, relative to the mode in which the rate of the clock was deduced, with which the pendulums were compared in the observation of coincidences contained in Tables III., IV., V., and VI. The clock, which was made by the late Mr. George Bolton, was compared every day at 12 P.M. with two other clocks, which were regarded as standards of comparison: one of these was the time-piece by Cumming, noticed in the Phil. Trans. for 1819, part III. p. 41; the other was a clock recently made by Molyneux on the same principle as Cumming's, and in Mr. Browne's estimation, is not inferior to it in performance. The standard clocks were regularly compared with astronomical time; the dependence placed on them being that of keeping an uniform rate from one transit observation to another. The rate of Bolton, entered in Table VII., is on mean time, as separately inferred from the comparison with each of the standard clocks.

TABLE VII.

1823.	Rate of Bolton deduced from		Mean.	1823.	Rate of Bolton deduced from		Mean.
	Cumming	Molyneux			Cumming	Molyneux	
12 P.M.	Gaining. s.	Gaining. s.	Gaining. s.	12 P.M.	Gaining. s.	Gaining. s.	Gaining. s.
Feb. 27 – ,, 28	0.43	0.43	0.43	March 14 – ,, 15	0.5	0.4	0.45
,, 28 – March 1	0.75	0.73	0.74	,, 15 – ,, 16	0.6	0.4	0.5
March 1 – ,, 2	0.3	0.2	0.25	,, 16 – ,, 17	0.5	0.5	0.5
,, 2 – ,, 3	0.45	0.6	0.52	,, 17 – ,, 18			
,, 3 – ,, 4	0.55	0.4	0.47	,, 18 – ,, 19	0.6	0.6	0.6
,, 4 – ,, 5	0.38	0.4	0.39	,, 19 – ,, 20	0.7	0.6	0.65
,, 5 – ,, 6	0.38	0.2	0.29	,, 20 – ,, 21	0.8	0.9	0.85
,, 6 – ,, 7	0.6	0.6	0.6	,, 21 – ,, 22	0.5	0.8	0.65
,, 7 – ,, 8	0.8	0.8	0.8	,, 22 – ,, 23	0.3	0.3	0.3
,, 8 – ,, 9	0.8	0.8	0.8	,, 23 – ,, 24	0.1	0.1	0.1
,, 9 – ,, 10	0.3	0.2	0.25	,, 24 – ,, 25	0.2	0.2	0.2
,, 10 – ,, 11	0.3	0.2	0.25	,, 25 – ,, 26	0.2	0.1	0.15
,, 11 – ,, 12	0.00	0.00	0.00		Losing.	Losing.	Losing.
,, 12 – ,, 13	0.35	0.35	0.35	April 4 – ,, 5	0.15	0.15	0.15
,, 13 – ,, 14	0.00	0.00	0.00	,, 5 – ,, 6	0.15	0.15	0.15

EXPERIMENTS IN 1824.

The experiments in London in the spring of 1823, the particulars of which have been just related, were made when the pendulums had been landed from the Pheasant on their return from the equatorial stations, and before their embarkation in the Griper, for the Arctic Circle. On the return of the Griper in the winter of 1823-1824, it became necessary to repeat the trial of the pendulums a third time in London, for the purpose of shewing that the attention given to their safe preservation had been as effectual in the second voyage, as it had been found to have been in the preceding one. There was also a second purpose, essential to the strict comparison of the experiments at the several stations with each other, which remained to be accomplished on the final return of the pendulums to England. Embracing climates so widely dissimilar, the range of temperature, at which the various results had been obtained, exceeded fifty degrees; it became, therefore, an object of primary importance, to determine experimentally, and with the utmost exactness of which the experiment should be capable, the expansion of the pendulums corresponding to the measures of heat. Two methods of proceeding in the attainment of this object presented themselves; one, by immersing the pendulums successively in fluids of different temperatures, and measuring their intermediate expansion, by means of a microscopical apparatus, which Captain Kater had devised on a similar occasion, and of which he was so kind as to offer me the use, as well as his own most valuable assistance in the operation; and the other, by ascertaining the effect of the expansion on the rate of the pendulums, when vibrating in temperatures, of which the difference should equal the extreme range which had occurred in the course of the experiments. The latter method bore the more immediate relation to the purpose for which the expansion was required; but as I was more aware of the difficulties which would oppose a sufficiently precise determination of the rate and temperature during the vibration, than of those which are attendant upon exact microscopical measurement, I should have preferred the adoption of the former method, in reliance on the skilfulness and experience of Captain Kater, had I not possessed advantages, through the kindness of Mr. Browne, in the use of his most excellent clocks, and in his very accurate determination of their rates, which encouraged me to an attempt, wherein I could otherwise have scarcely hoped to have succeeded; and I was further induced by the consideration, that if I should fail in determining with sufficient exactness the alteration of rate due to differences of temperature, I should at least obtain a rate of the pendulums in London, which would compare with the results in 1821 and 1823.

On my arrival in Portland-place in December, 1823, I found a clock of Mr. Browne's, made by Arnold, being the same which Captain Kater had used at the stations of the trigonometrical survey, in occupation of the recess beneath the pendulum support, and keeping, as Mr. Browne informed me, a tolerably good mean rate. Being anxious to take advantage of the cold weather which then prevailed, and which was the coldest of the season, in order to obtain the rate of the pendulums at the lowest temperature which natural circumstances would enable, I determined to proceed immediately in the observation of coincidences with the clock which was already stationed; hoping that by comparing it very carefully with Cumming and Molyneux at short intervals, any deviation which might take place from its mean rate might be detected and allowed for. By keeping one of the windows of the clock-room constantly open, and the shutters closed, and by discontinuing a fire in the adjoining room, the temperature was lowered to little more than that of the external atmosphere, and its fluctuations were reduced within very small limits. As the observations proceeded, however, I had the mortification to perceive, that the rate of the clock varied from hour to hour, so much, and so continually, as to make it doubtful whether the vibrations of the pendulums could be deduced from it with sufficient accuracy; the irregularity was shewn by the discordances in the partial results with the detached pendulums, which form the severest test to which the uniformity of a clock's going can be subjected, because they detect variation in smaller intervals, than those in which it can be discovered by the comparison with other clocks. I persevered, however, until ten results with each pendulum had been obtained; but finding on examination that they contained differences with each other, amounting to a whole vibration per diem, I was induced to reject them altogether, and to undertake a fresh series with Cumming, which Mr. Browne consented to remove into the recess for that purpose,

A further short delay took place, in furnishing the mercurial pendulum of the clock with a small plate bearing a disk of the same diameter with those on the pendulums of the other clocks; the disk was of silver, and was contrasted as usual by a coating of black varnish on the plate. The door of the clock-case happening to be larger than those of the other clocks of Mr. Browne's which had been used in the recess, the frame of wood to which Captain Kater's arc, measuring the extent of the vibration of the detached pendulums, was fixed, was too small to fit into the opening of the door-way. I had hitherto always used Captain Kater's arc in London, because the frame of the arc belonging to the Board of Longitude had been made to fit the clock which I had employed at the other stations, and was much too large for Mr. Browne's clocks; on this occasion, however, I had the frame reduced to fit the door of Cumming, and happening to compare the arcs when thus accidentally brought together, I perceived that I had too confidently presumed their radius to be the same, in consequence of their having been made for pendulums of equal length. The expansion of the pendulums, however, requiring all my attention, I postponed for the time the inquiry into which was in error, and into the exact amount of their difference.

From the time which necessarily elapsed in these previous arrangements, it was not until late in March that I was able to resume the observation of the coincidences, the particulars of which are contained in Tables VIII. and IX.

TABLE VIII. LONDON, 1824.——COINCIDENC

DATE.	Barometer.	Vibrations of Cumming per Diem.	No. of Coincidence.	Temperature.	Time of Disappearance.	Time of Re-appearance.	True Time of Coincidence.	Arc of Vibration.	Mean Temperature.	Mean Interval.	Correction for the Arc.	Vibrations in 24 hours.	Reduction to a mean Temperature.	Reduced Vibrations at 44°.66.
1824.	IN.	s.		°	M. S.	M. S.	H. M. S.	°	°	s.	s.			
April 1 A.M.	29.88	86401.1	1	41.7	53 16	53 17	11 53 16.5	1.31	42.35	704.15	+ 1.51	86157.21	−0.97	86156.24
			11	43	50 37	50 39	1 50 38	0.65						
,, 1 P.M.	29.80	86401.1	1	43.2	8 22	8 24	2 8 23	1.38	43.6	702.35	1.71	86156.81	−0.45	86156.36
			11	44	5 26	5 27	4 5 26.5	0.7						
,, 2 A.M.	29.15	86401.1	1	45	21 34	21 39	10 21 36.5	1.2	45.1	701.35	1.28	86156.02	+0.18	86156.20
			11	45.2	18 26	18 34	12 18 30	0.6						
,, 2 P.M.	29.30	86401.1	1	45.2	32 06	32 10	1 32 08	1.24	44.5	701.3	1.36	86156.10	+0.07	86156.17
			11	43.8	28 59	29 03	3 29 01	0.62						
,, 3 A.M.	30.05	86400.8	1	42.5	19 42	19 44	9 19 43	1.36	43.4	702.7	1.64	86156.52	−0.53	86155.99
			11	44.3	16 48	16 52	11 16 50	0.68						
P.M.	30.08	86400.8	1	44.7	32 52	32 55	1 32 53.5	1.4	44.9	700.75	1.74	86155.96	+0.10	86156.06
			11	45.1	29 39	29 43	3 29 41	0.7						
,, 4 A.M.	30.26	86400.8	1	44.8	03 37	03 40	10 03 38.5	1.24	45.4	701.45	1.36	86155.84	+0.31	86156.15
			11	46	00 32	00 34	12 00 33	0.62						
,, 4 P.M.	30.26	86400.8	1	46	08 41	08 44	2 08 42.5	1.37	46.5	699.7	1.62	86155.46	+0.77	86156.23
			11	47	05 16	05 23	4 05 19.5	0.67						
,, 5 A.M.	30.44	86400.8	1	44.1	07 51	07 53	9 07 52	1.3	44.9	701.5	1.5	86155.98	+0.10	86156.08
			11	45.7	04 46	04 48	11 04 47	0.64						
,, 5 P.M.	30.44	86400.8	1	45.7	11 11	11 13	2 11 12	1.37	45.95	700.15	1.62	86155.62	+0.54	86156.16
			11	46.2	07 52	07 55	4 07 53.5	0.67						
Means. . .	29.97								44.66			86156.15		86156.15

The correction for Buoyancy is +6.25, making 86162.40 Vibrations in vacuo; and 44°.66 of the scale of the registering Thermometer is equivalent to 44°.38 of Fahrenheit, being the temperature of the Pendulum.

TABLE IX. LONDON, 1824.——COINCIDENCES OBSERVED with PENDULUM No. 4; the Clock *(Cumming)* making 86401.3 Vibrations in a Mean Solar Day.

DATE.	Barometer.	No. of Coincidence.	Temperature.	Time of Disappearance.	Time of Re-appearance.	True Time of Coincidence.	Arc of Vibration.	Mean Temperature.	Mean Interval.	Correction for the Arc.	Vibrations in 24 hours.	Reduction to a Mean Temperature.	Reduced Vibrations at 47°.61.
1823.	IN.		°	M. S.	M. S.	H. M. S.	S.	°	S.	S.			
Mar. 24 A.M.	29.95	1	51	43 09	43 14	9 43 11.5	1.32	50.8	721.45	+ 1.55	86163.35	+1.34	86164.69
		11	50.6	43 24	43 28	11 43 26	0.66						
„ 24 P.M.	29.96	1	50.2	58 24	58 28	1 58 26	1.35	50	722.8	1.59	86163.84	+1.00	86164.84
		11	49.8	58 50	58 58	3 58 54	0.66						
„ 25 A.M.	30.09	1	48.2	14 46	14 48	10 14 47	1.35	48.35	724.8	1.59	86164.49	+0.30	86164.79
		11	48.5	15 33	15 37	12 15 35	0.66						
„ 25 P.M.	30.06	1	48.5	23 39	23 43	2 23 41	1.36	47.95	724.8	1.60	86164.50	+0.14	86164.64
		11	47.4	24 26	24 32	4 24 29	0.66						
„ 26 A.M.	30.00	1	46.9	33 46	33 50	9 33 48	1.32	47.4	725.85	1.51	86164.75	−0.09	86164.66
		11	47.9	34 44	34 49	11 34 46.5	0.64						
„ 26 P.M.	30.00	1	48	56 14	56 17	1 56 15.5	1.32	48.1	725.6	1.51	86164.66	+0.20	86164.86
		11	48.2	57 10	57 13	3 57 11.5	0.64						
„ 27 P.M.	29.83	1	46	35 28	35 32	9 35 30	1.36	46.5	727.2	1.60	86165.28	−0.47	86164.81
		11	47	36 40	36 44	11 36 42	0.66						
„ 28 A.M.	29.88	1	44	46 50	46 53	8 46 51.5	1.38	44.5	728.95	1.67	86165.93	−1.31	86164.62
		11	45	48 18	48 24	10 48 21	0.68						
„ 28 P.M.	29.88	1	45	55 39	55 43	1 55 41	1 32	44.9	729.05	1.51	86165.81	−1.14	86164.67
		11	44.8	57 08	57 15	3 57 11.5	0.64						
Means . . .	29.96							47.61			86164.73		86164.73

The correction for Buoyancy is +6.2, making 86170.93 Vibrations in vacuo; and 47° 61 of the scale of the registering Thermometer is equivalent to 47°.24 of Fahrenheit, being the temperature of the Pendulum.

The next procedure was to raise the temperature of the clock-room by artificial means, and to keep it steadily at an height which should exceed 80 degrees, the mean heat in the neighbourhood of the equator; for that purpose a stove was placed in the apartment beneath, and the pipe brought up through the floor into a part of the room most distant from the pendulum recess; the pipe was then bent in a right angle, about a foot and a half above the floor, and carried across the room into a hole in the chimney-board made to receive it; a skreen of gauze was spread horizontally a few inches above the pipe, to prevent the immediate ascent of the heated air, and to diffuse it more extensively in the lower stratum; the windows and shutters were closed, excepting when a part of one of the shutters was opened to admit the light required in the observations; the temperature of the adjoining room was raised by fires to between 70 and 80 degrees, so that when the door of communication was opened for the purpose of entering or quitting the clock-room, the temperature of the room might not be disturbed; the fire was kept up in the stove without intermission, and two days were suffered to elapse before the observation of coincidences commenced, so as to allow the walls, as well as every part of the apparatus, to become thoroughly warmed. Besides the usual register of the temperature at the first and eleventh coincidences, three intermediate observations were made at equal intervals, in order to obtain a more exact mean: by these precautions, aided by the admirable going of the clock, the partial results with each pendulum differed only in the hundredths of a vibration per diem from their respective means; the details are contained in the following Tables, X. and XI.

TABLE X. LONDON, 1824.——COINCIDENCES OBSERVED with PENDULUM No. 3; the Clock *(Cumming)* making 86399.92 Vibrations in a Mean Solar Day.

DATE.	Barometer.	No. of Coincidence.	Temperature.	Time of Disappearance.	Time of Re-appearance.	True Time of Coincidence.	Arc of Vibration.	Mean Temperature.	Mean Interval.	Correction for the Arc.	Vibrations in 24 hours.	Reduction to mean Temperature.	Reduced Vibrations at 84°.56.
1824.	IN.		°	M. S.	M. S.	H. M. S.	°	°	S.	°			
		1	84	7 26	7 30	1 7 28	1.24						
			84.2	. . .	. . .	1 34				+			
April 11 P.M.	29.37		84.9	. . .	. . .	2 01		84.55	661.6	1.55	86140.27	. . .	86140.27
			85	. . .	. . .	2 28							
		11	84.7	57 40	57 48	2 57 44	0.73						
		1	84.7	23 52	23 56	3 23 54	1.2						
			83.9	. . .	. . .	3 51							
„ 11 P.M.	29.37		82.6	. . .	. . .	4 18		84	662.4	1.45	86140.51	−0.23	86140.28
			84.9	. . .	. . .	4 45							
		11	83.9	14 12	14 24	5 14 18	0.7						
		1	86.7	42 05	42 07	8 42 06	1.24						
			85.2	. . .	. .	9 09							
„ 12 A.M.	29.42		85.6	. . .	. .	9 36		85.64	660.3	1.55	86139.77	+0.45	86140.22
			84.5	. . .	. . .	10 03							
		11	86.2	32 06	32 12	10 32 09	0.73						
		1	86.4	39 56	40 00	11 39 58	1.27						
			85.2	. . .	. . .	12 07							
„ 12 Noon.	29.46		84.6	. . .	. . .	12 34		84.36	661.5	1.62	86140.34	−0.08	86140.26
			83.2	. . .	. .	1 02							
		11	82.4	30 11	30 15	1 30.13	0 74						
		1	82.8	38 26	38 29	2 38 27.5	1.32						
			82.6	. . .	. .	3 06							
„ 12 P.M.	29.48		82.6	. . .	. . .	3 33		82.4	663.3	1.75	86141.15	−0.91	86140.24
			81.4	. . .	. .	4 01							
		11	82.5	28 58	29 03	4 29 00.5	0.77						
		1	85.3	44 12	44 17	7 44 14.5	1.2						
			87.2	. . .	. . .	8 12							
„ 13 A.M.	29.74		87.2	. . .	. .	8 40		86.42	659.85	1.45	86139.49	+0.78	86140.27
			85.6	. . .	. . .	9 07							
		11	86.8	34 10	34 16	9 34 13	0.7						
Means. . . .	29.47 = 29.38 at 53°							84.56			86140.26		86140.26

The correction for Buoyancy is +5.65, making 86145.91 Vibrations in vacuo; and 84°.56 of the scale of the registering Thermometer is equivalent to 83°.49 of Fahrenheit, being the temperature of the Pendulum.

TABLE XI. LONDON, 1824.——COINCIDENCES OBSERVED with PENDULUM No. 4; the Clock *(Cumming)* making 86399.9 Vibrations in a Mean Solar Day.

DATE.	Barometer.	No. of Coincidence.	Temperature.	Time of Disappearance.	Time of Re-appearance.	True Time of Coincidence.	Arc of Vibration.	Mean Temperature.	Mean Interval.	Correction of the Arc.	Vibrations in 24 hours.	Reduction to a mean Temperature.	Reduced Vibrations at 84°.69.
	IN.		°	M. S.	M. S.	H. M. S.	°	°	S.	S.			
		1	84	53 51	53 56	8 53 53.5	1.2						
			84.2	. . .	. . .	9 21 0				+			
Apr. 14 A.M.	29.90		84	. . .	. . .	9 50 0		84.52	687.5	1.45	86150.01	−0.07	86149.94
			84.4	. .	. . .	10 19 0							
		11	86	48 24	48 33	10 48 28.5	0.7						
		1	86	2 12	2 16	11 2 14	1.23						
			86.2	. . .	. . .	11 30 0							
„ 14 Noon	29.90		86	. . .	. . .	11 58 0		87.04	684.4	1.55	86148.95	+0.99	86149.94
			87.5	. . .	. . .	12 27 0							
		11	89.5	56 14	56 22	12 56 18	0.73						
		1	82.4	48 31	48 36	2 48 33.5	1.27						
			82.1	. . .	. . .	3 16 0							
„ 14 P.M.	29.90		85.7	. . .	. . .	3 44 0		86.4	684.85	1.63	86149.18	+0.72	86149.90
			91	. . .	. . .	4 13 0							
		11	90.8	42 38	42 46	4 42 42	0.74						
		1	82.2	52 09	52 13	8 52 11	1.26						
			81.6	. . .	. . .	9 20 0							
„ 15 A.M.	29.80		81.25	. . .	. . .	9 59 0		82.49	689.1	1.61	86150.79	−0.92	86149.87
			83.8	. . .	. . .	10 28 0							
		11	83.6	46 58	47 06	10 47 02	0.74						
		1	83.2	8 58	9 04	2 9 01	1.29						
			83	. . .	. . .	2 37 0							
„ 15 P.M.	29.80		83.3	. . .	. . .	3 05 0		83.7	687.6	1.68	86150.26	−0.42	86149.84
			84.5	. . .	. . .	3 34 0							
		11	84.5	3 32	3 42	4 3 37	0.74						
		1	82.2	9 35	9 40	9 9 37.5	1.34						
			84	. . .	. . .	9 38 0							
„ 16 A.M.	29.30		83.4	. . .	. . .	10 06 0		83.06	688.4	1.76	86150.66	−0.68	86149.98
			82.5	. . .	. . .	10 35 0							
		11	83.2	4 15	4 28	11 4 21.5	0.78						
		1	84	35 49	35 56	1 35 52.5	1.22						
			86	. . .	. . .	2 03 0							
„ 16 P.M.	29.30		84.8	. . .	. . .	2 32 0		85.62	686.25	1.51	86149.59	+0.39	86149.98
			87	. .	. . .	3 01 0							
		11	86.3	30 08	30 22	3 30 15	0.74						
Means . .	29.70 = 29.61 at 53°.							84.69			86149.92		86149.92

The correction for Buoyancy is +5.7, making 86155.62 Vibrations in vacuo; and 84°.69 of the scale of the registering Thermometer is equivalent to 83°.62 of Fahrenheit, being the temperature of the Pendulum.

The results then of the experiments to ascertain the effect of differences of temperature on the vibrations of the pendulums, are as follows:

			°				°
Pendulum 3	86162.40	Vibrations at	44.38	Pendulum 4	86170.93	Vibrations at	47.24
	86145.91	. . .	83.49		86155.62	. . .	83.62
Differences .	16.49	. .	39.11	Differences .	15.32	. .	36.38

Hence it appears, that by the experiments with No. 3, a degree of Fahrenheit is equivalent to 0.4216 parts of a vibration in twenty-four hours, and by those with No. 4, to 0.4208 parts; the mean, or 0.4212, corresponding to an expansion of the plate-brass of which the pendulums were composed, of 0.021125 parts of an inch per foot in 180 degrees, may be taken as the final deduction; the separate results are, of pendulum 3, 0.021117, and of pendulum 4, 0.021133.

As the figure in the fourth place of decimals in the number 0.4212 amounts only to one hundredth of a vibration per diem in 50 degrees of temperature, its consideration may safely be dropped, and 0.421 taken as the equivalent to a degree of Fahrenheit, in the reduction of the experiments to a general mean temperature.

The following notice was received by me from Mr. Browne, relative to the rate of the clock used in obtaining the expansion of the pendulums:

" The rates from the 24th of March to the 16th of April were deduced immediately from the observations taken with the transit instrument applied to Cumming, as from the admirable going of this clock it was not thought necessary to use the medium of any other: it cannot but be remarked, however, that the difference in these rates in so short an interval is greater than might have been expected from the general character of the clock: the circumstance of its having been removed and put up in its new place only on the 23d may account for the small change of rate between the 24th of March and the 5th of April; but the change which is observable from the 10th to the 16th, when the heated pipe was introduced, is greater than can be properly due to any defect of compensation, and appears to be an effect of the artificial heat, totally distinct from temperature, and arising from the excessive dryness caused by it in the surrounding atmosphere: how the dryness acted upon the clock, I must confess myself at a loss to explain, but I believe myself perfectly correct in ascribing it to that cause, as the new clock made for me by Molyneux upon the same principle, with which Cumming's was regularly compared, was affected in pecisely the same manner." H. B.

Before the rates of the pendulums which had been thus obtained could be compared with those of the preceding year, it was necessary to inquire into the cause and amount of the difference which had been noticed in the scale of the arcs belonging to Captain Kater and to the Board of Longitude; for that purpose both the arcs were referred for re-examination to Mr. Jones, by whom they had been made, when it appeared that the length of the degrees in Captain Kater's arc was correct, each degree measuring one inch and five hundredths, corresponding to a radius of five feet and about half an inch; but that the arc belonging to the Board of Longitude had either been inadvertently graduated for a different radius, or on a wrong calculation, if made designedly for the pendulum which it accompanied, as the length of a degree was not more than 0.975 parts of an inch*.

The following Table exhibits the reduction of the arcs registered in Tables VIII., IX., X., and XI. by the erroneous scale, into the true arcs in which the pendulums vibrated; the latter being in the inverse proportion to the former of 42 to 39: it contains also the corrections to be applied to the number of vibrations per diem, calculated both for the registered and for the true arcs.

* The division extended for two degrees on each side of the vertical; the four degrees occupied four inches two tenths on Captain Kater's, and three inches nine tenths on the one belonging to the Board of Longitude.

PENDULUM No. 3.

	TABLE VIII.				TABLE IX.			
	Arcs Registered.		True Arcs.		Arcs Registered.		True Arcs.	
	°	Correc. s.	°	Correc. s.	°	Correc. s.	°	Correc. s.
	1.31 0.65	1.51	1.22 0.6	1.30	1.24 0.73	1.55	1.15 0.68	1.34
	1.38 0.7	1.71	1.28 0.65	1.46	1.2 0.7	1.45	1.12 0.65	1.25
	1.2 0.6	1.28	1.12 0.56	1.11	1.24 0.73	1.55	1.15 0.68	1.34
	1.24 0.62	1.36	1.15 0.58	1.18	1.27 0.74	1.62	1.18 0.69	1.40
	1.36 0.68	1.64	1.26 0.63	1.41	1.32 0.77	1.75	1.23 0.71	1.51
	1.4 0.7	1.74	1.30 0.65	1.51	1.2 0.7	1.45	1.12 0.65	1.25
	1.24 0.62	1.36	1.15 0.58	1.18	. . . .	. .	. . . .	. .
	1.37 0.67	1.62	1.27 0.62	1.40	. . . .	. .	. . . .	. .
	1.3 0.64	1.50	1.21 0.60	1.28	. . . .	. .	. . . .	. .
	1.37 0.67	1.62	1.27 0.62	1.40	. . . .	. .	. . . .	. .
Means.		1.53		1.32		1.56		1.35
Difference.		0.21				0.21		

PENDULUM No. 4.

	TABLE X.				TABLE XI.			
	Arcs Registered.		True Arcs.		Arcs Registered.		True Arcs.	
	°	Correc. s.	°	Correc. s.	°	Correc. s.	°	Correc. s.
	1.32 0.66	1.54	1.23 0.61	1.33	1.2 0.7	1.45	1.12 0.65	1.25
	1.35 0.66	1.59	1.25 0.61	1.38	1.23 0.73	1.55	1.14 0.68	1.33
	1.35 0.66	1.59	1.25 0.61	1.38	1.27 0.74	1.62	1.18 0.69	1.40
	1.36 0.66	1.60	1.26 0.61	1.39	1.26 0.74	1.61	1.17 0.69	1.39
	1.32 0.64	1.51	1.23 0.60	1.31	1.29 0.74	1.68	1.20 0.69	1.45
	1.36 0.66	1.60	1.26 0.61	1.39	1.34 0.78	1.76	1.24 0.72	1.53
	1.38 0.68	1.67	1.28 0.63	1.44	1.22 0.74	1.51	1.13 0.69	1.33
	1.32 0.64	1.51	1.23 0.60	1.31	. . . .	. .	. . . .	. .
	. . . .	. .	. . . .	. .	. . . .	. .	. . . .	. .
	. . . .	. .	. . . .	. .	. . . .	. .	. . . .	. .
Means.		1.58		1.37		1.59		1.38
Difference.		0.21				0.21		

It is shown by this table, that in consequence of the arcs of vibration having been registered by an erroneous scale, the corrections applied in Tables VIII., IX., X., and XI., to reduce the number of vibrations per diem performed in circular arcs into the equivalent number in arcs indefinitely small, exceeds the corrections which were actually due by the same amount in each instance, *viz.* by 0.21 parts of a vibration; and that the rate of the pendulum in each of those Tables, which is given as the mean result of the experiments contained in them, ought, in strictness, to be diminished by 0.21 parts of a vibration per diem.

As, however, the deductions which ought thus to be made, would be to the same amount in each of the four instances, the *differences* in the rates of the pendulums in high and low temperatures, obtained by the comparison of the results with each other, remain the same, whether the deductions be made, or whether they be omitted.

The same remark extends to every purpose for which the comparative rate only, and not the absolute rate, of the pendulums is required; that is to say, to every purpose contemplated in these experiments.

By pursuing a similar investigation to the one contained in the preceding table, in every series in which the erroneous scale was used, (which comprehends the whole of the experiments at every station, excepting those at London in 1821 and 1823,) it is found, that 0.21 is a constant expression of the value of the difference between the corrections due to the registered and to the actual arcs.

Omitting therefore, for the moment, the consideration of the experiments in London in 1821 and 1823, no inconvenience whatsoever is occasioned by an adherence to the original register, and to the rates of the pendulums as they now appear in the several Tables: it being always remembered that, if the *absolute* rate of either of the pendulums should be required at any of the stations at which they were employed, 0.21 parts of a vibration per diem should be deducted from the tabular rate.

With respect to the experiments of 1821 and 1823, it is obvious that their results may equally be brought into just comparison with those on all other occasions, (and consequently with the results in London in 1824,) whether the number 0.21 be added to the former, or deducted from the latter.

In the following comparative view of the rates of the pendulums in London, as obtained in 1821, 1823, and 1824, the difference of the scale on which the respective arcs of vibration were measured, is compensated by the addition of 0.21 to the results in 1821 and 1823.

	Date.	Vibrations per Diem.	Temperature.	Compensation for difference of Arc.	Reduction to a Mean Temperature.	Reduced Vibrations at 62°.	Difference of the partial Results from the Mean.
Pendulum 3	1821, 1823	With Captain Kater's Planes of Suspension*.					
	1823	86158.47	52.97	+0.21	−3.79	86154.89	−0.05
	1824	86162.40	44.38	. . .	−7.42	86154.98	+0.04
	1824	86145.01	83.49	. . .	+9.05	86154.96	+0.02
Mean						86154.94	
Pendulum 4	1821	86161.78	68.13	+0.21	+2.58	86164.57	−0.08
	1823	86168.71	51.88	+0.21	−4.26	86164.66	+0.01
	1823	86167.57	54.4	+0.21	−3.2	86164.58	−0.07
	1824	86170.93	47.24	. . .	−6.22	86164.71	+0.06
	1824	86155.62	83.62	. . .	+9.10	86164.72	+0.07
Mean						86164.65	

* The experiments with Pendulum 3, in 1821, and in the first series in 1823, which were made with Captain Kater's planes of suspension, are not introduced into the table, on account of the effect which the employment of different planes was found to produce on the rate of that pendulum, and which interfered to prevent the comparison of the results obtained on those occasions, with those of the subsequent experiments on the planes belonging to the pendulum itself; they were, however, strictly comparative with regard to each other, and their accordance has been already adduced in page 192, to show that no change whatsoever had taken place in the pendulum between the first experiments in 1821, and those of 1823.

The correspondence of the results in the preceding Table is much too remarkable to be passed unnoticed, and the occasions are far too numerous to admit of their accordance being attributed to accident, or viewed otherwise than as a consequence of the method of experiment, and as an evidence of the refinement of which it is capable: in fact, under circumstances which leave no uncertainty as to the temperature of a detached pendulum, its rate may be determined to the utmost extent of the precision to which the rate of the clock is known, with which the pendulum is compared. It will be remembered, that the rate of the pendulum in twenty-four hours is obtained from its comparison with the clock during an interval which does not exceed in duration one twelfth part of the period for which the rate of the pendulum is inferred; and as it is not possible to determine the definite rate of a clock for so short an interval as that of two hours, either by astronomical observation or by its comparison with other clocks, it becomes necessary to rely on an uniform performance in an interval of sufficient length to enable the mean gain or loss to be ascertained. The degree of uniformity which is required in the clock's performance may be appreciated by the consideration, that a departure from the mean rate, amounting to the one hundred and twentieth part of a second in the two hours in which the coincidences are continued, will make a difference of one tenth of a vibration in the deduction of the rate of the detached pendulum, which has been shown to be a very important quantity in these experiments. The limit which the absence of a maintaining power occasions in the period for which the oscillations of the pendulum of experiment will continue, and in which consequently its rate must be determined, throws the principal responsibility towards precise deduction on the performance of the clock; and in proportion as its regularity is maintained in shorter intervals than are usually the objects of attention, may the results of the coincidences in successive distinct experiments be expected to be consistent. It is in this respect that the experiments in

London, the account of which has occupied the preceding pages, have had a peculiar advantage, in the employment and comparison with Mr. Browne's clocks, of which those in particular by Cumming and Molyneux are probably unequalled in the preservation of a constant and uniform rate; it is to their excellence in this qualification, that the very remarkable agreement in the results which is under notice may essentially be attributed.

The accordance of the results in the different years affords the best practical proof that can be given that the pendulums had not sustained injury from use or accident, from the commencement to the close of the operations in which they have been employed. After the last experiments, however, had been concluded, and before I had had leisure to compare the several results, I requested Mr. Browne and Dr. Wollaston to do me the favour of examining the knife edges of the pendulums; when neither by the eye, nor by a microscope, could the slightest effect of wear, or injury of any sort, be perceived on the parts of the knife edges which rest upon the planes.

I have deferred a statement of the reasons which induced me to prefer the method of observing coincidences which I have adopted, *viz.*, by taking a mean of the times of disappearance and of the reappearance of the disk, to that of observing the disappearances only and considering them as times of coincidence, until the detail of the observations with the detached pendulums had been gone through; because the difference of the methods, and their respective influence on the strict relation of the several results to each other, will perhaps be better understood by illustration than by description; and it is important that the subject should be understood, because it concerns the experiments of others, as well as those of mine.

If the oscillation of a detached pendulum could be really performed in a vacuum, and if the motion at the point of suspension were perfectly free, the vibration would continue indefinitely in an arc, of which the magnitude would be determined by the impulse which first gave motion to the pendulum, and would be thenceforward permanent.

In such case, the intervals comprised between successive disappearances of the disk would be all of equal duration, whilst the rates of the clock and pendulum remained the same; and each would bear the strict proportion of an interval between coincidences, to the difference between the rates of the clock and pendulum.

If an alteration be supposed to take place in the arc of the clock during the vibration of the detached pendulum in a vacuum, the effect on the intervals between successive disappearances (independently of the influence which the alteration might have on the rate of the clock) would be, to render the one interval, during which the alteration took place, erroneous; but the original duration would be restored in subsequent intervals.

If the rate either of the clock or pendulum were to undergo a change, the interval between coincidences would change correspondingly to the difference of the rates. In this case, also, the interval between successive disappearance, in which the change took place, would be rendered erroneous by reason of the alteration in the relative velocities of the pendulums in their respective arcs, the effect of which would be equivalent to an alteration in the magnitude of the arc of the pendulum that underwent the change of rate; but the succeeding interval would be of the correct duration, and it would be successively maintained, whilst the relative rates of the clock and pendulum were constant; and would shew the exact period of time in which the pendulum of the clock gained two complete vibrations on the detached pendulum, oscillating in a circular arc of certain dimension; or in which the latter gained two complete vibrations on the former, according as the detached pendulum might have been constructed to vibrate more or less frequently than that of the clock.

If the case which has been thus supposed, of the vibration taking place in a vacuum, could occur in practice, the times of successive disappearance might be considered as those of coincidence, and the rate of the pendulum be deduced from the intervals between them, without producing more than occasional irregularity; because the intervals would be of the same duration as those obtained by a more strict method of determining the times of coincidence, excepting when changes occurred in the rate either of the clock or of the pendulum, or in the arc of the clock, in which instances a single interval only would be vitiated.

The pendulum, however, does not oscillate in a vacuum, but in a resisting medium, which causes the arc, originally communicated, gradually to diminish until the pendulum arrives at rest: the consequences of the progressive diminution of the arc of the pendulum are, first, as affects the actual rate of the pendulum itself, which continually accelerates as the retardation lessens due to the vibration in circular arcs and in proportion to their magnitude; and second, as the intervals between successive disappearances are affected, independently of the rate of the pendulum which they are designed to measure; for as the arcs diminish, the pendulum moves with diminished velocity, occasioning the number of seconds in which the disk passes the field of the telescope in entire obscuration, to augment in successive coincidences; and as the true time of coincidence, *i. e.*, when both pendulums are simultaneously at the lowest point of their respective arcs, is the middle time between the disappearance of the disk and its re-appearance, the successive intervals deduced from the observation of the times of disappearance only will differ from those deduced from actual coincidences, by half the amount which the time of entire obscuration augments from one coincidence to the next.

The progressive increase in the rate of the pendulum occasions the interval between successive coincidences to augment as the arcs diminish. The increase in the time of obscuration occasions the interval between successive disappearances, on the other hand, to diminish; and according to the amount of this diminution (which is proportioned to half the increase in the period of obscuration in successive coincidences) does the rate of the pendulum, deduced from the intervals between times of disappearance, differ from the true rate of the pendulum*.

The object which is sought in these experiments is not necessarily the absolute length of the pendulum at different stations, but its acceleration; and the acceleration may be obtained with equal correctness from nominal rates, in which a constant difference from the actual rates is maintained, as from the actual rates themselves; if, therefore, the increase in the period of obscuration depended solely on the ratio of diminution in the arc of the pendulum, and was therefore on all occasions the same, the method of deduction from the times of disappearance might be substituted for those of more strictly assured coincidences, without occasioning error: but the period in which the disk passes the field of the telescope without being visible, is governed by a variety of considerations, amongst the least influential of which, are those that depend on the relation which the rates of the clock and pendulum bear to each other, or on that of the respective velocities of the pendulums.

* In the usual practice of observing eleven coincidences in succession, in which the vibration is commenced in an arc whose dimension is between a degree and a degree and a half, and terminates in one of six or seventh-tenths of a degree, the rate of the pendulum accelerates by reason of the diminution of the arc, and is about one vibration and three-tenths per diem faster at the close than at the commencement; whence the interval between the 10th and 11th coincidences should be longer than that between the 1st and 2d by nearly four seconds.

In the detached pendulum which I have employed, the period of obscuration of the disk has varied on different occasions in the 1st coincidences from one to eight seconds, and in the 11th from eight to thirty seconds: as these numbers are adduced at present solely for the sake of illustration, they may be supposed to average respectively four seconds and twenty seconds. In substituting, therefore, the times of disappearance for those of true coincidence, the first disappearance would take place two seconds, and the last ten seconds, before the times when the two pendulums were in both cases strictly coincident; making a difference, occasioned by the substitution, of eight seconds in the period due to the ten intervals; whence the rate of the pendulum obtained from the disappearances only, would be about three-tenths of a vibration per diem less than the actual rate. With pendulums of which the difference in rate on that of the clock with which they are compared is not so great as in mine, the difference between the actual rates and those so deduced would be much more considerable.

In the case of a pendulum oscillating in a resisting medium, and therefore in a progressively diminishing arc, the effect of an alteration in the relative velocities of the pendulums in their respective arcs, on the intervals between successive disappearances, differs from that which would take place if the pendulum oscillated in a vacuum and in a constant arc, in this respect, that it is not merely the single interval in which the alteration occurs which is affected, but that every succeeding interval is influenced as well as the first; and that the period of obscuration augments as the arc of the pendulum diminishes, with greater or less rapidity, according to the relation which the velocities bear to each other.

In order, therefore, that a constant difference might obtain between the actual and the deduced rates of the pendulum on all occasions, it would be necessary that the relative velocities should be strictly maintained at the different stations; which would require that both the detached pendulum, and that of the clock, should vibrate every where in the same respective arcs, and that the difference in the rates of the two pendulums should be every where the same; towards which, if the inevitable imperfection of instruments could be entirely removed, a still more serious embarrassment would present itself, in the necessity of preserving an uniform temperature at every station, and on every occasion.

The only case, therefore, in which the method of obtaining the intervals by observing the disappearance only could be rigorously correct in principle, even as far as merely theoretical considerations are involved, is one with conditions, which, it is probable, will prevent its ever being of practical occurrence.

The causes by which the period of obscuration is found in experience to be principally affected, independent of, and unconnected with the rates or velocities of the pendulums, are such as can scarcely fail to introduce irregularity and error in every case in which the intervals of disappearance only are employed.

It may readily be conceived, that if disks should be employed, of which the diameters should not be precisely of the same magnitude, whilst all other circumstances should remain the same, whatever variation might take place in the time of obscuration of the disks of different dimension in the first coincidences, would be greatly augmented in the eleventh coincidences, because the value which a certain definite space of the clock's arc bears to the whole arc of the pendulum, increases as the latter becomes smaller; thus, the rate of the pendulum deduced from intervals of disappearance only, would be made to vary, according as disks of different magnitudes were employed, whereas by the supposition the rate is constant, and should appear so, by a perfect method of observation.

It may as readily be conceived that the effect would be the same, whether the alteration in the magnitude of the disk were real, or apparent.

The apparent magnitude of the disk (as judged by the effect) is influenced by three considerations; of which two interfere with the comparison of observations made by different individuals, and the third with the comparison of observations made at different times by the same individual.

First.—The length of the period of obscuration is different, *ceteris paribus*, with different eyes; those persons who are what is usually termed short-sighted, retain the view of the disk longer, and perceive its re-appearance earlier than others.

Second.—The same remark applies, but in a much greater degree, to eyes which have more or less practice in observation.

Third.—The duration of the period will vary to the same eye, according to the quantity of light admitted into the room. In the experiments recorded in this volume, a difference of fifteen seconds has been frequently experienced in the time during which the disk was invisible when passing the telescope, on occasions when the sole cause of the difference was the greater or less portion of light by which it was rendered visible.

I proceed to exemplify, by instances drawn from the observations in the preceding pages, some of the points which I have endeavoured to explain.

The difference which takes place in the period for which the disk is obscured, in the observations of persons equally practised, and where all circumstances are correspondent, excepting the greater or less portion of light by which the disk is rendered visible, may bewell illustrated by the two earliest observations of coincidences recorded in this volume: those of the afternoon of the 6th of July, 1821, observed by Mr. Browne, and those of the following morning, observed by myself. To prevent the inconvenience of a reference, the particulars are repeated, and results are deduced from them by both the methods which are in question; by which the influence of the method of observation on the comparability of results may be further judged:

LONDON, 1821.——COINCIDENCES OBSERVED with PENDULUM No. 3.

DATE.	Observer.	No. of Coincidence.	Temperature.	Time of Disappearance.	Time of Re-appearance.	True Time of Coincidence.	Arc of Vibration.	Mean Temperature.	Mean Interval.	Correction for the Arc.	Vibrations in 24 hours.	Reduction to a mean Temperature.	Reduced Vibrations at 65°.66.
			°	M. S.	M. S.	H. M. S.	°	°	S.	S.			
July		1	66.2	27 52	27 59	2 27 55.5	1.22			+			
6 P.M.	Mr. Browne							66.35	676.2	1.35	86145.23	+0.29	86145.52
		11	66.5	20 25	20 50	4 20 37.5	0.64						
		1	65	25 38	25 43	11 25 40.5	1.26						
7 A.M.	Capt. Sabine							65.65	676.65	1.46	86145.52	. . .	86145.52
		11	66 3	18 22	18 32	1 18 27	0.66						
		1	66.2	27 52	. . .	. . . . .	1.22						
6 P.M.	Mr. Browne							66.35	675.3	1.35	86144.91	+0.29	86145.20
		11	66.5	20 25	. . .	. . . . .	0.64						
		1	65	25 38	. . .	. . . . .	1.26						
7 A.M.	Capt. Sabine							65.65	676.4	1.46	86145.44	. .	86145.44
		11	66.3	18 22	. .	. . . .	0.66						

It is here seen that the time of obscuration in the first coincidence in the afternoon series, exceeded by two seconds the corresponding time in the noon series; but that in the eleventh coincidences the excess had augmented to no less than fifteen seconds. It happened that each observation was the first complete series of eleven coincidences which either Mr. Browne or I had ever observed, and therefore in respect of practice we were equal; no part of the effect can be attributed to a difference of eyes, since it was found on the contrary that in alternately observing successive coincidences, when the circumstances, including those of light, were similar, Mr. Browne invariably made the time of obscuration shorter than I did; yet so, that the intervals deduced from a mean of the times of disappearance and re-appearance corresponded, notwithstanding the change of the observer, whilst those resulting from the disappearances only did not correspond; that, in fact, Mr. Browne saw a minuter portion of the disk, when following the pendulum previously to obscuration, and again when preceding it in re-appearance, than was perceptible by me, in consequence of a natural difference in our power of vision. The effect, however, at present under consideration is of an opposite character, as the periods of obscuration were longer in Mr. Browne's coincidences than in mine, and as it may be presumed, would have been still longer, had our eyes been alike in conformation; the effect was caused by a difference in the strength of the light, at the respective times of observation, in the room in which the experiments were made.

On a further examination of the table, it will be seen that when in these observations a mean is taken between the times of disappearance and re-appearance for the times of coincidence, the results deduced from the intervals are identical notwithstanding the difference in the periods of obscuration: secondly, that when the times of disappearance are considered as those of coincidence, the results which by the previous method were shewn to be identical, would appear to differ 0.24 parts of a vibration per diem; and consequently, that results, obtained by the method of disappearances only are not strictly comparative unless the equality of light can be assured: thirdly, that if different methods of observation be used on different occasions, the employment of the results as comparative may involve errors of still greater amount; if, for instance, the first result be deduced by the method of disappearances, and the second by that of the mean, they will appear to differ 0.32 parts of a vibration per diem: and fourthly, that in experiments to obtain, not the relative, but the absolute length of a pendulum, it is not indifferent which method be employed, for if the one be correct, the other must be incorrect, since the rates deduced from the mean of the two observations are shewn to differ 0.2 parts of a vibration per diem, according to the method by which they are derived.

Being desirous of obtaining a still more marked example of the influence of light on the period of obscuration, and on the respective intervals deduced by the two methods, I took occasion, at Spitzbergen, to submit the light, by which successive coincidences were observed, to considerable changes, which I was enabled to accomplish by the division of the roof of the pendulum-house into compartments, each of which was removable at pleasure, so as to augment or diminish the aperture, by which the light, passing through the canvass of the tent, was admitted into the room: the compartments were opened about two minutes preceding a coincidence, and closed immediately after it; and a day was selected for the experiment, in which the temperature of the interiors of the house and tent was nearly the same, and differed but little from that of the external atmosphere, so as to be likely to remain steady.

SPITZBERGEN. 1823, *July 15th*, P.M.——COINCIDENCES with PENDULUM 4.									
No. of Coincidence.	Tempe-rature.	Arc of Vibra-tion.	Times of Disap-pearance.	Inter-vals.	Times of Coincidence.	Inter-vals.	Times of Re-ap-pearance.	Inter-vals.	Compart-ments open.
	°	S.	M. S.	S.	H. M. S.	S.	M. S.	S.	
First . .	33.5	1.28	48 27		7 48 30		48 33		Three.
				728		728.5		729	
Second .	. .	. .	00 35		8 00 38.5		00 42		Three.
				731		729		727	
Third. .	. .	. .	12 46		8 12 47.5		12 49		Six.
				729		730		731	
Fourth .	. .	. .	24 55		8 24 57.5		25 00		Six.
				726		729.5		733	
Fifth . .	. .	. .	37 01		8 37 07		37 13		Two.
				729		730		731	
Sixth . .	. .	. .	49 10		8 49 17		49 24		Two.
				735		731		727	
Seventh.	. .	. .	01 25		9 01 28		01 31		Eight.
				727		731.5		736	
Eighth .	. .	. .	13 32		9 13 39.5		13 47		Two.
				726		731		736	
Ninth. .	. .	. .	25 38		9 25 50.5		25 03		Two, {and the porch door open.
				740		731.5		723	
Tenth. .	. .	. .	37 58		9 38 02		38 06		Eight.
				725		731		737	
Eleventh	33.4	0.68	50 03		9 50 13		50 23		Three.
Means	33.45			729.6		730.3		731	

This table appears to require no other explanation, than that the door of the porch being open, in which the telescope for the observation of coincidences was placed, and the consequent admission of light into the porch, produced the same effect as a decrease of light in the room; and that a circumstance of practical occurrence which it illustrates, is the influence of the direction, as it regards the telescope, in which light enters the room at different stations; whether behind the observer when seated for observation, or through a side window from whence it may have direct access to the object-glass of the telescope.

The influence which a greater or less degree of practice in different observers will have on the period for which the disk is obscured, may be exemplified by its amount in the coincidences observed by Mr. Renwick and myself at New York; in which it may be remarked that the period in the eleventh coincidences was invariably longer with pendulum 3, and in every instance except one with pendulum 4, in Mr. Renwick's observations than in mine; whilst in the first coincidences the period differed much less, and was occasionally longer in mine than in Mr. Renwick's.

NUMBER of SECONDS during which the DISK was OBSCURED in the COINCIDENCES at NEW-YORK.

	PENDULUM 3.				PENDULUM 4.			
	Mr. Renwick.		Captain Sabine.		Mr. Renwick.		Captain Sabine.	
	1st Coin.	11th Coin.	1st Coin.	11th Coin.	1st Coin.	11th Coin.	1st Coin.	11th Coin.
	s.	s.	s.	s.	s.	s.	s.	s.
	6	19	4	13	7	31	..	..
	4	20	3	15	2	27	6	17
	7	21	4	11	5	24	8	14
	4	21	4	15	7	18	3	18
	..	..	5	12	5	26	8	16
	..	..	..	..	6	20	..	..
	..	..	..	..	7	22	..	..
	..	..	..	..	1	21	..	..
Means ..	5.2	20.5	4	13.2	5	24	7.2	16.2

The error, which may arise in comparing the results of the observations of differently-practised individuals deduced from the times of disappearance only, may be usefully illustrated by means of the first five results obtained with pendulum 4 at New York, in page 130, wherein the circumstances were as nearly similar, as can well be conceived; and as this series was the second in which Mr. Renwick had been engaged, he could not be considered as entirely an unpractised observer: three of these results, being the second, third, and fourth in the page, were from the observations of Mr. Renwick; the first and fifth were from mine.

Results Deduced.	From the Disappearances.		From true Coincidences.		From the Re-appearances.	
From Mr. Renwick's Observations . .	86127.64		86128.06		86128.48	
	86127.70	86127.70	86128.14	86128.11	86128.58	86128.52
	86127.76		86128.12		86128.48	
From Capt. Sabine's Observations . .	86127.81	86128.00	86127.97	86128.10	86128.13	86128.20
	86128.19		86128.23		86128.27	
Difference between the Observers . .	. . . . . .	0.30	. . . . . .	0.01	. . . . . .	0.32

It is here seen, that the results obtained by Mr. Renwick and myself, which agreed within one-hundredth of a vibration per diem when correct intervals of coincidence were employed in the deduction, would have appeared to differ no less than three-tenths of a vibration, if the times of disappearance only had been observed, and regarded as those of true coincidence.

If therefore the observations had been made by Mr. Renwick and myself, seperately, and at different stations, a comparison of the results, deduced from the disappearances only, would have involved an error of three-tenths of a vibration; whilst the results obtained from the more correct intervals would have been strictly comparable, notwithstanding the inequality of experience in the observers.

Further, if the observations of Mr. Renwick at New York had been the commencement of a series of comparative experiments at different stations, to be carried into execution throughout by himself, it is reasonable to infer that as his practice would have increased, the periods of obscuration would have become less, in so far as they are dependant on a more or less experienced eye; thus, by employing the disappearances only, the error of three-tenths of a vibration might be equally involved in the comparison of the earlier and later stations of the same individual, as it has been shewn to be in the comparison of results obtained at the same stations by observers with different degrees of practice: whereas, by adopting the more correct method of a mean between the disappearances and re-appearances, the earlier and later results of the same observer are rendered as strictly comparable, as are the results of different observers at the same station who may be unequally experienced.

This example affords also a still stronger illustration, than the former one drawn from the observations of Mr. Browne and myself, of the preference which should be given to the latter method, in ascertaining the rate of a pendulum of measured length, for purposes wherein it is essential that the rate ascertained should be due precisely to the length; as in the case of an experimental pendulum employed in the establishment of a national standard. In this determination, it is the first importance that the method of proceeding should ensure, as far as may be possible, the attainment of identical results when conducted by different experimentors; and for that purpose that it should be as independent as it can be rendered of individual skill and accidental circumstance. In the two instances which have been adduced, wherein the same pendulum was used in alternate observations by different persons, the one method is shewn to have fulfilled the condition of agreement, the results being strictly identical in the one case, and differing only the 100th of a vibration in the other; whilst by the other method, the disagreement, occasioned entirely by the method of observation, amounted in the one case to two-tenths, and in the other to three-tenths of a vibration in the rate, equivalent to between 2 and 3,000ths of an inch in the length of the seconds' pendulum.

As an exemplification of the inaccuracy of the previously-received method of observing coincidences is important towards the establishment of the practice of a more correct method hereafter, as well as to an estimation of the probable error which may have obtained in former experiments, it may be useful to collect in one view the duration of the periods of obscuration in the first and eleventh coincidences, in the several observations in this volume. It is probable that the variation in the length of the periods thus exhibited may be much within its extent in general occurrence; since as the same clock was used at all the stations, the arc was of the same dimension throughout: and as no alteration was made in the length of the pendulum of the clock at different stations, its rate underwent the same changes from the variations of gravity, as those of the detached pendulums: and, lastly, as the light was endeavoured to be equalized at the several stations, by regulating the space of its admission, although, by the method of observation which was practised, its disturbing influence was counteracted. It is reasonable to suppose that the variation in the length of the period of obscuration may be far more considerable on occasions, when different clocks are employed of which the arcs may not be the same; when the rate of the clock is reduced to keep mean time at each station, whereby the length of the interval between the coincidences must be greatly changed; when care is not given to make the commencing and concluding arcs of the detached pendulums the same at each station; when the observations are made by persons variously practised; and, finally, when precautions are not adopted to avoid inequality of light.

PENDULUM 3. TROPICAL STATIONS.

SIERRA LEONE.				St. THOMAS.		ASCENSION.		BAHIA.		MARANHAM.		TRINIDAD.		JAMAICA.	
1 Coin.	11 Coin.	1 Coin.	11 Coin.	1 Coin.	11 Coin.	1 Coin.	11 Coin.	1 Coin.	11 Coin.	1 Coin.	11 Coin.	1 Coin.	11 Coin.	1 Coin.	11 Coin.
s.	s.	s.	s.	s.	s.	s.	s.	s.	s.	s.	s.	s.	s.	s.	s.
2	9	2	9	1	9	2	10	2	8	1	10	4	14	2	13
3	13	2	9	2	10	1	9	4	15	3	9	2	8	4	12
2	18	2	4	5	9	1	11	4	13	1	9	3	12	4	16
1	10	1	11	2	8	4	7	2	10	5	9	5	20	5	11
8	18	4	12	4	12	2	11	5	11	1	11	3	10	5	15
1	11	2	11	2	11	2	10	3	9	3	10	6	15	4	15
3	9	3	10	2	11	2	12	3	12	1	10	3	15	1	12
.	..	.	..	4	9	1	10	4	11	2	11	3	15	2	17
.	..	.	..	.	..	.	..	.	..	3	8	5	14	5	16
.	..	.	..	.	..	.	..	.	..	3	9	.	..	.	..
Means. Coincidences, { 1st, 2".6; 11th, 11"				2 8	10	2	10	3.4	11	2.3	9.6	3.8	13.7	3.6	14.1
Intervals, 643"				644"		636"		642"		647"		639"		642"	
Obscuration in the 1st Coincidences 3"; in the 11th Coincidences 11".3.															

PENDULUM 3. NORTHERN STATIONS.

HAMMERFEST.				SPITZBERGEN.				GREENLAND.				DRONTHEIM.			
1 Coin.	11 Coin.	1 Coin.	11 Coin.	1 Coin.	11 Coin.	1 Coin.	11 Coin.	1 Coin.	11 Coin.	1 Coin.	11 Coin.	1 Coin.	11 Coin.	1 Coin.	11 Coin.
s.	s.	s.	s.	s.	s.	s.	s.	s.	s.	s.	s.	s.	s.	s.	s.
3	12	4	25	5	9	5	10	4	9	2	13	5	16	5	13
2	23	5	14	5	12	3	12	5	11	5	17	4	16	5	16
4	13	1	17	5	13	3	19	7	16	2	16	1	16	2	16
1	15	2	17	5	13	5	11	6	10	3	16	5	18	5	31
2	15	4	18	5	11	4	12	5	11	7	15	5	14	4	21
5	16	5	15	4	11	4	13	2	16	4	17	5	16	7	21
1	18	5	12	5	12	3	12	.	..	.	..	5	16	1	18
.	..	.	..	3	11	1	13	.	..	.	..	5	25	3	19
.	..	.	..	.	..	.	..	.	..	.	..	3	16	.	..
Means. Coincidences, { 1st, 3".1; 11th, 16".5				1st, 4".1; 11th, 12".1				1st, 4".3; 11th, 14"				1st, 4".1; 11th, 18"			
Intervals, 675"				690"				686"				684"			
Obscuration in the 1st Coincidences 3".9; in the 11th Coincidences 15".1.															

PENDULUM 4. TROPICAL STATIONS.

	SIE. LEONE.		St. THOMAS.		ASCENSION.		BAHIA.		MARANHAM.		TRINIDAD.		JAMAICA.	
	1 Coin.	11 Coin.	1 Coin.	11 Coin.	1 Coin.	11 Coin.	1 Coin.	11 Coin.	1 Coin.	11 Coin.	1 Coin.	11 Coin.	1 Coin.	11 Coin.
	s.	s.	s.	s.	s.	s.	s.	s.	s.	s.	s.	s.	s.	s.
	6	14	5	13	5	12	5	12	1	14	5	15	6	20
	1	12	4	13	1	12	1	13	2	13	5	14	5	18
	4	13	3	15	2	14	1	12	3	12	4	16	7	19
	3	13	4	14	1	15	7	18	4	12	4	14	3	17
	5	13	5	12	2	14	4	13	3	14	4	12	2	14
	3	13	3	12	3	15	5	13	6	12	5	18	3	24
	6	18	6	16	5	12	2	13	2	12	4	18	4	16
	6	15	4	13	5	16	4	23	5	10	6	15	5	20
	3	14	7	16	..	..	4	15	6	12	6	19	..	..
	5	14	1	10	..	..	..	..	..	..	..	..	..	..
	5	17	..	..	..	..	..	..	..	..	..	..	..	..
Means . .	4.3	14	4.2	13.4	3	14	3.7	15	3.5	12.3	5	16	4.4	19
	Intervals	666″	667″		655″		671″		672″		665″		662″	
	Obscuration in the 1st Coincidences 4″; in the 11th Coincidences 14″.8.													

PENDULUM 4. NORTHERN STATIONS.

HAMMERFEST.				SPITZBERGEN.				GREENLAND.				DRONTHEIM.			
1 Coin.	11 Coin.	1 Coin.	11 Coin.	1 Coin.	11 Coin.	1 Coin.	11 Coin.	1 Coin.	11 Coin.	1 Coin.	11 Coin.	1 Coin.	11 Coin.	1 Coin.	11 Coin.
s.	s.	s.	s.	s.	s.	s.	s.	s.	s.	s.	s.	s.	s.	s.	s.
5	30	7	26	8	17	7	22	8	23	4	21	6	24	8	26
3	19	7	22	1	18	6	20	8	14	4	16	4	18	5	24
7	19	7	14	7	23	7	22	8	15	2	19	8	25	5	20
5	22	8	17	8	18	7	19	5	22	1	16	8	27	6	23
5	16	6	17	6	20	7	17	2	19	5	17	7	23	3	18
3	17	4	19	8	18	5	19	7	17	1	17	9	22	7	27
7	15	..	..	8	14	5	17	3	17	..	..	5	18	3	13
..	..	..	..	8	17	..	..	..	..	..	..	..	..	..	.
Means. Coincidences { 1st, 5″.3; 11th, 19″.5				1st, 6″.5; 11th, 18″.7				1st, 4″.4; 11th, 18″				1st, 6″; 11th, 22″			
Intervals . . 713″				725″				716″				711″			
Obscuration in the 1st Coincidences 5″.5; in the 11th Coincidences 19″.5.															

In these tables, the influence of the length of the interval between coincidences on the period of obscuration, will be found to receive a double illustration; for which purpose the experiments with each pendulum have been kept distinct, and have been arranged in two divisions, one comprehending the tropical stations or those of high temperature, and the other the northern stations, wherein the temperature was on the average forty degrees lower, or equivalent to an increase of about seventeen seconds per diem in the rates of the detached pendulums, (which increase did not take place in the pendulum of the clock in consequence of its compensation,) and to a consequent augmentation exeeding forty seconds in the interval between coincidences; the circumstances in the two divisions were strictly similar in all other respects, excepting possibly in occasional inequalities in the light, notwithstanding the precautions which were adopted. It is seen that the period of obscuration was augmented in every case in the longer intervals by about one third of the average amount at the tropical stations.

It may be remembered that the number of vibrations made by pendulum 4, exceeded those of pendulum 3, by rather less than ten vibrations in the twenty-four hours; its coincidences with the clock were consequently less frequent than those of No. 3; the difference of their respective intervals amounted, on the average of the stations, to rather less than thirty seconds; being the joint effect of the vibrations of pendulum 4 having taken place at a somewhat lower mean temperature than those of No. 3, and of the actual difference in the length of the two pendulums arising from their original construction. In this instance also, the augmentation in the interval between coincidences is shewn to have produced an increase in the period of obscuration, corresponding to that occasioned by the augmented rate of both pendulums from temperature.

In both instances, the illustration is not confined to the deductions from the averages only, but may be traced in nearly its just proportion into the several experiments at every station*.

In all the cases which have been thus illustrated, in which errors would be involved by the comparison of results obtained by the method of

* The practical importance of this illustration will appear, by instancing the experiments with a similar pendulum, made at Madras by Mr. Goldingham, and in London by Captain Kater, and published in the Phil. Tr. for 1822, Part I., in which the interval of coincidences differed at the two stations not less than nine hundred seconds, being 725 seconds at Madras, and 1630 seconds in London. In the evidence of the very sensible effect produced by a difference of thirty or forty seconds, the liability to the introduction of error of very serious consideration may be inferred, in the employment of the method of disappearances where the intervals are so widely dissimilar. Mr. Goldingham appears to have adopted a mode of observing coincidences very nearly the same as that which has been practised in the experiments in this volume, but without being aware of the inaccuracies involved by the more usual method, or that the results obtained by different methods would not be comparable with each other. In general the errors of such comparison would be greater than when the same method, although defective, is employed on both occasions; in the latter case its amount is proportioned to the difference, at the two stations, in the excess of half the period of obscuration in the eleventh coincidences over half the period in the first coincidences; in the former case, to the whole excess at the station where the disappearances only are observed; that station in the present instance is, the one in which the interval is of the longest duration.

Viewing the exceeding and admirable care with which the experiments appear to have been conducted by Mr. Goldingham at Madras, it is much to be desired that the pendulum and its apparatus should be returned to England, and that a series of experiments, strictly comparable, should be made in London by the same method, and with the apparatus employed at Madras. It has been seen that the relation of experiments to each other may be destroyed by the vibration of the pendulum on different agate planes; and as other parts of the apparatus, used at Madras, were furnished by the same maker as those which have been shewn in the preceding pages to have been incorrect in their construction, it becomes the more desirable that the same apparatus should be rigorously employed at both stations; and that the thermometer in particular should be subjected to a careful examination. The accomplishment of such a series in London is not less desirable, in regard to the great pains which Mr. Goldingham has taken at Madras, than due to the work in which the experiments are published in their present incomplete state; for the rate of an invariable pendulum at one station, however correctly obtained, possesses no value except in its combination with strictly corresponding results elsewhere.

disappearances only, the irregular influence of the circumstances which occasioned error will counteract itself, when the re-appearances also are observed and a mean is taken for the true time of coincidence.

As it might be imagined that the re-appearances may not be seen with the same distinctness or certainty as the disappearances, and as a difference in this respect might be deemed on superficial consideration an objection to the method of which the re-appearances form a part of the observation, it may be desirable briefly to trace the effect which such a difference, supposing it to exist, would produce on the results.

The re-appearance in the first coincidence being much too decided to admit of uncertainty in the observation, the possibility of such an occurrence must be limited to the eleventh coincidence. Let therefore the times of true coincidence be considered (as when they are deduced from a mean) to take place when the centre of the disk passes the telescope in coincidence with the middle of the tail-piece of the detached pendulum; and let it be supposed that the re-appearance in the eleventh coincidence is observed one or even two seconds more distant from the true coincidence, than the interval which took place between the disappearance and coincidence; the effect will be that the registered time of the eleventh coincidence will be a second later than the true time, if the error of the observation be two seconds, and half a second later if it be only a single second; whence the deduced rate of the pendulum will be four-hundredths of a vibration per diem quicker than the actual rate in the one case, and two-hundredths in the other; but as the late observation of the re-appearance is supposed to be a constant effect, as occasioned by a less facility in noticing the first apppearance of an object than in following it until its disappearance, however the deduction of the absolute rate might be influenced, the relative rates would be every where similarly affected, and the acceleration obtained from them would be as rigorously correct, as if derived more strictly from the actual rates.

The observation of a coincidence is not however of the same nature as that of the immersion and emersion of a satellite, in which the object is viewed until it gradually becomes invisible, and in which its re-appearance takes place through the same graduation of indistinctness: the pendulum is visible only when passing the opening in the diaphragm of the telescope, and therefore disappears to the eye and is seen afresh in each alternate vibration; a reason does not readily present itself why a minute portion of the white disk should not be seen with equal facility, during the passage across, whether it be on the preceding or on the following side of the pendulum; nor so far as my individual experience is concerned, have I ever been able to perceive a difference.

With respect to errors arising from inadvertency in the observer, or from other similar causes depending on the individual, rather than a consequence of the mode of proceeding, the possibility of their occurrence may be supposed to be equal in each case: but their effect on the times of coincidence deduced from two observations would be reduced to half the amount, which would take place when the times depended on a single observation.

The instant of coincidence is also capable of a more precise determination, when it is the mean of two, than when dependant on a single observation; in the one case it is the second, and in the other the half second of coincidence which is determined.

There are several other practical advantages in the method of the double observation, of which an attentive observer would become sensible in the course of a series of experiments; but which I do not think it necessary to particularize, having already exceeded the limits which I had originally intended in this discussion, under a sense of the importance of a correct appreciation being made of the relative value of different methods of obtaining the rate of a free pendulum.

It will be remembered that in the various important purposes in which the pendulum is now employed, its peculiar value is as a very accurate measure of very minute quantities; and that the inquiry into the best method of procedure with it must in consequence be concerned in the discussion of differences which may appear as extremely small; but which are by no means insignificant quantities in the purposes to which the pendulum is applied.

RESULTS WITH THE DETACHED PENDULUMS.

The results obtained with the detached pendulums at the several stations are collected in one view in the subjoined Table.

The particulars of the first nine columns appear to require no additional explanation to that which they have received in the several places from whence they are collected.

The tenth, eleventh, and twelfth columns are occupied in a comparison of the results of the two pendulums at the several stations. Column ten shows the excess of the vibrations of pendulum 4 over those of pendulum 3; and in column eleven is inserted the mean excess, which, in consideration of the number of observations from which it is derived, may be deemed the precise amount due to the actual difference in the length of the two experimental pendulums; this amount ought consequently to have been constant, if the results given by the pendulums had been every where strictly correspondent; or, in other words, if the length of the seconds' pendulum deducible from the experiments with No. 3, had been always precisely the same with that deducible with No. 4. Column twelve exhibits the several deviations from such perfect identity; the amount of which, even in the extreme cases, appears so small, that it may almost be deemed an over-refinement to attribute its occurrence to any particular circumstance. There is, however, one cause so much more prominent than others, that if the deviations had been larger I should not have hesitated to have ascribed them chiefly to it, and it may be proper therefore to be noticed. It will be remembered that the rate of the clock with which the pendulums were compared was determined by observations of Zenith Distances or Transits for the whole interval through which the experiments were continued; and that the rate in the half intervals occupied by the separate experiments with each pendulum, although apportioned under the guidance of the observations and comparisons, must still be considered as having been less precisely determined by them than the rate for the whole period, in consequence of the comparatively short duration of the half intervals. Now if an inaccuracy be supposed in the division of the rate into the half periods, its effect will be to cause an apparent difference of double its own amount in the rates of the detached pendulums; because what is gained in the rate of the clock in the one half interval must be taken from the other. The mean difference in the results of the two pendulums (omitting the signs) is 0.16, and the extreme difference 0.37 parts of a vibration per diem, which are equivalent respectively to an incorrect assignment of the clock's rate in the half intervals (generally of five days) of 0.08, and 0.19 parts of a second per diem; which amounts are not merely within the limits of probable occurrence, but are so small that the supposition of their non-occurrence must be regarded as extremely improbable.

The effect, however, of incorrectness introduced, in the division of the rate into the half intervals, ceases, when the separate results with the two pendulums are united in a mean result, as in the succeeding column.

Column thirteen exhibits a mean between the vibrations of the two pendulums; or rate of an imaginary pendulum supposed to oscillate in a vacuum, and at a uniform temperature, at every station.

And finally, in column fourteen, are contained the respective lengths of the seconds' pendulum at the several stations, corresponding to the vibrations in the preceding column, and resting on the determination of the length of the pendulum vibrating seconds in Portland-place, determined by Captain Kater. These measures are expressed in parts of Sir George Shuckburgh's standard scale, at the temperature of 62° Fahrenheit.

RESULTS WITH THE DETACHED PENDULUMS.

STATIONS.	Pendulum.	Vibrations.	Barometer.	Temperature.		Buoyancy.	Reduction to a Mean Temperature.	Vibrations in Vacuo at 62°.	Excess in the Vibrations of Pendulum 4.			Vibrations strictly comparative.	Length of the Seconds' Pendulum.
				Registered.	Fahrenheit.								
1	2	3	4	5	6	7	8	9	10	11	12	13	14
			IN.	°	°	s.	s.						IN.
St. Thomas.	3	86010.73	30.08	82.1	81.1	5.79	+ 8.04	86024.56	9.69		0.06	86029.40	39.02069
	4	86020.02	30.08	83.1	82.05	5.79	+ 8.44	86034.25					
Maranham	3	86001.39	30.01	81.83	80.82	5.78	+ 7.92	86015.09	9.39		0.24	86019.78	39.01197
	4	86010.70	30.00	82.02	81	5.78	+ 8.00	86024.48					
Ascension	3	86014.77	30 13	81.47	80.47	5.81	+ 7.78	86028.36	9.51		0.12	86033.11	39.02406
	4	86023 72	30.15	82.87	81.83	5.80	+ 8.35	86037.87					
Sierra Leone	3	86009.82	29.84	81.3	80.3	5.75	+ 7.70	86023.27	9.74		0.11	86028.14	39.01954
	4	86019.37	29.86	81.75	80.74	5.75	+ 7.89	86033.01					
Trinidad	3	86007.55	30.01	84.45	83.4	5.75	+ 9.01	86022.31	10.00		0.37	86027.31	39.01879
	4	86018.00	30.02	83.35	82.3	5.76	+ 8.55	86032.31					
Bahia	3	86016.82	29.99	75.2	74.3	5.86	+ 5.18	86027.86	9.90		0.27	86032.81	39.02378
	4	86027.61	30.06	72.9	72.1	5.90	+ 4.25	86037.76					
Jamaica	3	86026.78	30.04	81.77	80.77	5.79	+ 7.90	86040.47	9.60	9.63	0.03	86045.27	39.03508
	4	86035.65	30.04	83.6	82.55	5.77	+ 8.65	86050.07					
New York	3	86117.90	30.21	35.66	35.56	6.42	− 11.13	86113.19	9.59		0.04	86118.48	39.10153
	4	86128.20	30.37	33.81	33.75	6.47	− 11.89	86122.78					
London	3	page 211	. . .	. .	62	. .	. . .	86154.91	9.71		0.08	86159.79	39.13908
	4				62	. .	. . .	86164.65					
Drontheim	3	86193.99	29.82	47.41	47.06	6.19	− 6.29	86193.89	9.27		0.36	86198.52	39.17428
	4	86203 61	29.49	46.73	46.38	6.13	− 6.58	86203.16					
Hammerfest	3	86213.15	29.93	55.05	54.47	6.11	− 3.17	86216.09	9.74		0.11	86221.46	39.19512
	4	86227.18	29.68	44.34	44.06	6.20	− 7.55	86225.83					
Greenland	3	86227.07	29.88	44	43.73	6.24	− 7.69	86225.62	9.65		0.02	86230.44	39.20328
	4	86237.86	29.91	41.14	40.93	6.28	− 8.87	86235.27					
Spitzbergen	3	86239.21	29.89	45.2	44.9	6.23	− 7.20	86238.24	9.38		0.25	86242.93	39.21464
	4	86251.68	29.79	37.51	37.36	6.31	− 10.37	86247.62					

SECTION II.

With attached Invariable Pendulums.

THE pendulums employed in the experiments of the present section were attached to the machinery of a clock, by which their tendency to arrive at rest in consequence of the resistance of the air was counteracted, the continuance of their oscillation maintained, and the number of their vibrations registered.

The principle of their construction, in respect to invariability in length, was the same as that of the detached pendulums; they could undergo no change but from the expansion of the metal of which they were composed in different degrees of temperature.

The rate of a pendulum of this description is obtained by comparing the clock to which it is attached, at the commencement and close of the intervals for which the rate is desired, either with celestial time, or with another clock which is itself duly compared with the heavens; and by observing, as frequently as may be thought necessary during the intervals, the temperature and density of the atmosphere, the temperature of the pendulum, and the arc in which it vibrates. The extent of the arc is shown by a portion of a graduated circle affixed to the clock-case immediately behind the lower extremity of the pendulum; and the temperature of the pendulum, by a thermometer within the clock-case, suspended as near the pendulum as it can conveniently be placed. From these data may be computed the number of vibrations which the pendulum, being of a certain temperature, would make in a vacuum, and in an arc indefinitely small.

The two pendulums, similar in construction, numbered 1 and 2, and the clock in which they were successively used, were the same with which I had been furnished in the Expeditions of Arctic Discovery, in 1818, and in 1819—1820; and have been particularly described in the account of the experiments made in these voyages, published in the *Phil. Tran.* for 1821; the following more brief notice may, therefore, suffice on the present occasion.

The pendulums were each cast in one piece of solid brass, and were furnished with knife-edges of hard steel, secured as in the detached pendulums; the vibration was performed on agate planes ground into portions of hollow cylinders, and imbedded in a brass support firmly secured to the clock-case. The clock was supported by a triangular wooden frame, and was fitted with all the necessary and proper adjustments.

The knife-edges of the pendulums had been slightly corroded on their return from the North, and it was apprehended that the injury might increase so as to interfere with the vibration. They were therefore ground afresh, and the clock cleaned and oiled previously to the commencement of the present experiments. I had hoped that these operations might have been completed in sufficient time to have permitted a trial of the going of the pendulums in the clock, before their embarkation in the Iphigenia; but from circumstances, which occasioned much vexation at the time, but which it is needless now to particularize, this measure could not be accomplished, and their first trial was at Sierra Leone.

SIERRA LEONE.

THE qualifications of a room suited to the reception of the pendulum clock differ in some respects from those required in the apartment adapted to the experiments with the detached pendulums; and it may be proper to notice the two principal particulars of difference, because they influence materially the greater facility with which accomodation can be procured for the clock than for the pendulums. The support of the pendulum-clock resting wholly on the ground, the nature and thickness of the walls are no otherwise of consideration than as they may conduce towards an uniformity of temperature: it is merely requisite that the flooring of the room should be firm and substantial, or which is preferable, that the room should be on the basement, and that the flooring should be removed so as to allow the support to rest on the ground beneath. As the comparison of the clock and registry of the arc and temperature can be accomplished by the light of a taper as well as by the daylight, the admission of the latter into the room is not a necessary qualification, and being prejudicial to steadiness of temperature may be better avoided.

The room in which the clock was set up at Sierra Leone was a kitchen beneath the Officers' Quarter assigned for the pendulum experiments; the floor was of brick, and the windows were permanently boarded up; the room was entered only at the times and for the purposes of comparison and registry.

As the same methods of comparing the clock and of observing the temperature and arc were pursued at the different stations with very little variation, the following description may be considered to apply generally, and the alterations only will be hereafter noticed.

The clock was compared at intervals of twelve hours of mean time with a chronometer or clock the rate of which was determined by astronomical observation. As soon as the comparison had been made, the door of the clock-case was opened, and the temperature indicated by the pendulum thermometer was read to degrees and tenths. It was then seen, by means of the indices of a self-registering thermometer stationed also in the clock-case, to what extent the temperature had varied in the preceding interval above and below the indication thus read; the extremes of temperature so obtained were entered in the table as corresponding to the interval, and as those which the pendulum thermometer would have shewn, had it possessed the means of registering its own extremes; the purpose of the self-registering thermometer was only to perform this latter office, without regard to the agreement or otherwise of its scale with that of the pendulum thermometer. The amplitude was next observed; and as the graduated arc on which it was read was a fixture, and did not admit the adjustment of its zero to the pendulum when at rest, the extent of the arc was noticed in each semi-vibration, and the mean was entered in the table as the angular deviation of the pendulum from the vertical. The indices of the self-registering thermometer were then brought down to the mercury, and the clock-case closed until the next period of observation.

The clock was suffered to go for three or four days before the registry was commenced, in order to allow the rate to become steady; the observations with each pendulum were continued generally from five to seven days.

The following tables contain the particulars of the going of the two pendulums at Sierra Leone; and appear to require no other explanation at present, than that the correction for the arc has been computed by the same formula as in the detached pendulums: and that the reduction to a mean temperature is in the proportion of 0.44 parts of a vibration per diem for each degree of Fahrenheit; the chronometer with which the clock was compared was No. 423 of which the rate on mean solar time is taken from page 21.

SIERRA LEONE.——VIBRATIONS of PENDULUM No. 1, in the PENDULUM CLOCK.
Mean Height of the Barometer 29.85 Inches.

DATE.	Chronometer.	Pendulum Clock.	Clock's loss on Chron.	Daily Rates. Chron.	Daily Rates. Clock.	Temperature.			Arc of Vibration.		Correction for the Arc.	Reduction to a mean Temp.	Vibrations per Diem at 81°.19.
1822.	H. M. S.	M. S.	S.	Gaining. S.	Losing. S.	°	°	°	°	°			
Mar. 9 P.M.		56 46.9							1.02				
						81.5, 80.7	81.1						
„ 10 A.M.		55 31.4	152.9	1.01	151.89			81.2	1.00	1.007	1.67	...	86249.78
						81, 81.6	81.3						
„ 10 P.M.		54 14							1.00				
						81.6, 80.6	81.25						
„ 11 A.M.		52 58	152.2	1.01	151.19			81.28	1.00	1.00	1.65	+0.04	86250.50
						81.7, 80.9	81.3						
„ 11 P.M.		51 41.8							1.00				
						81.1, 80	80.55						
„ 12 A.M.	10 00 00	50 26.3	152.1	1.01	151.09			80.9	0.98	0.993	1.63	−0.13	86250.41
						80.5, 82	81.25						
„ 12 P.M.		49 09.7							1.00				
						82, 80.2	81.1						
„ 13 A.M.		47 54.5	151.1	1.01	150.09			81.15	0.95	0.983	1.59	...	86251.58
						80.6, 81.8	81.2						
„ 13 P.M.		46 38.6							1.00				
						81.8, 81	81.4						
„ 14 A.M.		45 23.4	151.1	1.01	150.09			81.4	0.95	0.983	1.59	+0.09	86251.59
						81, 81.8	81.4						
„ 14 P.M.		44 07.5							1.00				
MEANS.					150.87			81.19		0.993	1.63		86250.76

Sierra Leone.——VIBRATIONS of PENDULUM No. 2 in the PENDULUM CLOCK.
Mean Height of the Barometer 29.85 Inches.

DATE.	Chronometer.	Pendulum Clock.	Clock's loss on Chron.	DAILY RATES. Chron.	DAILY RATES. Clock.	Temperature.	Arc of Vibration.	Correction for the Arc.	Reduction to a Mean Temp.	Vibrations per Diem at 81°.5.
1822.	H. M. S.	M. S.	s.	Gaining. s.	Losing. s.	° ° °	° °	s. +	s.	
Mar. 16 P.M.		23 22.5					1.10			
„ 17 A.M.		23 01	43	0.64	42.36	{82.2, 81.2} 81.7; {82.2, 80.8} 81.5; 81.6	1.07 } 1.09	2.00	+0.04	86359.68
„ 17 P.M.		22 39.5					1.07			
„ 18 A.M.		22 18	43.7	0.64	43.06	{80.8, 82.2} 81.5; {80.8, 81.2} 81; 81.25	1.06 } 1.06	1.87	−0.11	86358.70
„ 18 P.M.		21 55.8					1.05			
„ 19 A.M.		21 33.8	41.6	0.64	40.96	{81.2, 81.8} 81.5; {81, 81.2} 81.1; 81.3	1.05 } 1.05	1.83	−0.09	86360.78
„ 19 P.M.		21 14.2					1.05			
„ 20 A.M.		20 53.4	41.8	0.64	41.16	{81.8, 79.8} 80.8; {79.8, 82.2} 81; 80.9	0.96 } 1.02	1.72	−0.26	86360.30
„ 20 P.M.		20 32.4					1.05			
„ 21 A.M.		20 11.2	42.2	1.05	41.15	{82.2, 80.9} 81.55; {81.9, 81.5} 81.2; 81.4	1.02 } 1.03	1.76	−0.04	86360.57
„ 21 P.M.	10 00 00	19 50.2					1.02			
„ 22 A.M.		19 30	41.3	1.05	40.25	{81.5, 80.4} 80.95; {80.4, 82.2} 81.3; 81.12	0.99 } 1.01	1.68	−0.17	86361.26
„ 22 P.M.		19 08.9					1.02			
„ 23 A.M.		18 47	44.5	1.05	43.45	{82.2, 82.5} 82.35; {82.5, 81.7} 82.1; 82.22	1.04 } 1.03	1.76	+0.31	86358.62
„ 23 P.M.		18 24.4					1.03			
„ 24 A.M.		18 02.4	44.4	1.05	43.35	{81.6, 82.6} 82.1; {82.5, 81.7} 82.1; 82.1	1.04 } 1.04	1.79	+0.20	[illegible]
„ 24 P.M.		17 40					1.05			
„ 25 A.M.		17 19.3	41.1	1.05	40.05	{81.6, 81.8} 81.7; {82, 81.2} 81.6; 81.65	1.04 } 1.04	1.79	+0.06	86361.80
„ 25 P.M.		16 58.9					1.03			
„ 26 A.M.		16 39	40.1	1.05	39.05	{82.2, 80.8} 81.5; {82, 81} 81.5; 81.5	1.04 } 1.04	1.79	. . .	86362.74
„ 26 P.M.		16 18.8					1.05			
MEANS					41.48	81.5	1.04	1.79		86360.31

In the experiments which I had made with the same clock and pendulums, during the voyages for the Discovery of a North-West Passage, the action of the weight, transmitted through the wheels, had maintained an amplitude always exceeding a degree and six-tenths; whereas in these experiments it may be seen that the amplitude with each pendulum scarcely equalled a degree. As the weight was the same which had been used on the former occasions, the diminution of its force appeared far too great to be accounted for by any ordinary effect of the cleaning which the clock had undergone intermediately; but as this was the first time that the clock had been set up since it had been cleaned, I thought it possible that the effect might be, at least in part, occasioned by friction; arising either from a defective supply of oil, or from time being required from the oil to work into the pivots: I supplied, therefore, a very small quantity of prepared oil to the principal parts where its absence would have occasioned the most resistance; and having assured myself that all the adjustments of the clock were perfect, and that the small fixed level, by which the horizontality of the agate planes was shown, corresponded with the parts of the planes on which the knife edge rested, the oscillation of pendulum 1 was suffered to proceed, and subsequently of pendulum 2, as shewn in the preceding tables. At the close of the experiments with pendulum 2, the clock had gone nearly a month, and as no material alteration had taken place in the arc during that period, it had become evident that none was to be expected from time. The invariability of the sustaining force being a most important condition, and within limits even an essential one, towards the success of the method of experiment, I felt extremely anxious to ascertain the cause of the great variation which had appeared since the clock was cleaned; as if it arose from any accidental defect, or derangement of the machinery (which I could not but suspect) it might yet admit of being remedied; whereas if I should fail in tracing it to an ostensible cause, or if it should be one beyond my power to remedy, it was scarcely to be expected, that being of sufficient magnitude to have reduced the arc to nearly the half of its original dimension, it should not be productive of further fluctuations. Before the clock was re-packed for embarkation, therefore, the works were submitted to a very close and careful examination; but without my being able to perceive that any part of them was misplaced or injured. As the whole of my attention, during the short remainder of my stay at Sierra Leone, was occupied in a repetition of the experiments with the detached pendulums as already described in consequence of the results of the first series having differed so much from previous expectation, I was obliged to postpone the further examination of the pendulum clock until my arrival at another station, when I designed to set it up in a room in which light should be admitted, for the purpose of observing it, whilst in motion, with more advantage than I had been able to accomplish at Sierra Leone. The next station which I visited was the Island of St. Thomas: but in consequence of the difficulties which were experienced in obtaining permission to land the instruments, and in consideration of the irresponsible nature of the government of the Island, I did not think it prudent to land more instruments than could conveniently be re-embarked by a boat's crew at the shortest notice; and accordingly the pendulum clock was not landed.

ASCENSION.

The room in the barrack-square at Ascension, which Major Campbell was kind enough to give up to me, but of which the walls were not sufficiently substantial to support the frame belonging to the detached pendulums, possessed the qualifications of a brick floor and steady temperature which rendered it suitable for the experiments with the pendulum clock: it was accordingly set up, and was going on the afternoon of my arrival.

On examining the movements of the clock on the following morning under the advantage of a strong light, I had the satisfaction of discovering, and of removing, the impediment to the free motion of the pendulum which had caused so much anxiety, and which I shall endeavour to explain. The cylindrical groove in which the knife edge vibrated was closed at the end towards the back of the clock by the brass frame in which the planes were imbedded, but was open at the other end: in order therefore to ensure the replacement of the pendulum at all times in the same longitudinal position, a brass cheek projected by a spring from the clock plate, and acted as a termination to the groove at the open end, but without pressing against or touching the side of the planes; the length included between the cheek and the opposite termination of the groove was designed to be just sufficient to receive the pendulum, and to allow it to oscillate without touching either extremity; and such was the space included before the clock was cleaned; but in putting the works together after that operation had been performed, the brass cheek had not been replaced precisely in its original position; it had apparently escaped Mr. Arnold's notice, to whom the clock had been intrusted to be cleaned, that the ends of the wootz prism on which the pendulum vibrated were made to slope outwards commencing from the top, so that the length at the knife edge visibly exceeded that of the head-piece of the pendulum. In examining the vibration of the knife edge in the cylinder, the lower part of the end of the prism, just at the angle but not higher, appeared nearly if not quite in contact with the cheek, although the latter stood well clear of the upper part of the prism and of the head-piece. On pressing back the cheek towards the plate, the arc immediately began to augment, and in a short time resumed and maintained its former dimension of one degree and $\frac{7}{10}$ths. The contact was so slight that I was not fully assured of its being actually the case, until the effect which followed on the pressure being relieved removed all doubt; and I then observed that the vibration at Sierra Leone had slightly marked the brass. The two pendulums were tried successively, but when the spring was at its full extent, the space included was not sufficient for either to vibrate without touching at one extremity. Fortunately the remedy was simple and did not prevent the proper use of the cheek; by employing it at the full extent of the spring, when the clock was first set up, the proper position of the pendulum was ensured: that object being effected, its pressure on the pendulum was relieved, so that the oscillation might be perfectly free.

It is from Ascension therefore that the comparative rates of the attached pendulums commence. It appeared by subsequent comparison with other stations, and by taking into account the acceleration as shewn by the detached pendulums, that the rate of each pendulum at Sierra Leone was accelerated about twenty-three seconds by the diminution of the arc produced by the resistance to the free motion of the pendulum. It is remarkable that no considerable irregularity should have taken place in the daily going of the clock, under a disturbing influence of such magnitude, by which its existence might have been indicated: had it not been for my previous knowledge of the clock, I should certainly have had no reason to have suspected from its going at Sierra Leone, that the proper action of the weight was modified to so considerable an extent.

The height of the clock at Ascension was the same as that of the detached pendulums; the rate of the Chronometer No. 423, with which it was compared, is taken from page 48.

ASCENSION.——VIBRATIONS of PENDULUM No. 1, in the PENDULUM CLOCK.

• Mean Height of the Barometer 30.05 Inches.

DATE.	Chronometer.	Pendulum Clock.	Clock's loss on Chron.	DAILY RATES. Chron.	DAILY RATES. Clock.	Temperature.	Arc of Vibration.	Correction for the Arc.	Reduction to a mean Temp.	Vibrations per Diem at 80°.05.
1822.	H. M. S.	M. S.	S.	Gaining. S.	Losing. S.	° ° °	°	S.		
June 28 P.M.		39 32.7					1.75			
						83.5, 80.2 } 81.85		+		
,, 29 A.M.			174.5	2.57	171.93	} 81.18	1.75 } 1.747	5.02	+0.50	86233.59
						82.3, 78.7 } 80.5				
,, 29 P.M.		36 38.2					1.74			
						83, 78.2 } 80.6				
,, 30 A.M.			174.2	2.57	171.63	} 80.6	1.75 } 1.747	5.02	+0.24	86233.63
						83.8, 77.4 } 80.6				
,, 30 P.M.		33 44					1.75			
						84, 77.5 } 80.75				
July 1 A.M.			174	2.57	171.43	} 80	1.75 } 1.747	5.02	−0.02	86233.57
						81, 77.5 } 79.25				
,, 1 P.M.		30 50					1.74			
						82.7, 77.8 } 80.25				
,, 2 A.M.	2 00 00		174.7	2.57	172.13	} 80.27	1.74 } 1.743	5.00	+0.10	86232.97
						77.8, 82.8 } 80.3				
,, 2 P.M.		27 55.3					1.75			
						83.5, 77.4 } 80.45				
,, 3 A.M.			174.3	2.57	171.73	} 80.28	1.75 } 1.747	5.02	+0.10	86233.39
						77.4, 82.8 } 80.1				
,, 3 P.M.		25 01					1.74			
						83.8, 77.6 } 80.7				
,, 4 A.M.			174.5	2.57	171.93	} 79.35	1.74 } 1.743	5.00	−0.31	86232.76
						78.5, 77.5 } 78				
,, 4 P.M.		22 06.5					1.75			
						79.5, 75 } 77.25				
,, 5 A.M.			174.5	2.57	171.93	} 78.67	1.76 } 1.76	5.10	−0.61	86232.56
						77.7, 82.3 } 80				
,, 4 P.M.		19 12					1.77			
MEANS.					171.81	80.05	1.748	5.03		86233.21

ASCENSION.——VIBRATIONS of PENDULUM No. 2, in the PENDULUM CLOCK.
Mean Height of the Barometer 30.07 Inches.

DATE.	Chronometer.	Pendulum Clock.	Clock's loss on Chron.	DAILY RATES. Chron.	DAILY RATES. Clock.	Temperature.		Arc of Vibration.		Correction for the Arc.	Reduction to a Mean Temp.	Vibrations per Diem at 79°.44.
1822.	H. M. S.	M. S.	S.	Gaining. S.	Losing. S.	° °	°	°	°	S.	S.	
July 5 P.M.	9 55 00	4 38				75.5, 78.5 — 77		1.75		+		
„ 6 A.M.		4 06.2	64.4	2.53	61.82	76.7, 83.5 — 80.1	78.55	1.75	1.75	5.04	−0.39	86342.83
„ 6 P.M.		3 33.6				77, 78.5 — 77.75		1.75				
„ 7 A.M.		3 01.8	64.8	2.58	62.22	78.5, 83.8 — 81.15	79.45	1.75	1.75	5.04	. . .	86342.82
„ 7 P.M.		2 28.8				79, 77.5 — 78.25		1.75				
„ 8 A.M.		1 57.2	64.3	2.58	61.72	78.5, 84 — 81.25	79.75	1.75	1.75	5.04	+0.14	86343.46
„ 8 P.M.		1 24.5				79.4, 78.4 — 78 9		1.75				
„ 9 A.M.		0 52.9	64.6	2.58	62.02	78.5, 83.8 — 81.15	80.02	1.75	1.75	5.04	−0.25	86343.27
„ 9 P.M.		0 19.9						1.75				
MEANS						79.44		1.75		5.04		86343.09

BAHIA.

The clock was set up in the same room at Bahia in which the experiments with the detached pendulums were made: the floor was boarded, but as the clock was situated in a part of the room which was not approached except for the purposes of comparison and registry, and as the joists, on which the floor was laid, rested directly on the foundation of the house, the support was regarded as sufficiently firm.

The rate of the Chronometer No. 423 from the 23d of July to the afternoon of the 2d of August is taken from the observations of which the results are collected in page 66; the rate from the 3d to the afternoon of the 5th is deduced from the zenith distances on the afternoon of the 2d in page 65, and those which are subjoined, observed on the afternoon of the 5th.

Bahia.——OBSERVATIONS to DETERMINE the RATE of the Chronometer 423, by ZENITH DISTANCES of the Sun, observed with a Repeating Circle.—Latitude 12° 59′ 21″ S.

August 5th P.M.; Barometer 30.05; Thermometer 73°, ☉'s U.L.

	Chronometer.	Level.		Readings, &c.			Chronometer.	Level.		Readings, &c.	
	H. M. S.				° ′ ″		H. M. S.				° ′ ″
	6 25 59.2	+2	+1	First Vernier	24 54 45		7 01 28.4	−2	−4	First Vernier	96 34 05
	6 27 22	+8	+8	Second ,,	54 20		7 02 54	+6	+5	Second ,,	33 45
	6 29 16.1	−5	−6	Third ,,	55 00		7 04 10	−2	−1	Third ,,	34 20
	6 30 52.8	+2	+2	Fourth ,,	54 20		7 06 17.2	0	0	Fourth ,,	33 30
	6 32 40.4	+3	+2	Mean . .	24 54 36		7 07 40.4	+2	+1	Mean . . .	96 33 55
	6 34 13.6	+2	+2	Index . . +	360 00 00 0 00 08.5		7 08 50	+2	+1	Index . . .	+335 05 24
Mean. .	6 30 04.07	+12	+9	Level . . .	+10.5	Mean. . .	7 05 18.33	+6	−1	Level . . .	2.5
True time.	3 56 41.13	+10.5			384 54 55	True time.	5 31 55.3	+2.5			431 39 21
Chron. fast	2 33 22.94			Observed Z.D.	64 09 09	Chron. fast	2 33 23.03			Observed Z.D.	71 56 33
				Ref. and Paral.	+1 46					Ref. and Paral.	+2 40
				Semidiam . .	+15 48		360° − 24° 54′ 36″ = 335° 05′ 24″			Semidiam . .	+15 48
				True Z.D. . .	64 26 43					True Z.D. . .	72 15 01

Chronometer, Fast { H. M. S. 2 33 22.94 / 2 33 23.03 } H. M. S. 2 33 22.98

		H. M. S.
August 2d P.M.	Chronometer Fast	2 33 14.1
,, 5th P.M.	Chronometer Fast	2 33 22.98
	Chronometer's Gain in 3 Days	8.88 = 2.96 seconds per Diem.

Bahia.—VIBRATIONS of PENDULUM No. 1, in the PENDULUM CLOCK.
Mean Height of the Barometer 29.93 Inches.

DATE.	Chronometer.	Pendulum Clock.	Clock's loss on the Chron.		DAILY RATES. Chron.	DAILY RATES. Clock.	Temperature.			Arc of Vibration.		Correction for the Arc.	Reduction to a Mean Temp.	Vibrations per Diem at 73°.27
1822.	H. M. S.	M. S.	S.	S.	Gaining.	Losing.	°	°	°	°	°	S.	S.	
July 23 A.M.		30 06.7								1.74				
			85.7		s.	s.	72.3 75.4	73.85				+		
„ 23 P.M.		28 41		171.1	2.68	168.42			73.47	1.74	1.737	4.97	+0.09	86236.64
			85.4				74.7 71.5	73.1						
„ 24 A.M.		27 15.6								1.73				
			85.8				72 75	73.5						
„ 24 P.M.		25 49.8		171.9	2.68	169.22			73.35	1.73	1.727	4.91	+0.04	86235.73
			86.1				74.4 72	73.2						
„ 25 A.M.		24 23.7								1.72				
			84.9				72.2 75	73.6						
„ 25 P.M.	10 00 00	22 58.8		871.2	2.68	168.52			73.2	1.74	1.73	4.93	−0.03	86236.38
			86.3				74.3 71.4	72.8						
„ 26 A.M.		21 32.5								1.73				
			85.7				71.4 74.6	73						
„ 26 P.M.		20 06.8		171	2.68	168.32			72.9	1.72	1.727	4.91	−0.16	86236.43
			85.3				73.5 72.1	72.8						
„ 27 A.M.		18 41.5								1.73				
			85.7				72.7 75	73.9						
„ 27 P.M.		17 15.8		171.5	2.68	168.82			73.45	1.73	1.73	4.93	+0.08	86236.19
			85.8				71.4 74.6	73						
„ 28 A.M.		15 50								1.73				
MEANS						168.66	73.27			1.73		4.93		86236.27

Bahia.——VIBRATIONS of PENDULUM No. 2, in the PENDULUM CLOCK.
Mean Height of the Barometer 30.00 Inches.

DATE.	Chronometer.	Pendulum Clock.	Clock's loss on Chron.	DAILY RATES. Chron.	DAILY RATES. Clock.	Temperature.	Arc of Vibration.	Correction for the Arc.	Reduction to a Mean Temp.	Vibrations per Diem at 71°.72.
1822.	H. M. S.	M. S.	S. S.	Gaining. S.	Losing. S.	° ° °	° °	S. +	S.	
July 29 A.M.	11 00 00	13 06.2					1.71			
,, 29 P.M.		12 35.8	30.4, 29.8 } 60.2	2.68	57.52	{71.5, 72.8} 72.15; {71.8, 70.2} 71; 71.58	1 70; 1.703	4.79	−0.06	86347.21
,, 30 A.M.		12 06					1.70			
,, 30 P.M.		11 35.4	30.6, 30 } 60.6	2.68	57.92	{70.5, 72.8} 71.65; {72.5, 70.5} 71.5; 71.58	1.70; 1.70	4.77	−0.06	86346.79
,, 31 A.M.		11 05.4					1.70			
,, 31 P.M.		10 35	30.4, 30 } 60.4	2.68	57.72	{71, 72.3} 71.65; {71.7, 70.4} 71.05; 71.35	1.70; 1.703	4.79	−0.16	96346.91
Aug. 1 A.M.		10 05					1.71			
,, 1 P.M.		09 34.7	30.3, 30.2 } 60.5	2.68	57.82	{70.8, 72.3} 71.55; {72.2, 70.6} 71.4; 71.47	1.71; 1.71	4.83	−0.11	86346.90
,, 2 A.M.		09 04.5					1.71			
,, 2 P.M.		08 33.9	30.6, 20.3 } 60.9	2.68	58.22	{70.8, 73.5} 72.15; {73.3, 71.7} 72.5; 72.32	1.71; 1.71	4.83	+0.26	86346.87
,, 3 A.M.		08 03.6					1.71			
,, 3 P.M.		07 33.6	30, 30.6 } 60.6	2.96	57.64	{71.7, 73.2} 72.45; {72.3, 71} 71.65; 72.05	1.71; 1.71	4.83	+0.14	86347.33
,, 4 A.M.		07 03					1.71			
,, 4 P.M.		06 33	30, 30.6 } 60.6	2.96	57.64	{71, 72.5} 71.75; {72.3, 70.2} 71.25; 71.5	1.71; 1.71	4.83	−0.10	86347.09
,, 5 A.M.		06 02.4					1.71			
,, 5 P.M.		05 32.2	30.2, 30.2 } 60.4	2.96	57.44	{70.4, 73.3} 71.85; {72.7, 71.3} 72; 71.92	1.71; 1.71	4.83	+0.09	86347.48
,, 6 A.M.		05 02					1.71			
MEANS.					57.74	71.72	1.707	4.81		86347.07

MARANHAM.

The clock was set up at Maranham in the same room in which the detached pendulums were used, the floor of which was paved with stone. The rate of the Chronometer 423 is taken from page 83.

Maranham.—VIBRATIONS of PENDULUM No. 2, in the PENDULUM CLOCK.
Mean Height of the Barometer 29.92 Inches.

DATE.	Chronometer.	Pendulum Clock.	Clock's loss on Chron.		Daily Rates. Chron.	Daily Rates. Clock.	Temperature.			Arc of Vibration.		Correction for the Arc.	Reduction to a Mean Temp.	Vibrations per Diem at 81°.03.
1822.	H. M. S.	M. S.	S.	S.	Gaining. S.	Losing. S.	°	°	°	°	°	S. +	S.	
Aug. 30 P.M.		10 12.7								1.75				
			39.9				81 / 80.6	80.8						
„ 31 A.M.		9 32.8		80	2.7	77.3			80.8	1.75	1.75	5.04	−0.10	86327.64
			40.1				80.4 / 81.2	80.8						
„ 31 P.M.		8 52.7								1.75				
			39.9				81.2 / 80.7	80.95						
Sept. 1 A.M.		8 12.8		80.2	2.7	77.5			80.95	1.75	1.75	5.04	−0.03	86327.51
			40.3				80.7 / 81.2	80.95						
„ 1 P.M.		7 32.5								1.75				
			40				81.2 / 80.6	80.9						
„ 2 A.M.	12 00 00	6 52.5		79.9	2.7	77.2			80.92	1.75	1.75	5.04	−0.03	86327.81
			39.9				80.6 / 81.3	80.95						
„ 2 P.M.		6 12.6								1.75				
			39.3				81.5 / 81.1	81.3						
„ 3 A.M.		5 33.3		78.7	2.7	76			81.22	1.75	1.75	5.04	+0.08	86329.12
			39.4				81.1 / 81.2	81.15						
„ 3 P.M.		4 53.9								1.75				
			39.4				81.5 / 81.2	81.35						
„ 4 A.M.		4 14.5		79	2.7	76.3			81.25	1.75	1.75	5.04	+0.10	86328.84
			39.6				81 / 81.3	81.15						
„ 4 P.M.		3 34.9								1.75				
MEANS						76.86			81.03		1.75	5.04		86328.18

MARANHAM.——VIBRATIONS of PENDULUM No. 1 in the PENDULUM CLOCK.
Mean Height of the Barometer 29.93 Inches.

DATE.	Chronometer.	Pendulum Clock.	Clock's loss on Chron.		DAILY RATES. Chron.	DAILY RATES. Clock.	Temperature.			Arc of Vibration.		Correction for the Arc.	Reduction to a mean Temp.	Vibrations per Diem at 81°.39.
	H. M. S.	H. M. S.	S.	S.	Gaining.	Losing.	°	°	°	°	°	S.	S.	
Aug. 23 A.M.		12 01 51.8								1.71				
			95		s.	s.	81.6 82.4	82						
" 23 P.M.		12 00 16.8		189.6	2.7	186.9			82.07	1.72	1.713	4.85	+0.30	86218.25
			94.6				82.4 81.9	82.15						
" 24 A.M.		11 58 42.2								1.71				
			95.2				81.9 82.2	82.05						
" 24 P.M.		11 57 07		189.8	2.7	187.1			82.1	1.71	1.71	4.83	+0.31	86218.04
			94.6				82.6 81.7	82.15						
" 25 A.M.		11 55 32.4								1.71				
			94.8				81.4 81.7	81.55						
" 25 P.M.		11 53 57.6		189.4	2.7	186.7			81.68	1.71	1.71	4.83	+0.13	86218.26
			94.6				82.1 81.5	81.8						
" 26 A.M.		11 52 23								1.71				
			94.8				81 81.3	81.15						
" 26 P.M.	12 00 00	11 50 48.2		189.2	2.7	186.5			81.3	1.71	1.71	4.83	−0.04	86218.26
			94.4				81.7 81.2	81.45						
" 27 A.M.		11 49 13.8								1.71				
			94.3				81.4 81	81.2						
" 27 P.M.		11 47 39.5		188.8	2.7	186.1			81.1	1.71	1.71	4.83	−0.13	86218.60
			94.5				81.2 80.8	81						
" 28 A.M.		11 46 05								1.71				
			94.5				80.8 81	80.9						
" 28 P.M.		11 44 30.5		189	2.7	186.3			80.85	7.71	1.71	4.83	−0.24	86218.29
			94.5				81 80.6	80.8						
" 29 A.M.		11 42 56								1.71				
			94.7				80.4 80.7	80.55						
" 29 P.M.		11 41 21.3		189.4	2.7	186.7			80.65	1.7	1.703	4.79	−0.33	86217.76
			94.7				80.9 80.6	80.75						
" 30 A.M.		11 39 46.6								1.7				
MEANS						186.6			81.39		1.709	4.81		86218.21

TRINIDAD.

The station of the pendulum clock at Port Spain was in the vestry of the Protestant church, which was paved with stone. The rate of the chronometer 423 from the 23d of September to the 4th of October is taken from page 97: and from the 4th, to the evening of the 8th of October, the rate is deduced from the zenith distances on the morning of the 4th in page 96, and the subjoined observations on the morning of the 10th.

Trinidad.——OBSERVATIONS to DETERMINE the RATE of the Chronometer No. 423, by ZENITH DISTANCES of the Sun, observed with a Repeating Circle.—Latitude 10° 38′ 43″ N.

October 10th A.M.; Barometer 30.00; Thermometer 81°; ☉'s L.L.

Chronometer.	H. M. S.	Level.		Readings, &c.	° ′ ″
	11 25 07.2	+4	+3	First Vernier	52 08 20
	11 26 32.4	+8	+6	Second „	08 00
	11 28 26.8	+2	+1	Third „	08 30
	11 29 20.8	−2	−4	Fourth „	08 00
	11 30 57.6	0	0	Mean . . .	52 08 12.5
	11 31 52.8	−3	−5	Index . . +{	360 00 00 0 00 08.5
Mean. . .	11 28 42.93	+9	+1	Level . . .	+5
True time .	7 20 51.55	+5			412 08 26.5
Chron. fast	4 07 51.38			Observed Z.D.	68 41 24.5
				Ref. and Paral.	+2 10
				Semidiam . .	−16 04
				True Z.D. . .	68 27 31

Chronometer.	H. M. S.	Level.		Readings, &c.	° ′ ″
	11 35 55.6	+7	+6	First Vernier	88 25 40
	11 37 38.4	+6	+5	Second „	25 00
	11 38 58.8	+16	+16	Third „	25 50
	11 40 32.6	−3	−5	Fourth „	25 00
	11 41 55.2	−3	−4	Mean . . .	88 25 22.5
	11 43 16	0	−1	Index . . . +	307 51 47.5
Mean. . .	11 39 42.93	+23	+17	Level . . .	+20
True time.	7 31 52	+20			396 17 40
Chron. fast	4 07 50.93			Observed Z.D	66 02 50
				Ref. and Paral.	+1 58
$360^{\circ}-52^{\circ}\ 08'\ 12''.5=307^{\circ}\ 51'\ 47''.5$				Semidiam . .	−16 04
				True Z.D. . .	65 48 48

Chronometer, Fast $\left\{\begin{matrix} \text{H. M. S.} \\ 4\ 07\ 51.38 \\ 4\ 07\ 50.93 \end{matrix}\right\}$ H. M. S. 4 07 51.15

		H. M. S.
October 4th A.M.	Chronometer Fast	4 07 34.53
„ 10th A.M.	Chronometer Fast	4 07 51.15
	Chronometer's Gain in 6 Days	16.62 = 2.8 seconds per Diem.

TRINIDAD.——VIBRATIONS of PENDULUM No. 1, in the PENDULUM CLOCK.
Mean Height of the Barometer 29.92 Inches.

DATE.	Chronometer.	Pendulum Clock.	Clock's loss on Chron.	DAILY RATES. Chron.	DAILY RATES. Clock.	Temperature.	Arc of Vibration.	Correction for the Arc.	Reduction to a mean Temp.	Vibrations per Diem at 82.98.
1822.	H. M. S.	M. S.	S.	Gaining. S.	Losing. S.	° ° °	° °	S. +	S.	
Sept. 23 P.M.	9 15 00	38 33					1.76			
" 24 A.M.			182.4	3.19	179.21	84, 81.4 } 82.7; 83, 84.1 } 83.55; } 8 2	1.76 } 1.76	5.10	+0.11	86226.00
" 24 P.M.		35 30.6					1.76			
" 25 A.M.			183	3.19	179.81	81.7, 84.4 } 83.05; 84, 83.2 } 83.6; } 83.32	1.76 } 1.76	5.10	+0.15	86225.44
" 25 P.M.		32 27.6					1.76			
" 26 A.M.			182.2	3.19	179.01	81.7, 84 } 82.85; 83.6, 83.1 } 83.35; } 83.1	1.76 } 1.763	5.12	+0.05	86226.16
" 26 P.M.		29 25.4					1.77			
" 27 M.			182.2	3.19	179.01	81.9, 83.5 } 82.7; 83.5, 82.3 } 82.9; } 82.8	1.76 } 1.76	5.10	−0.08	86226.01
" 27 P.M.		26 23.2					1.75			
" 28 A.M.			182.4	3.19	179.21	81.3, 82.8 } 82.05; 82.4, 83.8 } 83.1; } 82.58	1.75 } 1.75	5.04	−0.18	86225.65
" 28 P.M.		23 20.8					1.75			
MEANS					179.25	82.98	1.759	5.09		86225.85

TRINIDAD.——VIBRATIONS of PENDULUM No. 2, in the PENDULUM CLOCK. Mean Height of the Barometer 29.94 Inches.

DATE.	Chronometer.	Pendulum Clock.	Clock's loss on Chron.	DAILY RATES.		Temperature.		Arc of Vibration.		Correction for the Arc.	Reduction to a Mean Temp.	Vibrations per Diem at 82°.82.
				Chron.	Clock.							
1822.	H. M. S.	M. S.	S.	Gaining. S.	Losing. S.	°	°	°	°	S. +	S.	
Sep. 29 P.M.		32 27.3						1.72				
„ 30 A.M.		.....	72.6	3.19	69.41	82.8 / 83.6	83.2	1.72	1.72	4.86	+0.17	86335.62
„ 30 P.M.		31 14.7						1.72				
Oct. 1 A.M.		. ...	72.6	3.19	69.41	83.8 / 82.5	83.15	1.74	1.737	4.96	+0.15	86335.70
„ 1 P.M.		30 02.1						1.75				
„ 2 A.M.		.....	72.7	3.19	69.51	84.4 / 82.9	83.65	1.76	1.76	5.10	+0.36	86335.95
„ 2 P.M.		28 49.4						1.77				
„ 3 A.M.		.....	73	3.19	69.81	84.4 / 82	83.2	1.77	1.77	5.16	+0.17	86335.52
„ 3 P.M.		27 36.4						1.77				
„ 4 A.M.	2 30 00	.....	71.9	2.8	69.1	84 / 81.6	82.8	1.77	1.77	5.16	...	86336.06
„ 4 P.M.		26 24.5						1.77				
„ 5 A.M.		.....	71.9	2.8	69.1	82.3 / 83.7	83	1.77	1.77	5.16	+0.08	86336.14
„ 5 P.M.		25 12.6						1.77				
„ 6 A.M.		.....	72	2.8	69.2	82.7 / 81.8	82.25	1.77	1.77	5.16	−0.25	86335.71
„ 6 P.M.		24 00.6						1.77				
„ 7 A.M.		.....	71.9	2.8	69.1	82.8 / 81.9	82.35	1.77	1.77	5.16	−0.21	86335.85
„ 7 P.M.		22 48.7						1.77				
„ 8 A.M.		.....	72	2.8	69.2	82.3 / 81.3	81.8	1.77	1.77	5.16	−0.44	86335.52
„ 8 P.M.		21 36.7						1.77				
MEANS					69.32	82.82		1.76		5.10		86335.79

JAMAICA.

THE pendulum clock was set up at Port Royal in a room adjoining the one in which the experiments with the detached pendulums were made; it was on the same level, but had the advantage of a brick floor; the windows were darkened, and the room entered only for the purposes of comparison and registry.

The rate of the Chronometer No. 423 from the 22d to the 30th of October is taken from page 109; from the 30th of October to the 5th of November the rate is deduced from the Zenith Distances on the morning of the 30th in page 108, and those which are subjoined observed on the morning of the 5th of November.

JAMAICA.—OBSERVATIONS to DETERMINE the RATE of the Chronometer No. 423, by ZENITH DISTANCES of the Sun, observed with a Repeating Circle.—Latitude 17° 55′ 55″ N.

November 5th A.M.; Barometer 30.00; Thermometer 81°; ☉'s L.L.

	Chronometer.	Level.		Readings, &c.			Chronometer.	Level.		Readings, &c.	
	H. M. S.				° ′ ″		H. M. S.				° ′ ″
	0 28 49.6	−5	−6	First Vernier	34 25 10		1 05 52.4	+3	+2	First Vernier	346 45 50
	0 31 06	0	0	Second „	25 10		1 07 02.8	+5	+4	Second „	45 30
	0 32 36.4	0	0	Third „	25 40		1 08 38.4	−5	−4	Third „	46 00
	0 33 59.2	0	0	Fourth „	25 00		1 09 48.4	+4	+3	Fourth „	45 40
	0 35 46.4	+2	+3	Mean . . .	34 25 15		1 11 15.2	0	0	Mean . . .	346 45 45
	0 37 06.8	−1	−2	Index . . +	360 00 00 42 15 41		1 12 16	+4	+3	Index . . .	43 46 28
Mean. . .	0 33 14.07	−4	−5	Level . . .	−5	Mean. . .	1 09 08.87	+11	+8	Level . . .	+9
True time .	7 22 30.47	−1.5			436 40 51	True time .	7 58 25.8	+9.5			390 32 22
Chron. fast	5 10 43.6			Observed Z.D.	72 46 49	Chron. fast	5 10 43.07			Observed Z.D.	65 05 24
				Ref. and Paral.	+2 48					Ref. and Paral.	+ 1 50
360°−317° 44′ 19″=42° 15′ 41″				Semidiam . .	−16 10	360°−316° 13′ 32″=43° 46′ 28″				Semidiam . .	−16 10
				True Z.D.. .	72 33 27					True Z.D. . .	64 51 04

Chronometer, Fast { H. M. S. 5 10 43.6 / 5 10 43.07 } H. M. S. 5 10 43.33

		H. M. S.
October 30th A.M.	Chronometer Fast	5 10 23.83
November 5th A.M.	Chronometer Fast	5 10 43.33
	Chronometer's Gain in 6 Days . . .	19.5 = 3.25 seconds per Diem.

JAMAICA.——VIBRATIONS of PENDULUM No. 1, in the PENDULUM CLOCK.
Mean Height of the Barometer 29.96 Inches.

DATE.	Chronometer.	Pendulum Clock.	Clock's loss on Chron.	DAILY RATES. Chron.	DAILY RATES. Clock.	Temperature.		Arc of Vibration.		Correction for the Arc.	Reduction to a Mean Temp.	Vibrations per Diem at 80°.96.
1822.	H. M. S.	M. S.	S.	Gaining. s.	Losing. s.	°	°	°	°	S. +	S.	
Oct. 22 A.M.		20 21.4						1.74				
,, 22 P.M.		. . .	164.8	4.14	160.66	79 / 81.5	80.25		1.735	4.98	−0.31	86244.01
,, 23 A.M.		17 36.6						1.73				
,, 23 P.M.		. . .	165	4.14	160.86	81.3 / 79.3	80.3	. .	1.725	4.92	−0.29	86243.77
,, 24 A.M.		14 51.6						1.72				
,, 24 P.M.	1 50 00	.	165.1	4.14	160.96	81.8 / 79.8	80.8	. .	1.725	4.92	−0.07	86243.89
,, 25 A.M.		12 06.5						1.73				
,, 25 P.M.		. .	164.7	4.14	160.56	80.2 / 82.6	81.4	. .	1.72	4.89	+0.19	86244.52
,, 26 A.M.		09 21.8						1.71				
,, 26 P.M.		. . .	165.8	4.14	161.66	83.6 / 80.5	82.05	. .	1.72	4.89	+0.48	86243.71
,, 27 A.M.		06 36						1.73				
MEANS					160.94	80.96		1.725		4.92		86243.98

JAMAICA.——VIBRATIONS of PENDULUM No. 2, in the PENDULUM CLOCK.
Mean Height of the Barometer 29.96 Inches.

DATE.	Chronometer.	Pendulum Clock.	Clock's loss on Chron.	DAILY RATES. Chron.	DAILY RATES. Clock.	Temperature.		Arc of Vibration.		Correction for the Ac.	Reduction to a Mean Temp.	Vibrations per Diem at 82°.28.
1822.	H. M. S.	M. S.	S.	Gaining. S.	Losing. S.	°	°	°	°	S.	S.	
Oct. 28 A.M.	6 00 00	57 14.6						1.70				
„ 28 P.M.		.....	55.6	4.14	51.46	80.5 / 83.7	82.1	..	1.705	4.80	−0.08	86353.26
„ 29 A.M.		56 19						1.71				
„ 29 P.M.		.....	54.7	4.14	51.56	81.6 / 83.8	82.7	..	1.715	4.86	+0.18	86353.48
„ 30 A.M.		55 24.3						1.72				
„ 30 P.M.		.....	54.5	3.25	51.25	80.8 / 83.8	82.3	..	1.715	4.86	+0.01	86353.64
„ 31 A.M.		54 29.8						1.71				
„ 31 P.M.		.....	54.5	3.25	51.25	81 / 83.4	82.2	..	1.71	4.83	−0.04	86353.56
Nov. 1 A.M.		53 35.3						1.71				
„ 1 P.M.		.....	54.5	3.25	51.25	81.4 / 83.4	82.4	..	1.71	4.83	+0.05	86353.65
„ 2 A.M.		52 40.8						1.71				
„ 2 P.M.		.....	54.5	3.25	51.25	80.8 / 83.8	82.3	..	1.71	4.83	+0.01	86353.61
„ 3 A.M.		51 46.3						1.71				
„ 3 P.M.		.....	54.5	3.25	51.25	83.5 / 80.9	82.2	..	1.71	4.83	−0.04	86353.56
„ 4 A.M.		50 51.8						1.71				
„ 4 P.M.		.....	54.6	3.25	51.35	80.4 / 83.7	82.05	..	1.71	4.83	−0.10	86353.40
„ 5 A.M.		49 57.2						1.71				
MEANS					51.31		82.28		1.71	4.83		86353.52

NEW YORK.

The experiments with the pendulum clock at Columbia College were made in a room adjoining the library of the college, in the story beneath the detached pendulums: the height above the sea was 53 feet.

The rate of the Chronometer No. 423 from the 22d of December to the 3d of January is taken from the observations of which the results are collected in page 127; the rate from the 16th of December to the 22d was obtained by Zenith Distances of the sun, west of the meridian, on the 16th (which are subjoined,) compared with the corresponding observations on the 22d, detailed in page 121.

New York.——OBSERVATIONS to DETERMINE the RATE of the Chronometer No. 423, by ZENITH DISTANCES of the Sun, West of the Meridian.—Latitude 40° 42′ 43″ N. Longitude 74° 03′.5 W.

December 16th P.M.; Barometer 30.00; Thermometer 32°; ☉'s U.L.

Chronometer.		Level.		Readings, &c.		Chronometer.		Level.		Readings, &c.	
	H. M. S.				° ′ ″		H. M. S.				° ′ ″
	7 59 29.6	0	0	First Vernier	281 33 20		8 12 38	−8	0	First Vernier	34 00 00
	8 00 53.6	+8	−2	Second „	33 15		8 14 05.2	−2	+7	Second „	33 59 40
	8 02 35.2	0	0	Third „	33 50		8 15 44.8	−2	−11	Third „	34 00 00
	8 03 54.4	0	0	Fourth „	33 20		8 17 15.6	+9	+1	Fourth „	33 59 40
	8 05 51.8	+8	+1	Mean . . .	281 33 26		8 20 05.6	−3	−12	Mean . .	33 59 50
	8 08 00.8	−9	0	Index . . .	+180 13 00		8 21 48.4	−3	+6	Index . . +	360 00 00 78 26 34
Mean . .	8 03 28.07	+17	−11	Level . . .	+03	Mean .	8 16 56.27	−41	+23	Level . .	−09
True time .	3 02 19.4	+3			461 46 29	True time .	3 15 48.23	−9			472 26 15
Chron. fast	5 01 08.64			Observed Z.D.	76 57 45	Chron. fast	5 01 08.04			Observed Z.D.	78 44 22.5
				Ref. and Paral.	+4 08					Ref. and Paral.	+4 48
360 − 179 47 00 = 180 13 00				Semidiam. . .	+16 17	360 − 281 33 26 = 78 26 34				Semidiam. . .	+16 17
				True Z.D. . .	77 18 10					True Z.D. . .	79 05 27

Chronometer Fast {H. M. S. 5 01 08.64 / 5 01 08.04} H. M. S. 5 01 08.34

		H. M. S.
December 16th P.M.	Chronometer Fast	5 01 08.34
December 22d P.M.	Chronometer Fast	5 01 01.65
	Chronometer's Loss in 6 Days . . .	6.69 = 1.12 seconds per Diem.

The intervals at which the clock was compared, and the arc and temperature registered, were of twenty-four hours duration instead of twelve, in consequence of my residence at New York being at some distance from the College.

New York.——VIBRATIONS of PENDULUM No. 1, in the PENDULUM CLOCK.
Mean Height of the Barometer 30.20 Inches.

DATE.	Chronometer.	Pendulum Clock.	Clock's loss on Chron.	Daily Rates. Chron.	Daily Rates. Clock.	Temperature.	Arc of Vibration.	Correction for the Arc.	Reduction to a Mean Temp.	Vibrations per Diem at 37°.95.
1822.	H. M. S.	M. S.	S.	Losing. S.	Losing. S.	° °	° °	S. +	S.	
Dec. 16 A.M.		09 14.4					1.73			
			68.4	1.12	69.52	{37.4, 39.4} 38.4	1.73	4.93	+0.20	86335.61
„ 17 A.M.		08 06					1.73			
			68.7	1.12	69.82	{38, 39.7} 38.85	1.73	4.93	+0.40	86335.51
„ 18 A.M.		06 57.3					1.73			
			69.3	1.12	70.42	{39.7, 42.3} 41	1.73	4.93	+1.34	86335.85
„ 19 A.M.		05 48					1.73			
			71.4	1.12	72.52	{41.9, 43.6} 42.75	1.725	4.91	+2.11	86334.50
„ 20 A.M.		04 36.6					1.72			
			71.4	1.12	72.52	{41.2, 43.7} 42.45	1.72	4.88	+1.98	86334.34
„ 21 A.M.*		03 25.2					1.72			
	7 10 00		69.6	1.12	70.72	{42.8, 39.8} 41.3	1.715	4.86	+1.47	86335.61
„ 22 A.M.		02 15.6					1.71			
			68.6	2.62	71.22	{42, 37.2} 39.6	1.70	4.77	+0.75	86334.30
„ 23 A.M.		01 07					1.68			
			65.8	2.62	68.42	{36.2, 31} 33.6	1.685	4.68	−1.91	86334.35
„ 24 A.M.		00 01.2					1.68			
			65.7	2.62	68.32	{31.2, 36.2} 33.7	1.68	4.64	−1.87	86334.45
„ 25 A.M.		58 55.5					1.68			
			65.	2.62	67.62	{31.4, 34.7} 33.05	1.68	4.64	−2.16	86334.86
„ 26 A.M.		57 50.5					1.68			
			64.3	2.62	66.92	{31, 34.4} 32.7	1.68	4.64	−2.31	86335.41
„ 27 A.M.		56 46.2					1.68			
MEANS					69.82	37.95	1.707	4.80		86334.98

New York.——VIBRATIONS of PENDULUM No. 2, in the PENDULUM CLOCK.
Mean Height of the Barometer 30.26 Inches.

DATE.	Chronometer.	Pendulum Clock.	Clock's gain on Chron.	DAILY RATES. Chron.	DAILY RATES. Clock.	Temperature.	Arc of Vibration.	Correction for the Arc.	Reduction to a Mean Temp.	Vibrations per Diem at 33°.88.
1822.	H. M. S.	M. S.	S.	Losing. S.	Gaining. S.	° °	° °	S. +	S.	
Dec. 28 A.M.		42 50.5				31.2	1.68			
„ 28 P.M.		. . . .	46	2.62	43.38	32.95	. . 1.68	4.65	−0.41	86447.62
						34.7				
„ 29 A.M.		43 36.5				32	1.68			
„ 29 P.M.		. . .	44.9	2.62	42.28	32.75	. . 1.68	4.65	−0.50	86446.43
						33.5				
„ 30 A.M.		44 21.4				35.3	1.68			
„ 30 P.M.		. . . .	45.6	2.62	42.98	32.3	. . 1.68	4.65	−0.69	86446.94
						29.3				
„ 31 A.M.		45.07				33.4	1.68			
„ 31 P.M.	7 10 00	. . .	45.7	2.62	43.08	31.6	. . 1.69	4.71	−1.00	86446.79
1823.						29.8				
Jan. 1 A.M.		45 52.7				36	1.70			
„ 1 P.M.		. .	45.3	2.62	42.68	33.5	. . 1.70	4.77	−0.15	86447.30
						31				
„ 2 A.M.		46 38				37.3	1.70			
„ 2 P.M.		. . . .	43.2	2.62	40.58	36	. . 1.70	4.77	+0.93	86446.28
						34.7				
„ 3 A.M.		47 21.2				36.8	1.70			
„ 3 P.M.		. . . .	43.6	2.62	40.98	38.05	. . 1.69	4.71	+1.81	86447.53
						39.3				
„ 4 A.M.		48 04.8					1.68			
MEANS				. .	42.28	33.88	1.69	4.70		86446.98

HAMMERFEST.

THE framed house constructed for the pendulums at the Northern Stations, and conveyed in the Griper, was of sufficient size to contain the pendulum clock as well as the detached pendulums, and to permit them to be set up at the same time without the interference of any part of their respective apparatus. The pendulum clock stood on the same side of the house as the astronomical clock, and the triangular frame by which it was supported rested on the ground or rock beneath the flooring of the room.

The comparisons and registry were repeated at intervals of twelve hours as usual. The Rate of the Chronometer No. 649, with which the clock was compared, is taken from the observations of which the results are collected in page 144.

HAMMERFEST.——VIBRATIONS of PENDULUM No. 1, in the PENDULUM CLOCK.
Mean Height of the Barometer 29.93 Inches.

DATE.	Chronometer.	Pendulum Clock.	Clock's gain on Chron.	DAILY RATES. Chron.	DAILY RATES. Clock.	Temperature.	Arc of Vibration.	Correction for the Arc.	Reduction to a Mean Temp.	Vibrations per Diem at 51°.9.
1822.	H. M. S.	M. S.	S. S.	Gaining. S.	Gaining. S.	° ° °	° °	S. +	S.	
June 9 A.M.	8 55 00	41 36	14.2			41.7, 51.7 → 46.7	1.67			
„ 9 P.M.		41 50.2	13.5 → 27.7	1.43	29.13	53.2, 50.2 → 51.7 → 49.2	1.68 → 1.687	4.68	−1.18	86432.63
„ 10 A.M.		42 03.7	12			53.2, 59.7 → 56.45	1.71			
„ 10 P.M.		42 15.7	13.3 → 25.3	1.43	26.73	56.7, 47.7 → 52.2 → 54.35	1.70 → 1.70	4.76	+1.08	86432.57
„ 11 A.M.		42 29	13.6			47.7, 54.7 → 51.2	1.69			
„ 11 P.M.		42 42.6	13.2 → 26.8	1.43	28.23	51, 54.4 → 52.7 → 51.95	1.67 → 1.68	4.64	...	86432.87
„ 12 A.M.		42 55.8	11.5			61.5, 53.3 → 57.4	1.68			
„ 12 P.M.		43 07.3	13.7 → 25.2	1.43	26.63	53.3, 48.3 → 50.8 → 54.1	1.68 → 1.677	4.62	+0.97	86432.22
„ 13 A.M.		43 21	13.8			49, 52.2 → 50.6	1.67			
„ 13 P.M.		43 34.8	14.2 → 28	1.43	29.43	50.7, 47.8 → 49.25 → 49.92	1.66 → 1.667	4.56	−0.88	86433.11
„ 14 A.M.		43 49					1.67			
MEANS					28.03	51.9	1.682	4.65		86432.68

HAMMERFEST.——VIBRATIONS of PENDULUM No. 2, in the PENDULUM CLOCK.
Mean Height of the Barometer 29.70 Inches.

DATE.	Chronometer.	Pendulum Clock.	Clock's gain on Chron.		DAILY RATES. Chron.	DAILY RATES. Clock.	Temperature.		Arc of Vibration.		Correction for the Arc.	Reduction to a Mean Temp.	Vibrations per Diem at 43°.
1823.	H. M. S.	M. S.	S.	S.	Gaining. S.	Gaining. S.	°	°	°	°	S. +	S.	
June 14 P.M.		25 49.4							1.67				
			68.4				56.1, 49 } 52.55						
„ 15 A.M.		26 57.8		138.3	1.38	139.68		50.12	1.62	1.59	4.16	+ 3.13	86546.97
			69.9				49, 46.4 } 47.7						
„ 15 P.M.		28 07.7							1.48				
			70.3				48, 44 } 46						
„ 16 A.M.		29 18		140.5	1.38	141.88		46.3	1.64	1.59	4.16	+ 1.45	86547.49
			70.2				48, 45.2 } 46.6						
„ 16 P.M.		30 28.2							1.64				
			71.6				45.2, 42.8 } 44						
„ 17 A.M.		31 39.8		142.4	1.38	143.78		43.88	1.44	1.54	3.88	+ 0.39	86548.05
			70.8				42.5, 45 } 43.75						
„ 17 P.M.		32 50.6							1.53				
			72.7				42.5, 39.3 } 40.9						
„ 18 A.M.		34 03.3		143.9	1.38	145 28		39.95	1.4	1.48	3.60	− 1.34	86547.54
			71.2				39.3, 38.7 } 39						
„ 18 P.M.	8 55 00	35 14.5							1.52				
			71.1				38.6, 39.6 } 39.1						
„ 19 A.M.		36 25.6		141.3	1.38	142.68		39.95	1.67	1.63	4.34	− 1.34	86545.68
			70.2				39.6, 42 } 40.8						
„ 19 P.M.		37 35.8							1.70				
			70				41, 41.5 } 41.25						
„ 20 A.M.		38 45.8		139.8	1.38	141.18		41.12	1.69	1.69	4.70	− 0.83	86545.05
			69.8				42, 40 } 41						
„ 20 P.M.		39 55.6							1.67				
			71.2				40.3, 37.8 } 39.05						
„ 21 A.M.		41 06.8		141.1	1.38	142.48		39.2	1.63	1.67	4.58	− 1.67	86545.39
			69.9				38.5, 40.2 } 39.35						
„ 21 P.M.		42 16.7							1.70				
			70				39.3, 41.8 } 40.55						
„ 22 A.M.		43 26.7		139.8	1.38	141.18		43.6	1.71	1.71	4.83	+ 0.26	86546.27
			69.8				41.8, 51.5 } 46.65						
„ 22 P.M.		44 36.5							1.72				
MEANS	. . . .					142.27	43		1.61		4.28		86546.55

SPITZBERGEN.

At Spitzbergen the pendulum clock was compared directly with the astronomical clock, at precise intervals of twelve hours of mean solar time as shewn by the Chronometer No. 649. The rate of the astronomical clock is taken from the transits and zenith distances, of which the results are collected in page 155.

Spitzbergen.——VIBRATIONS of PENDULUM No. 1, in the PENDULUM CLOCK.
Mean Height of the Barometer 29.90 Inches.

DATE.	Chronometer.	Astronomical Clock.	Pendulum Clock.	Pend. Clock's loss on Astr. Cl.	DAILY RATES. Astr. Cl.	DAILY RATES. Pend. Cl.	Temperature.	Arc of Vibration.	Correction for the Arc.	Reduction to a Mean Temp.	Vibrations per Diem at 44°.07.
1823.	H. M. S.	H. M. S.	M. S.	S.	Gaining. S.	Gaining. S.	° ° °	° °	S. +	S.	
July 7 P.M.		2 52 43.8	35 24.6					1.69			
„ 8 A.M.		. . .	. .	36.1	88.02	51.92	{48.2, 43.8} 46; {43.8, 45.6} 44.7; 45.35	1.68 } 1.68	4.65	+0.56	86457.13
„ 8 P.M.		2 54 09.8	36 14.5					1.67			
„ 9 A.M.		. . . .	. . .	35.2	88.25	53.03	{47.6, 41.4} 44.5; {40.8, 43.4} 42.1; 43.3	1.68 } 1.677	4.63	−0.34	86457.32
„ 9 P.M.		2 55 38.5	37 08					1.68			
„ 10 A.M.		. . .	. .	36.1	88.23	52.13	{43.4, 47.3} 45.35; {45.7, 41.7} 43.7; 44.52	1.69 } 1.69	4.71	+0.20	86457.04
„ 10 P.M.	3 00 00	2 57 05.6	37 59					1.70			
„ 11 A.M.			. . .	37.5	88.7	51.20	{45.7, 48.5} 47.1; {45.7, 50.7} 48.2; 47.65	1.69 } 1.697	4.75	+1.57	86457.52
„ 11 P.M.		2 58 31.8	38 47.7					1.70			
„ 12 A.M.		. .	.	36.5	88.68	52.18	{51, 42.3} 46.65; {41.4, 43.5} 42.45; 44.55	1.68 } 1.69	4.71	+0.21	86457.10
„ 12 P.M.		3 00 00.2	39 39.6					1.69			
„ 13 A.M.		. . .	. .	34	88.73	54.73	{38.1, 43.5} 40.8; {36.5, 38.1} 37.3; 39.05	1.67 } 1.68	4.65	−2.21	86457.17
„ 13 P.M.		3 01 29.6	40 35					1.68			
MEANS				. . . . .		52.53	44.07	1.686	4.69		86457.21

SPITZBERGEN.——VIBRATIONS of PENDULUM No. 2, in the PENDULUM CLOCK.
Mean Height of the Barometer 29.80 Inches.

DATE.	Chronometer.	Astronomical Clock.	Pendulum Clock.	Pend. Clock's gain on Astr. Cl.	DAILY RATES.		Temperature.	Arc of Vibration.	Correction for the Arc.	Reduction to a Mean Temp.	Vibrations per Diem at 37°.63.
					Astr. Cl.	Pend. Cl.					
1823.	H. M. S.	H. M. S.	M. S.	S.	Gaining.	Gaining.	° ° °	° °	S.		
July 14 P.M.		3 02 58	34 20.6					1.68			
					S.	S.	{36.5, 34} 35.25		+		
„ 15 A.M.		....	...	77.8	88.65	166.45	34.48	1.67 } 1.677	4.63	−1.39	86569.69
							{34, 33.4} 33.7				
„ 15 P.M		3 04 26	37 06.4					1.67			
							{34, 32.6} 33.3				
16 A.M.		...	...	78.4	88.64	167.04	33.6	1.67 } 1.673	4.61	−1.77	86569.88
							{32.6, 35.2} 33.9				
„ 16 P.M.		3 05 56.9	39 55.7					1.68			
							{35.2, 37.2} 36.2				
„ 17 A.M.		...	...	77	88.65	165.65	37.1	1.69 } 1.687	4.69	−0.23	86570.11
							{37.4, 38.6} 38				
„ 17 P.M.		3 07 26.6	42 42.4					1.69			
							{38, 36.7} 37.35				
„ 18 A.M.	3 00 00	.. .	...	76.9	88.70	165.6	37.4	1.69 } 1.69	4.71	−0.11	86570.20
							{36.3, 38.6} 37.45				
„ 18 P.M.		3 08 56.7	45 29.4					1.69			
							{38.6, 37.4} 38				
„ 19 A.M.		. .	...	75.9	88.70	164.6	38.95	1.69 } 1.69	4.71	+0.58	86569.89
							{37.4, 42} 39.9				
„ 19 P.M.		3 10 27	48 15.6					1.69			
							{42, 39.5} 40.75				
„ 20 A.M.		. ..	...	75.1	88.70	163.8	40.75	1.69 } 1.69	4.71	+1.37	86569.88
							{39.5, 42} 40.75				
„ 20 P.M.		3 11 56.8	51 00.5					1.69			
							{42.6, 40.6} 41.6				
„ 21 A.M.		....	...	75	88.70	163.7	41.12	1.69 } 1.69	4.71	+1.54	86569.95
							{40, 41.3} 40.65				
„ 21 P.M.		3 13 26.3	53 45					1.69			
MEANS						165.29	37.63	1.686	4.68		86569.97

GREENLAND.

THE pendulum clock was set up in the pendulum house at Greenland, as at the two preceding stations.

The rate of the Chronometer No. 423, with which the clock was compared, is taken from the transits of which the results are collected in page 167.

GREENLAND.——VIBRATIONS of PENDULUM No. 1, in the PENDULUM CLOCK.
Mean Height of the Barometer 29.90 Inches.

DATE.	Chronometer.	Pendulum Clock.	Clock's gain on Chron.	DAILY RATES. Chron.	DAILY RATES. Clock.	Temperature.	Arc of Vibration.	Correction for the Arc.	Reduction to a Mean Temp.	Vibrations per Diem at 37°.7.
1823.	H. M. S.	H. M. S.	S.	Gaining. S.	Gaining. S.	° ° °	° °	S. +	S.	
Aug. 26 P.M.		8 53 43.9					1.68			
						45.1, 32.7 } 38.9				
„ 27 A.M.		8 54 02.4	37.7	7.72	45.42	} 38.3	1.68 } 1.68	4.65	+0.26	86450.33
						32.7, 42.7 } 37.7				
„ 27 P.M.		8 54 21.6					1.68			
						42.3, 34 } 38.15				
„ 28 A.M.		8 54 40.6	37.6	7.21	44.81	} 38.45	1.66 } 1.667	4.57	+0.33	86449.71
						43.5, 34 } 38.75				
„ 28 P.M.	9 05 00	8 54 59.2					1.66			
						34.7, 43.1 } 37.9				
„ 29 A.M.		8 55 19.6	39.5	6.17	45.67	} 38.17	1.68 } 1.673	4.61	+0.21	86450.49
						42.2, 34.7 } 38.45				
„ 29 P.M.		8 55 38.7					1.68			
						30.4, 40.6 } 35.5				
„ 30 A.M.		8 55 58.7	39.7	6.67	46.37	} 35.8	1.68 } 1.68	4.65	−0.84	86450.18
						30.6, 41.6 } 36.1				
„ 30 P.M.		8 56 18.4					1.68			
MEANS					45.57	37.7	1.675	4.62		86450.18

GREENLAND.——VIBRATIONS of PENDULUM No. 2, in the PENDULUM CLOCK.
Mean Height of the Barometer 29.90 Inches.

DATE.	Chronometer.	Pendulum Clock.	Clock's gain on Chron.	DAILY RATES. Chron.	DAILY RATES. Clock.	Temperature.	Arc of Vibration.	Correction for the Arc.	Reduction to a Mean Temp.	Vibrations per Diem at 39°.
1823.	H. M. S.	H. M. S.	s.	Gaining. s.	Gaining. s.	° ° °	° °	s. +	s.	
Aug. 20 P.M.		9 42 57.8					1.67			
„ 21 A.M.		9 44 12	148.6	6.61	155.21	{33.1, 42.1} 37.6; {34.6, 44.6} 39.6; 38.6	1.68 } 1.67	4.59	−0.18	86559.62
„ 21 P.M.		9 45 26.4					1.66			
„ 22 A.M.		9 46 40.5	147.8	6.34	154.14	{36, 42} 39; {41, 44.2} 42.6; 40.8	1.66 } 1.66	4.53	+0.79	86559.46
„ 22 P.M.		9 47 54.2					1.66			
23 A.M.	9 05 00	9 49 09	149.2	5.60	154.80	{34.5, 40.5} 37.5; {36.6, 46.1} 41.35; 39.42	1.67 } 1.667	4.57	+0.19	86559.56
„ 23 P.M.		9 50 23.4					1.67			
„ 24 A.M.		9 51 37.8	150	5.72	155.72	{34, 37 4} 35.7; {37.4, 44.6} 41; 38.35	1.67 } 1.67	4.59	−0.29	86560.02
„ 24 P.M.		9 52 53.4					1.67			
„ 25 A.M.		9 54 08	149.6	5.06	155.56	{33, 37.4} 35.2; {37.4, 43.6} 40.5; 37.85	1.67 } 1.67	4.59	−0.31	86559.04
„ 25 P.M.		9 55 23					1.67			
MEANS					155.08	39	1.667	4.57		86559.66

DRONTHEIM.

At Drontheim the pendulum clock stood in the same room and by the side of the astronomical clock, with which it was compared at the usual intervals: both clocks rested independently on the ground beneath the flooring of the room which was removed for that purpose.

The rate of the astronomical clock is taken from page 177.

Drontheim.——VIBRATIONS of PENDULUM No. 2, in the PENDULUM CLOCK.
Mean Height of the Barometer 29.82 Inches.

DATE.	Chronometer, No. 649.	Astron. Clock.	Pendulum Clock.	Pend. Cl. on Ast. Cl. Slow.	Pend. Cl. on Ast. Cl. Gaining.	Daily Rates. Astr. Cl.	Daily Rates. Pend. Cl.	Temperature.	Arc of Vibration.	Correction for the Arc.	Reduction to a Mean Temp.	Vibrations per Diem at 47°.4.
1823.	H. M. S.	M. S.	M. S.	M. S.	S.	Gaining. S.	Gaining. S.	° ° °	°	S. +	S.	
t. 19 A.M.	7 00 00	9 49.6	57 23.5	12 26.1					1.69			
„ 19 P.M.		10 12	58 22.9	. . .	73.5	45.44	118.94	{44, 46} 45; {44.7, 46} 45.35; 45.17	1.69 }1.69	4.71	−0.98	86522.67
„ 20 A.M.		10 34.5	59 21.9	11 12.6					1.69			
„ 20 P.M.		10 56.7	00 21.1	. . .	74	45.45	119.45	{44.2, 46.5} 45.35; {44.3, 46.4} 45.35; 45.35	1.69 }1.69	4.71	−0.88	86523.28
„ 21 A.M.		11 18.7	01 20.3	9 58.6					1.69			
„ 21 P.M.		11 40.7	02 19.5	. . .	74.4	45.45	119.85	{44.3, 45.4} 44.85; {47.8, 44.8} 46.3; 45.57	1.70 }1.697	4.75	−0.81	86523.79
„ 22 A.M.		12 02.8	03 18.6	8 44.2					1.70			
„ 22 P.M.		12 24.2	04 16.2	. . .	71.8	45.45	117.25	{47.8, 50.8} 49.3; {50, 51} 50.5; 49.9	1.69 }1.697	4.75	+1.10	86523.10
„ 23 A.M.		12 45.7	05 13.3	7 32.4					1.70			
„ 23 P.M.		13 07.5	06 12	. . .	71.6	45.47	117.07	{50.8, 51.8} 51.3; {50.1, 51.3} 50.7; 51	1.70 }1.70	4.77	+1.58	86523.42
„ 24 A.M.		13 29.6	07 08.8	6 20.8					1.70			
MEANS							118.51	47.4	1.695	4.74		86523.25

DRONTHEIM.——VIBRATIONS of PENDULUM No. 1, in the PENDULUM CLOCK.

Mean Height of the Barometer 29.50 Inches.

DATE.	Chronometer, No. 649.	Astron. Clock.	Pendulum Clock.	Pend. Cl. on Ast. Cl. Slow.	Pend. Cl. on Ast. Cl. Loss.	Daily Rates. Ast. Cl.	Daily Rates. Pen. Cl.	Temperature.			Arc of Vibration.		Correction for the Arc.	Reduction to a Mean Temp.	Vibrations per Diem 44°.7.
	H. M. S.	M. S.	M. S.	M. S.	S.	Gaining. s.	Gaining. s.	°	°	°	°	°	s	s.	
Oct. 25 P.M.		14 34.8	6 23.4	8 11.4				44.7, 48	46.35		1.73		+		
„ 26 A.M.		14 56.9	6 27.2	. . .	37.1	45.49	8.39	48, 50	49	47.67	1.72	1.73	4.93	+1.31	86414.
„ 26 P.M.		15 18.4	6 29.9	8 48.5				49.7, 51.7	50.7		1.74				
„ 27 A.M.		15 39.9	6 32.1	. . . .	39.2	45.49	6.29	51.2, 52	51.6	51.15	1.74	1.74	4.99	+2.84	86414.
„ 27 P.M.		16 00.7	6 33	9 27.7				49.6, 51.3	50.45		1.75				
„ 28 A.M.		16 21.1	6 34.7	. . . .	37.7	45.49	7.79	49, 50	49.5	49.98	1.75	1.75	5.05	+2.32	86415.1
„ 28 P.M.		16 42.6	6 37.2	10 05.4				49.3, 46.4	47.85		1.75				
„ 29 A.M.		17 04.5	6 40.5	. . . .	37.2	45.58	8.38	46, 48	47	47.42	1.75	1.74	4.99	+1.20	86414.5
„ 29 P.M.		17 25.9	6 43.3	10 42.6				47.4, 42	44.7		1.72				
„ 30 A.M.	7 00 00	17 47.6	6 46.9	. .	35.3	45.59	10.29	42.2, 43.2	42.7	43.7	1.72	1.72	4.86	−0.44	86414.7
„ 30 P.M.		18 09.2	6 51.3	11 17.9				42.2, 42.6	42.4		1.72				
„ 31 A.M.		18 30.8	6 55.7	. . . .	35.5	45.58	10.08	41.8, 43.4	42.6	42.5	1.72	1.72	4.86	−0.97	86413.9
„ 31 P.M.		18 52.4	6 59	11 53.4				43.5, 39.5	41.5		1.72				
Nov. 1 A.M.		19 14.5	7 03.6	. . . .	34	45.59	11.59	37.2, 39.8	38.5		1.72	1.72	4.86	−2.07	86414.3
„ 1 P.M.		19 36.5	7 09.1	12 27.4				36, 39	37.5		1.72				
„ 2 A.M.		19 58.6	7 14	. .	34	45.48	11.48	39, 41	40	38.75	1.72	1 72	4.86	−2.64	86413.7
„ 2 P.M.		20 20.7	7 19.3	13 01.4				40.8, 41	40.9		1.72				
„ 3 A.M.		20 41.7	7 23.2	. . .	34.7	45.48	10.78	40.4, 41.6	41	40.95	1.72	1.72	4.86	−1.65	86413.9
„ 3 P.M.		21 02.6	7 26.5	13 36.1							1.72				
MEANS							9.45	44.7			1.729		4.91		86414.3

LONDON.

THE pendulum clock was set up in Mr. Browne's library in Portland-place, which is on the same level with the room in which the experiments with the detached pendulums were made. The rate was obtained by comparison with Mr. Browne's clock by Cumming.

LONDON.——VIBRATIONS of PENDULUM No. 1, in the PENDULUM CLOCK.—Mean Height of the Barometer 29.50 Inches.

DATE.	Clock's Rate on Mean Solar time.	Temperature.	Arc of Vibration.	Barometer.	Correction for the Arc.	Reduction to a Mean Temp.	Vibrations per Diem at 54°.1.
1823.	s.	°	° °	IN.	s.	s.	
February 21			1.75		+		
	35.34	52.8 / 54.6 → 53.7	1.745		4.98	−0.18	86369.46
„ 22			1.74				
	35.24	53.2 / 54.8 → 54	1.745		4.98	−0.04	86369.70
„ 23			1.75				
	35.44	53 / 55.6 → 54.3	1.74		4.95	+0.09	86369.60
„ 24			1.73				
	36.04	53.8 / 55.4 → 54.6	1.735	29.33	4.92	+0.22	86369.10
„ 25			1.74				
	35.84	52.4 / 55 → 53.7	1.745		4.98	−0.18	86368.96
„ 26			1.75				
	35.34	52.4 / 55 → 53.7	1.745		4.98	−0.18	86369.46
„ 27			1.74				
	35.24	51.6 / 54.6 → 53.1	1.745		4.98	−0.44	86369.30
„ 28			1.75				
1824.							
January 21			1.78				
	34.4	53.2 / 53.8 → 53.5	1.78		5.20	−0.26	86370.54
„ 22			1.78				
	34.6	53 / 54 → 53.5	1.775		5.17	−0.26	86370.31
„ 23			1.77				
	35.6	55 / 56 → 55.5	1.77	29.65	5.14	+0.62	86370.16
„ 24			1.77				
	35	53 / 55 → 54	1.765		5.11	−0.04	86370.07
„ 25			1.76				
	35.6	55 / 56 → 55.5	1.76		5.07	+0.62	86370.09
„ 26			1.76				
MEANS.	35.31	54.1	1.754		5.04		86369.75

LONDON.—VIBRATIONS of PENDULUM No. 2, in the PENDULUM CLOCK.
Mean Height of the Barometer 29.53 Inches.

DATE.	Cumming.	Pendulum Clock.	Clock's gain on Cumm.	DAILY RATES.		Temperature.		Arc of Vibration.		Correction for the Arc.	Reduction to a Mean Temp.	Vibrations per Diem at 53°.9.
				Cumm.	Pend.Cl.							
1823.	H. M. S.	M. S.	S.	Losing. S.	Gaining. S.	°	°	°	°	S. +	S.	
Mar. 1 P.M.		07 42.6						1.72				
,, 2 A.M.		....	75.3	0.24	75.06	52.6, 54.2	53.4		1.72	4.83	−0.21	86479.68
,, 2 P.M.		08 57.9						1.72				
,, 3 A.M.		....	74.3	0.24	74.06	53.8, 55.2	54.5		1.72	4.83	+0.26	86479.15
,, 3 P.M.		10 12.2						1.72				
,, 4 A.M.		.....	74	0.24	73.76	53.4, 54.9	54.15		1.725	4.86	+0.10	86478.62
,, 4 P.M.		11 26.2						1.73				
,, 5 A.M.		.....	74.2	0.24	73.96	53.3, 55.9	54.6		1.73	4.89	+0.31	86479.16
,, 5 P.M.		12 40.4						1.73				
,, 6 A.M.	2 00 00	.....	74.3	0.24	74.06	53.8, 55.2	54.5		1.73	4.89	+0.26	86479.21
,, 6 P.M.		13 54.7						1.73				
,, 7 A.M.		.....	75	0.24	74.76	53.2, 54.4	53.8		1.73	4.89	−0.04	86479.61
,, 7 P.M.		15 09.7						1.73				
,, 8 A.M.		.....	75.5	0.24	75.26	52.8, 54.2	53.5		1.73	4.89	−0.17	86479.98
,, 8 P.M.		16 25.2						1.73				
,, 9 A.M.		.....	75.3	0.24	75.06	52.7, 53.3	53		1.73	4.89	−0.40	86479.55
,, 9 P.M.		17 40.5						1.73				
,, 10 A.M.		.....	75.3	0.24	75.06	52.8, 54.2	53.5		1.73	4.89	−0.17	86479.78
,, 10 P.M.		18 55.8						1.73				
MEANS					74.56		53.9		1.728	4.87		86479.43

RESULTS WITH THE ATTACHED PENDULUMS.

THE results with the attached pendulums at the several stations are collected in one view in the subjoined table.

The reduction of the vibrations to a general mean temperature of 62° in column 7 is in the ratio of 0.44 parts of a vibration per diem for each degree of Fahrenheit; correspondent to an expansion of the cast brass of which the pendulums were composed of 0.0220 parts of an inch per foot in 180 degrees.

In column 8 are inserted the vibrations finally corrected of each pendulum at the several stations of experiment; and in column 9 the excess of the vibrations of pendulum No 2 over those of pendulum No 1; whence it appears that supposing the effect of the sustaining force of the clock to have been the same on both pendulums, the actual difference in their length was equivalent to 109.97 vibrations per diem; and that the greatest deviation from identity, in the ultimate deduction from the separate results with each pendulum, at any one of the stations, corresponds to a difference of 0.41 parts of a vibration per diem, and the mean deviation (omitting the signs) to a difference of 0.17 parts.

The tenth column exhibits a mean between the vibrations of the two pendulums; or the rate of an imaginary pendulum supposed to have oscillated in the clock in a vacuum and at an uniform temperature at every station. These vibrations are consequently the final results of the experiments with attached pendulums; and were the method of experiment certain, and the execution sufficiently exact, the several lengths of the seconds' pendulum should be to each other in the duplicate ratio of the numbers in this column.

The first of these preliminary questions, namely, that which concerns the sufficiency of the method of experiment, is much the more important consideration; it is dependant on the assurance that can be entertained, that the influence on the rate of the pendulum, of the force applied to sustain and register its oscillations, has been uniform at every station.

When, in the year 1818, it was determined to take advantage of the opportunity afforded by the expedition of Northern Discovery then in preparation, to extend the inquiry into the figure of the earth, by means of the pendulum, into the latitudes of the Arctic Circle, the method of attached pendulums was chosen as that of procedure, by the committee of the Royal Society to whom the consideration of the subject was referred; and the clock and pendulums which have been now experimented with were prepared under the direction of that committee: my part on that occasion was to obtain the results which the method thus selected might produce, with the utmost correctness which the nature of the experiment should permit.

The publication of the experiments made in the voyages of 1818, 1819, and 1820, revived the objection to the employment of pendulums attached to clocks in the prosecution of the inquiry, which several years antecedently, had induced in France the substitution of free pendulums. The following summary of the objection may not, perhaps, be deemed an incorrect statement. It was admitted that no very decisive reason could be assigned why the action of the weight, transmitted through the wheels, might not be a constant force; but it was urged that no sufficient proof of the affirmative existed; and that, as, in the absence of such proof, had the compression deduced by the clocks deviated widely from the result of methods not liable to the objection, or from received opinion, there would have been no hesitation in attributing the difference to an effect of the sustaining force, and in rejecting altogether the results of the attached pendulums, so, in the existing case, which happened to be that of near accordance, the results could not be considered as entitled to the weight of an independant authority, in confirmation of the compression towards which they approximated.

The question of the invariability, or otherwise, of the sustaining force, could obviously receive no other than an experimental solution: the opportunities which my subsequent prosecution of the inquiry with detached pendulums presented, of exhibiting the acceleration deduced by pendulums vibrating in clocks, in comparison with that of free pendulums, promised to be sufficiently extensive, to enable a fair practical decision on a point of considerable interest in the history, and in the employment of clocks. It was accordingly undertaken, and has been carried through, in the operations which are now recorded.

There is one indication afforded in the going of a pendulum attached to a clock, by which an inference may be drawn *a priori*, and independently of the comparison with free pendulums, as to the constancy of the sustaining force. The resistance of the air to the motion of the pendulum is the impediment which the action of the clock has to overcome, and the vibration is therefore maintained in an arc of such dimension, that the force and the resistance are exactly in equilibrio: the steadiness of the arc is consequently an evidence of the preservation of their relative proportion to each other, and its variation, of changes in the proportion, occasioned by an alteration of either: now the alterations to which the resistance of the air is liable, are necessarily of very small amount, and their causes are equally cognizable by other means; if then a difference in the arc takes place which exceeds in amount that which may reasonably be ascribed to the variation of the resistance, or if it happens when the resistance must be presumed to have been unchanged, the difference must be regarded as indicating an irregularity in the sustaining force; and in like manner, the steadiness of the arc, or the confinement of its fluctuations within the small limits by which it may be supposed to have been affected by variations in the resistance, will indicate a regularity of the sustaining force.

It may be seen by the tables containing the details at the several stations, that the differences in the arc of vibration at the eleven stations, of which the results are collected in the subjoined table, were comprehended between 1°.76 and 1°.61; those of pendulum No. 1 varying from 1°.76 to 1°.67, and those of pendulum No. 2 from 1°.76 to 1°.61 *; and further, that this small diminution of the arc obtained progressively from the equatorial to the polar stations. Now, the effect of the increased velocity of the pendulum, due to the increase in the force of gravity in proceeding from the equator towards the pole, would be to augment the resistance of the air, and consequently to contract the arcs; the increased density of the atmosphere in the colder latitudes would operate in like manner to augment the resistance, and would tend still more to contract the arcs; to the joint operation of these two causes, modified by the very effect which they tended to produce, may doubtless be ascribed the small and apparently systematic variation of the arcs; leaving the inference that the proper action of the clock was constant throughout the experiments.

The influence which the inconsiderable variation that took place in the arcs of vibration may have had on the respective rates of the pendulum, can scarcely be supposed to have been otherwise than very small; it must, however, be admitted that no very certain authority can be adduced for estimating the correction to be applied in compensation of the difference: there can be little doubt that the effect would vary in dependence

* 1°.61, in a single case only, *i. e.*, at Hammerfest: omitting that instance, the lowest arc to which No. 2 was reduced was between 1°.66 and 1°.67. The irregularity at Hammerfest was occasioned by one of the legs of the clock-frame having rested on a fragment of rock which had been disunited from the general mass, but remained imbedded in its original position: part of the foundation of the pendulum-house rested also on this fragment, near one of its extremities; and during the violent gales which were experienced, communicated a slight tremulous motion to the whole fragment. As soon as the cause of the irregular action of the clock was discovered, the contact of the house with the slab on which the leg of the clock rested was relieved, when the arc was immediately restored to its original dimension, which was subsequently maintained. To the same cause may be attributed the unusual discordances of the partial results with pendulum 2 at Hammerfest.

on the causes by which the difference in the arcs might be occasioned; and that the same causes would produce different effects in different clocks, according to the mode in which the action of the sustaining force is applied to the pendulum; or in other words, according to the nature of the escapement.

The corrections for the arc which have been applied in the tables to the rates of the attached pendulums at each station, and which have had for their object the reduction to the supposed corresponding rates in arcs infinitely small, have been in all cases those which would be due to the difference between the vibration in a circular and in a cycloidal arc of the same dimension as the observed arcs. This correction for an arc of 1.76, is + 5.10, and for 1.66 + 4.53; the uncertainty which affects the results of these experiments, is only as to the value of the difference between these corrections, namely, whether the quantity 0.57 should be more or less. Were it of consequence to pursue the investigation, it might possibly be shewn from the detail of the clocks going, that if the quantity 0.57 is erroneous, it is probably so in defect, that the retardation increased in the larger arcs in a ratio somewhat greater than the difference between the vibration in circular and cycloidal arcs; but the quantity is altogether so small, that the evidence of probability could not be very satisfactory.

Before the comparison of the results with those of the detached pendulums is proceeded in, it may be proper to notice another source of inexactness, by which the precision of the former may have been slightly affected. In the estimated value of the corrections applied to the rates at the several stations to reduce them to a general mean temperature, errors of small amount may have obtained in two ways; first, in the method pursued of registering the temperature, which was not a strictly comparative one at stations widely differing in climate; and, secondly, in the assignment of the equivalents to the effects produced on the rates by differences of temperature. The mean of the extremes taken twice in twenty-four hours, may afford a strict comparative temperature, whilst the observations are confined to adjoining latitudes or similar climates; but as a mean so taken does not bear precisely the same proportion to the true mean temperature in all climates, and as the observations included a very extensive range, a nearer approximation by a more frequent registry would have been requisite, had the utmost attainable accuracy of the method of experiment been sought; as at Melville Island for instance, where the temperature in the clock case was registered every hour for six weeks; but the object now sought by the attached pendulums, in relation to the ultimate purpose to which the acceleration was to be applied, was at the utmost, a general corroboration of the results obtained with the detached pendulums; and as no hope could be entertained that the utmost devotion of time or attention could have enabled the former method to compete in minute accuracy with the latter, the registry was confined to a sufficient frequency, to obtain the temperatures at the several stations comparable within limits, by which the general conclusion might not be affected. The same consideration influenced the adoption of an expansion of the pendulums by heat, drawn from the general result obtained by former experimenters on the expansion of cast brass (of which the pendulums were composed), instead of an attempt to determine it by special experiment, as was done in the case of the detached pendulums. The determination, by the rate of vibration in extreme temperatures would have been much more difficult of precise accomplishment with pendulums vibrating in clocks; and if artificial temperatures had been employed, the conclusion would have been far less certain in its practical application, than when the results were unembarrassed by the machinery and action of a clock. Under any probable supposition, the assumed expansion will not occasion error of greater amount than one tenth of a vibration per diem.

Having premised the causes, occurring in the execution, by which the precision of the results with the attached pendulums may have been interfered with, I proceed to their comparison with those of the free pendulums, which is exhibited in the 10th, 11th, 12th, 13th, and 14th columns of the subjoined table.

The 10th column has been already described as presenting a mean of the vibrations of the two attached pendulums, or the rate of a supposititious pendulum, oscillating in a clock in a vacuum, and at an uniform temperature, at every station.

The 11th column contains the corresponding vibration of a supposititious detached pendulum, oscillating also in a vacuum, and at an uniform temperature, brought forward from the 13th column of the table in p. 236.

In column 12 is inserted the excess of the vibrations at each station of the attached over thoso of the detached pendulum ; and in column 13 the mean excess. Had the length of the seconds' pendulum deducible from the results of the two methods of experiment been everywhere strictly identical, the values in column 12 would have corresponded in every instance with the mean excess in column 13. The deviations from identity are inserted in the 14th, or final column.

The deviations are obviously greater on some occasions than can be ascribed to inaccuracy in the corrections for the arc and temperature: when all circumstances are duly considered, there can be no hesitation in believing them to have been occasioned by accidental and temporary affections of thc action of the weight, in its transmission to the pendulum, through the machinery of the clock.

When the various processes are borne in mind, which the clock underwent in its removal from station to station, the production of occasional irregularities in the action of the machinery would seem to be scarcely avoidable; and when it is further considered that the account of its going was usually commenced on the second or third day after it had been set up, and concluded before it had gone a fortnight, the small extent of the limits, within which the amount of the deviations are comprised, will appear a remarkable testimony of the excellence of the clock.

That the irregularities were occasioned by accidental and temporary causes, and that the influence of the sustaining force on the rate of the pendulums underwent no permanent change during the operations, may be shewn by the usual process of grouping several results into a mean, whereby the *accidental* irregularities, with which, what will then become the partial results were affected, will counteract each other, and cease to embarrass the comparison. In column 14, the results at the five stations within 20° on either side of the equator (which were also the first obtained in the order of time), are thus collected into one group; and the remaining six stations, between the latitudes of 40° and 80°, into a second group; and it is then seen, that although the acceleration at a single station obtained by attached pendulums, may be in error (inferred from the comparison with free pendulums) to an amount, in an extreme case, of two seconds per diem, yet, if the stations of experiment are sufficiently multiplied, to extinguish, in an average result, the influence of the partial irregularities introduced by the machinery of the clock, the two methods of experiment produce an identical result.

Thus then, the objects designed by the employment of the attached pendulums appear to have been most satisfactorily effected; first, as a method of experimenting upon the figure of the earth, it is seen to be less exact in single determinations, but equally so, in extensively multiplied operations, as is the method with detached pendulums; secondly, in regard to the authority of the present experiments, the agreement of two methods, the processes of which are totally distinct, and which have only in common a reference to the same determination of astronomical time, can scarcely be deemed less than evidence of proof, of the correctness of the general result in which they agree.

RESULTS WITH THE ATTACHED PENDULUMS.

STATIONS.	Pendulum.	Vibrations.	Barometer.	Thermometer.	Correction for Buoyancy.	Reduction to a Mean Temp.	Vibrations in a Vacuum at 62°.	Excess in the Vibr. of Pend. 2.	Vibr. strictly comparative. Att. Pend.	Vibr. strictly comparative. Det. Pend. (page 236.)	COMPARISON of the METHODS.			
1	2	3	4	5	6	7	8	9	10	11	12	13	14	
			IN.	°	S.	S.								
aranham	1	86218.21	29.93	81.39	5.80	+ 8.53	86232.54	109.81	86287.44	86019.78	167.66		−0.7	
	2	86328.18	29.92	81.03	5.80	+ 8.37	86342.35							
scension	1	86233.21	30.05	80.05	5.82	+ 7.94	86246.97	109.64	86301.79	86033.11	168.68		+0.3	
	2	86343.09	30.07	79.44	5.85	+ 7.67	86356.61							
inidad	1	86225.85	29.92	82.98	5.77	+ 9.23	86240.85	109.88	86295.79	86027.31	168.48		+0.1	+0.04
	2	86335.79	29.94	82.82	5.78	+ 9.16	86350.73							
ahia	1	86236.27	29.93	73.27	5.89	+ 4.96	86247.12	110.15	86302.19	86032.81	169.38		+1.0	
	2	86347.07	30.00	71.72	5.92	+ 4.28	86357.27							
maica	1	86243.98	29.96	80.96	5.80	+ 8.34	86258.12	110.11	86313.17	86045.27	167.90		−0.5	
	2	86353.52	29.96	82.28	5.79	+ 8.92	86368.23							
ew York	1	86334.98	30.20	37.95	6.33	− 10.58	86330.73	110.38	86385.92	86118.48	167.44	168.4	−1.0	
	2	86446.98	30.26	33.88	6.50	−12.37	86441.11							
ondon	1	86369.73	29.50	54.1	6.11	− 3.48	86372.36	109.68	86427.20	86159.79	167.41		−1.0	
	2	86479.43	29.53	53.9	6.07	− 3.46	86482.04							
rontheim	1	86414.36	29.50	44.7	6.15	− 7.61	86412.90	110.14	86467.97	86198.52	169.45		+1.0	
	2	86523.25	29.82	47.4	6.21	− 6.42	86523.04							−0.04
ammerfest	1	86432.68	29.93	51.9	6.12	− 4.44	86434.46	109.95	86489.43	86221.46	167.97		−0.4	
	2	86546.55	29.70	48	6.22	− 8.36	86544 41							
reenland	1	86450.18	29.90	37.7	6.35	−10.69	86445.84	109.97	86500.81	86230.44	170.37		+2.0	
	2	86559.66	29.90	39	6.27	−10.12	86555.81							
pitzbergen	1	86457.21	29.90	44.07	6.24	− 7.89	86455.56	110.02	86510.57	86242.93	167.64		−0.8	
	2	86569.97	29.80	37.63	6.33	−10.72	86565.58							

There is yet a third source from which a corroboration may be adduced of the general correctness of the results with the detached pendulums; and which it may be interesting therefore briefly to examine. It is by their comparison with the acceleration shewn by the astronomical clock.

The pendulum of this clock was on the principle which is usually termed the gridiron compensation; it had the ordinary contrivance for regulating its length by means of a screw and circular nut at its lower extremity; the ball of the pendulum rested on the nut, which was divided and figured round its circular rim, corresponding to fractions of a second of time; the vibration was performed on a knife edge, working in a cylindrical groove ground in agate; presuming the compensation to have been perfect in temperatures fifty degrees apart, and the position of the nut unchanged, this pendulum was strictly invariable in its length.

The exact relation of the rate of this pendulum to the force of gravity at the several stations may have been affected, first, by irregularities inherent in the attachment to machinery; secondly, by defective compensation; and thirdly, by variations in the arc of vibration. As no intention was entertained of using the rate of the astronomical clock for any other than its immediate purpose at each station, no registry was made of the extent of the arcs: all that can be said of them now therefore is, that they certainly did not vary to any considerable amount, but that they probably underwent the small alterations due to the variable resistance of the atmosphere; and that no correction has been attempted for them.

The length of the pendulum, as dependant on the regulating screw, was kept the same at all the stations, except at Ascension, where the nut was accidentally turned, and was not discovered to be so until the clock was in motion, when it was not deemed of consequence to be rectified; the rate at Ascension, therefore, is not included in the comparison.

The columns in the subjoined table exhibit respectively, the rate of the astronomical clock brought forward from preceding pages; the corrections for buoyancy, and the rates corrected; the comparative vibration of the detached pendulums from column 13 in the table in page 236; the differences between the clock and the detached pendulums; the mean difference; and finally the discordances.

STATIONS.	Vibrations of the Ast. Clock.	Correction for Buoyancy.	Vibrations per Diem. Ast. Clock.	Vibrations per Diem. Detached Pendulums.	Difference.	Mean Difference.	Discordances.	
St. Thomas .	86277.23	5.79	86283.02	86029.40	253.62		−0.5	
Maranham . .	86266.83	5.78	86272.61	86019.78	252.83		−1.3	
Sierra Leone .	86276.8	5.75	86282.55	86028.14	254.41		+0.3	+0.4
Trinidad . . .	86276.1	5.75	86281.85	86027.31	254.54		+0.4	
Bahia	86283.77	5.88	86289.65	86032.81	256.84		+2.7	
Jamaica . .	86254.5	5.78	86300.28	86045.27	255.01	254.08	+0.9	
New York .	86367.37	6.45	26373.82	86118.48	255.34		+1.2	
London . . .	86408.20	6.00	86414.20	86159.79	254.41		+0.3	
Drontheim . .	86445.46	6.16	86451.62	86198.52	253.10		−1.0	−0.5
Hammerfest .	86468.50	6.16	86474.66	86221.46	253.20		−0.9	
Greenland . .	86477.94	6.26	86484.2	86230.44	253.76		−0.3	
Spitzbergen .	86488.55	6.27	86494.82	86242.93	251.89		−2.2	

The evidence, which is afforded by this comparison, of partial irregularities in the rate of pendulums attached to clocks, and of general agreement with detached pendulums, when the stations are sufficiently multiplied to reduce the effect of accidental interferences, is very similar to that furnished by the corresponding table in page 281.

The number of stations required to produce a mean result, in which the influence of partial irregularities may be neutralized, must depend on the amount of error which may be liable to occur in an extreme case; it may be inferred from the comparisons that both the average and the extreme irregularity was about one-fourth or one-fifth greater in the astronomical than in the pendulum clock, and it might be expected therefore that the same grouping of stations, which produced a correspondence between the detached and solid pendulums, would not be sufficient to extinguish altogether the influence of extreme cases in the astronomical clock; and accordingly the effects of the discordancies at Bahia and Spitzbergen are not wholly destroyed by the five stations combined with each, but are still visible in the means.

In the following table are collected in one view the discordancies of the results of the detached pendulums at the several stations, 1st, with those of the attached solid pendulums; 2dly, with those of the attached compensated pendulum; and 3dly, with the mean of the three distinct methods, ascribing to each for the moment an equal weight. The object of this table will be more fully apprehended when the accidental irregularities in the force of gravity, as evidenced by these experiments, shall be under consideration; its purpose is to exhibit the *utmost* error which could be attributed to the results with the detached pendulums at each station, were each of the distinct and separate methods employed of equal authority*; and the consequent limit, within which they must be presumed, on such concurrent testimony, to be exact; it will be seen in the sequel that the utmost errors fall far short of the anomalies which will be manifested in the application of the results, and consequently that it

* It is hardly necessary, but it may be proper to state expressly, that it is not intended to represent the methods as being really equal in authority; on the contrary, the results with the detached pendulums must be esteemed as far more exact than even the mean of the three methods. It is hoped that the details of the experiments throughout are sufficiently ample, to enable every person to form his own estimate of the probability of error in each determination.

must be regarded as extremely improbable that the causes of the anomalies should be in the experiments.

STATIONS.	The Detached Pendulums in excess or defect, by			STATIONS.	The Detached Pendulums in excess or defect, by		
	The Solid Pendulums.	The Compensated Pendulum.	The Mean of the three Methods.		The Solid Pendulums.	The Compensated Pendulum.	The Mean of the three Methods.
	s.	s.					
St. Thomas .	. . .	+0.5	+0.2	New York .	+1.0	−1.2	−0.1
Maranham .	+0.7	+1.3	+0.6	London . .	+1.0	−0.3	+0.2
Ascension .	−0.3	. . .	−0.1	Drontheim .	−1.0	+1.0	0.0
Sierra Leone	. . .	−0.3	−0.1	Hammerfest	+0.4	+0.9	+0.4
Trinidad . .	−0.1	−0.4	−0.2	Greenland .	−2.0	+0.3	−0.5
Bahia . . .	−1.0	−2.7	−1.3	Spitzbergen	+0.8	+2.2	+0.9
Jamaica . .	+0.5	−0.9	−0.1				

The following table, in which the going of the pendulums in their employment during the Expedition of Discovery in 1819 and 1820, is compared with their subsequent going at the same or corresponding stations recorded in this volume, appears of sufficient interest to merit insertion, from its bearing on some points of the preceding discussion.

The particulars in the first nine columns would seem to require no other explanation, than that the vibrations of the pendulums in 1819 and 1820 are brought forward from the details published in the *Phil. Trans.* for 1821 ; two clocks were then employed, in both of which the pendulums were tried, with results of which the means coincided within 0.2 of a vibration per diem ; the vibrations in the clock which was then considered to deserve a preference, for reasons adduced in the memoir, are those introduced in the present table ; the weights of the two clocks were of the same amount.

It has been noticed that on the return of the pendulums from the north in 1820, their knife edges were ground afresh, in consequence of injury which they had received from rust; it is to that cause that the differences in column 13 are due, (presuming that as the arcs were equal, the influence of equal weights on the rate was uniform also); the pendulums appear to have been lengthened by the operation, an amount equivalent to about eight vibrations per diem; and No. 2 rather less than No. 1.

The small correction for ellipticity in column 11, is the difference of vibration between the latitudes of Melville Island and Greenland, which would be occasioned by an ellipticity of $\frac{1}{288}$th; it is introduced for the purpose of comparing the acceleration between London and Melville Island, with that between London and Greenland, which in conformity to a compression of $\frac{1}{288}$th should differ 0.5 per diem. It is seen by the final column that the comparative acceleration obtained in 1819—1820, and 1823, differed 0.54 parts of a vibration per diem; being much within the limit of the differences to which the results of attached pendulums have been shewn to be liable, from accidental irregularities in the sustaining force.

The most interesting column, however, is that which exhibits the arcs of vibration; it is seen that on both the occasions, in which the clock and pendulums were taken from the latitude of London to that of between seventy-four and seventy-five degrees, the arcs diminished to the same amount; presenting a strong testimony of the systematic nature of the cause of the diminution, and confirming the probability, that it has been correctly ascribed to the increased velocity of the pendulum in the higher latitudes, and to the consequently increased resistance of the atmosphere.

In the uncertainty which is involved in the comparison of the vibrations of attached pendulums performed in different arcs, it might, perhaps, be desirable, should they be again employed in this or similar inquiries, to employ a variable, instead of a fixed weight; and to proportion the sustaining force in such manner, that the arcs might always be maintained of the same dimension.

Stations.	Latitudes.	Dates.	Pendulum.	Arc of Vibration.	Temperature.	Vibrations.	Excess of Pendulum No. 2.	Mean Vibration.	Corrections.		Comparative Vibrations.	Differences of Pend. in 1820 & 1823.	Difference between the Acceleration in 1820 & 1823.
									Temp.	Ellip.			
London . . .	51 31 08.4	1819	1	1.76	45°	86388.10	108.89	86442.54	−7.4	. .	86435.14	7.94	0.54
			2	1.73	45	86496.99							
		1823	1	1.75	62	86372.36	109.68	86427.20	. .	. .	86427.20		
			2	1.73	62	86482.04							
Melville Island.	74 47 12.4	1820	1	1.65	45	86462.53	109.33	86517.19	−7.4	−0.5	86509.29	8.48	
			2	1.67	45	86571.86							
Greenland. . .	74 32 19	1823	1	1.67	62	86445.84	109.97	86500.81	. .	. .	86500.81		
			2	1.67	62	86555.81							

LATITUDES OF THE PENDULUM STATIONS.

THE instruments employed in determining the latitudes were a repeating circle of six inches diameter; a repeating reflecting circle of eight inches diameter; and a sextant of eight inches radius.

The repeating circle was made by the direction, and at the expense of the Board of Longitude, for the purpose of exemplifying the efficacy of the principle of repetition when applied to a circle of so small a diameter as six inches, carrying a telescope of seven inches focal length and one inch aperture; and of practically ascertaining the degree of accuracy which might be retained, whilst the portability of the instrument should be increased, by a reduction in the size to half the amount, which had been previously regarded by the most eminent artists as the extreme limit of diminution to which repeating circles, designed for astronomical purposes, ought to be carried.

The excellent workmanship of Mr. Dollond, and the many ingenious and useful contrivances which he has applied to the repeating circles of his latest construction, rendered this little instrument extremely complete, and by no means inconvenient in use; the arrangement, in particular, of the screws was such, that each could be managed without the liability of interfering with or of being mistaken for others, and with full room for the fingers even under very unfavourable circumstances; as at New York where the temperature was frequently below 20°, and obliged the use of gloves.

It has already been noticed that the diminution in the size of this instrument brought the several particulars of an observation, being the contact, the levels, and the time, within the command of a single observer;

and that the advantages gained in those respects are scarcely of less importance than the increased portability. The observations with the circle, detailed in the following pages, were made without an assistant.

The practical value of the six inch circle may be estimated by comparing the differences of the partial results from the mean at each station, with the correspondence of any similar collection of observations made with a circle on the original construction and of large dimension; such for instance as the latitudes of the stations of the French Arc recorded in the *Base du Systéme Metrique;* when if due allowance be made for the extensive experience and great skill of the distinguished persons who conducted the French observations, the comparison will scarcely appear to the disadvantage of the smaller circle, even if extended generally through all the stations of the present volume; but if it be particularly directed to Maranham and Spitzbergen, at which stations the partial results were more numerous than elsewhere, and obtained with especial regard to every circumstance by which their accuracy might be affected, the performance of the six inch circle will appear fully equal to that of circles of the larger dimension; the comparison with the two stations, at which a more than usual attention was bestowed, is the more appropriate, because it was essential to the purposes for which the latitudes of the French stations were required, that the observations should always be conducted with the utmost possible regard to accuracy.

It would appear, therefore, that in a repeating circle of six inches, the disadvantages of a smaller image enabling a less precise contact or bisection, and of an arch of less radius admitting of a less minute subdivision, may be compensated by the principle of repetition; whilst the advantage is obtained, of a less pressure on the centre work, and of a more free and independent motion of the several parts of the instrument, in consequence of the reduction in size; which advantage is of much practical consideration.

The repeating reflecting circle will be particularly described in a subsequent part of the volume, in an account of the Longitudes of the several stations, in which determination it was principally employed.

The sidereal chronometer used in noting the distance of stars from the meridian, corresponding to their zenith distances observed with the repeating circle, was made by Molyneux, and was lent me by Captain Frederic Marryatt of the Royal Navy; its rate appeared, by the comparisons at the several stations, to be losing about 5 seconds on a sidereal day, and to be tolerably steady.

The corrections for astronomical refraction, employed in the following calculations, are taken from Dr. Thomas Young's table, published in the *Nautical Almanac* for 1822, and subsequent years; the temperature of the table being considered 48° instead of 50°

The corrections for aberration and nutation applied to the stars, of which the true apparent places are not inserted in the *Nautical Almanac*, have been computed by Mr. Groombridge's *Universal Tables*, published in the first volume of the *Transactions of the Astronomical Society of London.*

SIERRA LEONE.

Place of Observation.—The West Bastion of Fort Thornton.

March 8th, 1822. Barom. 29.90; Ther. 80°. The Chron. 423 fast 48′ 50″. Sirius, AR. 6^h 37′ 19″.4, on the Meridian at 8^h 22′ 17″ by the Chronometer. The Zenith Distances observed with a Repeating Circle.

Chronometer.	Horary Angles.	N. V. Sines.	Level.	
H. M. S.	M. S.			
8 10 10	12 07	1397	−3	−4
8 13 51	8 26	677	+3	+2
8 16 22	5 55	333	+3	+2
8 20 01	2 16	49	+5	+4
8 24 24	2 07	43	−6	−7
8 26 25	4 08	163	+4	+3
8 28 53	6 36	415	+8	+8
8 31 26	9 09	797	−5	−6
Means . . .		484.2	+5.5	

Lat. 8° 29′ 30″	Cosine	9.9952127
Dec. 16° 29′	Cosine	9.9817744
Z.D. 24° 58′	Cosecant . . .	10.3745940
Log. Sine 1″	A.C.	5.3168000
Log. 484.2 (+4)	. . . . .	6.6850248
Correction 3′ 45′.6	Log.	2.3534059

Readings, &c.

		° ′ ″
Previous	First Vernier . .	90 25 00
	Second „ . . .	24 48
	Third „ . . .	25 15
	Fourth „ . . .	24 40
	Mean	90 24 56
Final. .	First Vernier . .	290 37 10
	Second „ . . .	37 00
	Third „ . . .	37 50
	Fourth „ . . .	37 10
	Mean	290 37 17.5
	Index	−90 24 56
	Level	+5.5
		200 12 27
	Observed Z.D. . .	25 01 33.4
	Refraction +0′ 27″.2	
	Barometer. . −0.1	+0 25.5
	Thermometer −1.6	
	Correction . . .	−3 45.6
	True Z.D. . . .	24 58 13.3
	Star's Declin. . .	16 28 45.4
	Latitude North . .	8 29 27.9

St. THOMAS.

Place of Observation.—The Mansion House of Fernandilla; Man of War Bay.

June 7th, 1822. Bar. 30.10; Ther. 78°. The Chron. 423 slow, 29′ 56″· α Crucis (AR. 12^h 16′ 52″.4) on the Meridian at 6^h 44′ 33″, by the Chronometer. The Altitudes observed with a Repeating Reflecting Circle, and a Mercurial Horizon.

Chronometer.	Horary Angles.	N.V. Sines.	Readings, &c.	
H. M. S.	M. S.			° ′ ″
6 45 20	0 47	6	Arc passed through	110 00 30
6 49 18	4 45	215	Apparent Altitude	27 30 07.5
Mean		110.5	Refraction − 1′ 51″.5 } Barometer . . − 0.4 } Thermometer + 6.4 }	−1 45.5
Lat. 0° 25′ Cosine .		9.999988	Correction	+0 12.1
Dec. 62° 07′ Cosine .		9.669942		27 28 34.1
Alt. 27° 28′ Secant .		10.051939	Star's South Polar Distance, 1st Jan. 1818 .	27 54 32 }
Log. Sine 1″ A.C. .		5.316800	Prec. Aber. and Nut.	− 1 15.6 }
Log. 110.5(+4) . .		6.043362	Latitude North	0 24 42.3
Correction 12″.1 = .		1.082031		

ST. THOMAS.——June 10th, 1822.—Bar. 30.10; Ther. 80°; the Chron. 423 slow, 29′ 50″; Arcturus (AR. 14^h 07′ 35″.5) on the Meridian at 8^h 23′ 15″.5 by the Chron. The Altitudes observed by a Repeating Reflecting Circle, and a Mercurial Horizon.

		° ′ ″
Meridian Double Altitude		140 36 44
Apparent Altitude		70 18 22
Refraction . . −20″.9		
Barometer . . −0.1		−19.9
Thermometer . +1.2		
True Altitude		70 18 02.1
North Polar Distance		69 53 22
Latitude North		0 24 40.1

RECAPITULATION.

June 7th, α Crucis	0 24 42.3
June 10th, Arcturus	0 24 40.1
Latitude	0 24 41.2 North.

ASCENSION.

Place of Observation.—The Barrack Square.

June 26th, 1822. Bar. 30.14; Ther. 80°. The Chron. No. 423 fast 55′ 08″. α Centauri (AR. 14^h 28′ 13″), on the Meridian at 9^h 05′ 45″, by the Chron. The Altitudes observed with a Sextant and Mercurial Horizon. Index Correc. —1′.

Chronometer.	Horary Angles.	N.V. Sines.	Observed Double Altitudes.
H. M. S.	M. S.		
8 57 20	8 25	674	75 39 20
8 59 33	6 12	366	75 40 20
9 01 45	4 00	152	75 41 20
9 03 20	2 25	56	75 42 00
9 04 23	1 22	18	75 42 20
9 08 27	2 42	69	75 41 55
9 09 44	3 59	151	75 40 50
9 11 05	5 20	271	75 39 40
Means		219.6	75 40 58

Lat. 7° 56′ Cosine . . .	9.9958235
Dec. 60° 07′ Cosine	9.6974347
Alt. 37° 49′ Secant	10.1023857
Log. Sine 1″ A.C.	5.3168000
Log. 219.6 (+4)	6.3416323
Correction 28″.5 = Log.	1.4540762

Deduction.

		° ′ ″
Observed Double Altitude . .		75 40 58
Index		− 1 00
Apparent Altitude		37 49 59
Refraction	−1 15	
Barometer	− 0.3	. . . . − 1 11
Thermometer	+ 4.3	
Correction		+ 0 28.5
True Altitude		37 49 16.5
South Polar Distance, 1st Jan. 1818		29 54 33
Prec. Aber. and Nut.		− 1 05.6
Latitude South		7 55 49.1

ASCENSION.—July 2d, 1822. Bar. 30.15; Therm. 80°. The Chron. 423, fast 55′ 22″. α Centauri (AR. 14ʰ 28′ 13″) on the Meridian at 8ʰ 42′ 20″ by the Chron. The Altitudes observed with a Sextant and Mercurial Horizon. Index Correc. − 1′.

Chronometer.	Horary Angles.	N. V. Sines.	Observed double Altitudes.
H. M. S.	M. S.		° ′ ″
8 32 42	9 38	883	75 38 10
8 34 17	8 03	617	75 39 35
8 35 40	6 40	423	75 40 00
8 36 57	5 23	276	75 41 05
8 38 10	4 10	165	75 41 25
8 39 29	2 51	77	75 41 30
8 40 29	1 51	33	75 41 50
8 41 49	0 31	3	75 42 10
8 42 57	0 37	4	75 41 50
8 44 12	1 52	33	75 41 30
8 45 26	3 06	92	75 41 00
8 47 41	5 21	272	75 40 25
8 48 46	6 26	394	75 39 50
8 49 54	7 34	545	75 39 05
8 51 12	8 52	748	75 38 45
Means . .		304.3	75 40 32.7

Deduction.

Latitude 7° 56′	Cosine . .	9.9958235
Declination 60° 07′	Cosine . . .	9.6974347
Altitude 37° 49′	Secant . . .	10.1023857
Log. Sine 1″	A.C. . . .	5.3168000
Log. 304.3 (+4)	. . . . . .	6.4833020
Correction 39″.4	Log.	1.5957459

	° ′ ″
Observed double Altitude . . .	75 40 32.7
Index Correction	− 1 00
Apparent Altitude	37 49 46.3
Refraction − 1′ 14″.9; Barometer − 0.3; Thermometer + 4.5	− 1 10.7
Correction	+ 0 39.4
True Altitude.	37 49 15
South Polar Dist., 1st Jan. 1818. .	29 54 33
Prec. Aber. and Nut.	− 1 05.6
Latitude, South	7 55 47.6

ASCENSION.——July 6th, 1822. Bar. 30.15; Therm. 80°. The Chron. 423, fast ·55′ 32″. α Centauri (AR. 14^h 28′ 13″) on the Meridian at 8^h 26′ 46″ by the Chron. The Zenith Distances observed with a Repeating Circle.

Chronometer.	Horary Angles.	N. V. Sines.	Level.	
H. M. S.	M. S.			
8 23 32	3 14	99	−7	−9
8 26 45	0 01	0	+4	+2
8 30 24	3 38	126	+2	0
8 33 10	6 24	390	0	−3
8 35 47	9 01	774	+6	+6
8 38 42	11 56	1355	+5	+2
Means		457.3	+4	

Lat. 7° 56′	Cosine	9.9958235
Dec. 60° 07′	Cosine .	9.6974347
Z.D. 52° 11′	Cosecant . . .	10.1023857
Log. Sine 1″	A.C.	5.3168000
Log. 457.3 (+4)		6.6602012
Correction 59″.2	Log.	1.7726451

Readings, &c.

			° ′ ″
Final. .	First Vernier . . .		313 03 05
	Second „		02 50
	Third „		03 30
	Fourth „		03 30
	Mean.		313 03 14
	Index		+ 08.5
	Level		+ 4
			313 03 26.5
	Observed Z.D.		52 10 34.4
	Refraction + 1′ 14″.9		
	Barometer	+ 0.3	+ 1 10.7
	Thermometer	− 4.5	
	Correction		− 0 59.2
	True Z.D. . . .		52 10 45.9
	Star's Decl. 1 Jan. 1818.	60 05 27	
	Prec. Aber. and Nut. .	+ 1 05.6	
	Latitude, South. . .		7 55 46.7

RECAPITULATION.

	°	′	″
June 26, Arcturus	7	55	49.1
July 2, α Centauri	7	55	47.6
July 6, α Centauri	7	55	46.7
	7	55	47.8 South.

BAHIA.

Place of Observation.—At Mr. Pennell's House at Vittoria. The Zenith Distances observed with a Repeating Circle.

July 23d, 1822. Barom. 30.04; Therm. 71°. The Chron. 423 fast $2^h\ 32'\ 47''$, (page 56); α Lyræ (AR. $18^h\ 30'\ 57'.8$) on the Meridian at $12^h\ 59'\ 00''$ by the Chronometer.

Chronometer.	Horary Angles.	N. V. Sines.	Level.	
H. M. S.	M. S.			
12 46 15	12 45	1547	0	− 4
12 51 46	7 14	498	+ 3	+ 1
12 53 50	5 10	254	− 6	− 5
12 56 34	2 26	56	+ 1	− 2
13 00 13	1 13	14	− 2	− 5
13 02 30	3 30	117	+ 3	+ 1
13 05 12	6 12	366	+10	+ 7
13 07 11	8 11	637	− 6	− 3
13 10 32	11 32	1266	+15	+15
13 13 37	14 37	2033	−11	− 8
Means . .		678.8	+ 2	

Readings, &c.

		° ′ ″
Final .	First Vernier .	156 18 50
	Second „ . . .	18 55
	Third „ . .	19 35
	Fourth „ . . .	19 00
	Mean	156 19 05
	Index	+360 00 08.5
	Level	+2
		516 19 15.5
	Observed Z.D. .	51 37 55.6
	Refraction +1′ 13″.3; Barometer. +0.1; Thermometer −3.2	+1 10.2
	Correction . .	−2 16.7
	True Z.D. . . .	51 36 49.1
	Star's Declin. .	38 37 29.7
	Latitude South .	12 59 19.4

Lat. 12° 59′	Cosine . . .	9.9887531
Dec. 38° 37′.5	Cosine	9.8927885
Z.D. 51° 37′	Cosecant . . .	10.1057537
Log. Sine 1″	A.C.. . .	5.3168000
Log. 678.8 (+4). .	. .	6.8317418
Correction 2′ 16″.7	Log .	2.1358371

Bahia.——July 26th, 1822. Barom. 29.98; Therm. 73°. The Chron. 423 fast 2^h 32′ 55″, (page 59); α Lyræ (AR. 18^h 30′ 57″.8) on the Meridian at 12^h 47′ 19″ by the Chronometer.

Chronometer.	Horary Angles.	N. V. Sines.	Level.	
H. M. S.	M. S.			
12 35 26	11 53	1344	− 5	− 8
12 37 40	9 39	886	−10	−12
12 40 15	7 04	475	− 1	− 4
12 42 15	5 04	244	+ 4	+ 1
12 44 27	2 52	78	+ 4	+ 1
12 46 17	1 02	10	− 1	− 4
12 48 16	0 57	8	− 8	− 5
12 51 33	4 14	171	− 6	− 3
12 53 53	6 34	410	0	− 3
12 56 26	9 07	791	− 3	0
12 58 55	11 36	1281	− 5	− 3
13 00 43	13 24	1709	− 1	− 4
Means . .		617.3	− 5	

Readings, &c.		° ′ ″
Final. .	First Vernier . .	259 33 05
	Second „ . . .	33 00
	Third „ . . .	33 35
	Fourth „ . .	33 00
	Mean . . .	259 33 10
	Index . . .	+360 00 08.5
	Level	−5
		619 33 13.5
	Observed Z.D. . .	51 37 46.1
	Refraction +1′ 13″.3; Barometer . . . . .; Thermometer −3.4	+1 09.9
	Correction . . .	−2 04.3
	True Z.D. . . .	51 36 51.7
	Star's Declin. . .	38 37 30.5
	Latitude South . .	12 59 21.2

Lat. 12° 59′	Cosine . .	9.9887521
Dec. 38° 37′.5	Cosine	9.8927885
Z.D 51° 37′	Cosecant . . .	10.1057537
Log. Sine 1″	A.C..	5.3168000
Log. 617.3 (+4)	. . . . .	6.7904963
Correction 2′ 04″.3	Log .	2.0945916

BAHIA.——July 31st, 1822. Barom. 30.05; Therm. 71°. The Chron. 423, fast 2h 33′ 09″, (page 63); α Pavonis (AR. 20h 11′ 36″) on the Meridian, at 14h 08′ 14″ by the Chronometer.

Chronometer.	Horary Angles.	N. V. Sines.	Level.	
H. M. S.	M. S.			
14 02 58	5 16	264	−4	−6
14 04 52	3 22	108	+5	+2
14 07 15	0 59	9	−1	0
14 08 47	0 33	3	+3	+1
14 11 15	3 01	87	+5	+2
14 12 52	4 38	204	+9	+6
14 16 20	8 06	624	+5	+2
14 20 49	12 35	1507	+2	−1
14 23 21	15 07	2175	+3	0
14 27 49	19 35	3648	+2	0
Means		862.9	+17.5	

Lat. 12° 59′ Cosine . .		9.9887531
Dec. 57° 17′ Cosine . .		9.7327837
Z.D. 44° 18′ Cosecant . .		10.1558863
Log. Sine 1″ A.C. . .		5.3168000
Log. 862.9 (+4)		6.9359605
Correction 2′ 14″.9	Log.	2.1301836

Readings, &c.

		° ′ ″
Final . .	First Vernier . . .	83 15 10
	Second „	14 40
	Third „	15 15
	Fourth „	14 40
	Mean	83 14 56
	Index . . . . +	360 00 08.5
	Level	+ 17.5
		443 15 22
	Observed Z.D.	44 19 32.2
	Refraction + 0′ 56″.8; Barometer + 0.1; Thermometer − 2.5	+ 54.4
	Correction	− 2 14.9
	True Z.D.	44 18 11.7
	Mean Decl. Jan. 1, 1818.	57 18 23
	Prec. Aber. and Nut. .	− 0 48.9
	Latitude, South. . .	12 59 22.4

RECAPITULATION.

	° ′ ″
July 23, α Lyræ	12 59 19.4
July 26, α Lyræ	12 59 21.2
July 31, α Pavonis	12 59 22.4
	12 59 21 South Latitude.

MARANHAM.

Place of Observation.—In Mr. Hesketh's House.

The observations were made with the six inch repeating circle belonging to the Board of Longitude, and were designed to afford a fair example of the accuracy of which that Instrument is capable. The circle was supported on the window-sills, and was always suffered to remain several minutes after the adjustments were perfect, before the observations were commenced. The temperature was registered by a thermometer freely suspended near the circle, and its height was observed before the mercury had risen in consequence of the approach of the lamp, by which it was read. The screws for slow motion, of the circle, level, and telescope, were turned in opposite directions, in successive pairs of observation.

August 28th, 1822. Barometer 29.95; Thermometer 80°. The Chron. 423, fast 2h 56′ 59″, (page 76); α Lyræ (AR. 18h 30′ 57″.4) on the Meridian, at 8h 04′ 35″ Mean Time, and at 11h 01′ 34″ by the Chronometer.

Chronometer.	Horary Angles.	N.V. Sines.	Level.	
H. M. S.	M. S.			
10 49 06	12 28	1479	+ 3	+ 1
10 52 40	8 54	754	− 2	− 4
10 55 50	5 44	313	− 4	− 6
10 57 44	3 50	140	− 1	− 2
11 00 49	0 45	5	− 8	− 10
11 03 42	2 08	43	+ 4	+ 2
11 06 52	5 18	267	− 8	− 10
11 09 17	7 43	567	+ 7	+ 5
Means		446	− 16.5	

Readings, &c.		° ′ ″
Previous	First Vernier . . .	167 11 50
	Second „	11 30
	Third „	12 10
	Fourth „ . . .	11 40
	Mean.	167 11 47.5
Final . .	First Vernier . . .	136 35 00
	Second „	34 30
	Third „	35 30
	Fourth „	35 00
	Mean	136 35 00
	Index	+192 48 12.5
	Level	− 16.5
		329 22 56
	Observed Z D. . . .	41 10 22
	Refraction +1′ 50″.8 } Barometer − 0.1 } Thermometer − 3.0 }	+ 0 47.7
	Correction	− 1 49.7
	True Z.D. . . .	41 09 20
	Star's Decl.	38 37 37.6
	South Latitude . .	2 31 42.4

Lat. 2° 32′ Cosine . . .	9.9995753
Dec. 38° 37′ Cosine	9.8928395
Z.D. 41° 09′ Cosecant. . . .	10.1817526
Log .Sine 1″ A.C.	5.3168000
Log. 446 (+4)	6.6493349
Correction 1′ 49″.7 Log.	2.0403023

MARANHAM.——August 29, 1822. Bar. 30.00; Therm. 80°. The Sidereal Chron. No. 702, slow on Sidereal Time 8^h 23′ 38″. α Lyræ (AR. 18^h 30′ 57″.4) on the Meridian at 10^h 7′ 19″ by the Chronometer.

Chronometer.	Horary Angles.	N. V. Sines.	Level.	
H. M. S.	M. S.			
9 48 35	18 44	3339	−3	−1
9 51 00	16 19	2533	+4	+2
9 54 20	12 59	1604	−2	−4
9 57 04	10 15	1000	0	−2
9 59 45.5	7 31.5	539	+4	+1
10 02 30.5	4 48.5	220	−2	−4
10 06 11.5	1 07.5	12	+5	+3
10 08 43	1 24	19	−4	−2
10 11 34	4 15	172	−7	−9
10 14 08.5	6 49.5	443	0	−2
10 17 33.5	10 14.5	998	0	−3
10 20 05	12 46	1551	−3	0
Means . . .		1035.8	−14.5	

Readings, &c.		° ′ ″
Previous	First Vernier	116 26 10
	Second „	25 50
	Third „ . . .	26 20
	Fourth „	25 55
	Mean	. 116 26 04
Final	First Vernier . . .	. 250 59 50
	Second „ . .	250 59 30
	Third „	. 251 00 10
	Fourth „	. 250 59 20
	Mean	. 250 59 42.5
	Index	+243 33 56
	Level	− 14.5
		494 23 24
	Observed Z.D. . .	41 12 47
	Refraction +0′ 50″.9; Barometer ...; Thermometer − 3.0	+ 0 47.9
	Correction . .	− 4 13.4
	True Z.D. . . .	41 09 21.5
	Star's Declination . .	38 37 37.7
	South Latitude . . .	2 31 43.8

Lat. 2° 32′	Cosine	9.9995753
Dec. 38° 37′	Cosine . . .	9.8928395
Z.D. 41° 09′	Cosecant . . .	10.1817526
Log. Sine 1″ A.C.	. . . .	5.3144251
Log. 1035.8 (+4)	. . . . .	7.0152759
Correction 4′ 13″.4	Log.	2.4038684

MARANHAM.——August 29th, 1822. Barom. 30.00; Therm. 80°. The Sidereal Chron. 702 slow on Sidereal Time 8^h 23′ 39″. α Pavonis (App. AR. 20^h 11′ 36″) on the Meridian at 11^h 47′ 57″ by the Chronometer.

Chronometer.	Horary Angles.	N. V. Sines.	Level.	
H. M. S.	M. S.			
11 39 28	8 29	685	− 1	− 3
11 42 54.5	5 02.5	242	0	− 3
11 46 55	1 02	10	− 1	− 3
11 49 12	1 15	15	+ 5	+ 2
11 54 45	6 48	410	+11	+ 9
11 56 42	8 45	729	− 9	−11
11 58 58	11 01	1155	− 3	− 5
12 01 11	13 14	1667	− 2	0
12 06 56	18 59	3428	+ 5	+ 6
12 08 51	20 54	4155	+ 3	+ 1
Means . . .		1252.6	−0.5	

Lat. 2° 32′ Cosine . . .		9.9995753
Dec. 57° 17′ Cosine . .		. 9.7327837
Z.D. 54° 46′ Cosecant .		. 10.0878793
Log. Sine 1″ A.C. . .		5.3144251
Log. 1252.6 (+4) . . .		7.0978124
Correction 2′ 50″ 8	Log .	2.2324758

Readings, &c.

		° ′ ″
Previous	First Vernier .	250 59 50
	Second „	250 59 30
	Third „ .	251 00 10
	Fourth „	250 59 20
	Mean . . .	250 59 42.5
Final . .	First Vernier .	78 54 20
	Second „ .	54 20
	Third „ . .	54 40
	Fourth „ . .	54 10
	Mean	78 54 22.5
	Index . . . +	109 00 17.5
		360 00 00
	Level .	+0.5
		547 54 40.5

Observed Z.D. . .		54 47 28
Refraction +1′ 22″.4	}	+1 17 6
Barometer . .		
Thermometer −4.8		
Correction		−2 50.8
True Z.D. . . .		54 45 54.8
Star's Dec. 1 Jan. 1818	57 18 23	}
Prec. Aber. and Nut.	−0 43.7	
Latitude South . .		2 31 44.5

MARANHAM.——August 31st, 1822. Bar. 29.97; Ther. 80°. The Sidereal Chron. slow on Sidereal Time 8h 23′ 50″. α Lyræ (AR. 18h 30′ 57″.3) on the Meridian at 10h 7′ 7″ by the Chronometer.

Chronometer.	Horary Angles.	N.V. Sines.	Level.		Readings, &c.		
H. M. S.	M. S.						° ′ ″
9 52 35	14 32	2010	− 3	− 5	Previous	First Vernier . . .	280 39 10
9 55 46	11 21	1226	0	0		Second „	39 20
9 58 43	8 24	672	+12	+10		Third „	39 40
10 00 48.5	6 18.5	379	− 3	− 5		Fourth „	39 00
10 04 12	2 55	81	− 1	− 3		Mean	280 39 17.5
10 06 11.5	0 55.5	8	+ 5	+ 2			
10 09 37.5	2 30.5	60	− 4	− 6	Final	First Vernier . . .	55 09 25
10 11 37.5	4 30.5	194	− 1	− 3		Second „	09 15
10 16 05	8 58	765	+ 9	+ 7		Third „ . . .	09 50
10 18 14	11 07	1176	0	0		Fourth „	09 10
10 21 03.5	13 56.5	1850	+ 2	0		Mean	55 09 25
10 25 17	18 10	3140	− 8	−10		Index . . . . +	79 20 42.5 / 360 00 00
Means		963.4	− 2.5			Level	− 2.5
							494 30 05
						Observed Z.D. . . .	41 12 30.4
						Refraction +0′ 50″.9 / Barometer ... / Thermometer − 3.0	+ 0 47.9
						Correction	− 3 55.7
						True Z.D.	41 09 22.6
						Star's Declination . .	38 37 38
						South Latitude . .	2 31 44.6

Lat. 2° 32′	Cosine	9.9995753
Dec. 38° 37′	Cosine	9.8928395
Z.D. 41° 09′	Cosecant . .	10.1817526
Log. Sine 1″ A.C.	. . . .	5.3144251
Log. 963.4 (+4)	. . . . .	6.9838066
Correction 3′ 55″.7	Log.	2.3723991

MARANHAM.——August 31st, 1822. Barom. 29.97; Therm. 80°; the Sidereal Chron. slow on Sidereal Time 8^h 23′ 50″; α Cygni (AR. 20^h 35′ 25″) on the Meridian at 12^h 11′ 35″ by the Chronometer.

Chronometer.	Horary Angles.	N.V. Sines.	Level.	
H. M. S.	M. S.			
11 57 46.5	13 48.5	1814	− 5	− 7
12 00 03.5	11 31.5	1264	+ 4	+ 2
12 02 26.5	9 08.5	795	0	0
12 04 33.5	7 01.5	470	− 4	− 6
12 07 59.5	3 35.5	123	+ 2	0
12 10 12	1 23	18	− 2	− 4
12 13 09	1 34	23	0	0
12 16 03.5	4 28.5	190	+ 4	+ 1
12 20 31	8 56	760	+11	+ 9
12 22 40	11 05	1169	+ 6	+ 4
12 25 45.5	14 10.5	1912	− 7	− 9
12 27 19	15 44	2355	− 2	− 4
Means . .	. . .	907.7	−3.5	

Readings, &c.

		° ′ ″
Previous	First Vernier . .	55 09 25
	Second „ . .	09 15
	Third „ . .	09 50
	Fourth „ .	09 10
	Mean .	55 09 25
Final . .	First Vernier	261 45 20
	Second „ .	45 00
	Third „ . .	45 40
	Fourth „ . .	44 40
	Mean . . .	261 45 10
	Index . .	+304 50 35
	Level . .	−3.5
		566 35 41.5
	Observed Z.D.	47 12 58.5
	Refraction +1 02.9, Barometer, Thermometer −3.7	+0 59.2
	Correction	−3 01.4
	True Z.D.	47 10 56.3
	Star's Declin.	44 39 14.2
	South Latitude	2 31 42.1

Lat. 2° 32′ Cosine	9.9995753
Dec. 44° 39′ Cosine	9.8521218
Z.D. 47° 11′ Cosecant	10.1345808
Log. Sine 1″ A.C.	5.3144251
Log. 907.7 (+4)	6.9579423
Correction 3′ 01″.4 . . Log.	2.2586453

MARANHAM.——September 2d, 1822. Barom. 29.95; Therm. 77°; the Sidereal Chronom. 702 slow on Sidereal Time 8^h 24′ 02″; α Gruris (App. AR. 21^h 57′ 03″) on the Meridian at 13^h 33′ 01″ by the Chronometer.

Chronometer.	Horary Angles.	N.V. Sines.	Level		Readings, &c.	
H. M. S.	M. S.					° ′ ″
13 18 50	14 11	1914	+ 2	+ 1	Final . First Vernier . .	183 45 40
13 21 44.5	11 16.5	1210	+ 6	+ 5	Second „ . .	45 40
13 25 08	7 53	592	− 5	− 7	Third „ .	46 20
13 27 33.5	5 27.5	283	+ 5	+ 3	Fourth „ . .	45 45
13 31 15	1 46	30	− 8	−10	Mean	183 45 51
13 33 45	0 44	5	+ 4	+ 2	Index . . .	+360 00 08.5
13 35 59	2 58	84	− 4	− 6	Level	+18
13 39 18	6 17	376	− 1	− 3		543 46 17.5
13 41 24	8 23	669	+ 8	+ 6	Observed Z.D. .	45 18 51.5
13 43 23	10 22	1023	+ 5	+ 3	Refraction +58″.7	
13 47 32	14 31	2005	+18	+13	Barometer − 0″.1	+0 55.4
13 49 26.5	16 25.5	2567	0	− 1	Therm. . − 3″.2	
Means		896.5	+18			

Lat. 2° 32′ Cosine	9.9995753	Correction . . .	−2 54.6
Dec. 47° 49′ Cosine	9.8270493	True Z.D. . .	45 16 52.3
Z.D. 45° 17′ Cosecant	10.1483780	Star's Dec. 1 Jan. 1818	47 49 59
Log Sine 1″ A.C.	5.3144251	Prec. Aber. Nut. .	−1 24.5
Log 896.5 (+4)	6.9525503		
Correction 2′ 54″.6 . . . Log.	2.2419780	Latitude South .	2 31 42.2

RECAPITULATION.

August 28,	α Lyræ	2	31	42.4
August 29,	α Lyræ	2	31	43.8
August 29,	α Pavonis	2	31	44.5
August 31,	α Lyræ	2	31	44.6
August 31,	α Cygni	2	31	42.1
Septem. 2,	α Gruris	2	31	42.2
		2	31	43.3 South.

TRINIDAD.

Place of Observation.—In Colonel Young's House, in the second ground-lot West of the Protestant Church in Port Spain. The Zenith Distances were observed with a Repeating Circle.

September 30th, 1822.—Barom. 29.98; Therm. 77°. The Sidereal Chron. 702 slow of Sidereal Time 7^h 17′ 30″. Achernar (Apparent AR. 1^h 31′ 07″.5) on the Meridian at 18^h 13′ 37″.5 by the Chronometer.

Chonometer.	Horary Angles.	N.V. Sines.	Level.	
H. M. S.	M. S.			
18 09 01.5	4 36	201	− 2	− 1
18 11 18.5	2 19	51	+ 9	+ 7
18 13 48.5	11	0	− 2	− 1
18 15 51.5	2 14	47	+ 3	+ 1
18 17 49	4 11.5	167	0	0
18 19 41.5	6 04	350	− 5	− 7
18 22 08.5	8 31	690	+ 3	+ 1
18 24 23.5	10 46	1103	− 2	− 4
18 26 37.5	13 00	1608	− 2	− 3
18 28 46.5	15 09	2184	+ 3	+ 1
18 31 53.5	18 16	3175	− 2	0
18 34 09.5	20 32	4011	+ 5	+ 3
Means. . . .		1132.25	+ 2	

Lat. 10° 39′ Cosine		9.9924539
Dec. 58° 08′ Cosine		9.7225881
Z.D. 68° 47′ Cosecant		10.0304823
Log. Sine 1″ A.C.		5.3144251
Log. 1132.25 (+4)		7.0539400
Correction 2′ 10″ .	. Log.	2.1138894

Readings, &c.

			° ′ ″
Previous	First Vernier	.	192 17 40
	Second „ .	.	17 35
	Third „ . .		17 55
	Fourth „ . .	.	17 30
	Mean .	. .	192 17 40
Final	First Vernier.	. .	297 40 20
	Second „ . .	.	40 00
	Third „ . .	.	40 20
	Fourth „ . .	.	40 00
	Mean . .	.	297 40 10
	Index . .	+	167 42 20 / 360 00 00
	Level .	.	+ 2
			825 22 32
	Observed Z.D. . .		68 46 52.7
	Refraction + 2.28.8 / Barometer . . . / Thermometer − 8.1	+	2 20.7
	Correction .	.	− 2 10
	True Z.D. . . .		68 47 03.4
	Star's Dec. 1 Jan. 1818.		58 09 51
	Prec. Aber. and Nut.		— 1 43.7
	Latitude North . .		10 38 56.1

TRINIDAD.——October 3d, 1822. Barom. 29.98; Therm. 78°. The Sidereal Chron. 702 slow of Sidereal Time $7^h\ 17'\ 47''$; α Gruris (Apparent AR. $21^h\ 57'\ 03''$) on the Meridian at $14^h\ 39'\ 16''$ by the Chronometer.

Chronometer.	Horary Angles.	N. V. Sines.	Level.	
H. M. S.	M. S.			
14 20 19	18 57	3416	+1	0
14 22 38	16 38	2632	+4	+2
14 25 05	14 11	1914	0	0
14 26 49	12 27	1475	+2	+1
14 29 13	10 03	961	−4	−5
14 32 14	7 02	471	+1	0
14 35 01	4 15	172	−7	−8
14 37 32	1 44	29	0	0
14 39 52	0 36	3	+3	+2
14 41 35	2 19	51	+5	+3
14 43 51	3 35	122	+4	+2
14 45 41	6 25	392	+5	+3
14 48 51	9 35	874	−4	−6
14 51 07	11 51	1336	−3	−5
Means . .		989.1	−2	

Lat. 10° 39′	Cosine	9.9924539
Dec. 47° 49′	Cosine	9.8270493
Z.D. 58° 27′	Cosecant . . .	10.0694667
Log. Sine 1″	A.C.	5.3144251
Log. 989.1 (+4)		6.9952402
Correction 2′ 38″	Log.	2.1986352

Readings, &c.		° ′ ″
Previous	First Vernier . .	328 20 20
	Second „	20 10
	Third „	20 40
	Fourth „	20 10
	Mean	328 20 20
Final . .	First Vernier . .	67 10 30
	Second „	01 00
	Third „	01 40
	Fourth „	01 10
	Mean	67 01 20
	Index . . . . +	31 39 40
		720 00 00
	Level	− 2
		818 40 58
	Observed Z.D. . .	58 28 38.5
	Refraction +1′ 34″.8	
	Barometer . . .	+1 29.3
	Thermometer −5.5	
	Correction . .	−2 38
	True Z.D.	58 27 29.8
	Star's Dec. 1 Jan. 1818	47 49 59
	Prec. Aber. and Nut.	− 1 21.4
	Latitude North . .	10 38 52.2

Trinidad.——October 4th, 1822. Barom. 29.96; Therm. 76°. The Sidereal Chron. 702 slow of Sidereal Time 7h 17′ 53″.5. Achernar (App. AR. 1h 31′ 07″.6) on the Meridian at 18h 13′ 14″ by the Chronometer.

Chronometer.	Horary Angles.	N. V. Sines.	Level.	
H. M. S.	M. S.			
17 42 19	30 55	9070	−1	−2
17 45 19.5	27 54.5	7405	+6	+5
17 48 15	24 59	5936	−6	−7
17 50 42	22 32	4829	−1	−2
17 53 15	19 59	3799	−5	−7
17 55 13	18 01	3088	+4	+2
17 57 41	15 33	2301	0	0
17 59 05.5	14 08.5	1903	−5	−6
18 02 20.5	10 53.5	1130	−5	−6
18 04 17.5	8 56.5	761	0	0
18 07 02.5	6 11.5	365	−4	−5
18 08 41.5	4 32.5	196	+8	+6
18 11 35.5	1 38.5	25	−7	−9
18 13 09	0 05	0	+7	+5
Lamp trimmed.				
18 23 19	10 05	968	−6	−7
18 28 47.5	15 33.5	2303	+7	+5
18 30 54.5	17 40.5	2970	+2	+1
18 33 35	20 21	3940	−1	0
18 39 17	26 03	6453	+6	+5
18 41 03.5	27 49.5	7360	+7	+5
Means . . .		3240.1	−5.5	

Lat. 10° 39′	Cosine.	9.9924539
Dec. 58° 08′	Cosine. . . .	9.7225881
Z.D. 68° 47′	Cosecant . . .	10.0304823
Log. Sine 1″	A.C.	5.3144251
Log. 3240.1 (+4)	. . .	7.5105584
Correction 6.12	Log .	2.5705078

Readings, &c.

		° ′ ″
Previous	First Vernier . . .	211 41 25
	Second „ . . .	41 15
	Third „ . .	41 55
	Fourth „ . . .	41 05
	Mean	211 41 25
Final. .	First Vernier .	148 41 10
	Second „ . .	40 50
	Third „	41 20
	Fourth „ . . .	40 50
	Mean	148 41 02.5
	Index . . . +	148 18 35 1080 00 00
	Level	−5.5
		1376 59 32
	Observed Z.D. . .	68 50 58.6
	Refraction +2′ 29″.3 Barometer . . . Thermometer −7.8	+2 21.5
	Correction . . .	−6 12
	True Z.D. . . .	68 47 08.1
	Star's Dec. 1 Jan. 1818	58 09 51
	Prec. Aber. and Nut.	−1 42.2
	Latitude North . .	10 38 59.3

RECAPITULATION.

			° ′ ″
September 30th,	Achernar	. . . .	10 38 56.1
October 3d,	α Crucis	. . . .	10 38 52.2
„ 4th,	Achernar	. . . .	10 38 59.3
			10 38 56 North.

JAMAICA.

Place of Observation.—Fort Charles. The Zenith Distances observed with a Repeating Circle.

October 25th, 1822. Bar. 30.01; Therm. 78° The Sidereal Chron. slow on Sidereal Time 6^h 18′ 18″. Polaris (AR. 58′ 14″.2) on the Meridian at 18^h 39′ 56″ by the Chron.

Chronometer.	Horary Angles.	N. V. Sines.	Level.	
H. M. S.	M. S.			
18 08 52	31 04	9178	+3	+1
18 11 45	28 11	7552	+1	0
18 14 38	25 18	6087	−5	−3
18 16 40	23 16	5149	−2	−3
18 19 01	20 55	4162	0	0
18 21 54	18 02	3093	−1	0
18 23 55	16 01	2441	+6	+5
18 26 04	13 52	1830	−1	0
18 28 43	11 13	1197	0	0
18 30 58	8 58	765	−2	0
18 33 40	6 16	374	+1	0
18 35 42	4 14	171	0	0
18 37 49	2 07	43	+3	+2
18 39 33	0 23	1	0	0
18 42 32	2 36	64	0	0
18 46 22	6 26	394	0	0
18 48 00	8 04	619	+3	+2
18 50 04	10 08	977	−2	0
18 52 04	12 08	1401	+3	0
18 53 45	13 49	1817	+3	+2
18 55 27	15 31	2291	0	0
18 56 56	17 00	2750	0	0
18 58 22	18 26	3233	0	0
18 59 26	19 30	3618	+2	+1
19 01 25	21 29	4390	+2	+1
19 02 56	23 00	5031	−5	−3
19 04 25	24 29	5701	0	0
19 05 08	25 12	6039	−3	−1
19 07 33	27 37	7251	+8	+6
19 09 27	29 31	8282	0	0
Means . . .		3183.4	+12	

Readings, &c.

Lat. 17° 56′ Cosine . . .		9.9783702
Dec. 88° 22′ Cosine		8.4548934
Z.D. 70° 26′ Cosecant		10.0258327
Log. Sine 1″ A.C.		5.3144251
Log. 3183.4 (+4) . . .		7.5028912
Correction 18″.9	Log.	1.2764126
		° ′ ″
Previous	First Vernier	117 31 25
	Second „ . . .	31 10
	Third „	31 35
	Fourth „ . .	31 00
	Mean	117 31 17.5
Final . .	First Vernier . . .	69 21 05
	Second „ . . .	20 40
	Third „ . . .	20 50
	Fourth „ . . .	20 25
	Mean	69 20 45
	Index . . . . +	242 28 42.5 1800 00 00
	Level	+ 12
		2111 49 39.5
	Observed Z.D. . .	70 23 39.3
	Refraction + 2′ 42″.3 Barometer . . . Thermometer − 9.2	+ 2 33.1
	Correction . . .	− 0 18.9
	True Z.D.	70 25 53.5
	Star's Decl.	88 22 02.1
	Latitude North . . .	17 56 08.6

JAMAICA.——November 3d, 1822. Bar. 30.00; Therm. 78°. The Sidereal Chron. slow on Sidereal Time 6^h 19′ 06″.4. Polaris (AR. 58′ 12.″4) on the Meridian at 18^h 39′ 06″ by the Chronometer.

Chronometer.	Horary Angles.	N. V. Sines.	Level.	
H. M. S.	M. S.			
17 53 18.5	45 48	19902	−2	−4
17 56 36	42 30	17144	0	0
17 59 41	39 25	14754	0	0
18 02 15	36 51	12899	+5	+4
18 04 50	34 16	11157	+5	+4
18 07 09	31 57	9702	0	0
18 09 11	29 55	8508	0	0
18 10 56	28 10	7543	−2	−4
18 13 47	25 19	6093	−7	−9
18 15 55	23 11	5112	−2	−4
18 18 44	20 22	3946	+1	+3
18 20 26.5	18 40	3315	+5	+3
18 22 30.5	16 36	2622	+3	+2
18 24 36	14 30	2001	−2	−4
18 26 42	12 24	1463	0	0
18 28 11	10 55	1134	+5	+3
18 30 16	8 50	743	+3	+1
18 31 46	7 20	512	+4	+2
18 33 31.5	5 35	297	−1	−[illegible]
18 35 12	3 54	145	0	0
18 37 18	1 48	31	0	0
18 38 42	0 24	2	0	0
18 40 19	1 13	14	−2	−4
18 41 39.5	2 33	62	+6	+5
18 43 32	4 26	187	0	0
18 45 00.5	5 54	331	+3	+1
18 47 29	8 23	669	0	0
18 50 01	10 55	1134	0	0
18 52 25.5	13 20	1692	+2	−0
18 54 07.5	15 01	2146	0	0
Means		4508.7	+10	

Readings, &c.

Lat. 17° 56′	Cosine	9.9783702
Dec. 88° 22′	Cosine	8.45489_{34}
Z.D. 70° 26′	Cosecant	10.0258327
Log. Sine 1″	A.C.	5.3144251
Log. 4508.7 (+4)	. . . . . .	7.6540513
Correction 26″.8	Log.	1.4275727

		° ′ ″
Previous	First Vernier . .	5 48 40
	Second ,, ...	48 10
	Third ,, ...	48 50
	Fourth ,, ...	48 00
	Mean	5 48 25
Final . .	First Vernier	317 44 25
	Second ,, ...	44 10
	Third ,, ...	44 40
	Fourth ,, ...	44 00
	Mean.	317 44 19
	Index . . +	354 11 35 1440 00 00
	Level	+ 10
		2111 56 04
	Observed Z.D.	70 23 52.1
	Refraction + 2′ 42″.3 Barometer . . . Thermometer − 9.9	+ 2 33.1
	Correction	− 0 26.8
	True Z.D.	70 25 58.4
	Star's Decl.	88 22 05
	Latitude, North. . . .	17 56 06.6

RECAPITULATION.

	° ′ ″
October 25, Polaris	17 56 08.6
November 3, Polaris	17 56 06.6
	17 56 07.6 North.

NEW YORK.

Place of Observation.—The Cupola of Columbia College. The Zenith Distances were observed with a Repeating Circle.

December 24th, 1822. Barom. 30.40; Therm. 20°.5. The Chronometer 423 fast 5h 00′ 57″ (page 123); Sun on the Meridian at 23h 59′ 53″ Mean Time, 5h 00′ 50″ by the Chronometer.

Chronometer.	Time from Noon.	N.V. Sines.	Level.	
H. M. S.	M. S.			
4 45 27	15 23	2252	0	0
4 49 06	11 44	1310	0	0
4 51 50	9 00	771	+3	− 7
4 57 23	3 27	113	+2	− 7
4 59 40	1 10	13	0	−10
5 03 10	2 20	52	+9	0
5 05 25	4 35	200	+7	− 3
5 07 10	6 20	382	0	0
5 09 10	8 20	661	+6	− 4
5 10 38	9 48	914	+7	− 3
5 12 54	12 04	1386	+8	− 3
5 15 00	14 10	1910	+8	− 2
Means. After Noon	1 23	830.3	+5.5	

	H. M.
App. Greenwich time at Noon . . .	4 56.2
Observation later than Noon . . .	1.4
App. Greenwich time, corresponding to the Mean Z.D.	4 57.6

Lat. 40° 42′ 43″	Cosine . . .	9.8796683
Dec. 23° 26′ 34″	Cosine . . .	9.9625861
Z.D. 64° 09′ 15′	Cosecant . .	10.0457718
Log. Sine 1″	A.C.. . . .	5.3144251
Log. 830.3 (+4)	. . . . . .	6.9192350
Correction 2′ 12″.3 . . .	Log.	2.1216863

Readings, &c.

		° ′ ″
Previous	First Vernier .	328 26 20
	Second „ . . .	26 20
	Third „ . . .	26 50
	Fourth „ . . .	26 00
	Mean	328 26 22.5
Final . .	First Vernier . . .	21 34 50
	Second „ . . .	34 30
	Third „ . . .	35 10
	Fourth „ . . .	34 30
	Mean	21 34 45
	Index . . . +	31 33 37.5 720 00 00
	Level	+5.5
		773 08 28
	Observed Z.D. .	64 25 42.3
	Refraction +2′ 01″.1 Barometer . . +1.6 Thermometer . +6.7 Parallax . . . −7.9	+2 01.5
	Semidiam	−16 17.7
	Correction . .	2 12.3
		64 09 13.8
	Decl. at 4h 57′.6 . . App. Greenwich time	23 26 33.7
	North Latitude . .	40 42 40.1

New York.—December 24th, 1822. Barom. 30.40; Therm. 21°. The Sidereal Chron. 702 slow on Sidereal Time $6^h\ 33'\ 55''$; Polaris (AR. $0^h\ 57'\ 46''.3$) on the Meridian at $18^h\ 23'\ 51''$ by the Sidereal Chronometer.

Chronometer.	Horary Angles.	N.V. Sines.	Level.	
H. M. S.	M. S.			
18 09 48	14 03	1879	0	0
18 13 13	10 38	1076	+11	0
18 15 32	8 19	658	0	0
18 16 52	6 59	464	0	0
18 19 02	4 49	221	−11	− 1
18 20 48	3 03	89	+10	+ 1
18 23 12	0 39	4	+ 8	− 4
18 25 14	1 23	18	− 9	− 1
18 30 37	6 46	436	0	0
18 33 00	9 09	797	+ 9	+ 2
18 35 08	11 17	1212	+ 8	− 4
18 37 09.5	13 18	1683	+ 9	− 2
Means . . .		711.4	+ 13	

Lat. 40° 42′ 43″	Cosine . . .	9.8796683
Dec. 88° 22′ 18″	Cosine . . .	8.4535622
Z.D. 49° 39′ 30″	Cosecant . . .	10.1312705
Log. Sine 1″	A.C.	5.3144251
Log. 711.4 (+4)	. . . . . .	6.8521139
Correction 0′ 04″.3 . . .	Log.	0.6310400

Readings, &c.		° ′ ″
Previous	First Vernier . .	228 56 30
	Second „ . .	56 20
	Third „ . .	57 00
	Fourth „ . . .	56 30
	Mean	228 56 35
Final . .	First Vernier . . .	80 37 40
	Second „ . . .	37 00
	Third „ . . .	38 10
	Fourth „ . .	37 30
	Mean	80 37 35
	Index . . . . +	131 03 25
		360 00 00
	Level	+ 13
		571 41 13
	Observed Z.D. . .	47 38 26.1
	Refraction +1′ 03″.8; Barometer +0.9; Thermometer +3.5	+ 1 08.2
	Correction . . .	− 0 04.3
	True Z.D. . . .	47 39 30
	Polar Distance . .	1 37 41.1
	Co. Latitude . . .	49 17 11.1
	North Latitude . .	40 42 48.9

NEW YORK.——December 31st, 1822. Barom. 30.76; Therm. 26°. The Chron. 423 fast 5^h 00′ 40″; the Sun on the Meridian at 0^h 03′ 19″ Mean time, 5^h 03′ 59″ by the Chronometer.

Chronometer.	Time from Noon.	N. V. Sines.	Level.		Readings, &c. Sun's L.L.		
H. M. S.	M. S.						° ′ ″
4 49 13	14 46	2075	0	0	Previous	First Vernier . . .	128 22 40
4 52 12	11 47	1321	− 9	0		Second „ . . .	22 20
4 54 26	9 33	868	+ 7	− 2		Third „ . .	23 10
4 56 45	7 14	498	+12	+ 3		Fourth „ . .	22 20
4 59 20	4 39	206	+ 9	0		Mean	128 22 37.5
5 00 47	3 12	97	+ 6	− 4			
5 02 45	1 14	14	− 3	−12	Final	First Vernier . . .	177 31 50
5 04 44	0 45	5	0	0		Second „ . . .	31 40
5 07 00	3 01	87	+ 3	− 7		Third „ . . .	32 00
5 08 32	4 33	197	0	− 9		Fourth „ . . .	31 20
5 10 52	6 53	451	+10	0		Mean . . .	177 31 42.5
5 13 24	9 25	844	+10	0		Index . . . +	231 37 22.5 / 360 00 00
Means. Before Noon	1 39	555.25	+ 7			Level . .	+7
							769 09 12

	H. M.
Apparent Greenwich time at Noon . .	4 56.2
Observation earlier than Noon . . .	1.7
Appt. Greenwich time corresponding to the Mean Z.D.	4 54.5

Lat. 40° 42′ 43″ Cosine	9.8796683
Dec. 23° 07′ 17″ Cosine	9.9636344
Z.D. 63° 50′ 00″ Cosecant . . .	10.0469582
Log. Sine 1″ A.C. . . .	5.3144251
Log. 555.25 (+4) . . .	6.7444886
Correction 1′ 28″.9 . . . Log.	1.9491746

Observed Z.D. L.L. .		64 05 46
Refraction	+1′ 59″.3	+1 59.7
Barometer . .	+3.0	
Thermometer	+5.3	
Parallax . .	−7.9	
Semidiam		−16 17.8
Correction . .		−1 28.9
True Z.D. . . .		63 49 59
Decl. at 4^h 54′.5 . . App. Greenwich time		23 07 17.6
North Latitude .		40 42 41.4

New York.——January 3d, 1823. Barom. 30.20; Therm. 36°. The Sidereal Chron. 702 slow on Sidereal Time 6^h 34′ 55″. β Ursæ Minoris (AR. 14^h 51′ 17″.5) on the Northern Meridian at 20^h 16′ 22″ by the Chronometer.

Chronometer.	Horary Angles.	N. V. Sines.	Level.	
H. M. S.	M. S.			
20 05 40	10 42	1090	0	0
20 08 18	8 04	619	+ 3	− 6
20 11 00	5 22	274	0	0
20 12 32	3 50	140	0	0
20 16 36.5	0 14	1	−10	0
20 20 57	4 35	200	− 9	0
20 24 14.5	7 52	589	−10	0
20 26 12.5	9 50	920	+ 2	− 7
20 28 02	11 40	1295	+10	0
20 29 53.5	13 31	1739	+ 3	− 6
20 32 06.5	15 44	2355	0	0
20 34 55	18 33	3274	− 9	0
Means		1041.3	−19.5	

Readings, &c.

		° ′ ″
Previous	First Vernier	219 43 00
	Second „	43 10
	Third „ . . .	43 25
	Fourth „	42 50
	Mean	219 43 06
Final .	First Vernier . .	272 08 50
	Second „ . .	08 50
	Third „ .	09 30
	Fourth „ .	08 30
	Mean	272 08 55
	Index . . . . +	140 16 54 360 00 00
	Level	− 19.5
		772 25 29
	Observed Z.D. . .	64 22 07.4
	Refraction +2′ 00″.7 Barometer +0.8 Thermometer +2.9	+2 04.4
	Correction. .	+0 47.1
	True Z.D. . . .	64 24 58.9
	Polar Distance .	15 07 41.2
	Co. Latitude . .	49 17 17.7
	Latitude North .	40 42 42.3

Lat. 40° 42′ 43″ Cosine . .		9.8796683
Dec. 74° 52′ 19″ Cosine .		. 9.4166026
Z.D. 64° 25′ 00″ Cosecant .		10.0448136
Log. Sine 1″ A.C.		5.3144251
Log. 1041.3 (+4) . .		7.0175750
Correction 0′ 17″.1	Log.	1.6730855

RECAPITULATION.

	° ′ ″
December 24, the Sun	40 42 40.1
December 24, Polaris	40 42 48.9
December 31, the Sun	40 42 41.4
January 3, β Ursæ Minoris, S.P.	40 42 42.3
	40 42 43.2 North.

HAMMERFEST.

Place of Observation.—At Fugleness. The Zenith Distances observed with a Repeating Circle.

June 12th, 1823. Bar. 29.90; Ther. 61° The Chron. No. 649, slow 1h 33′ 44″.5. Sun on the Meridian at 23h 59′ 15″ Mean Time, and at 22h 25′ 30″.5 by the Chron.

Chronometer.	Horary Angles.	N.V. Sines.	Level.	
H. M. S.	M. S.			
22 07 28	18 02.5	3096	− 1	− 5
22 08 55	16 35.5	2620	+ 1	− 2
22 10 38	14 52.5	2105	+ 6	+ 2
22 13 07	12 23.5	1461	− 3	0
22 14 45	10 45.5	1101	0	− 3
22 16 08	8 22.5	668	0	0
22 32 48	7 17.5	506	− 5	− 8
22 33 50	8 19.5	660	0	0
22 35 08	9 37.5	881	+ 6	+10
22 36 27	10 56.5	1140	0	0
22 38 05	12 34.5	1505	0	0
22 39 26	13 55.5	1845	+ 5	+ 1
22 41 15	15 44.5	2357	0	0
22 42 13	16 42.5	2656	+ 5	+ 1
22 44 00	18 29.5	3253	0	0
22 45 34	20 03.5	3827	− 4	0
Means. After Noon. . 3 14		1855	+ 3	

Apparent Greenwich time at Noon	22 25	
Observations later than Noon . .	3.2	
Apt. Greenwich time, corresponding to the Mean Zenith Distances. .	22 28.2	
Lat. 70° 40′	Cosine	9.5199112
Dec. 23° 08′	Cosine	9.9635957
Z.D. 47° 32′	Cosecant . .	10.1321377
Log. Sine 1″	A.C.	5.3144251
Log. 1855 (+4)	. . . . . .	7.2683439
Correction 2′ 37″.9	Log.	2.1984136

Readings, &c. Sun's L.L.

		° ′ ″
Previous	First Vernier . . .	266 10 20
	Second „	9 45
	Third „	10 20
	Fourth „	9 40
	Mean	266 10 00
Final . .	First Vernier . . .	311 26 10
	Second „	26 20
	Third „	26 40
	Fourth „	25 45
	Mean	311 26 14
	Index . . . . . +	93 50 00 360 00 00
	Level	+ 3
		765 16 17
	Observed Z.D. L.L. .	47 19 16
	Refraction + 1′ 04″.2 Barometer . − 0.2 Thermometer − 1.4 Parallax . − 6.3	+ 0 56.3
	Semidiam	− 15 46.4
	Correction	− 2 37.9
	True Z.D.	47 32 18
	Declination at 22h 28′ (June 11th) Appt. Greenwich time . .	23 07 47.3
	Latitude North . . .	70 40 05.3

SPITZBERGEN.

Place of Observation.—At the Observatory on the Inner Norway Island. The Zenith Distances were observed with a Repeating Circle.

July 5th, 1823. Bar. 30.10; Therm. 41°. The Chron. 649, slow 44′ 51″.5, (page 152); the Sun on the Southern Meridian at 0ʰ 04′ 00″ Mean Time, and at 23ʰ 19′ 08 .5 by the Chronometer.

Chronometer.	Times from Noon.	N. V. Sines.	Level.	
H. M. S.	M. S.			
22 55 38	23 30.5	5256	0	0
22 57 54	21 14.5	4292	+10	0
23 00 00	19 08.5	3486	+ 6	− 3
23 02 00	17 08.5	2796	+ 6	− 3
23 04 13	14 55.5	2119	+ 7	− 2
23 05 45	13 23.5	1707	0	0
23 08 00	11 08.5	1182	+ 3	− 6
23 11 04	8 04.5	620	+ 2	− 8
23 13 17	5 51.5	327	+ 2	− 7
23 16 16	2 52.5	79	0	0
23 18 38	0 30.5	3	+ 2	− 7
23 20 28	1 19.5	17	+ 6	− 4
23 22 26	3 17.5	103	0	0
23 24 02	4 53.5	228	+ 6	− 3
23 25 36	6 27.5	397	+ 7	− 3
23 27 08	7 59.5	608	+ 6	− 3
23 28 47	9 38.5	885	−10	− 1
23 31 48	12 39.5	1525	+ 8	− 1
23 33 52	14 43.5	2063	+ 7	− 3
23 35 45	16 36.5	2625	+ 3	− 7
Before Noon . . 3 01		1515.9	+ 5	

	H. M.
Apparent Greenwich time at Noon	23 13.3
Observations before Noon .	− 3
Apparent Greenwich Time corresponding to the Mean Z.D. . . .	23 10.3

Readings, &c. Sun's L.L.

Lat. 79° 50′ Cosine .	9.2467746
Dec. 22° 52′ Cosine . . .	9.9644537
Z.D. 56° 58′ Cosecant	0.0765728
Log. Sine 1″ A.C.	5.3144251
Log. 1515.9 (+4)	7.1806706
Correction 1′ 00″.7 Log.	1.7828968

		° ′ ″
Previous	First Vernier . . .	39 18 20
	Second ,, ...	18 10
	Third ,, . . .	18 30
	Fourth ,, . .	18 10
	Mean . . .	39 18 17.5
Final . .	First Vernier . .	103 39 05
	Second ,, . . .	38 45
	Third ,, ...	39 30
	Fourth ,, ...	38 50
	Mean . . .	103 39 02.5
	Index . . . . . +	320 41 42.5 720 00 00
	Level	+ 5
		1144 20 50
	Observed Z.D. L.L. .	57 13 02.5
	Refraction + 1′ 30″.3 Barometer + 0.3 Thermometer + 1.3 Parallax − 7.4	+ 1 24.5
	Semidiam. . . .	−15 45.5
	Correction	− 1 00.7
	True Z.D.	56 57 40.8
	Decl. at 23ʰ 10′.3 App. Greenwich Time	22 52 15.3
	North Latitude . . .	79 49 56.1

SPITZBERGEN.——July 6th, 1823. Bar. 29.90; Therm. 38°.5. The Chronometer 649, slow 44′ 51″ (page 152). The Sun on the Northern Meridian at 12^h 04′ 16″ Mean Time, and at 11^h 19′ 25″ by the Chronometer.

Chronometer.	Time from Midnight.	N. V. Sines.	Level.	
H. M. S.	M. S.			
11 05 26	13 59	1861	0	0
11 07 35	11 50	1333	0	0
11 10 07	9 18	823	0	0
11 12 00	7 25	524	+10	− 2
11 14 47.5	4 37.5	203	1+0	+ 1
11 17 07	2 18	51	+10	+ 1
11 20 45.5	1 20.5	17	+10	+ 1
11 22 44	3 19	105	+ 2	− 7
11 25 06	5 41	307	+10	+ 3
11 27 37.5	8 12.5	641	−10	− 2
11 30 18	10 53	1127	+11	+ 3
11 32 17	12 52	1575	+ 2	− 6
Means. Before Mid.	0.35	714	+13.5	

Readings, &c. Sun's U.L.

		° ′ ″
Previous	First Vernier .	284 25 20
	Second „ .	25 10
	Third „ .	25 50
	Fourth „ . .	25 00
	Mean	284 25 20
Final . .	First Vernier . .	129 38 00
	Second „	37 50
	Third „ . . .	38 25
	Fourth „ .	38 05
	Mean . .	129 38 05
	Index . . +	75 34 40
		720 00 00
	Level	+ 13.5
		925 12 58.5
	Observed Z.D. U.L.	77 06 04.9
	Refraction +4′ 09″.4; Barometer −0 00.8; Thermometer +0 04.9; Parallax −0 08.6	+ 4 04.9
	Semidiam. . .	+ 15 45.5
	Correction . .	+ 0 24.6
	True Z.D.	77 26 19.9
		102 33 40.1
	Decl. at 11^h 12′.7 App. Greenwich Time	22 43 44.2
	North Latitude . . .	79 49 55.9

	H. M.
App. Greenwh. Time at Midnight .	11 13.3
Observations earlier than Midnight .	0.6
App. Greenwich Time corresponding to the Mean Z.D.	11 12.7

Lat. 79° 50′ Cosine	9.2467746
Dec. 22° 44′ Cosine . .	9.9648785
Z.D. 77° 26′ Cosecant .	0.0105308
Log. Sine 1″ A.C.	5.3144251
Log. 714 (+4)	6.8536982
Correction 0′ 24″.6 Log.	1.3903072

Spitzbergen.——July 7th, 1823. Barom. 29.74; Therm. 34°. The Chron. 649 slow 44′ 49″.5 (page 253); the Sun on the Northern Meridian at 12ʰ 04′ 26″ Mean Time, and at 11ʰ 19′ 36″.5 by the Chronometer.

Chronometer.	Time from Midnight.	N.V. Sines.	Level.	
H. M. S.	M. S.			
10 52 43	26 53.5	6876	0	0
10 56 03	23 33.5	5278	+ 6	− 3
10 57 56	21 40.5	4469	0	0
11 00 12	19 24.5	3584	+13	+ 4
11 02 57	16 39.5	2640	+ 2	− 7
11 04 27	15 09.5	2187	+13	+ 3
11 07 38	11 58.5	1365	0	0
11 10 32	9 04.5	783	+ 7	− 3
11 12 35	7 01.5	470	+ 2	− 7
11 14 00	5 36.5	300	0	0
11 15 48	3 48.5	138	+ 2	− 7
11 17 15	2 21.5	53	0	0
11 19 08	0 28.5	2	+ 8	− 2
11 20 36	0 59.5	10	−12	− 3
11 30 27	10 50.5	1118	+ 6	− 3
11 31 51	12 14.5	1426	+ 3	− 7
11 34 08	14 31.5	2008	+ 2	− 7
11 35 18	15 41.5	2342	+ 9	− 1
11 37 09	17 32.5	2928	+ 2	− 7
11 38 36	18 59.5	3431	+10	+ 1
Means. Before Midnight	3 38	2070.4	+ 12	

Readings, &c. Sun's U.L.

		° ′ ″
Previous	First Vernier . .	203 00 00
	Second ,, . .	202 59 40
	Third ,, . .	203 00 30
	Fourth ,, . .	202 59 50
	Mean . . .	203 00 00
Final . .	First Vernier . .	306 47 20
	Second ,, . .	47 05
	Third ,, .	47 30
	Fourth ,,	46 50
	Mean .	306 47 11.2
	Index . . +	157 00 00
		1080 00 00
	Level . .	+12
		1543 47 23.2
	Observed Z.D. U.L.	77 11 22.2
	Refraction +4′ 11″; Barometer . −2.2; Thermometer +7.3; Parallax −8.5	+4 07.6
	Semidiam . .	+15 45.5
	Correction	+ 1 11.3
	True Z.D. . .	77 32 26.6
		102 27 33.4
	Decl. at 11ʰ 09′.7 App. Greenwʰ Time	22 37 34.8
	North Latitude .	79 49 58.6

	H. M.
Apparent Greenwich Time at Midnight .	11 13.3
Observation earlier than Midnight . .	3.6
Apparent Greenwich Time corresponding to the Mean Z.D.	11 09.7

Lat. 79° 50′ Cosine	9.2467746
Dec. 22° 38′ Cosine	9.9651953
Z.D. 77° 32′ Cosecant	10.0103624
Log. Sine 1″ A.C.	5.3144251
Log. 2070.4 (+4) . . .	7.3160543
Correction 1′ 11″.3 . . . Log.	1.8528117

SPITZBERGEN.——July 9th, 1823. Barom. 29.93; Therm. 37°. The Chron. 649, slow 44′ 48″; the Sun on the Northern Meridian at 12ʰ 04′ 45″ Mean Time, and at 11ʰ 19′ 57″ by the Chronometer.

Chronometer.	Time from Midnight.	N. V. Sines.	Level.	
H. M. S.	M. S.			
11 05 25	14 32	2010	+ 2	− 6
11 06 53	13 04	1625	+ 6	− 2
11 09 07	10 50	1117	+ 1	− 7
11 10 29	9 28	653	+ 8	− 1
11 12 27	7 30	535	+ 6	− 2
11 14 10	5 47	318	0	− 9
11 16 07	3 50	140	0	0
11 17 19	2 38	66	0	0
11 19 10	0 47	6	+ 5	− 3
11 21 30	1 33	23	+ 7	− 2
11 24 00	4 03	156	+ 7	− 2
11 25 28	5 31	289	+12	+ 4
11 29 07	9 10	800	0	0
11 30 46	10 49	1114	+ 6	− 3
11 32 42	12 45	1547	0	0
11 34 23	14 26	1982	+ 3	− 5
Means. Before Mid.	0 38	786.3	+ 12.5	

Readings, &c. Sun's U.L.

		° ′ ″
Previous	First Vernier . . .	176 08 55
	Second „ . . .	09 00
	Third „ .	09 30
	Fourth „ . .	08 50
	Mean . . .	176 09 03.7
Final	First Vernier . . .	334 57 40
	Second „ . . .	57 50
	Third „ . . .	58 00
	Fourth „ . .	57 25
	Mean . . .	334 57 43.7
	Index . . . +	183 50 56.3
		720 00 00
	Level	+12.5
		1238 48 52.6
	Observed Z D. U.L. .	77 25 33.3
	Refraction +4′ 15″; Barometer −0.6; Thermometer +5.8; Parallax −8.5	+4 11.7
	Semidiam . .	+15 45.6
	Correction	+0 27.1
	True Z.D.	77 45 57.7
		102 14 02.3
	Decl. at 11ʰ 12″.7 App. Greenwich Time	22 24 03
	North Latitude . .	79 49 59.3

	H. M.
Apparent Greenwich Time at Midnight	11 13.3
Observation earlier than Midnight .	0.6
App. Greenwich Time corresponding to the Mean Z.D.	11 12.7

Lat. 79° 50′ Cosine	9.2467746
Dec. 22° 24′ Cosine . .	9.9659285
Z.D. 77° 46′ Cosecant. . .	0.0099753
Log. Sine 1″ A.C.	5.3144251
Log. 786.3 (+4) . .	6 8955883
Correction 0′ 27″.1 . . Log.	1.4326918

SPITZBERGEN.——July 10th, 1823. Barom. 30.00; Therm. 45°. The Chron. 423 fast of No. 649, 3′ 25″, slow of Mean Time 41′ 23″. The Sun on the Northern Meridian at 12^h 04′ 54″ Mean Time, and at 11^h 23′ 31″ by 423.

Chronometer.	Time from Midnight.	N. V. Sines.	Level.	
H. M. S.	M. S.			
10 54 42	28 49	7894	0	0
10 56 56	26 35	6719	+ 7	− 2
10 59 54	23 37	5305	0	0
11 02 51	20 40	4063	+ 2	− 7
11 05 25	18 06	3117	+ 5	− 3
11 07 28	16 03	2451	+11	+ 2
11 10 15	13 16	1675	+ 3	− 5
11 12 28	11 03	1162	0	0
11 15 15	8 16	650	+ 2	− 7
11 17 10	6 21	384	+ 3	− 5
11 19 21	4 10	165	+ 6	− 2
11 22 00	1 31	22	+ 6	− 2
11 25 48	2 17	50	+ 3	− 5
11 27 45	4 14	171	+ 7	− 2
11 30 42	7 11	491	+ 2	− 6
11 32 31	9 00	771	+ 2	− 7
11 34 47	11 16	1208	0	− 8
11 36 52	13 21	1696	+ 9	+ 1
11 39 17	15 46	2365	0	0
11 40 40	17 09	2799	+ 8	− 1
Means. Before Midnight.	4 55	2157.9	+ 8.5	

		H. M.
Appt. Greenwich Time at Midnight	.	11 13.3
Observation earlier than Midnight	. .	4.9
Appt. Greenwich Time corresponding to the Mean Zenith Distance	. .	11 08.4

Lat. 79° 50′ Cosine	9.2467746
Dec. 22° 17′ Cosine . . .	9.9662920
Z.D. 77° 53′ Cosecant . .	10.0097845
Log. Sine 1″ A.C.	5.3144251
Log. 2157.9 (+4) . . .	7.3340313
Correction 1′ 14″.3 Log.	1.8713075

Readings, &c. Sun's U.L.

		° ′ ″
Previous.	First Vernier .	334 57 40
	Second „ . . .	57 50
	Third „ . .	58 00
	Fourth „ . .	57 25
	Mean	334 57 43.75
Final . .	First Vernier . .	85 40 50
	Second „ . .	40 20
	Third „	41 00
	Fourth „ . .	40 20
	Mean	85 40 37.5
	Index . +	25 02 16.25
		1440 00 00
	Level .	+ 8.5
		1550 43 02.25
	Observed Z.D. U.L.	77 32 09.1
	Refraction +4′ 18″; Barometer . 0; Thermometer +1.6; Parallax . −8.5	+ 4 11.1
	Semidiam . . .	+ 15 45.6
	Correction . . .	+ 1 14.3
	True Z.D. . .	77 53 20.1
		102 06 39.9
	Decl. at 11^h 08′.4 App. Greenwh. time	22 16 44.1
	North Latitude . .	79 49 55.8

SPITZBERGEN.——July 12th, 1823. Barom. 29.94; Therm. 37°.5. The Chron. No. 649 slow 44′ 48″. The Sun on the Southern Meridian at 00^h 05′ 06″.5 Mean Time, 23^h 20′ 18″.5 by the Chronometer.

Chronometer.	Time from Noon.	N.V. Sines.	Level.	
H. M. S.	M. S.			
23 15 10	5 08.5	252	0	0
23 17 03	3 15.5	101	+ 7	− 1
23 18 50	1 28.5	20	+ 3	− 5
23 20 25	0 06.5	0	0	0
23 22 22	2 03.5	40	+ 7	− 2
23 23 32	3 13.5	99	+ 1	− 8
23 25 04	4 45.5	216	+ 3	− 6
23 27 33	7 14.5	499	0	0
23 29 00	8 41.5	719	+ 2	− 7
23 31 35	11 16.5	1210	+ 7	− 2
23 33 34	13 15.5	1673	+ 1	− 8
23 36 10	15 51.5	2392	+ 3	− 6
Means. After Noon	4 42	601.8	−5.5	

	H. M.
App. Greenwich Time at Noon . .	23 13.3
Observation after Noon	4.7
App. Greenwich Time, corresponding to the Mean Z.D.	23 18

Lat. 79° 50′ Cosine	9.2467746
Dec. 22° 05′ Cosine	9.9669101
Z.D. 57° 45′ Cosecant . . .	10.0727694
Log. Sine 1″ A.C.	5.3144251
Log. 601.8 (+4)	6.7794522
Correction 00′ 24″ . . Log. .	1.3803314

Readings, &c. Sun's U.L.

		° ′ ″
Previous	First Vernier . . .	205 55 40
	Second „ . . .	55 40
	Third „ . . .	56 15
	Fourth „ . . .	55 20
	Mean	205 55 43.75
Final . .	First Vernier . . .	175 35 15
	Second „ . .	35 15
	Third „ . . .	35 40
	Fourth „ . . .	34 50
	Mean	175 35 15
	Index . . . . +	154 04 16.2 360 00 00
	Level	−5.5
		689 39 25.7
	Observed Z.D. U.L.	57 28 17.1
	Refraction +1′ 31″.2 Barometer . . −0.2 Thermometer . +1.9 Parallax . . . −7.4	+1 25.5
	Semidiam.	+15 45.8
	Correction . .	−00 24
		57 45 04.4
	Decl. at 23^h 18′. . . App. Greenwich Time	22 04 57.1
	North Latitude . .	79 50 01.5

RECAPITULATION.

	°	′	″
July 5, Sun on the Southern Meridian	79	49	56.1
July 6, Sun on the Northern Meridian	79	49	55.9
July 7, Sun on the Northern Meridian	79	49	58.6
July 9, Sun on the Northern Meridian	79	49	59.3
July 10, Sun on the Northern Meridian	79	49	55.8
July 12, Sun on the Southern Meridian	79	50	01.5
	79	49	57.8 North.

GREENLAND.

Place of Observation.—At the Observatory on the Inner Pendulum Island.

August 21st, 1823. Bar. 29.90; Ther. 39°. The Altitudes observed with a Sextant and Mercurial Horizon. The Chron. No. 423 fast 1ʰ 23′ 45″ (page 168); the Sun on the Meridian at 0ʰ 03′ 03″ Mean Time, and at 1ʰ 26′ 48″ by the Chron.

Chronometer.	Time from Noon.	N.V. Sines.	Altitudes.
H. M. S.	M. S.		
1 18 00	8 48	737	56 07 58 ☉̄
1 19 00	7 48	582	55 04 50 ☉̲
1 33 40	6 52	449	55 04 30 ☉̲
1 35 00	8 12	640	56 07 45 ☉̄
Means. Before Noon	23	602	55 36 16

	H. M.
App. Greenwich Time at Noon . .	1 15.3
Observation earlier than Noon	0.4
Apparent Greenwich Time corresponding to the Mean Altitude	1 14.9

Lat. 74° 32′ Cosine . .	9.4259867
Dec. 12° 19′ Cosine . . .	9.9898873
Alt. 27° 46′ Secant	10.0531293
Log. Sine 1″ A.C.	5.3144251
Log. 602 (+4)	6.7795965
Correction 36″.6 Log.	1.5630249

Deduction.

	° ′ ″
Observed Double Altitude	55 36 16
Index	− 1 20
	55 34 56
Apparent Altitude . .	27 47 28
Refraction 1′ 50″.1; Barometer +0.4; Thermometer −2; Parallax +7.7 . . .	− 1 44
Correction . .	+ 0 36.6
True Altitude	27 46 20.6
Zenith Distance	62 13 39.4
Decl. at 1ʰ 14′.9 app. Greenʰ. Time	12 18 39
North Latitude	74 32 18.4

GREENLAND.——August 22nd, 1823. Bar. 29.95; Therm. 39°. The Altitudes observed with a Repeating Reflecting Circle of six inches diameter, and a Mercurial Horizon. The Chron. No. 423, fast 1h 23′ 51″, page 168. The Sun on the Meridian at 0h 02′ 48″ Mean Time, and at 1h 26′ 39″ by the Chronometer.

Chronometer.	Time from Noon.	N. V. Sines.
H. M. S.	M. S.	
1 20 30	6 09	360
1 21 50	4 49	221
1 32 20	5 41	307
1 33 24	6 45	434
1 35 08	8 29	685
1 36 30	9 51	923
Means. After Noon.	3 17	488.3

	H. M.
Apparent Greenwich Time at Noon	1 15.3
Observation later . . .	3.3
App. Greenwh Time corresponding to the Mean Altitude	1 18.6

Lat. 74° 32′ Cosine .	9.4259867
Dec. 11° 59′ Cosine	9.9904312
Alt. 27° 26′ Secant .	10.0518084
Log. Sine 1″ A.C. .	5.3144251
Log. 488.3 (+4) .	6.6886867
Correction, 0′ 29″.6 Log.	1.4713381

Deduction. (The Limbs observed alternately.)

		° ′ ″
Arc passed through		329 31 10
		54 55 11.7
Apparent Altitude		27 27 35.85
Refraction − 1′ 51″.6	}	
Barometer . + 0.2	}	− 1 45.85
Thermometer − 2.1	}	
Parallax . + 7.7	}	
Correction		+ 0 29.6
True Altitude		27 26 19.6
Zenith Distances		62 33 40.4
Declin. at 1h 18′.6 App. Greenwich Time	. . .	11 58 36.2
North Latitude		74 32 16.6

GREENLAND.——August 23d, 1823. Barom. 29.90; Therm. 37°.5; the Altitudes observed with a Repeating Reflecting Circle of six inches diameter, and a Mercurial Horizon. The Chronometer No. 423 fast 1^h 23′ 56″ (page 168). The Sun on the Meridian at 0^h 02′ 33″ Mean Time, and at 1^h 26′ 29″ by the Chronometer.

Chronometer.	Time from Noon.	N.V. Sines.
H. M. S.	M. S.	
1 12 07	14 22	1964
1 14 45	11 44	1310
1 18 20	8 09	632
1 19 50	6 39	421
1 35 15	9 16	817
1 38 15	11 46	1318
1 41 15	14 46	2075
1 42 50	16 21	2544
Means. After Noon	1 24	1385

	H. M.
Apparent Greenw^h. Time at Noon	1 15.3
Observations later	1.4
App. Green. Time corresponding to Mean Alt.	1 16.7

Deduction.—(The Limbs observed alternately.)

Lat. 74° 32′ Cosine	9.4259867
Dec. 11° 38′ Cosine	9.9909859
Alt. 27° 06′ Secant	10.0505062
Log. Sine 1″ A.C.	5.3144251
Log. 1385 (+4)	7.1414498
Correct. 1′ 23″.8	Log. 1.9233537
	° ′ ″
Arc passed through	433 44 00
	54 13 00
Apparent Altitude	27 06 30
Refraction −1′ 53″.3; Barometer + 0.4; Thermometer − 2.4; Parallax + 7.7	− 1 47.6
Correction	+ 1 23.8
True Altitude	27 06 06.2
Zenith Distance	62 53 53.8
Decl. at 1^h 16′.7 app. Greenwich Time	11 38 26.2
Latitude North	74 32 20

GREENLAND.——August 25th, 1823. Barom. 29.72; Therm. 38°. Zenith Distances with a Repeating Circle. The Chron. 423 fast 1h 24′ 09″ (page 168); the Sun on the Meridian at 02′ 02″ Mean Time, and at 1h 26′ 11″ by the Chronometer.

Chronometer.	Time from Noon.	N. V. Sines.	Level.		Readings, &c. Alternately upper and lower Limbs.	
H. M. S.	M. S.					° ′ ″
1 13 10	13 01	1612	+ 3	− 6	Previous: First Vernier . .	179 58 00
1 15 45	10 26	1036	+ 6	− 3	Second „ . . .	57 50
1 17 45	8 26	677	+10	+ 2	Third „ . . .	58 20
1 20 10	6 01	345	+ 6	− 3	Fourth „ . . .	58 00
1 31 45	5 34	295	0	0	Mean	179 58 02.5
1 33 32	7 21	514	+ 2	− 6	Final: First Vernier . . .	95 37 30
1 35 30	9 19	826	+ 8	0	Second „ . . .	37 20
1 37 12	11 01	1155	+10	+ 2	Third „ . . .	38 00
1 38 45	12 34	1503	+ 3	− 6	Fourth „ . . .	37 20
1 40 28	14 17	1941	+ 3	− 6	Mean	95 37 32.5
					Index . . . . +	180 01 57.5 / 360 00 00
Means. After Noon	2 13	990.4	+ 12.5		Level	+12.5
						635 39 42.5

	H. M.
Apparent Greenwich Time at Noon .	1 15.3
Observation earlier than Noon . .	2.2
App. Greenwich Time corresponding to the Mean Z.D.	1 17.5

	° ′ ″
Observed Z.D. . . .	63 33 58.25
Refraction +1′ 56″.5; Barometer −1.1; Thermometer +2.4; Parallax . . −7.9	+1 49.9
Correction . . .	−0 59.7
True Z.D. . . .	63 34 48.45
Decl. at 1h 17′.5 . App. Greenwich Time	10 57 31.5
North Latitude . .	74 32 19.95

Lat. 74° 32′ Cosine	9.4259867
Dec. 10° 58′ Cosine	9.9919956
Z.D. 63° 35′ Cosecant	10.0478945
Log. Sine 1″ A.C. . . .	5.3144251
Log. 990.4 (+4)	6.9958106
Correction 0′ 59″.7 . . Log.	1.7761125

GREENLAND.——August 26th, 1823. Barom. 29.74; Therm. 44°. Zenith Distances observed with a Repeating Circle. The Chron. 423 fast 1^h 24′ 17″ (page 168); Sun on the Meridian at 01′ 46″ Mean Time, and at 1^h 26′ 03″ by the Chronometer.

Chronometer.	Time from Noon.	N. V. Sines.	Level.	
H. M. S.	M. S.			
1 7 33	18 30	3256	− 4	+ 3
1 10 07	15 56	2416	+13	+ 7
1 12 10	13 53	1834	− 4	−11
1 14 40	11 23	1233	0	+ 6
1 17 16	8 47	734	− 5	+ 1
1 19 22	6 41	425	− 1	+ 5
1 30 57	4 54	229	− 6	0
1 32 45	6 42	427	− 1	+ 5
1 34 54	8 52	748	− 6	0
1 36 50	10 47	1107	− 1	+ 5
1 38 33	12 30	1487	− 2	+ 4
1 40 18	14 15	1932	− 7	0
Means. Before Noon	1 26	1319	+ 1	

	H. M.
Apparent Greenwich Time at Noon	1 15.8
Observation earlier than Noon .	1.4
App. Greenwich Time corresponding to the Mean Z.D.	1 13.9

Lat. 74° 32′ Cosine	9.4259867
Dec. 10° 37′ Cosine . . .	9.9925013
Z.D. 63° 55′ Cosecant	10.0466485
Log. Sine 1″ A.C.	5.3144251
Log 1319 (+4) .	7.1202448
Correction 1′ 19″.4 . . . Log.	1.8998064

Readings, &c. Limbs alternately observed.

		° ′ ″
Previous	First Vernier . .	95 37 30
	Second „ . . .	37 20
	Third „ . . .	38 00
	Fourth „ . . .	37 20
	Mean	95 37 32.5
Final . .	First Vernier . .	142 36 40
	Second „ . . .	36 30
	Third „ . . .	37 00
	Fourth „ . . .	36 30
	Mean . . .	142 36 40
	Index . . . +	264 22 27.5 360 00 00
	Level	+1
		766 59 08.5
	Observed Z.D. . .	63 54 55.7
	Refraction +1′ 58″.4 Barometer −1.0 Thermometer +1.0 Parallax −7.9	+1 50.5
	Correction . .	−1 19.4
	True Z.D. . . .	63 55 26.8
	Decl. at 1^h 13′.9 App. Greenw^h Time	10 36 51.1
	North Latitude .	74 32 17.9

RECAPITULATION.

					° ′ ″
August	21st,	Sun on the Southern Meridian.		Sextant	74 32 18.4
„	22d,	„	„	Repeating Reflecting Circle	74 32 16.6
„	23d,	„	„	Repeating Reflecting Circle	74 32 20
„	25th,	„	„	Repeating Circle . . .	74 32 19.9
„	26th,	„	„	Repeating Circle . .	74 32 17.9
				North Latitude . .	74 32 18.6

DRONTHEIM.

Place of Observation.—Mr. Hans Wentzel's Villa. The Zenith Distances were observed with a Repeating Circle.

October 16th, 1823. Barom. 29.46; Therm. 41°. The Chron. 649 slow 39′ 35″.5. The Sun on the Meridian at 11ʰ 06′ 08″ by the Chronometer.

Chronometer.	Time from Noon.	N. V. Sines.	Level.	
H. M. S.	M. S.			
10 48 15	17 53	3043	−6	0
10 49 15	16 23	2554	+6	0
10 52 01	14 07	1896	+4	−2
10 53 21	12 47	1555	+2	−1
10 55 25	10 43	1093	0	−6
10 56 44	9 24	841	+6	−0
10 58 41	7 27	528	+2	−4
11 01 01	5 07	249	+4	−2
11 02 58	3 10	95	+2	−4
11 04 11	1 57	36	+8	+3
11 09 01	2 53	79	0	0
11 10 11	4 03	156	+9	+3
11 12 05	5 57	337	+8	+2
11 13 11	7 03	473	+5	−2
11 15 06	8 58	765	+2	−4
11 16 04	9 56	939	+9	+3
11 17 56	11 48	1325	0	0
11 19 04	12 56	1592	+6	0
11 20 51	14 43	2061	+4	−2
11 22 08	16 00	2436	+7	+2
Means. Before Noon.	0.16	1102.65	+30	

Lat. 63° 26′	Cosine	9.6505395
Dec. 8° 41′	Cosine	9.9919933
Z.D. 72° 07′	Cosecant	10.0215073
Log. Sine 1″ A.C.		5.3144251
Log. 1102.65 (+4)		7.0424370
Correction 1′ 45″.6	Log.	2.0239022

Readings, &c. Limbs alternately observed.

		° ′ ″
Previous	First Vernier	210 21 00
	Second „	20 30
	Third „	21 20
	Fourth „	20 25
	Mean	210 20 49
Final	First Vernier	212 19 40
	Second „	19 30
	Third „	20 10
	Fourth „	19 20
	Mean	212 19 40
	Index	+ { 149 39 11 / 1080 00 00 }
	Level	+ 30
		1441 39 21
	Observed Z.D.	72 05 58

Refraction	+2′ 58″.6	+2 50.4
Barometer	−3.3	
Thermometer	+3.3	
Parallax	−8.2	
Correction		−1 45.6
True Z.D.		72 07 02.8
Declination South		8 41 11.5
Latitude North		63 25 51.3

Drontheim.——November 10th, 1823. Bar. 30.44; Therm. 30°. The Chron. No. 649 slow 38′ 53″. α Ursæ (Apparent AR. 10^h 52′ 47″) on the Meridian below the Pole at 6^h 57′ 17″ by the Chronometer.

Chronometer.	Horary Angles.	N.V. Sines.	Level.		Readings, &c.		
H. M. S.	M. S.						° ′ ″
6 50 45	6 32	406	+ 8	− 3	Previous	First Vernier . .	209 22 40
6 53 06	4 11	167	+ 7	− 3		Second „ . .	22 30
6 54 56	2 21	53	0	0		Third „ . . .	23 10
6 57 30	0 13	0	+10	− 1		Fourth „ . .	22 20
6 59 32	2 15	48	+ 4	− 6		Mean	209 22 40
7 01 02	3 45	134	+ 6	− 4			
7 03 23	6 06	354	+ 6	− 4	Final	First Vernier . .	135 27 00
7 04 42	7 25	524	+ 8	− 3		Second „ . .	26 30
7 06 30	9 13	809	+ 2	− 8		Third „ . .	27 20
7 08 17	11 00	1152	+ 8	− 2		Fourth „ . .	26 40
7 11 03	13 46	1804	+ 2	− 8		Mean . . .	135 26 52.5
7 14 40	17 23	2875	0	0		Index . . +	{150 37 20; 360 00 00}
						Level . .	+ 9.5
Means.		694	+ 9.5				646 04 22
						Observed Z.D. . .	53 50 21.8
						Refraction + 1′ 19″.5; Barometer +1.2; Thermometer +3.2	+ 1 23.9
						Correction . . .	+ 0 36.3
						True Z.D. . . .	53 52 22
						Altitude	36 07 38
						Apparent North P.D.	27 18 19.2
						Latitude North . .	63 25 57.2

Lat. 63° 26′ Cosine	9.6505395
Dec. 62° 42′ Cosine . . .	9.6614810
Z.D. 53° 52′ Cosecant	10.0927784
Log. Sine 1″ A.C.	5.3144251
Log. 694 (+4)	6.8413595
Correction 36″.35 . . . Log.	1.5605835

RECAPITULATION.

		° ′ ″
October 16,	The Sun	63 25 51.3
November 10,	α Ursæ S.P.	63 25 57.2
		63 25 54.2 North.

APPLICATION OF THE OBSERVED VARIATION IN THE LENGTH OF THE SECONDS' PENDULUM TO THE DETERMINATION OF THE FIGURE OF THE EARTH.

THE elements, required towards the determination of the figure of the earth, are the ratios of the length of a pendulum vibrating equal portions of time, at the level of the sea, in different latitudes.

The values, which the operations recorded in the preceding pages have experimentally determined, are the lengths of the pendulum vibrating seconds of mean solar time, at stations of ascertained latitude, but elevated, in all the instances, more or less considerably above the sea. In order, therefore, to render the results applicable to the proposed determination, it is necessary that each should receive a small correction proportionate to the elevation at which it was obtained.

The value of the corrections which may be actually due in the several cases, is, unfortunately, not susceptible of a very exact determination, either by calculation or by experiment. Were the surface of the earth an unbroken plain, of uniform density, and were the space between its level and that of the station of experiment unoccupied by matter, the reduction of the length of the pendulum at the upper level to that of the lower level, would be strictly proportioned to the squares of their respective distances from the earth's centre: but the materials composing the eminence on which the pendulum is placed, as well as those which are adjacent, will influence the vibration by virtue of their own attraction; whence the difference in the length of a pendulum required to vibrate in equal times at the level of the sea, and at an elevation, must in all cases be less than would be due to a variation of gravity proportioned to the squares of the distances; and as the existing arrangement and disposition of the materials at the surface of the earth is one of much irregularity, both in figure and density, and as it is obviously impossible to calculate with exactness the peculiar attraction due to each locality, the value of the employed corrections must necessarily be assigned in some measure on arbitrary assumption, and must, therefore, be deficient in that precise experimental determination, of which all other parts of the operation appear to be capable.

The uncertainty, to which the results, to be adopted in the general conclusion, are liable from this source, may however be altogether avoided, or reduced within limits of inconsiderable amount, by the selection of stations but little removed from the level of the sea: but as stations cannot always be obtained in which this important advantage may be combined to its fullest extent with the other necessary qualifications, it may be proper to shew the limit of elevation at which the uncertainty produces a sensible effect on the results, as well as the extent, to which the correction for elevations exceeding that limit, may be considered as uncertain.

The co-efficient of a formula, which should correctly represent the modification, which the decrease of gravity at elevations proportioned to the square of the distance from the earth's centre undergoes, by reason of the attraction of the masses which surround and on which the operations take place, must vary in its amount in relation both to their external configuration and density: with respect to the first consideration, it has been stated by Dr. Young in the *Philosophical Transactions* for 1819, that if a station be situated on a tract of table land of two thirds the mean density of the earth, its attraction would equal half the diminution of gravity occasioned by receding from the earth's centre; and that in almost any country which could be chosen for the experiment, wherein the inequalities of surface might be excessive, the correction for elevation would not equal three fourths of the amount deducible from the duplicate proportion of the distances from the centre: now as the general disposition of the surface is much more conformable to the first supposition, than to the second extreme, (which is that of a station raised on a sphere), and as stations of experiment are rarely chosen in situations deviating much from what may be deemed a level surface broken by occasional small irregularities; if the co-efficient, due to an average superficial density, be assumed at $\frac{6}{10}$ths, it is probable that the correction would be in no instance in error, from the circumstances of figure, more than $\frac{1}{10}$th of the amount deducible from the squares of the distances, excepting in a very extreme case, when a special allowance might be made.

The lengths of the pendulum at the several stations determined by these experiments, are given to the fifth place of decimals, and may be presumed to be correct in their relation to each other, as far as the figure in the fourth place, corresponding to tenths of a second in the daily rate; if the elevation be under twenty feet, an uncertainty amounting to $\frac{1}{10}$th of the correction due to the squares of the distances, will not affect even the figure in the fifth place of decimals, and it is not until the figure in the fourth place is affected, that the presumed correctness of the experimental determinations is interfered with.

In the view that has been thus taken, the variation of the co-efficient due to the *form* of the eminences which rise above the general level of the earth's surface, has alone been taken into the account; it remains to consider the variation which may be occasioned by the different *densities* of the materials of which the eminences are composed. The first difficulty that would present itself, in an attempt to vary the co-efficient in this relation, would be found in the estimation of the density itself, or of its proportion to the average superficial density; of this, the pendulum may be considered to furnish the best evidence which is attainable; and however inexpedient it may appear, to derive from an effect produced, a correction of that effect proportioned to itself, the evidence which the pendulum affords of the influence exerted by the peculiar attraction of a locality, in modifying the mean force of gravity, ascribed to the parallel by a combination of experiments in different localities, may present the best practical means of approximating towards a true estimation, where the object of the determination requires an especial accuracy. A second difficulty would exist in appreciating the effect of different densities on the co-efficient; if a judgment may be formed from the present experiments, the bearing of which on the point in question will be discussed in the sequel, the local variations of gravity are influenced far more considerably by the density of the masses on which the pendulum is immediately placed, than on the general disposition of the surface.

To attempt a modification of the co-efficient of $\frac{6}{10}$ths, in consideration of the presumed variations of local attraction, either from form or density, at the stations of the present experiments, would be a refinement beyond the occasion; at four of the stations only, an error of twice the amount of the uncertainty, supposed in reference to the form of the eminence, would affect by a single unit the figure which has been presumed to have been accurately determined at the station; and at the remaining nine stations the possible errors may be regarded as wholly insignificant.

In the subjoined table the corrections have been inserted, both as derived from the squares of the distances, and as modified by a constant co-efficient of 0.6. The utmost facility is thus afforded for the substitution of any other co-efficient which may be deemed preferable, either generally, or in individual cases; the modified corrections are those which have been employed in reducing the lengths of the pendulum at the several stations, to the supposed corresponding lengths at the level of the sea, inserted in the final column.

STATIONS.	Latitudes.	Height above the Sea.	Length of the Seconds' Pendulum.	Corrections for Elevation.		Deduced Pendulum at the level of the Sea.
	° ′ ″	FT.	IN.			
St. Thomas	0 24 41 N.	21	39.02069	.00008	.00005	39.02074
Maranham	2 31 43 S.	77	39.01197	.00029	.00017	39.01214
Ascension	7 55 48 S.	17	39.02406	.00006	.00004	39.02410
Sierra Leone	8 29 28 N.	190	39.01954	.00071	.00043	39.01997
Trinidad	10 38 56 N.	21	39.01879	.00008	.00005	39.01884
Bahia	12 59 21 S.	213	39.02378	.00079	.00047	39.02425
Jamaica	17 56 07 N.	9	39.03508	.00003	.00002	39.03510
New York	40 42 43 N.	67	39.10153	.00025	.00015	39.10168
London . . .	51 31 08 N.	92.5	39.13908	.00035	.00021	39.13929
Drontheim	63 25 54 N.	121.5	39.17428	.00046	.00028	39.17456
Hammerfest	70 40 05 N.	29	39.19512	.00011	.00007	39.19519
Greenland	74 32 19 N.	31.5	39.20328	.00012	.00007	39.20335
Spitzbergen	79 49 58 N.	21	39.21464	.00008	.00005	39.21469

In order to obtain from the lengths of the pendulum contained in the final column of the antecedent Table, the Ellipticity, which may represent in the best possible manner their combined indication, when considered as expressing the direct ratios of gravitation in the respective parallels of latitude, it is desirable to employ the method of least squares in the deduction; by that method, the values of the equatorial pendulum, and of the total increase of gravitation between the Equator and the Pole, are determined in such manner, that the variation in the lengths of the pendulum deduced from the combination (on the principle that the length varies as the square of the sine of the latitude,) being compared with the observed variation, the sum of the squares of the several differences, may be less, than by any other possible determination.

In the ensuing calculation, the values of the equatorial pendulum, and of the total increase of gravitation, are represented by x and y; and the differences between the partial and the combined experimental determinations at each station by D^1, D^2, D^3, *&c.* The sum of the first series of conditional equations, divided by 13, and made equal to zero, expresses the equation of minimum in respect to x; the sum of the second series (which are the first thirteen equations, severally multiplied by the co-efficient of y in each equation) divided by 13, and made equal to zero, expresses the equation of minimum in respect to y.

	° ′ ″	
St. Thomas . . .	0 24 41 ;	$39.02074 - x - 0.0000515.y = D^1$
Maranham	2 31 43 ;	$39.01214 - x - 0.0019464.y = D^2$
Ascension	7 55 48 ;	$39.02410 - x - 0.0190338.y = D^3$
Sierra Leone . . .	8 29 28 ;	$39.01997 - x - 0.0218023.y = D^4$
Trinidad	10 38 56 ;	$39.01884 - x - 0.0341473.y = D^5$
Bahia	12 59 21 ;	$39.02425 - x - 0.0505201.y = D^6$
Jamaica	17 56 07 ;	$39.03510 - x - 0.0948286.y = D^7$
New York	40 42 43 ;	$39.10168 - x - 0.4254385.y = D^8$
London	51 31 08 ;	$39.13929 - x - 0.6127966.y = D^9$
Drontheim	63 25 54 ;	$39.17456 - x - 0.7999544.y = D^{10}$
Hammerfest . . .	70 40 05 ;	$39.19519 - x - 0.8904120.y = D^{11}$
Greenland	74 32 19 ;	$39.20335 - x - 0.9289304.y = D^{12}$
Spitzbergen . . .	79 49 58 ;	$39.21469 - x - 0.9688402.y = D^{13}$
		$39.09107 - x - 0.3729771.y = 0$

$$
\begin{array}{rcrcr}
- \; 0.002012 & + & x.0.0000515 & + & y.0.0000000 \\
- \; 0.075934 & + & x.0.0019464 & + & y.0.0000038 \\
- \; 0.742778 & + & x.0.0190338 & + & y.0.0003623 \\
- \; 0.850725 & + & x.0.0218023 & + & y.0.0004753 \\
- \; 1.332390 & + & x.0.0341473 & + & y.0.0011660 \\
- \; 1.971510 & + & x.0.0505201 & + & y.0.0025523 \\
- \; 3.701643 & + & x.0.0948286 & + & y.0.0089924 \\
- \; 16.635365 & + & x.0.4254385 & + & y.0.1809980 \\
- \; 23.984428 & + & x.0.6127966 & + & y.0.3755200 \\
- \; 31.337865 & + & x.0.7999544 & + & y.0.6399270 \\
- \; 34.899873 & + & x.0.8904120 & + & y.0.7928336 \\
- \; 36.417184 & + & x.0.9289304 & + & y.0.8629118 \\
- \; 37.992760 & + & x.0.9688402 & + & y.0.9386515 \\
\\
- \; 14.611113 & + & x.0.3729771 & + & y.0.2926457 = 0
\end{array}
$$

From the equations of minimum the value of x is found $= 39.01568$ inches, the pendulum at the equator; and $y = 0.20213$, the increase of gravitation between the equator and pole. The ellipticity corresponding to these values, is $\frac{1}{288.4}$ of the equatorial diameter.

The following table exhibits in the second column the lengths of the pendulum in the several latitudes, computed from the preceding values of x and y, or those corresponding generally with the experiments; the third column contains the lengths actually observed at each station; the fourth column, the excess or defect of the individual results on those of the combined determination (or the values respectively of D^1, D^2, D^3, *&c.*); and the fifth column, the number of vibrations per diem corresponding to the excess or defect in the preceding column.

STATIONS.	$x+ y.\text{Sin.}^2\text{Lat.}$	Lengths individually determined.	Individual determinations in excess or defect.	The excess or defect in Vibr.	GEOLOGICAL CHARACTERS.
St. Thomas .	39.01568	39.02074	+.00506	+5.58	Basaltic rock.
Maranham .	39.01607	39.01214	−.00393	−4.34	Alluvial.
Ascension . .	39.01953	39.02410	+.00457	+5.04	Compact volcanic rock.
Sierra Leone	39.02009	39.01997	−.00012	−0.12	A soft and rapidly disintegrating granite.
Trinidad . .	39.02258	39.01884	−.00374	−4.12	Alluvial.
Bahia . . .	39.02589	39.02425	−.00164	−1.80	A deep soil on a sandstone basis.
Jamaica . .	39.03485	39.03510	+.00025	+0.28	Calcareous rock.
New York .	39.10167	39.10168	+.00001	0.00	A stratum of 100 feet of sand, on serpentine.
London . . .	39.13954	39.13929	−.00025	−0.28	Gravel and chalk.
Drontheim .	39.17738	39.17456	−.00282	−3.10	Argillaceous soil on mica slate.
Hammerfest .	39.19566	39.19519	−.00047	−0.52	Mica slate.
Greenland . .	39.20344	39.20335	−.00009	−0.08	Sandstone.
Spitzbergen .	39.21151	39.21469	+.00318	+3.50	Quartz.

The most remarkable circumstance which this table presents to the view, is the extensive range in the amount of the differences between the individual and the combined experimental results; indicating either errors of experiment far more considerable than those which have hitherto been brought in question, or actual irregularities in gravitation much greater than have been previously evidenced.

That the differences are not altogether occasioned by errors of experiment, and that, in fact, the utmost portion of them must be very small, which, without an extreme violation of probability, can be attributed to that source, may be affirmed from the strong support which the individual results receive in the correspondence of a second, and totally distinct method of experiment, in that of the attached pendulums. The instances of extreme irregularity are, in defect at Maranham and Trinidad, and in excess at Spitzbergen, Ascension, and St. Thomas. At Maranham and Trinidad, the vibrations of the detached pendulums appear in defect no ess than 4.34 and 4.12 seconds per diem respectively; the attached pendulums (as evidenced in the comparative table, in page 281) shew in like manner a defect at Maranham of 5.04, and at Trinidad of 4.22 seconds. At Spitzbergen and Ascension, the detached pendulums are in excess 3.50 and 5.04 seconds respectively; the attached are also in excess 2.70 seconds at Spitzbergen, and 5.34 seconds at Ascension. At St. Thomas's, the solid pendulums were not employed; but in their absence a corroborative testimony of the same nature is afforded by the rate of the astronomical clock, relatively to its rate at all the other stations: the excess of vibration shewn by the detached pendulums is 5.58 seconds, and by the astronomical clock 5.08 seconds. An objection might be raised to the full authority of the astronomical clock as an independent corroboration, in the possibility, although extreme improbability, of a mutual influence having subsisted from proximity, or by communication through the respective supports; but no such possibility can be supposed between the detached and the solid pendulums; nor have the two methods a single point of connexion in which a common error could obtain, except in being referred to the same determination of astronomical time, which determination rests on observations much too extensively varied and multiplied to be questionable. When it is considered that the differences include a range between extreme cases of 10 seconds per diem, and that they are manifested alike in every instance (the discordances being absolutely insignificant in the comparison) by two decidedly distinct methods of procedure, and even by a third, of which the claim to be considered as an independent authority will not be refused by those who carefully examine the details, the conclusion that the irregularities do not originate in the experiments, but in the natural phenomena which are the objects of experiment, appears inevitable.

Viewing the differences, then, as indicating the existence of irregularities in gravitation itself, and as measures of the local excess or defect at each station, over the mean force in the respective parallels corresponding to the experiments generally, it is desirable to inquire into the relation which they may appear to bear, to the peculiarities of the superficial strata, in form and density.

The three stations, at which the force of gravitation would appear in principal excess, are St. Thomas, Ascension, and Spitzbergen; the character of these stations, in regard to the disposition of the attractive matter near the surface, is similar and peculiar: they are situated on islands of small extent but considerable elevation, the sides of which, both above and below the water, are abrupt and almost precipitous: they may be considered, therefore, as resembling stations on the declivity of an eminence, intermediate between the summit and the foot; and to be especially opposed to stations on an extensive tract of table-land. Now the effect of such a locality on the sum of the attractions, derived from considerations of form alone, should be, to produce a weaker force than the mean gravitation of the parallel, whereas the experiments indicate an increased force. Again, Maranham and Trinidad, the stations where the force was in principal defect, are also bordering on the sea, but being situated near the mouths of extensive rivers, the coasts continue shallow at a great distance from the land; those stations, therefore, have little inequality of elevation, and may be regarded as approximating very nearly to that state of the surface, in which the influence of form is at a maximum in augmenting the attraction: here, again, the effect evidenced by the experiments is of a totally opposite description.

The conclusion is far otherwise, however, when the respective densities of the materials near the surface are viewed in connexion with the excess or defect of local gravitation, and regarded as the circumstances of principal influence. It is fortunately not required for this purpose, that the estimation of the density should be very precise, as, independently of the pendulum, it would not be easy to be ascertained; it is sufficient to compare the particulars of the column indicating the irregularities of gravitation, with the geological characters of the several stations, to perceive their general, and almost to trace their individual, connexion In arranging the stations agreeably to the order of the densities evidenced by experiment, the compact volcanic rock of Ascension, the still more compact basalt of St. Thomas, the quartz of Spitzbergen, and the alluvial soils of Maranham, and Port Spain, in Trinidad, are found in their appropriate places at the opposite extremities of the succession; whilst of the intermediate stations (of which the correctness of the arrangement, relatively to each other, does not admit of the same positive testimony), not one can be said to differ materially from the position which would be assigned it, from the best estimation that can be formed of the density of the strata situated immediately beneath the pendulums.

STATIONS.	Excess or defect of Vibr.	Scale of density.	
St. Thomas . .	+5.58	100	A compact and very weighty basalt.
Ascension . .	+5.04	94	A compact volcanic rock.
Spitzbergen . .	+3.50	79	An extensive and deep bed of quartz.
Jamaica . . .	+0.28	45	Calcareous rock.
New York . .	0.00	43	A stratum of 100ft. of sand, resting on serpentine.
Greenland . .	−0.08	43	The debris of a compact sandstone rock.
Sierra Leone .	−0.12	42	A stratum of several feet of earth, resting on soft and rapidly disintegrating granite.
London . . .	−0.28	41	Gravel and chalk.
Hammerfest .	−0.52	37	Mica slate on a peninsula nearly surrounded by deep water.
Bahia	−1.80	26	Several feet of soil resting on sandstone.
Drontheim . .	−3.10	12	An argillaceous soil resting on rocks of mica slate.
Trinidad . . .	−4.12	2	Alluvial soil and sand.
Maranham . .	−4.34	1	Alluvial soil and sand.

Amongst the many interesting inferences which may be drawn from the view that has been thus presented, the following may deserve to be especially noticed:

1. If the irregularities in the force of gravitation are principally owing to the different densities of the materials near the surface, and if the influence of exterior configuration is so inconsiderable in comparison, as not to be recognisable in the results of experiments, the assumption of a co-efficient, for reducing the vibration at heights to that at the level of the sea, varied in relation to the form alone, cannot be supposed to meet the difficulties attendant upon a correct assignment; so far otherwise indeed, that in estimating the counteracting effect of the attraction of the eminence on which the experiment is made, on the regular decrease of gravitation in receding from the earth's surface, proportioned to the squares of the distances from the centre, it would appear that the consideration of its form may be safely neglected.

2. If a clock, or pendulum, is liable to vary 10 seconds in the same latitude, according to the nature of the materials on which it rests (including such only as are commonly found at the surface of the earth,) the length of the seconds' pendulum at the level of the sea, correctly determined at two places on land in the same parallel, may differ as much as .01 of an inch; and as the force of gravitation at alluvial stations may be supposed to exceed that which prevails over the extent, and at the surface, of the ocean, nearly as much as it falls short of the force at the stations of greatest local density, the actual variation of gravity in the same parallel may be considered as not less than equivalent to 20 seconds per diem, or to $\frac{1}{10}$th nearly of the difference between gravity at the pole and at the equator due to a compression of $\frac{1}{288}$, or to $\frac{1}{2000}$th of the whole attraction of the earth.

3. To obtain the force of gravitation corresponding to a parallel, to be

employed in the deduction of the total increase between the equator and the pole, due to the Ellipticity of the Earth, it is requisite, therefore, that several stations in or near the parallel should be grouped, so as to produce a mean result, in which the irregularities that render single stations unavailing in the deduction, may mutually destroy each other; it is desirable also that stations in either extreme of local density should be avoided as far as may be possible; or that if accidentally included, that an equal number of stations in each extreme should be comprehended in a group: thus, in the present experiments, St. Thomas and Maranham, Ascension and Trinidad, Drontheim and Spitzbergen, are respectively opposed to each other; it is preferable however, to confine the experiments to stations, at which the differences from the mean may be less considerable; as a general guide, perhaps, to where the specific gravity of the superficial strata may be between 2.25 and 2.75.

4. If the length of the pendulum assigned to a particular latitude, by the combined results of the experiments at the thirteen stations of this volume, be regarded as an approximate representation of the mean gravitation in the part of the parallel which is occupied at the surface by land; and if its amount, over the part which is occupied by the ocean, be supposed less than at alluvial stations, by more than half the difference between the stations of greatest and of least local density; and, if the parallel be equally occupied by land and ocean, the true mean pendulum of the latitude will be even less than the shortest of the individual deductions: thus, the equatorial pendulum, or the length representing the mean gravitation at the equator, will be less than 39.01568 (the value of x in page 334,) which is its length in situations only where the disposition and density of the materials near the surface correspond with the general average of the stations; and less even than 39.01175, which is the length deducible from Maranham, the station of least local attrac-

tion. It follows also, that although the ratio of gravitation in different latitudes may be determinable, by multiplying sufficiently the experiments in and near the respective parallels,—as well as the total increase between the equator and the pole, by making the groups sufficiently distant from each other,—the multiplication of stations on the land alone will not approximate towards a knowledge of the true equatorial pendulum. Fortunately, this length is not required to be very accurately known for the determination of the figure of the earth; since, if 39.01 were substituted for 39.01568 as the value of x, its combination with $y = 0.20213$ (the total increase between the equator and the pole,) in page 334, would produce an ellipticity of $\frac{1}{288.3}$; the difference between which deduction and $\frac{1}{288.4}$, resulting from $x = 39\ 01568$ and $y =$ as before, is too small to be significant in the present state of our knowledge.

5. The scale afforded by the pendulum for measuring the intensities of local attraction, appears to be sufficiently extensive, to render it an instrument of possible utility in inquiries of a purely geological nature. It has been seen* that the rate of a pendulum may be ascertained by proper care to a single tenth of a vibration per diem; whilst the variation of rate, occasioned by the geological character of stations, has amounted in extreme cases to nearly ten vibrations per diem; a scale of 100 determinable parts is thus afforded, by which the local attraction, dependant on the geological accidents, may be estimated.

The lengths of the seconds' pendulum determined by Captain Kater, at the principal stations of the trigonometrical survey of Great Britain, in

* *Vide*, the Table in page 211.

pursuance of an address of the House of Commons to the King, and published in a memoir in the *Philosophical Transactions* for 1819, may be connected and compared with the thirteen results contained in this volume, by means of the station in London, which is common to both series.

The experiments, by which Captain Kater's determinations were effected, were made with an invariable detached pendulum, and by a procedure similar in all respects to that which has been adopted in the operations recorded in this volume, excepting in the mode of observing coincidences; the rate of the pendulum used at the stations of the survey, having been deduced throughout from the intervals between the disappearances only; except in this one circumstance, the united series of nineteen results (London being a common station) are (it is believed) strictly comparable.

The following table contains the names of the stations of the Survey, with their respective latitudes, elevations, and pendulums, determined at the station; the corrections for elevation in two columns, as in the similar table in page 333; exhibiting, first, the corrections due to the duplicate proportion of the distances from the earth's centre, and, second, the corrections reduced by a co-efficient of 0.6 *; and in the final column, the

* Captain Kater has employed a co-efficient, varying from five to seven-tenths according to what he supposed might have been the influence of the *form* of the eminences on which the pendulum was placed, and without regard to the variations of density. The constant co-efficient of 0.6, which is here employed in the reduction of his results to the level of the sea, is not introduced in preference, as presumed to represent better the actual attraction of the several eminences, but inasmuch as it was necessary towards the just comparison of Captain Kater's results with mine, either that mine should be reduced by a variable co-efficient also, or that all should be reduced by the same constant quantity. Now, as the influence of the form does not appear to be recognisable in the actual variations of local attraction, it would have been superfluous at the least to have varied the corrections in relation to it, whilst the consideration of the density, which is the really influential circumstance, is omitted.

deduced lengths of the pendulum in the respective latitudes at the level of the sea.

The height of the pendulums in Mr. Browne's house, in London, being here described as 92.5 feet above the level of the sea, whilst in Captain Kater's memoir, in the *Philosophical Transactions*, is stated to be 83 feet only, it is necessary to explain that Captain Kater's estimation of the height was founded, in part, on the understanding (on the authority of the Royal Society) that the elevation of their barometer at Somerset-house is 81 feet above *low-water mark;* but as the latter elevation has been since corrected by Mr. Bevan, who has determined it by levelling to be 90.5 feet above the *mean level,* the height of the pendulums must now be considered as 92.5 feet, and is so esteemed by Captain Kater. It may be proper also to notice that Captain Kater's elevations are occasionally measured from low water, whereas mine are invariably measured from the mean level of the sea. The difference, however, may be safely disregarded in the comparison.

STATIONS.	Latitudes.	Elevation.	Pendulums at the Stations.	Corrections for Elevation.		Deduced Pendulums at the level of the sea.
	° ′ ″					
Unst . . .	60 45 28	28	39.17145	.00010	.00006	39.17151
Portsoy . .	57 40 59	94	39.16140	.00035	.00021	39.16161
Leith . . .	55 58 41	68	39.15540	.00026	.00016	39.15556
Clifton . .	53 27 43	339	39.14517	.00127	.00076	39.14593
Arbury Hill .	52 12 55	737	39.14057	.00276	.00166	39.14223
London	51 31 08	92.5	39.13908	.00035	.00021	39.13929
Shanklin . .	50 37 24	242	39.13551	.00091	.00055	39.13606

The stations of the Survey are combined in the following page with those of the present volume, in producing their corresponding Ellipticity by the method of least squares:

	° ′ ″							
St. Thomas	0 24 41 ;	39.02074	−	x	−	0.0000515.y	=	D^1
Maranham	2 31 43 ;	39.01214	−	x	−	0.0019464.y	=	D^2
Ascension	7 55 48 ;	39.02410	−	x	−	0.0190338.y	=	D^3
Sierra Leone . . .	8 29 28 ;	39.01997	−	x	−	0.0218023.y	=	D^4
Trinidad	10 38 56 ;	39.01884	−	x	−	0.0341473.y	=	D^5
Bahia	12 59 21 ;	39.02425	−	x	−	0.0505201.y	=	D^6
Jamaica	17 56 07 :	39.03510	−	x	−	0.0948286.y	=	D^7
New York	40 42 43 ;	39.10168	−	x	−	0.4254385.y	=	D^8
Shanklin	50 37 24 ;	39.13606	−	x	−	0.5975163.y	=	D^9
London	51 31 08 ;	39.13929	−	x	−	0.6127966.y	=	D^{10}
Arbury Hill . . .	52 12 55 ;	39.14223	−	x	−	0.6246044.y	=	D^{11}
Clifton	53 27 43 ;	39.14593	−	x	−	0.6455676.y	=	D^{12}
Leith	55 58 41 ;	39.15556	−	x	−	0.6869473.y	=	D^{13}
Portsoy	57 40 59 ;	39.16161	−	x	−	0.7141988.y	=	D^{14}
Unst	60 45 28 ;	39.17151	−	x	−	0.7613667.y	=	D^{15}
Drontheim	63 25 54 ;	39.17456	−	x	−	0.7999544.y	=	D^{16}
Hammerfest . . .	70 40 05 ;	39.19519	−	x	−	0.8904120.y	=	D^{17}
Greenland	74 32 19 ;	39.20335	−	x	−	0.9289304.y	=	D^{18}
Spitzbergen . . .	79 49 58 ;	39.21469	−	x	−	0.9688402.y	=	D^{19}
		39.11036	−	x	−	0.4673107.y	=	0

− 0.00201	+	x.0.0000515	+	y.0.0000000
− 0.07593	+	x.0.0019464	+	y.0.0000038
− 0.74278	+	x.0.0190338	+	y.0.0003623
− 0.85072	+	x.0.0218023	+	y.0.0004753
− 1.33239	+	x.0.0341473	+	y.0.0011660
− 1.97151	+	x.0.0505201	+	y.0.0025523
− 3.70164	+	x.0.0948286	+	y.0.0089924
− 16.63536	+	x.0.4254385	+	y.0.1809980
− 23.38443	+	x.0.5975163	+	y.0.3570258
− 23.98443	+	x.0.6127966	+	y.0.3755200
− 24.44841	+	x.0.6246044	+	y.0.3901308
− 25.27135	+	x.0.6455676	+	y.0.4167575
− 26.89780	+	x.0.6869473	+	y.0.4718966
− 27.96918	+	x.0.7141988	+	y.0.5100800
− 29.82389	+	x.0.7613667	+	y.0.5796793
− 31.33786	+	x.0.7999544	+	y.0.6399270
− 34.89987	+	x.0.8904120	+	y.0.7928336
− 36.41718	+	x.0.9289304	+	y.0.8629118
− 37.99276	+	x.0.9688402	+	y.0.9386515
− 18.30208	+	x.0.4673107	+	y.0.3436823 = 0.

Whence $x = 39.01566$, the Seconds' Pendulum at the Equator; $y = 0.20265$, the increase of gravitation from the Equator to the Pole; and the Ellipticity $= \frac{1}{289.5}$.

The following Table exhibits the pendulums, computed for the several latitudes of the stations from the values of x and y, as in the last page; with the differences between the individual and the combined experimental determinations, expressed in linear measure, and in the correspondent daily vibrations; the differences are presented in accompaniment with the geological characters of the stations, on which they are considered to depend.

STATIONS.	$x +$ $y.\text{Sin.}^2$ Lat.	Excess or defect of the individual Determinations.	Excess or defect, in Vibrations.	GEOLOGICAL CHARACTERS.
St. Thomas . .	39.01566	+.00508	+ 5.60	Basalt.
Maranham . . .	39.01605	−.00391	− 4.32	Alluvial.
Ascension . . .	39.01952	+.00458	+ 5.03	Computed volcanic rock.
Sierra Leone. . .	39.02008	−.00012	− 0.12	A soft and rapidly disintegrating granite.
Trinidad	39.02258	−.00374	− 4.12	Alluvial.
Bahia	39.02590	−.00165	− 1.81	A deep soil on a sandstone foundation.
Jamaica	39.03488	+.00022	+ 0.25	Calcareous rock.
New York . . .	39.10187	−.00019	− 0.20	100 feet of sand, resting on serpentine.
Shanklin	39.13675	−.00069	− 0.76	Chalk.
London	39.13984	−.00055	− 0.60	Gravel and chalk.
Arbury Hill . . .	39.14223	−.00000	0.00	Chalk, in the vicinity of primitive rocks.
Clifton . . .	39.14648	−.00055	− 0.60	Clay and shale.
Leith.	39.15487	+.00069	+ 0.76	Sandstone and scattered basaltic rock.
Portsoy	39.16039	+.00122	+ 1.32	Serpentine and granite.
Unst	39.16995	+.00156	+ 1.72	Serpentine.
Drontheim . .	39.17777	−.00321	− 3.54	Argillaceous earth, resting on mica slate.
Hammerfest . . .	39.19610	−.00091	− 1.00	Mica slate.
Greenland . . .	39.20391	−.00056	− 0.60	Sandstone.
Spitzbergen . . .	39.21200	+.00269	+ 3.00	Quartz rock.

The following arrangement exhibits a mode of grouping the stations into partial results corresponding to particular latitudes, from the subsequent comparison of which with each other, the Ellipticity of the earth may possibly receive even a more satisfactory elucidation, than from their general combination according to the method of least squares. The middle group comprises the six stations contained within the limits of England and Scotland, and assigns, from their several experimental determinations, the length of the pendulum in the latitude of 54 degrees. The first group comprises in like manner the five stations nearest the equator, and the third group the five most northern stations, from which are respectively assigned the pendulums corresponding to the latitudes of 5 degrees, and of 70 degrees.

STATIONS.	Latitudes.	Pendulums.	Mean Latitudes.	Corresponding reduced Pendulums. $y = 0.20245$.	
	° ′ ″		° ′ ″		
St. Thomas . .	0 24 41	39.02074		39.02227	
Maranham . .	2 31 43	39.01214		39.01329	
Ascension . . .	7 55 48	29.02410	5 00 00	39.02179	39.01758
Sierra Leone .	8 29 28	39.01997		39.01710	
Trinidad . . .	10 38 56	39.01884		39.01347	
Shanklin . . .	50 37 24	39.13606		39.14760	
London	51 31 08	39.13929		39.14774	
Arbury Hill . .	52 12 55	39.14223	54 00 00	39.14829	39.14832
Clifton . . .	53 27 43	39.14593		39.14774	
Leith	55 58 41	39.15556		39.14900	
Portsoy	57 40 59	39.16161		39.14953	
Unst	60 45 28	39.17151		39.19614	
Drontheim . .	63 25 54	39.17456		39.19138	
Hammerfest . .	70 40 05	39.19519	70 00 00	39.19370	39.19452
Greenland . . .	74 32 19	39.20335		39.19405	
Spitzbergen . .	79 49 58	39.21469		39.19732	

From the combination of the mean result of the first and second groups, the value of $y = 0.20210$, and the Ellipticity $\frac{1}{288.3}$.
From the first and third groups, $y = 0.20212$, and the Ellipticity $\frac{1}{288.4}$.
From the second and third groups, $y = 0.20218$, and the Ellipticity $\frac{1}{288.5}$.

I proceed to the comparison of the lengths of the seconds' pendulum at the stations which have been hitherto under notice, with the similar determinations at several points of the arc of the meridian comprised between Formentera and Unst, effected conjointly by MM. Biot, Arago, Bouvard, Mathieu, and Chaix, in a suite of operations undertaken at the instance of the Academy of Sciences of Paris, and carried on under the authority and support of the late and present governments of France, having commenced in 1807, and terminated in 1817. An account of the operations is published in detail at the conclusion of the third volume of the *Base du Système Métrique.*

The experiments of the French philosophers were made with an apparatus on the principle invented by Borda for the measurement of the seconds' pendulum at Paris, and by the process devised by that eminent philosopher; its distinctive peculiarity from the method employed at the stations of the British survey, and at those of this volume, is, that the absolute length of the pendulum is separately determined at each station, instead of the relative lengths to a particular station, serving as a common basis. If, however, the fundamental length on which the several determinations in the one mode of operation depend be correctly measured, and if the process by which the separate measurements at each station are effected in the other mode, be without an inherent error, the several results, when reduced to the same measure, ought to be strictly comparable, with the exception of the uncertainty which must prevail in the several reductions to the level of the sea.

The first of the two following tables comprises the names of the eight stations at which experiments were made by the French philosophers; the latitudes of the stations, and their elevation above the sea in metres; the lengths of the decimal pendulum observed at the station, expressed in millimetres, and the names of the experimentors. The second table exhibits the corresponding lengths of the sexagesimal pendulum, as well in millimetres as in parts of Sir George Shuckburgh's scale at the temperature of 62 degrees*; it contains also, the elevation in British feet, and the respective corrections to the level of the sea, computed first, by the duplicate proportion of the distances from the earth's centre, and second, by the same proportion with a co-efficient of 0.6; the pendulums at the level of the sea, as inserted in the final column, are obtained by employing the corrections computed with this co-efficient†.

STATIONS.	Latitudes.	Elevation.	Decimal Pendulum.	OBSERVERS.
	° ′ ″	Metres.	Millimetres.	
Unst	60 45 25	9	742.721034	Biot.
Leith	55 58 37	21	742.408533	Biot.
Dunkirk . .	51 02 10	4	742.07610	Biot, Mathieu.
Paris	48 50 14	70	741.90112	Biot, Mathieu, Bouvard.
Clermont	45 46 48	406	741.61059	Biot, Mathieu.
Bordeaux	44 50 26	17	741.60464	Biot, Mathieu.
Figeac	44 36 45	223	741.56033	Biot, Mathieu.
Formentera . . .	38 39 56	203	741.20540	Biot, Arago, Chaix.

* The metre is accounted 39.37079 inches of Sir George Shuckburgh's scale; the metre being at the temperature of melting ice, and the British scale at that of 62° of Fahrenheit.

† The allowance for the elevation of the stations used in the memoir in which the experiments are recorded, is in the first of the above proportions; the second is introduced here, solely with a view of rendering the results in the final column more strictly comparable with the others, in which the co-efficient of 0.6 has been employed.

STATIONS.	Sexagesimal Pendulum.		Elevation in feet.	Corrections for Elevation.		Reduced lengths of the Seconds' Pend. at the level of the Sea.
	In Millimetres.	In British Measure.				
Unst	994.94308	39.17170	30	.00011	.00007	39.17177
Leith	994.52445	39.15522	69	.00026	.00016	39.15538
Dunkirk . . .	994.07913	39.13769	13	.00004	.00002	39.13771
Paris	993.84473	39.12843	230	.00085	.00051	39.12894
Clermont . . .	993.45554	39.11313	1332	.00499	.00299	39.11612
Bordeaux . . .	993.44756	39.11282	56	.00021	.00013	39.11295
Figeac	993.38822	39.11048	732	.00274	.00164	39.11212
Formentera . .	992.91275	39.09176	606	.00248	.00149	39.09325

In the ensuing calculation, the stations of the French Arc are combined with the thirteen stations of this volume, in the deduction of the corresponding Ellipticity by the method of least squares.

	°	′	″	
St. Thomas . . .	0	24	41;	$39.02074 - x - 0.0000515.y = D^{1}$
Maranham . . .	2	31	43;	$39.01214 - x - 0.0019464.y = D^{2}$
Ascension . . .	7	55	48;	$39.02410 - x - 0.0190338.y = D^{3}$
Sierra Leone . .	8	29	28;	$39.01997 - x - 0.0218023.y = D^{4}$
Trinidad	10	38	56;	$39.01884 - x - 0.0341473.y = D^{5}$
Bahia	12	59	21;	$39.02425 - x - 0.0505201.y = D^{6}$
Jamaica	17	56	07;	$39.03510 - x - 0.0948286.y = D^{7}$
Formentera . . .	38	39	56;	$39.09325 - x - 0.3903417.y = D^{8}$
New York . . .	40	42	43;	$39.10168 - x - 0.4254385.y = D^{9}$
Figeac	44	36	45;	$39.11212 - x - 0.4932370.y = D^{10}$
Bordeaux . . .	44	50	26;	$39.11295 - x - 0.4972172.y = D^{11}$
Clermont	45	46	48;	$39.11612 - x - 0.5136118.y = D^{12}$
Paris	48	50	14;	$39.12894 - x - 0.5667721.y = D^{13}$
Dunkirk	51	02	10;	$39.13771 - x - 0.6045723.y = D^{14}$
London	51	31	08;	$39.13929 - x - 0.6127966.y = D^{15}$
Leith	55	58	37;	$39.15538 - x - 0.6869301.y = D^{16}$
Unst	60	45	25;	$39.17177 - x - 0.7613525.y = D$
Drontheim . . .	63	25	54;	$39.17456 - x - 0.7999544.y = D^{18}$
Hammerfest . . .	70	40	05;	$39.19519 - x - 0.8904120.y = D^{19}$
Greenland . . .	74	32	19;	$39.20335 - x - 0.9289304.y = D^{20}$
Spitzbergen . . .	79	49	58;	$39.21469 - x - 0.9688402.y = D^{21}$

$$39.10534 - x - 0.4458446.y = 0$$

$$
\begin{aligned}
&- \ 0.00201 + x.0.0000515 + y.0.0000000\\
&- \ 0.07593 + x.0.0019464 + y.0.0000038\\
&- \ 0.74278 + x.0.0190338 + y.0.0003623\\
&- \ 0.85072 + x.0.0218023 + y.0.0004753\\
&- \ 1.33239 + x.0.0341473 + y.0.0011660\\
&- \ 1.97151 + x.0.0505201 + y.0.0025523\\
&- \ 3.70164 + x.0.0948286 + y.0.0089924\\
&- 15.25973 + x.0.3903417 + y.0.1523667\\
&- 16.63536 + x.0.4254385 + y.0.1809980\\
&- 19.29154 + x.0.4932370 + y.0.2432827\\
&- 19.44764 + x.0.4972172 + y.0.2472250\\
&- 20.09049 + x.0.5136118 + y.0.2637970\\
&- 22.17718 + x.0.5667721 + y.0.3212305\\
&- 23.66157 + x.0.6045723 + y.0.3655076\\
&- 23.98443 + x.0.6127966 + y.0.3755200\\
&- 26.89702 + x.0.6869301 + y.0.4718730\\
&- 29.82353 + x.0.7613525 + y.0.5796577\\
&- 31.33786 + x.0.7999544 + y.0.6399270\\
&- 34.89987 + x.0.8904120 + y.0.7928336\\
&- 36.41718 + x.0.9289304 + y.0.8629118\\
&- 37.99276 + x.0.9688402 + y.0.9386515
\end{aligned}
$$

$$- \ 17.456817 + x.0.4458446 + y.0.30711115 = 0$$

Whence $x = 39.01510$ the equatorial pendulum; $y = 0.20227$ the total increase of gravitation; and the Ellipticity $= \frac{1}{288.7}$.

In the four first columns of the table in the following page, are collected in one view the whole of the stations which have been thus severally considered, their latitudes, observed pendulums, and the names of the observers; the values of x and y, which best correspond with their combined indication, are stated at the bottom of the page, the details of the calculation by the method of least squares being omitted, as the equations corresponding to each station have been already inserted in the preceding pages: the fifth column contains the pendulum in the respective latitudes computed from the values of x and y; and the sixth column, the discordances between the individual and combined experimental determinations, expressed in linear measure.

STATIONS.	Latitudes.	Experimental Pendulums.	Observers.	Computed Pendulums.	Experimental Pendulums in excess or defect.
	° ′ ″				
St. Thomas .	0 24 41	39.02074	Sabine	39.01520	+ .00554
Maranham . .	2 31 43	39.01214	Sabine	39.01559	− .00345
Ascension . . .	7 55 48	39.02410	Sabine	39.01905	+ .00505
Sierra Leone .	8 29 28	39.01997	Sabine	39.01961	+ .00036
Trinidad . . .	10 38 56	39.01884	Sabine	39.02211	− .00327
Bahia	12 59 21	39.02425	Sabine	39.02543	− .00118
Jamaica . . .	17 56 07	39.03510	Sabine	39.03440	+ .00070
Formentera . .	38 39 56	39.09176	Biot, Arago, Chaix. . .	39.09422	− .00246
New York . . .	40 42 43	39.10168	Renwick, Sabine . .	39.10133	+ .00035
Figeac . . .	44 36 45	39.11212	Biot, Mathieu	39.11506	− .00294
Bordeaux . . .	44 50 26	39.11295	Biot, Mathieu	39.11586	− .00291
Clermont . . .	45 46 48	39.11612	Biot, Mathieu . . .	39.11918	− .00306
Paris	48 50 14	39.12894	Biot, Mathieu, Bouvard	39.12994	− .00100
Shanklin . . .	50 37 24	39.13606	Kater	39.13617	− .00011
Dunkirk . . .	51 02 10	39.13771	Biot, Mathieu . . .	39.13760	+ .00011
London	51 31 08	39.13929	Kater, Sabine	39.13926	+ .00003
Arbury Hill . .	52 12 55	39.14223	Kater	39.14165	+ .00058
Clifton	53 27 43	39.14593	Kater	39.14590	+ .00003
Leith	55 58 39	39.15547	Biot . . . 39.15538 Kater . . . 39.15556	39.15427	+ .00120
Portsoy	57 40 59	39.16161	Kater	39 15979	+ .00182
Unst	60 45 26·5	39.17164	Biot . . . 39.17177 Kater . . . 39.17151	39.16934	+ .00230
Drontheim . .	63 25 54	39.17456	Sabine	39.17715	− .00259
Hammerfest . .	70 04 05	39.19519	Sabine	39.19546	− .00027
Greenland . .	74 32 19	39.20335	Sabine	39.20326	+ .00009
Spitzbergen . .	79 49 58	39.21469	Sabine	39.21134	+ .00335

Whence $x = 39.01520$, the equatorial pendulum : $y = 0.20245$ the increase of gravitation between the equator and the pole; and the Ellipticity $= \frac{1}{289.1}$.

In the following table are collected in one view the deductions from the several combinations which have been examined in the preceding pages.

	The Equatorial Pendulum, x.	The total increase of gravitation, y.	The Ellipticity.	
			x & y = as in the respective Columns.	y = as before, x = 39.01.
	IN.	IN.		
From the thirteen stations of this volume.	39.01568	0.20213	$\frac{1}{288\cdot4}$	$\frac{1}{288\cdot3}$
From the same, combined with the eight stations of the French Philosophers.	39.01516	0.20227	$\frac{1}{288\cdot7}$	$\frac{1}{288\cdot6}$
From the same, combined with the seven stations of the British Survey.	39.01566	0.20265	$\frac{1}{289\cdot5}$	$\frac{1}{289\cdot4}$
From the comparison of the pendulum of the Latitude of 5° deduced from the five stations nearest the Equator, with the Pendulum in the Latitude of 54° deduced from the six stations in England and Scotland.	39.01606	0.20210	$\frac{1}{288\cdot3}$	$\frac{1}{288\cdot2}$
From the comparison of the Pendulum of the Latitude of 5°, deduced as before, with the pendulum of the Latitude of 70° deduced from the five most Northerly stations.	39.01599	0.20212	$\frac{1}{288\cdot4}$	$\frac{1}{288\cdot3}$
From the comparison of the pendulum of the Latitude of 54°, with that of the Latitude of 70°, both deduced as above.	39.01599	0.20218	$\frac{1}{288\cdot5}$	$\frac{1}{288\cdot4}$
From the general combination of the stations of this volume, of the British Survey, and of the French Arc; in all twenty-five stations.	39.01520	0.20245	$\frac{1}{289\cdot1}$	$\frac{1}{288\cdot9}$
MEANS		0.20227	$\frac{1}{288\cdot7}$	$\frac{1}{288\cdot6}$

The attempt to determine the figure of the earth by the variations of gravity at its surface, has thus been carried into full execution, on an arc of the meridian of the greatest accessible extent; and the results which it has produced are seen to be consistent with each other, in combinations too varied to admit a probability of the correspondence being accidental. The Ellipticity to which they conform differs more considerably than could have been expected from $\frac{1}{306.75}$, which had been previously received on the authority of the most eminent Geometrician of the age, as the concurrent indication of the measurements of terrestrial degrees, of pendulum experiments, and of the lunar inequalities dependant on the oblateness of the earth. In further attestation of the irreconcilability of the variation of gravity now manifested, with the Ellipticity inferred in the memoir in which the Marquis de Laplace has discussed the results of previous observation and experiment, it may be noticed, that if each of the tropical stations which I have visited be severally combined with each of the stations within 45 degrees of the pole, no one result, amidst all the irregularities of local attraction, will be found to indicate so small a compression as that of previous reception.

The consideration, which the Ellipticity indicated by these experiments may receive from the public, as a final determination, will depend, first, on the conclusiveness which may attach to it, as the ultimate result of the method of experiment; and, second, on the conclusiveness of the method itself, in regard to the determination sought.

It is not easy to anticipate what accumulation, or what variety of experimental evidence will be required in a final determination of such magnitude as that of the exterior configuration of the earth; and (as a consequence) of the laws of density according to which the attractive matter of which it consists is distributed in its interior; the inquiry has been actively and unremittingly pursued for more than a century and a half, and has been deemed of sufficient importance to receive the co-operation of governments, where the undertakings have been beyond the power of individuals, or of associations, to accomplish;

and it may be confidently presumed that the subject will not be allowed to drop, until the object of the inquiry is deemed to be most thoroughly and satisfactorily ascertained.

The individual who has conducted the experiments, is peculiarly disqualified for anticipating the general opinion as to their conclusiveness, by reason of his intimate knowledge of the sources from whence error might have arisen, and of the efficacy of the means which were adopted to guard against its occurrence. The conviction which this knowledge produces on his own mind, cannot be imparted in its full force even by the most careful and extended detail: he cannot, therefore, anticipate what may be the impression on the minds of others, and the decision must remain with those, in whom maturity of judgment gives authority to opinion.

Should more evidence, however, appear to be yet desirable, in confirmation of the difference between the polar and equatorial gravitation shewn by the experiments which have been under notice, it may conduce to the completion of the inquiry, to point out by what measures it may be most satisfactorily procured.

It is presumed to have been sufficiently shewn, that single stations are quite unavailing in the assignment of the length of the pendulum due to a particular latitude; and that a group of not less than six or seven stations is requisite for that purpose: it may be further observed, that in the present advanced stage of the inquiry, no result ought to be admitted to have weight, which has not been obtained with due regard to all the minute circumstances by which error in experiment may be avoided. A reflection on the progress of the inquiry hitherto, will abundantly shew, that inexact experiments have tended but to perplex, and even to mislead: the irregularities of gravitation present a sufficient difficulty in the determination, without the additional embarrassment of irregularities of experiment.

1. There is already an experimental pendulum at Madras; were it conveyed to five or six stations of various geological character between the latitudes of Madras and the equator, and returned to England to have its corresponding rate in London correctly ascertained, with such arrangements as should manifest the intermediate invariability of the pendulum of comparison, a second tropical group would be obtained, doubling the evidence at the one extremity of the arc.

2. The seabord of the United States presents every requisite facility for the execution of a third group in the middle latitudes, varied almost at pleasure, in regard to the circumstances of locality. This group would possess the more interest, as the length of the pendulum on the average of the stations in the corresponding parallels in Europe, has hitherto fallen short of the general ratio in latitudes both to the north and to the south. It might be accomplished with great convenience by the Government of the United States, should a disposition be felt to co-operate with Great Britain and France in the prosecution of this important inquiry; and New York would, in such case, form a connecting station with the experiments in Great Britain, in France, within the Tropics, and in the Arctic Circle.

3. An additional group in the high latitudes would be comparatively a more arduous undertaking; the west coast of Greenland, and the northern shores of Baffin's Sea, would be suitable localities in which it might be accomplished, however, in a single season; the stations on the Greenland coast from 60° to 74° would consist of primitive rocks, and might be visited between April and July; those in still higher latitudes, which are accessible towards the close of July, and for which the months of August and September would remain, would consist of transition and secondary rocks. We know, on the authority of Baffin, who is yet our only authority for the northern limits of the sea which bears his name, that the experiments might be carried in that direction, at least as far as the latitude of 78°

It may be reasonably presumed that the three groups which have been thus suggested, combined with the four which have been already accomplished, would terminate the inquiry into the figure of the earth by means of the pendulum, as far at least as regards the northern hemisphere; by producing a result which would be decisive in the estimation even of the most cautious judgment; or by shewing (what the present experiments must render exceedingly improbable) that a decisive result is not attainable by it.

Having thus considered the deductions, in regard to the increase of gravitation between the equator and the pole, and the corresponding Ellipticity of the earth, furnished by the experiments at the twenty-five stations enumerated in page 351, as well in their general as in their partial combination, it may not be uninteresting, briefly to examine the correspondence of the mean ratio of gravitation at the surface of the earth, as it may be severally inferred from the experiments of the French philosophers, from those of Captain Kater, and from mine.

If the length of the pendulum in the latitude of forty-five degrees be deduced from the experimental lengths at each of the stations, by the aid of the co-efficient $y = 0.20227$, and if the stations of each distinct series of experiments be collected into a mean result, their correspondence or otherwise will be manifested on comparison: thus,

The Pendulum in the latitude of 45° corresponding to the experiments of	The French Philosophers at 8 Stations	=	39.11535	− .00119
	Capt. Kater at 7 Stations	=	39.11729	+ .00075
	Capt. Sabine at 6 Northern Stations	=	39.11664	+ .00010
	Capt. Sabine at 7 Southern Stations	=	39.11688	+ .00034
		Mean.	39.11654	

It is here seen that the mean ratio of gravitation deducible from the eight stations of the French philosophers is considerably in defect, when viewed in comparison either with Captain Kater's, or with mine; from Captain Kater's, especially, it differs not less than would influence the going of a clock two seconds in a day. It can scarcely be supposed that the discrepancy is occasioned by the different mode of experiment adopted by the French philosophers, and still less by any inexactness in the execution, where such consummate skill was displayed: that portion of it, then, which, on a due consideration of the circumstances at the several stations, shall appear beyond the utmost uncertainty which can attach to the respective conclusions, must be regarded as proceeding from, and evidencing an actual deficiency of gravitation on the average of the eight stations, when compared with the average of nearly twice their number; all the stations having been selected without reference to the peculiarities of local attraction.

Unless, therefore, some adequate cause for the discrepancy can be found in the experiments themselves, it must appear, that if the pendulum of a particular latitude be supposed to have been determined by the mean of *eight* measurements at different stations in the parallel indiscriminately selected,—and if the process be subsequently repeated at *eight* other stations as indifferently chosen, in the expectation of obtaining a mean result identical with the former,—the expectation may be defeated by a difference not less in amount than 002 parts of an inch. It is unnecessary to show by how much this difference may be exceeded, where single determinations only are employed.

The principal, and it may almost be said, the only source, from whence the individual results may have been liable to material inaccuracy, is in the reduction of the pendulums actually determined by the experiments at the stations, to the presumed corresponding lengths at the level of the sea; in which respect the magnitude of the possible inaccuracy may to a certain extent have been proportionate to the height: now the elevation of Clermont, one of the stations of the French arc, being 1332 feet, far exceeds that of any other of the stations with which it is included; and the doubt which must be allowed to exist, respecting the co-efficient which it might be most proper to use, in reducing the pendulum of that height to the corresponding one at the sea, renders the reduced result uncertain to a much greater amount than is involved elsewhere*. If therefore the influence of the experiments at Clermont be withdrawn from the pendulum of the latitude of forty-five degrees deduced from the French stations, the comparison may, perhaps, be deemed to take place under circumstances more strictly corresponding to each other; when it will be seen that some portion of the apparent discrepancy is removed.

The Pendulum of the latitude of 45° corresponding to the experiments of	The French Philosophers, (omitting Clermont)—7 Stations	=	39.11563	− .00098
	Capt. Kater—8 Stations	=	39.11729	+ .00068
	Capt. Sabine—6 Northern Stations	=	39.11664	+ .00003
	Capt. Sabine—7 Southern Stations	=	39.11688	+ .00027
		Mean.	39.11661	

It may be inferred, then, from the premises in the foregoing pages, that 39.1166 is the approximate mean length of the seconds' pendulum in the part of the parallel of 45 degrees which is occupied by land: that in consequence of the inequalities of local attraction, the length, correctly measured at individual stations in the parallel, may be liable to vary from 39.1221 to 39.1132: that if the several densities of the strata near the surface at the twenty-five stations of experiment, be

* It is probable that at such considerable heights, the value of the co-efficient, if one be used, should be varied in relation to the height, as well as to the disposition of the surface, and to the density of the substances composing the eminence.

supposed to be comprehended (as it is probable they may) between 1.8 and 3, the length of the pendulum over the sea, could it be there measured,—or which is the same thing, the ratio of gravitation over the surface of the ocean,—might possibly be found as low as 39.1042*: and on the further supposition that the parallel consists of equal surfaces of land and ocean, the mean pendulum of the latitude may not differ considerably from 39.11.

The same method of inference will assign 39.1391 as the mean length of the pendulum in the latitude of London (51° 31′ 08″.4), required to vibrate seconds on the land, 39.1267 on the ocean, and 39.1329 as the mean pendulum; also, that the length correctly measured at different stations in the parallel on land at the level of the sea, might vary from 39.1446 to 39.1357.

It will also assign 39 01 as at least an approximate mean length of the pendulum at the equator, corresponding to gravitation at all points of the circumference; whence the space which a body would fall through in one second of time is equal to 16.04223 feet; and the cen-

* Had circumstances permitted, it was my intention, whilst at Spitzbergen, to have obtained the rate of the pendulum clock on one of those vast accumulations of ice and snow which are occasionally met with, filling entire valleys, and presenting towards the ocean a front of five or six hundred feet, and sometimes even more, in perpendicular height. The summit of a Glacier is frequently a level surface, connecting the mountain ridges by which the valley is enclosed; and on one of the larger Glaciers a clock might be so placed, resting on pickets blunted at the end and driven into the snow, that the sides of the hills might be more distant from it than the bottom of the valley, and thus that no materials of a specific gravity greater than unity might be within five or six hundred feet of the clock. Its comparative going, when so stationed, and when supported at the same elevation above the sea on the adjacent land, might have afforded a more highly interesting illustration of the influence of superficial density on the general gravitation, than any which has been hitherto produced. Had the Griper commenced her voyage earlier in the season of 1823, this experiment would have been at least attempted; and it is now noticed, in the hope that it may yet be accomplished by some future voyager to Spitzbergen (which on account of its geographical position is occasionally visited for purposes of science), to whom time may be an object of less consequence than it was to me

trifugal force at the equator is $\frac{1}{288.032}$ of gravitation, or $\frac{1}{289.031}$ of gravity. The correspondence between these fractions, and that which has been found to express the Ellipticity of the Earth, or, in other words, between gravitation at the pole and at the equator in terms of the equatorial gravitation, and the radius of the earth at the pole and at the equator in terms of the equatorial radius, is too remarkable a coincidence to escape notice.

The success which has thus attended the attempt to carry into effect, under the conditions most favourable for the experiment, the method of investigating the figure of the earth by means of the pendulum, and the consistent and precise result, far exceeding previous expectation, which, under such circumstances, it has been found to afford, encourage the belief that an equally satisfactory conclusion, and one highly interesting in the comparison, might be obtained by the measurement of terrestrial degrees, performed also under the requisite conditions to give its due efficiency to the method of experiment. Experience has fully shewn, that no result of decisive character is to be expected from the repetition or comparison of measurements in the middle latitudes; and that it is only from operations carried on in portions of the meridian widely separated from each other, that such an event can be regarded as of probable accomplishment. The project of the original experimentors,—of those eminent men who, nearly two centuries ago, devised and executed corresponding measurements at the equator and at the arctic circle, —was of far more vigorous conception, than the steps of their successors have ventured to follow, even to the present period; and it is due to their memory to recognise that the failure on that occasion was not from insufficient extension of view, or from deficiency in the spirit of enterprise; but from the attempt having been made in the infancy of practical science, when the instruments were inferior, and the modes of their most advantageous employment less understood, than they have since been rendered.

The discordancies, which appear in the comparison of the measurements hitherto accomplished, are not so great as those which had resulted from the comparison of pendulum experiments, previously to the present attempt to give the latter method its full and efficient trial: it has been also seen that in proportion as the arcs have been enlarged, so as to include the continuous measurement of more extended portions of the meridian, and as the processes of operation have been conducted with improved means, and increased attention to accuracy, the anomalies have progressively diminished; the prospect, therefore, that they may be made wholly to disappear, by combining the interposition of the greatest interval between the measurements that the meridian of an hemisphere will admit, would seem sufficiently probable to justify and induce the undertaking.

Through the munificent liberality and splendid patronage of the East India Company, India already presents a determination of the arc contained between the 10th and 20th parallels: and as a consequence of the political changes which have recently taken place in South America, there is reason to hope, that the impediments to a measurement between the equator and the 10th degree, in the quarter of the globe best suited for the operation, will speedily be removed.

In regarding the polar extremities of the meridian, the attention is naturally directed in the first instance to Spitzbergen, as the land of highest convenient access in either hemisphere; its qualification, in that respect, is indeed far beyond comparison with other lands, and is a point of very principal importance; its high latitude and conveniency of access do not, however, form its only suitability; for, on due consideration, it will be found to possess many very peculiar advantages for the operations of a triangulation.

The general geological character of Spitzbergen is a group of islands of primitive rock, the ordinary hills of which are from 1000 to 2000 feet in height, commanding generally extensive views, and unencumbered with the vegetation which presents so great an obstacle to the connexion of stations in the more genial climates. The access to all parts of the interior is greatly facilitated by the extensive fiords, and arms of the sea, by which the land is intersected in so remarkable a manner; these, whether frozen over, as in the early part of the season, or open to navigation, as in the later months, form routes of communication suited to the safe conveyance of instruments either in sledges* or in boats; the fiord, in particular, which separates the western and eastern divisions of Spitzbergen, would be of great avail; it extends in a due north and south direction for above 120 miles, with a breadth varying from ten to thirty miles, and communicates at its northern extremity, by a short passage across the land, with the head of another fiord proceeding to meet it from the northern shores of the island, and affording similar facilities for carrying on either a triangulation, or a direct measurement, on the surface of the ice at the level of the ocean. It is hardly necessary to add, that the latter operation would be unembarrassed by the inequalities of surface, and uncertain temperature of the apparatus, which occasion so much trouble, and require so much precaution in the usual determination of a base.

The extent of the arc in the direction of the meridian, between the

* Sledges with rein-deer trained to draft, and the Fins by whom they are managed, may be hired for the season, at Hammerfest, in any number that might be required. Spitzbergen abounds more in the food of the rein-deer, and is more plentifully stocked with the animals themselves in their wild state, than any other arctic country which I have visited. The officers of the Griper killed more than fifty deer on the small islands which form the northern part of the harbour of Fairhaven.

southern shores of Spitzbergen and the islands on its northern coast in the eighty-first degree of latitude, is between four and five degrees. At the period of the celebrity of Spitzbergen as a fishing station, in the middle of the seventeenth century, when above 200 vessels, manned by 10 or 12,000 seamen, annually resorted to its vicinity, and frequented its harbours for the purposes of boiling oil, and when the harbours were divided by convention amongst the vessels in consequence of their numbers, according to the nation and towns to which they belonged, all parts of the coast were known to and visited by the hardy and enterprising Dutch and German seamen, by whom the fishery was then principally conducted. The whales have long since deserted the haunts which their kind had enjoyed for ages before in unmolested security, and have sought retreats less accessible to man; the graves, which occupy every level spot around the harbours, contain the only and in that climate the almost imperishable memorials of the once busy scene, which has reverted to its original solitude; even the accidental presence of a whaling ship in the western harbours is an event of rare occurrence*, and it is probable that more than half a century has elapsed since any vessel has passed to the North-eastern shores; it is not surprising, therefore, that the delineation of land, represented in the charts of the period when Spitzbergen was so greatly frequented as existing to the East of the seven islands, and to extend in a northerly direction far into the eighty-second parallel, should neither have been established nor disproved by modern authorities; those persons who have had opportunities of becoming acquainted, by examination on the spot, with the remarkable cor-

* During the Griper's stay of three weeks in the neighbourhood of the harbour of principal resort in earlier times, and in the middle of the fishing season, not a single whale fish or whaling ship were seen. The only vessels which now frequent the shores of Spitzbergen, are Norwegian sloops in quest of sea-horses and eider down. Their visits have been hitherto confined to the fiords and the islands on the southern and western coasts: they arrive early in March, and remain as late as November, making occasionally three voyages in a season.

rectness of the older charts in general, in the insertion and in the relative position (when not separated by much extent of ocean) of lands then recently discovered, will hesitate too hastily to reject their testimony, until it has been satisfactorily disproved; should land exist as represented in the charts of the period alluded to, even though not visible from Spitzbergen, its triangular connexion might be estabished on the surface of the ice, and latitudes yet unattained be included in the operations of the survey; nor would it be safe to assign too confidently the northern limit of such operations even in the absence of land, in our present ignorance of the facilities which the ice itself may afford for their extension towards the pole.

The measurement of a portion of the meridian in the higher latitudes is, however, one of the many experimental inquiries, beyond the reach of individual means to accomplish, for which the advancement of natural knowledge is delayed; if its accomplishment may be hoped for by that nation which has been most forward in exploring the regions of the north, —to whom its climates and its natural difficulties are familiar,—it must still await the existence of a channel in one of the departments of the state, through which the liberal disposition of the British Government to forward every undertaking worthy of a great nation, and by which it may occupy an additional page in history, shall be rendered available to other branches of scientific research, than those which are immediately conducive to the interests of navigation.

As the Pendulum experiments which have been related, bear in several points both theoretically and practically on the subject of a natural standard of linear measure, it may be useful to bring their connexion with it in such points directly under notice, rather than to leave it to be inferred incidentally.

In selecting a length in nature as a reference for a national linear scale, there are two qualities in particular the possession of which is essentially requisite ; first, it must be an invariable length ; and second, it must be also easily accessible.

The quadrant of the meridian, of which the French metre was designed to be the ten millionth part, possesses the first requisite ; but is inapplicable to its purpose, in consequence of the difficulties which impede the actual determination of its magnitude, and the great time and labour which would be required in the operation, even if it were possible that the impediments to its execution could be overcome. The relation of the metre to the length in nature to which it professes to bear a certain proportion, exists, therefore, only in the name ; and, if the measurement of the quadrant were hereafter to be actually accomplished, and the metre should be found to differ from its nominal proportion, it cannot be doubted, that it would be the proportion, and not the established scale, which would undergo the change.

Failing the actual determination of the quadrant of the meridian, its magnitude, and that of its aliquot part, the unit of the French scale, have been assigned from the measurement of an arc of the meridian, comprising about a tenth part of the quadrant, and by the assumption of a certain ellipticity : now, as the length of a definite portion of the circumference is as invariable as the circumference itself, the re-measurement of the arc which supplied the foundation of the scale, might serve equally for its recovery if lost, or for its restoration if injured by accident or wear, provided that the labour, time, and expense attendant on such an operation (admitting the certain identity of the result on repetition) did not render such a natural standard one that is not easily accessible ; which consideration has occasioned its practical abandonment as a reference, by the distinguished persons themselves whose lives have been engaged in the original measurement, and by whom the pendulum is now proposed as the means of defining and determining the metre.

In the act which passed the British Legislature in the session of 1824, entitled " An Act for ascertaining and establishing uniformity of Weights and Measures," the British imperial yard is declared to be in the proportion of 36 inches to 39.1393 (ten thousandths) of an inch, when compared with a pendulum vibrating seconds of mean time in the latitude of London, in a vacuum at the level of the sea. In thus designating and adopting the pendulum of a particular latitude as the natural standard of British measure, the act necessarily assumes that the length so adopted is of an uniform magnitude; namely, that the seconds' pendulum at the level of the sea and in a vacuum is of the same length every where in the same latitude; this assumption, however, is directly opposed by the evidence of the facts which have appeared in the course of the present experiments, and which is particularly summed up in the second, third, and fourth inferences in pages 339 and 340, and again in pages 358 and 359.

In the third report of the commissioners appointed by his Majesty to consider the subject of weights and measures, dated March, 1821, on which report the act of the legislature of 1824 was founded, it is recommended that the authentic legal standard of the British Empire should be identified, by declaring that 39.1393 inches of the standard, at the temperature of 62° of Fahrenheit, have been found equal to the length of a pendulum supposed to vibrate seconds in London, at the level of the sea, and in a vacuum. The recommendation of the commissioners is a nearer approximation than the act itself, to that more simple standard of determinate and determinable magnitude, and which is, in fact, the only experimental foundation of the provisions either of the act or of the report, the pendulum of a particular spot; and it is observable, that just inasmuch as the specification of the report departs from the simplicity of its foundation, does it fail in precision, and in substituting that which is supposititious and uncertain, for that which is susceptible of direct and experimental proof. This is seen, first, in the attribution of a certain definite length to the pendulum vibrating seconds in London, whilst the subject of the experiment was, more precisely, that of the pendulum vibrating seconds in a certain part of London; wherein the expression substituted is in strictness incorrect, except on the supposition that the seconds' pendulum is of the same length in all parts of London; which it would not be safe to assume even for meridians under the same parallel; and which is theoretically opposed (and doubtless also practically) to the sensible variation in the length of the pendulum in the northern and southern parts of London, due to the Ellipticity of the Earth: secondly, in the substitution of a supposititious pendulum by the reduction to the level of the sea, for the real pendulum measured at the spot, wherein two suppositions are involved, both open to question; namely, first, that the elevation was correctly ascertained, and secondly, that it was correctly allowed for. With respect to the elevation itself, it is now admitted (page 343) that the height was incorrectly assigned to the amount of several feet, by the error of one (and that the only one which has been subsequently examined) of the data on which it rested; and in regard to the correction, the present experiments have rather increased than diminished the uncertainty that previously prevailed as to the proper co-efficient to be employed in the reduction. Were the reference made to the pendulum which was actually the subject of experiment, with the understanding that all future repetitions, designed to produce identical results, should be made identically at the same spot, the accuracy or otherwise, both of the elevation and its correction, would be immaterial, because those particulars would be omitted in the specification, as superfluous; but in the case obviously contemplated by the report, and implied in its language, that the standard should be recoverable by measurements made elsewhere in London, the elevation and its just correction are essential, and require to be known with the same accuracy as the length of the measured pendulum itself.

In the twenty first Number of the "Journal of the Royal Institution (April, 1821), is a communication from Professor Schumacher to the Secretary of the British Board of Longitude, announcing the adoption, for the Danish standard of length, of the pendulum vibrating seconds of mean solar time in the latitude of 45° north, and in the meridian of Skaagen, at the level of the sea, and in a vacuum. If this specification is to be understood literally, the geographical position to which it refers is in Italy, not far from Mantua; and as its pendulum would require an arbitrary correction to reduce it to the level of the sea, in order to fulfil the conditions of the specification, it could not become the subject of a direct experimental determination. But it is more probable that the understanding should be, and that the intention is, to ascertain the length of the seconds' pendulum at the level of the sea at some spot in the Danish dominions, possibly at Skaagen itself, and to deduct a certain proportion of its length from the measurement, corresponding to the effect of the supposed Ellipticity of the Earth between the parallel of the experiment and of 45° If the latter understanding be correct, the Danish measure will be identified by the pendulum of the spot where the experiment is made, and the same spot must be recurred to for its recovery; and the Danish natural standard will be the pendulum of that spot, and not the pendulum of the latitude of 45°. The correspondence of the divisions of the scale with the aliquot parts of the supposed pendulum of 45°, will indeed establish a relation between them; but it will be like that of the metre to the quadrant of the meridian, a nominal relation to an inferred, but not determined length, having no practical superiority over an assumption more purely arbitrary. If it be designed that the Danish standard should be the representative of the mean ratio of gravitation between the equator and the pole, which is the probable intention, it does not necessarily follow from the proposed mode of determination that it should be so, even if the ellipticity were correctly known, because it would also be requisite that the materials near the surface at the place of experiment in the Danish dominions, should be a mean in the scale of the general superficial density *.

It is by France only that the experimental length itself, the measured pendulum of a spot, has been distinctly recognised as the subject of reference, and as the means of identifying the national scale; nor is it in that respect only, that France has advanced beyond other nations in the preliminary steps towards the establishment of a reference which may live through succeeding generations, and become available to distant posterity; she has repeated the measurement, which was supposed to have fixed in perpetuity the value of the metre; and the capability of the process of Borda to produce an identical result in the hands of other experimentors, has undergone a practical examination; the length of the sexagesimal pendulum at the observatory at Paris, by Borda's measurement, is 39.12776 inches of British measure, and by that of Messrs. Bouvard, Biot, and Mathieu, 39.12843; concerning which measurements, differing so considerably, M. Arago has remarked, that it would be difficult to pronounce to which the preference should be given. We are thus enabled to form a practical estimate of the extent to which the metre may be considered as identified, and may be capable of recovery on repetition by the present process of reference; the knowledge that the metre is not yet referred with certainty to the third place of decimals of a British inch, is an advance, in comparison with the erroneous opinion that might otherwise have been entertained, that being given to the fifth place it was correct in the fourth.

We are thus also furnished with evidence, if evidence were required for

* See page 358, where the mean seconds' pendulum in the part of the parallel of 45° which is occupied by land, is inferred from the mean of 24 stations, and the probable amount of difference at single stations or from fewer combinations is fully discussed.

conviction, how essential the experiment itself of repetition is to enable a correct judgment to be formed of what repetition will produce. It is seen that it is not sufficient, that a certain scale has been found to bear a certain proportion to a certain length in nature; it is also necessary, that it should be proved, that it will be so found again by repetition in other hands, in order that the purpose of identification, which is that of recovery, should be fulfilled.

There is reason to believe that the method for which science is indebted to the ingenuity and mechanical skill of Captain Kater, will be found capable of greater precision than that of Borda; it is on the proceedings of that method that the details of these experiments have much practical bearing, since what is true generally in regard to the accidents of experiment with the pendulums which I have used, is also applicable to the convertible pendulum. Thus the reasoning and experiments in pages 213 to 233, apply equally to the convertible as to the comparative pendulum, and shew, that according to the method that is practised of observing coincidences, will a pendulum of determinate and invariable length appear to possess different rates; and that unless the re-appearances of the disk be observed as well as its disappearances, neither the true rate corresponding to the length will be obtained, nor will the results of different experimentors be independent of individual peculiarity or accidental circumstance, and consequently that they will not be identical.

So also does the evidence, commented upon in pages 195 and 196, that the experimental rate of a pendulum may be influenced by an accidental peculiarity in the agate planes on which it vibrates discoverable only by a trial on other planes, bear on the convertible pendulum with more force, perhaps, than on the simple invariable pendulum; first, because it has two axes of suspension instead of one; and second, because the sliding weight is more likely to interfere with the uniform bearing of the knife edges on the planes. The existence of such and similar inaccuracies is best disproved by identity on repetition with different instruments; but they may certainly exist unsuspected in an unrepeated experiment.

It is in the same view that the comparison of different methods of ascertaining the length of the pendulum is highly important, and by consequence, the invention of new modes of procedure. It is understood that a third method has been proposed by Dr. Young, by means of a weight sliding on a rod, or bar, with a single axis of suspension, as a yet more convenient method of obtaining a correct standard, than the processes of Borda and Kater. It would be highly interesting to ascertain, by competent trial, the relative values of the three methods, and to examine the correspondence of their results; or, rather, to work at them until they should correspond, or until the reason of a difference should be apparent. The transmission of our measures to those distant times when our manufactured scales shall have perished (and such is the object of the reference to nature), is a purpose of such magnitude in all respects, as to require the utmost evidence which the ingenuity and labour of the age can supply: on its exact accomplishment depends the value to posterity of every attainment of the present age, in which linear measure is concerned; and it may be reasonably expected that the habits, in regard to accuracy in experiment, of the times when our proceedings shall be examined, for the purpose of recovering by them that which is lost, will be incomparably more precise than at the present period.

In selecting a spot, the pendulum of which is to supply an invariable length in perpetuity, it is expedient to avoid, as far as may be possible, the causes which may interfere with the permanency of local gravitation; for which reason great cities, or their vicinity, may be considered generally as objectionable stations. It may be reasonably doubted, for instance, whether the present pendulum, in any particular spot in London, is sensibly the same as it was before that part of the city was built; since an alteration in the density of the materials at the surface, equivalent to only 100th part of the natural differences which actually prevail in various localities (page 339), would be sufficient to influence the pendulum in the fourth place of decimals. We have recently seen the substitution attempted of masses of iron for the paving stones of London; and it is obviously impossible to anticipate the changes, which the ingenuity of man in constant operation may hereafter effect, to promote the interest or convenience of an immense population collected within a small compass. A station sufficiently distant from dwellings, and not likely to become their site, and yet not so remote as to be of inconvenient resort to foreigners, who may desire to compare the standards of their respective countries with that of Great Britain, would seem to be much preferable.

Even the changes which the accidents of nature may produce, admit of being provided against, by the well-established comparison of the standard pendulums of different countries with each other; whereby the means are furnished of recovering the standard of any particular country, even if the spot of its original determination should be destroyed.

GEOGRAPHICAL NOTICES.

LONGITUDE OF THE PENDULUM STATIONS.

The longitude of the Pendulum Stations was required to be known, in order to deduce the time at Greenwich corresponding to that at the several stations. For such purpose, I might have availed myself in some instances of the received longitude, as being probably sufficiently correct; but as original determinations would have been necessary in many cases, and as I was desirous that the memoir should be in all respects as complete in itself, and as independent as it could be made of the observations of others, it was deemed preferable to undertake an original ascertainment of the longitude at all the stations.

In the prosecution of this undertaking, it was conceived that a useful service might be rendered to navigation and geography, by affording an extensive practical exemplification of the value of lunar observations, when made with instruments of the best construction, and their results computed with due regard to what are usually termed the minute corrections.

It is well known that the instruments employed in the British navy and marine for observing lunar distances, are almost universally sextants, and that circular instruments are very rarely to be met with. Now, angular distances observed with sextants are liable to certain errors, caused by defects of construction, for which the circular completion of the arc enables a remedy: three sources of error may be specified in particular; first, from imperfect graduation; second, from the index and horizon glasses not being parallel when the index of the limb is at Zero, constituting what is termed index error; and third, from defective centring of the limb with respect to the arch. The great improvements which have been made of late years in the art and practice of graduation have rendered the errors arising from that source much less significant than they were formerly, but those who have carefully examined modern sextants, know, that this imperfection is by no means wholly removed.—The index error admits of its amount being determined, and when known, it may be allowed for by an equal increase or diminution of the angle read on the arch; in good sextants, and with due precaution, the index error is not very liable to change, but it requires constant examination, and occasions, therefore, at the best, much additional trouble.—The third source of error, however, that of the eccentricity of the limb with regard to the arc, is a much more serious evil, being extremely prevalent, and admitting of no very easy method of detection, or of having the value of the error at different parts of the arc ascertained*. The reflecting circle devised by Mr. Troughton, on which construction the very few circles which are met with in British ships are made, was designed expressly to obviate these defects; but its use has been found to be attended with so much practical inconvenience, as in

* The best mode with which I am acquainted of practically examining a sextant in this respect, is by observing the meridian altitudes of several stars of known declination at different altitudes at the same station. If the centring is correct, the latitudes deduced from the several observations will agree; if it is not so, the errors occasioned by it at different parts of the arc may be ascertained and allowed for in future observation, in addition to Index Error. By employing a mercurial horizon, and a telescope magnifying from ten to fourteen times, and by deducing the meridian altitudes from several observations made whilst the star is near the meridian, noting and correcting for the horary angles, this examination may be made with much exactness, by a tolerably practised observer. I may add that of many sextants which I have myself examined, of makers in most repute, I have met with only two, which had the same error at all points of the arc.

great measure to counterbalance the superiority of its principle, and to have impeded its general adoption: it may be sufficient to particularize, that the errors of imperfect graduation and defective centring are counteracted in that circle by a multiplication of verniers, making three distinct readings necessary in each observation; whilst in practice, the reading the arc, especially in night observations, is by far the most inconvenient and irksome part of the whole process of observation, the frequency of which it is most desirable to diminish, rather than to increase. A more convenient reflecting circle than Mr. Troughton's was therefore a desideratum of much practical importance; and it was particularly to be wished that the principle of repetition should be introduced, as a means of dispensing with the necessity of reading the arc, until the close of the several observations of a series.

In the spring of 1821, Mr. Dollond, to whom practical astronomy has so many obligations, was kind enough to show me the design of a repeating reflecting circle, which was then in progress of execution: as this instrument appeared, so far as could be judged from the design, to promise to supply precisely what was wanting, I requested him to make a second on my account, intending to give it an extensive trial. Without entering minutely into the details of the construction of Mr. Dollond's circle, it may be sufficient to notice, that it consists of two concentric circles in the same plane and nearly in contact, the one of which moves within the other: both circles are graduated, the outer into 720° subdivided into spaces of 10 seconds; the inner, at every tenth degree on both sides of Zero to 180°, referring for its subdivisions to those of the outer circle: the inner circle carries the telescope, horizon glass, and a vernier applying to the graduation on the outer circle, to which it clamps and is furnished with an apparatus for slow motion, placed with the clamp near the telescope: the index glass is carried on a limb moving freely round upon its own centre, having an apparatus for slow motion, a clamp by which it may be attached to the outer circle, and a vernier applying to the inner one. This part of the circle corresponds with, and is similar to the limb of a sextant, and may be used accordingly. To employ the circle in its more extended application as a repeating instrument, the vernier of the inner circle is clampt and read off, or set if it is preferred at the primary division; the telescope and horizon glass, which are both fixtures to this circle, are then directed to either of the objects, and the limb carrying the index glass is moved on its own centre, until the reflected image of the other object is in the field, when the limb is clampt, and the contact effected. So far the process is the same as in an ordinary sextant, with this advantage, that the angle may be observed on either side of the Zero, avoiding the necessity of the inversion of the instrument, (which at sea is frequently very inconvenient,) and the value of the angle thus measured may be read by the vernier of the limb, as already noticed. This reading, however, is an unnecessary step, where it is purposed to proceed in repetition, which is done as follows: the inner circle is now unclampt and moved round, (the limb with the index glass remaining fixed to the outer circle) until its vernier has passed through twice the angle which is measuring; it is then clampt, and the telescope and horizon glass being directed to the object which was reflected in the first observation, (or to the same as before, accompanied by the inversion of the circle, if it best suits the nature of the observation, or the convenience of the observer,) the objects are again perceived in the field, and their contact is effected as usual. The vernier of the circle becomes then charged with the sum of the two observed angles, which, if it is not wished to proceed further in the repetition, may be read off, and being divided by two, the quotient is the distance corresponding to the mean time between the observations. The angle so obtained is free from index error, but is still liable to be affected by those of imperfect graduation, and eccentricity, although their effect has been diminished by the process already gone through. In order wholly to extinguish these errors also, the repetition of the same double process must be continued, until the vernier of the circle has progressively completed the entire round, or as nearly so as the amount of the measured angle will admit; when a single reading, divided by the number of observations, will shew, as before, the angle corresponding to the mean time of observation.

When used as a sextant only, this instrument possesses the following advantages over sextants of the ordinary construction; first, it enables the angle to be measured alternately on each side of zero, whence the index error is compensated, and the liability to those of imperfect division and centring diminished; secondly, by clamping the vernier of the circle successively at primary divisions, about a third of the circle apart, in succeeding pairs of observation, the errors of centring may be destroyed; and thirdly, the angle which may be measured is not limited by the extent of the arc, but may be carried to the utmost amount in which the relative position of the glasses will admit of reflection.

When used as a circle, the following additional advantages are gained; the process of observation is shortened at least a half, by dispensing with the reading off and writing down the angle at each repetition. The errors which are frequently introduced in those operations are avoided; those of imperfect graduation and eccentricity are rendered insensible; and in night observations especially, the eye is spared the alternate reference to a strong artificial light, necessary for reading the arc, but extremely prejudicial to the most favourable state of the eye for observation.

An incidental advantage arising from thus shortening the process of observation is, that it places the whole operation within the power of an individual to accomplish by himself; whereas it previously consisted of too many distinct parts, and was consequently too laborious and fatiguing for accuracy. The subjoined observations were made (with very few exceptions) without an assistant, the times being noted by the beats of a chronometer. The satisfaction is great to an observer to have all the parts of an observation thus within his own command; it is convenient also, because assistance is not always to be obtained; and it is conducive to accuracy, because the attention of an assistant is rarely equal to that of the observer.

In the subjoined tabular abstract, the " Time by the Chronometer" is a mean of the number of observations expressed in column 4, the details of the time corresponding to each observation being omitted. The correction of the chronometer No. 423, to the mean time at the several stations inserted in column 3, is taken from the preceding pages of this volume; in column 5 is shewn the whole arc passed through by the vernier of the circle in the process of repetition; and in column 6, being the whole arc divided by the number of repetitions, is the apparent distance corresponding to the mean chronometer time in column 2. Columns 7 and 8 exhibit the apparent altitudes of the moon and sun, or star, calculated for the known apparent time at the station. The corrections for refraction, or the differences between the true and apparent altitudes, have been computed for the states of the atmosphere shewn in columns 9 and 10, by Dr. Young's table in the " Nautical Almanac" for 1822; much pains was taken to obtain the true temperature of the air, uninfluenced by radiation on the thermometer from the surfaces around; for which purpose the thermometer was enclosed in a highly-polished metal cylinder, pierced with holes in the top and bottom, and placed in the shade. The true distances in column 11 have been deduced by Dr. Maskelyne's method, published in the preface to " Taylor's Logarithms," with corrections introduced,—of the horizontal parallax on account of the Ellipticity of the Earth,—and of the distance, where the oblique semi-diameters were sensibly affected by refraction.

In deducing the time at Greenwich, corresponding to the true distances, from those inserted in the "Nautical Almanac" for every third hour, the second differences of the moon's motion in relation to the sun or star have been duly taken into the account. It sometimes happens that the second difference of the distances of the moon and stars inserted in the Nautical Almanac for every third hour, amounts to more than one minute of space; in such instances the correction due to the second difference will exceed six seconds of space during more than half the intermediate interval, and consequently, if neglected, will occasion an error of about three minutes of longitude in the deduction. This circumstance is thus specially adverted to, because its notice is omitted in the very useful summary of the minute corrections, requiring attention where precision is desired, published by the secretary of the British Board of Longitude in the Journal of the Royal Institution for July, 1820.

The circle with which the distances were observed was ten inches in diameter, and weighed five pounds; the telescope was furnished with a magnifying power of fourteen. The observations at Sierra Leone were not strictly its first employment, as I had observed sixty-four distances with it at Madeira, on the outward passage; with the exception of these, however, the use of the circle was new to me at Sierra Leone, and the awkwardness which attends the employment of a new instrument was still to be overcome.

The sixty-four distances at Madeira, of which forty were of Regulus west of the moon, and twenty-four of the Sun east of her, made the British Consul's house, at Funchal, in 16° 55′ 00″ W.*; the longitude of the Consul's garden has since been ascertained by the mean of sixteen chronometers, specially sent for the purpose, at the direction of the Commissioners of Longitude, and has been found 16° 54′ 45″. 3 W.

* Letter to Sir Humphry Davy, P.R.S., dated Goree, January, 1822, printed in the Journal of the Royal Institution, April, 1823.

AN ABSTRACT OF LUNAR OBSERVATIONS, MADE WITH A REPEATING REFLECTING CIRCLE OF TEN INCHES DIAMETER, AT THE SEVEN PENDULUM STATIONS WITHIN THE TROPICS.

SIERRA LEONE.——In the West Bastion of Fort Thornton. Latitude 8° 29′ 28″ N.

DATE.	Time by the Chron. 423.	423 on Mean Time Fast.	No. of Observations.	Arc passed through.	Apparent Distance of Limbs.	Apparent Altitudes.		Barom.	Therm.	True Distance.	Longitude.	PHENOMENA and REMARKS.
						Moon.	Sun or Star.					
1822.	H. M. S.	H. M. S.		° ′ ″	° ′ ″	° ′ ″	° ′ ″	IN.	°	° ′ ″	° ′ ″	
Feb. 28	2 31 29.8	0 48 48.6	12	1084 49 35	90 24 07.9	26 30 17	62 11 57	30.00	81	90 06 03.4	13 18 18 W.	Moon from Sun W. of her.
,,	2 51 52.1	0 48 48.6	12	1086 39 05	90 33 15.4	30 52 27	57 59 25	30.00	81	90 16 00.4	13 14 24	,, ,,
,,	3 14 03.9	0 48 48.6	12	1088 34 35	90 42 52.9	35 35 36	53 07 57	30.00	81	90 29 09.3	13 16 33	,, ,,
,,	3 30 45	0 48 48.6	12	1089 57 55	90 49 49.6	39 08 13	49 22 09	30.00	81	90 38 21	13 19 45	,, ,,
,,	3 48 29.4	0 48 48.6	18	1637 01 30	90 56 45	42 51 09	45 17 42	30.00	81	90 47 57.3	13 18 30	,, ,,
Mar. 1	3 39 17.6	0 48 49.5	12	1247 37 20	103 58 06.7	27 46 35	47 30 58	30.00	81	103 41 10	13 14 40.5	Moon from Sun W. of her.
,,	3 54 32.4	0 48 49.5	12	1248 57 40	104 04 48.3	31 00 06	43 59 12	30.00	81	103 49 26	13 16 51	,, ,,
,,	4 06 01.75	0 48 49.5	12	1249 52 50	104 09.24.2	33 25 58	41 17 48	30.00	81	103 55 29.6	13 13 50	,, ,,
,,	4 18 07.9	0 48 49.5	12	1250 51 25	104 14 17.1	35 58 49	38 26 15	30.00	81	104 02 03.2	13 15 30	,, ,,
,,	4 34 25.1	0 48 49.5	18	1878 07 20	104 20 24.4	39 23 11	34 34 28	30.00	81	104 10 37.3	13 08 49.5	,, ,,
2	4 50 29.2	0 48 51	18	2112 19 45	117 21 05.8	30 07 12	30 50 38	30.00	80	117 06 53.3	13 12 46.5	,, ,,
5	8 37 32.9	0 48 51.2	12	890 13 30	74 11 07.5	46 29 27	56 12 23	29.90	80	73 50 38.4	13 15 19.5	Moon from Aldebaran W. of her.
,,	9 06 23.2	0 48 51.2	18	1338 22 10	74 21 13.9	53 19 45	49 18 50	29.90	80	74 06 10.2	13 17 25	,, ,,
,,	9 33 27.6	0 48 51.2	12	894 00 40	74 30 03.3	59 31 43	42 49 44	29.90	80	74 20 32.8	13 13 51	,, ,,
,,	10 00 59.2	0 48 51.2	12	895 44 10	74 38 40.8	66 10 49	36 11 46	29.90	80	74 35 15.3	13 12 49.5	,, ,,
29	8 43 51.2	0 49 09.5	12	603 14 50	50 16 14.1	58 18 19	65 40 38	30.00	78	49 33 17.4	13 11 34.5	Moon from Regulus E. of her.
,,	9 04 17.6	0 49 09.5	12	601 44 10	50 08 40.8	54 25 28	70 35 57	30.00	78	49 21 24.5	13 14 59	,, ,,
April 2	8 24 29.25	0 49 12	12	485 24 45	40 27 03.7	61 53 57	67 35 51	30.00	78	40 21 17	13 18 45	Moon from Pollux W. of her.
,,	8 48 35.9	0 49 12	18	730 34 15	40 35 14.2	67 42 35	64 30 06	30.00	78	40 34 02.8	13 19 24	,, ,,
,,	9 11 52.8	0 49 12	18	732 50 20	40 42 47.8	73 20 54	60 45 56	30.00	78	40 46 16.2	13 17 10.5	,, ,,
,,	9 31 17.3	0 49 12	12	489 50 30	40 49 12.5	78 00 57	57 16 42	30.00	78	40 56 35.5	13 18 59.5	,, ,,
,,	11 41 14.9	0 49 12	12	595 32 15	49 37 41.25	70 08 18	58 41 38	30.00	78	49 05 08.7	13 11 49.5	Moon from Spica E. of her.
,,	12 04 27.2	0 49 12	18	890 57 05	49 29 50.3	64 29 35	63 06 42	30.00	78	48 52 48.2	13 13 12	,, ,,
										MEAN of 318 Distances . .	13 15 26.8	West Longitude.

St. Thomas.——In the Mansion House at Man-of-War Bay. Latitude 0° 24′ 41″ N.

DATE.	Time by the Chron. 423.	423 on Mean Time. Slow.	No. of Observations.	Arc passed through.	Apparent Distance of Limbs.	Apparent Altitudes. Moon.	Apparent Altitudes. Sun or Star.	Barom.	Therm.	True Distance.	Longitude.	PHENOMENA and REMARKS.
1822.	H. M. S.	H. M. S.		° ′ ″	° ′ ″	° ′ ″	° ′ ″	IN.	°	° ′ ″	° ′ ″	
May 29	8 45 46.3	0 30 08.5	12	786 50 20	65 34 11.6	64 04 30	44 10 23	30.10	82	65 00 46.8	6 46 23 E.	Moon from Antares E. of her.
,, ,,	9 04 38.1	0 30 08.5	12	785 36 35	65 28 03	59 25 53	47 48 02	30.10	82	64 51 01	6 41 33	,, ,,
,, ,,	9 22 31.8	0 30 08.5	12	784 28 05	65 22 20.4	55 03 29	51 04 29	30.10	82	64 42 02.4	6 46 21	,, ,,
,, ,,	9 34 37.2	0 30 08.5	12	783 37 35	65 18 07	52 05 25	53 09 04	30.10	82	64 35 45.1	6 41 39	,, ,,
,, ,,	9 48 43.7	0 30 08.5	18	1173 58 20	65 13 14.4	48 38 05	55 27 10.5	30.10	82	64 28 31.3	6 40 13.5	,, ,,
June 5	9 25 29.1	0 29 58.7	12	785 50 40	65 29 13.4	39 35 00	64 12 12	30.08	81	64 37 46	6 47 01.5	Moon from Spica W. of her.
,, ,,	10 06 49.8	0 29 58.7	18	1182 46 20	65 42 34.4	47 17 13	54 33 27	30.08	81	64 58 26.6	6 44 48	,, ,,
,, 10	7 05 59.4	0 29 51	12	1282 57 10	106 54 46	49 40 20	22 21 08	30.10	81	106 52 46.7	6 45 08	Moon from Sun E. of her.
,, ,,	7 25 02.8	0 29 51	18	1922 34 00	106 48 33.3	45 09 56.5	26 39 12	30.10	81	106 43 17	6 46 27	,, ,,
,, ,,	7 40 14.5	0 29 51	12	1280 39 50	106 43 19.1	41 32 31.5	30 03 47.5	30.10	81	106 35 40.6	6 46 22.5	,, ,,
,, ,,	7 53 04.5	0 29 51	12	1279 45 25	106 38 47.1	38 29 12	32 55 17.5	30.10	81	106 29 20.7	6 49 07.5	,, ,,
										Mean of 150 Distances . . .	6 45 00.4	East Longitude.

ASCENSION.——In the Barrack Square at the foot of Cross Hill. Latitude 7° 55′ 56″. S.

DATE.	Time by the Chron. 423.	423 on Mean Time. Fast.	No. of Observations.	Arc passed through.	Apparent Distance of Limbs.	APPARENT ALTITUDES.		Barom.	Therm.	True Distance.	Longitude.	PHENOMENA and REMARKS.
						Moon.	Sun or Star.					
1822.	H. M. S.	H. M. S.		° ′ ″	° ′ ″	° ′ ″	° ′ ″	IN.	°	° ′ ″	° ′ ″	
June 26	9 53 21.75	0 55 08.4	12	677 25 05	56 27 05.4	49 42 10	66 34 39	30.10	78	55 40 15.6	14 23 32 W.	Moon from Antares E. of her.
,, 27	9 07 12.9	0 55 11	10	444 23 15	44 26 19.5	70 28 23	59 12 02	30.10	80	43 57 41.2	14 20 39	Moon from Antares E. of her.
,, ,,	9 22 36.2	0 55 11	10	443 37 35	44 21 45.5	66 51 51	62 04 59.5	30.10	80	43 49 55.3	14 22 40.5	,, ,,
,, ,,	9 38 27.6	0 55 11	10	442 50 10	44 17 01	63 07 09	64 50 09	30.10	80	43 41 55.8	14 24 31.5	,,
,, 29	10 22 11.7	0 55 16	10	772 11 20	77 13 08	69 07 53.5	31 30 49.5	30.10	80	76 42 38.8	14 24 10.5	Moon from α Aquilæ E. of her.
,, ,,	10 51 03.9	0 55 16	10	771 11 00	77 07 06	63 42 06	38 24 09	30.10	80	76 30 33.1	14 24 10.5	,, ,,
,, ,,	11 02 24.3	0 55 16	10	770 46 00	77 04 36	61 24 28.5	41 04 27	30.10	80	76 25 47.2	14 24 36	,, ,,
July 2	10 57 41.2	0 55 23.5	10	621 48 10	62 10 49	67 07 12	38 39 48.5	30.10	80	62 15 00.1	14 25 52.5	Moon from Spica W. of her.
,, 3	8 52 57.9	0 55 25.5	10	737 05 00	73 42 30	34 18 47	68 27 13	30.10	80	73 15 08.9	14 28 03	Moon from Spica W. of her.
,, ,,	9 16 16	0 55 25.5	10	738 29 30	73 50 57	39 19 18.5	62 41 58.5	30.10	80	73 26 53.4	14 27 29	,, ,,
,, ,,	9 28 35	0 56 25.5	10	739 12 20	73 55 14	41 55 05	59 39 51	30.10	80	73 33 04.5	14 26 37.5	,, ,,
,, ,,	11 56 45.4	0 55 25.5	12	676 01 30	56 20 07.5	68 34 19	16 45 43	30.10	80	56 22 52	14 25 25.5	Moon from Fomalhaut E. of her.
,, 5	11 42 33.5	0 55 31	10	541 22 40	51 08 16	50 22 03	63 23 50.5	30.10	80	53 23 48	14 20 07.5	Moon from Antares W. of her.
,, ,,	11 55 29.4	0 55 31	10	542 04 40	51 12 28	53 11 54.5	61 04 06.5	30.10	80	53 30 28.2	14 18 54	,, ,,
,, 6	1 25 19.9	0 55 33.5	10	674 14 47	67 25 28.7	63 53 55	41 40 07	30.10	80	66 52 11.4	14 20 21	Moon from Antares W. of her.
,, ,,	1 39 48.3	0 55 33.5	10	675 00 13	67 30 01.3	67 06 13	38 28 57	30.10	80	66 59 49.7	14 20 05.5	,, ,,
									MEAN of 164 Distances . . .		14 23 35	West Longitude.

Bahia.——In the House of William Pennell, Esq., His Britannic Majesty's Consul, at Vittoria. Latitude 12 59′ 22″ S.

DATE.	Time by the Chron. 423.	423 on Mean Time. Fast.	No. of Observations.	Arc passed through.	Apparent Distance of Limbs.	Apparent Altitudes.		Barom.	Therm.	True Distance.	Longitude.	Phenomena and Remarks.
						Moon.	Sun or Star.					
1822.	H. M. S.	H. M. S.		° ′ ″	° ′ ″	° ′ ″	° ′ ″	IN.	°	° ′ ″	° ′ ″	
July 25	4 33 02.3	2 32 52.5	6	523 08 48	87 11 28	38 19 20	46 55 22	30 05	74	87 05 23.7	38 34 34.5W	Moon from Sun W. of her.
,, ,,	4 47 13.3	2 32 52.5	10	872 43 52	87 16 23.2	41 56 57	44 34 28	30.05	74	87 11 53	38 32 19.5	,, ,,
,, ,,	4 59 10.7	2 32 52.5	10	873 24 30	87 20 27	44 46 09	42 29 11	30.05	74	87 17 27.1	38 33 36	,, ,,
,, 26	5 35 37.4	2 32 54.5	6	592 01 25	98 40 14.2	42 56 14	35 45 10	29.97	74	98 34 52.7	38 29 16.5	,, ,,
,, ,,	5 41 52.5	2 32 54.5	6	592 14 45	98 42 27.5	44 26 55	34 31 45	29.97	74	98 37 54.8	38 35 00	,, ,,
,, ,,	5 48 16.6	2 32 54.5	6	592 26 20	98 44 23.3	46 02 46	33 12 43	29.97	74	98 40 40.1	38 29 48	,, ,,
,, ,,	11 12 48.1	2 32 55	10	801 48 10	80 10 49	56 20 50	43 34 38	29.98	72	79 27 23.4	38 33 36	Moon from α Aquilæ E. of her.
,, ,,	11 25 47.6	2 32 55	10	801 14 45	80 07 28.5	53 24 05	46 25 20	29.98	72	79 21 51.5	38 35 00	,, ,,
,, 31	11 34 24.9	2 33 09.5	10	831 57 40	83 11 46	65 34 26	26 05 09	30.05	70	83 10 33.6	38 33 51	Moon from Spica W. of her.
,, ,,	11 46 16.7	2 33 09.5	6	499 30 00	83 15 00	67 51 07	23 13 46.5	30.05	70	83 16 33.4	38 31 24	,, ,,
,, ,,	11 55 31.5	2 33 09.5	8	666 19 10	83 17 23.7	69 37 04	20 59 18	30.05	70	83 21 13.7	38 29 30	,, ,,
,, ,,	12 12 49.6	2 33 09.5	10	487 03 10	48 42 19	72 36 28	25 23 47	30.05	70	48 43 46	38 33 15	Moon from Fomalhaut E. of her.
,, ,,	12 24 37.2	2 33 09.5	10	486 36 00	48 39 36	74 23 07	27 49 57	30.05	70	48 38 36	38 31 54	,, ,,
,, ,,	12 41 03.7	2 33 09.5	20	972 00 58	48 36 02.9	76 19 58	31 30 13	30.05	70	48 31 15.1	38 34 45	,, ,,
										Mean of 128 Distances . . .	38 32 39	West Longitude.

MARANHAM.——In the House of Robert Hesketh, Esq., His Britannic Majesty's Consul. Latitude 2° 31′ 43″ S.

DATE.	Time by the Chron. 423.	423 on Mean Time. Fast.	No. of Observations.	Arc passed through.	Apparent Distance of Limbs.	Apparent Altitudes. Moon.	Apparent Altitudes. Sun or Star.	Barom.	Therm.	True Distance.	Longitude.	PHENOMENA and REMARKS.
1822.	H. M. S.	H. M. S.		° ′ ″	° ′ ″	° ′ ″	° ′ ″	IN.	°	° ′ ″	° ′ ″	
Aug. 24	11 26 02.2	2 56 49	8	508 45 20	63 35 40	45 39 52	71 01 13	29.95	78	62 44 01.2	44 21 33	Moon from α Aquilæ E. of her.
,, ,,	12 19 49.1	2 56 49	10	633 17 45	63 19 46.5	34 37 11	78 52 16	29.95	78	62 22 21.1	44 20 12	,, ,,
,, 25	6 48 36.6	2 56 52.5	10	1021 39 05	102 09 54.5	43 18 18	31 19 53	30.00	80	102 05 38.8	44 20 34.5	Moon from Sun W. of her.
,, ,,	6 57 08.3	2 56 52.5	10	1022 02 50	102 12 17	45 04 14	29 12 31	30.00	80	102 09 33.2	44 21 45	,, ,,
,, ,,	7 06 02.7	2 56 52.5	10	1022 27 10	102 14 43	46 39 10	27 06 33	30.00	80	102 13 34.6	44 21 06	,, ,,
,, ,,	13 00 05.2	2 56 53.5	10	744 06 25	74 24 38.5	36 23 01	44 51 55	29.95	78	73 40 28	44 21 24	Moon from Fomalhaut E. of her.
,, ,,	13 12 24.4	2 56 53.5	10	743 30 30	74 21 03	33 50 59	47 07 21	29.95	78	73 34 51.1	44 21 57	,, ,,
,, ,,	13 29 34.2	2 56 53.5	10	742 38 50	74 15 53	30 16 00	50 06 26	29.95	78	73 27 02.5	44 22 18	,, ,,
,, ,,	13 43 53.9	2 56 53.5	10	741 53 20	74 11 20	27 16 40.5	52 25 45	29.95	78	73 20 30.4	44 23 09	,, ,,
,, 27	11 22 35.7	2 56 57	10	528 30 30	52 51 03	66 01 05	26 59 45	29.95	78	52 44 23.4	44 22 19.5	Moon from Fomalhaut E. of her.
,, ,,	13 26 28.5	2 56 57	10	523 56 55	52 23 41.5	52 00 29	50 53 05	29.95	78	51 49 12	44 21 48	,, ,,
,, ,,	13 36 52.2	2 56 57	10	523 32 45	52 21 16.5	50 04 48	52 34 26	29.95	78	51 44 36	44 20 42	,, ,,
,, 31	12 26 37.85	2 57 08	10	817 11 00	84 43 06	57 37 02	30 16 24	29.95	78	84 36 07	44 21 00	Moon from Antares W. of her.
,, ,,	13 27 15.45	2 57 08	10	850 30 20	85 03 02	71 53 11	16 49 35	29.95	78	85 09 49.7	44 19 04	,, ,,
,, ,,	13 39 42.5	2 57 08	10	851 08 25	85 06 50.5	71 41 09	14 02 26	29.95	78	85 16 54.5	44 23 32.5	,, ,,
Sept. 2	12 45 51.05	2 57 15	10	1129 46 30	112 58 39	39 55 52	24 20 08	29.95	77	112 03 39.85	44 20 33	Moon from Antares W. of her.
MEAN of 158 Distances . .											44 21 25.5	West Longitude.

TRINIDAD.——In the second Ground-Lot, West of the Protestant Church, Port Spain. Latitude 10° 38′ 56″ N.

DATE.	Time by the Chron. 423.	423 on Mean Time. Fast.	No. of Observations.	Arc passed through.	Apparent Distance of Limbs.	Apparent Altitudes.		Barom.	Therm.	True Distance.	Longitude.	PHENOMENA and REMARKS.
						Moon.	Sun or Star.					
1822.	H. M. S.	H. M. S.		° ′ ″	° ′ ″	° ′ ″	° ′ ″	IN.	°	° ′ ″	° ′ ′	
Sept. 23	7 01 05.9	4 6 59.6	10	938 55 50	93 53 35	28 51 06	43 36 05	30.00	80	93 46 47.4	61 39 04.5	Moon from Sun W. of her.
,,	7 11 14	4 6 59.6	10	939 26 25	93 56 38.5	30 42 31	41 08 59	30.00	80	93 51 23.6	61 38 18	,, ,,
,,	7 22 51	4 6 59.6	10	940 00 30	94 00 03	32 47 00	38 19 44	30.00	80	93 56 40.1	61 37 27	,, ,,
,,	7 34 15.3	4 6 59.6	10	940 32 35	94 03 15.5	34 46 40	35 34 23.5	30.00	80	94 01 40.88	61 35 34.5	,, ,,
,,	12 55 38.2	4 7 00.2	10	677 15 45	56 26 18.75	37 48 09	41 03 51	29.97	78	55 51 41.3	61 33 50	Moon from Fomalhaut E. of her.
,,	13 11 15	4 7 00.2	10	563 40 50	56 22 05	35 00 31	42 58 59	29.97	78	55 44 44	61 35 00	,, ,,
,,	13 23 35.2	4 7 00.2	10	563 11 00	56 19 06	32 44 36	44 20 30	29.97	78	55 39 11	61 38 00	,, ,,
28	13 25 04	4 7 17.5	10	550 23 30	55 02 21	67 55 46	29 16 31	29.96	78	54 51 01.5	61 38 43	Moon from α Arietis E. of her.
,,	13 40 27.1	4 7 17.5	10	549 30 40	54 57 04	70 16 10	32 48 35	29.96	78	54 42 31.5	61 36 15	,, ,,
,,	13 57 47.3	4 7 17.5	10	548 28 35	54 50 51.5	72 19 02	36 48 34	29.96	78	54 32 42	61 34 24	,, ,,
,,	14 26 27.9	4 7 17.5	10	546 46 00	54 40 36	73 44 31	43 24 45	29.96	78	54 16 28	61 34 48	,, ,,
,,	14 39 25.4	4 7 17.5	10	546 00 00	54 36 00	73 25 21	46 26 16	29.96	78	54 09 11.6	61 33 32	,, ,,
29	14 32 52.4	4 7 21	10	586 27 50	58 38 47	76 23 49	41 34 09	29.97	78	58 45 10	61 36 43.5	Moon from α Aquilæ W. of her.
,,	14 44 28.1	4 7 21	10	586 53 20	58 41 20	78 11 09	38 42 54	29.96	78	58 50 34.6	61 35 37.5	,, ,,
30	13 43 18.9	4 7 23.5	10	702 49 00	70 16 54	56 34 59	52 48 39	29.96	77	70 02 00	61 35 17	,, ,,
,,	13 54 55.4	4 7 23.5	10	703 23 30	70 20 21	59 22 49	49 57 07	29.96	77	70 07 53.3	61 37 21	,, ,,
										MEAN of 162 Distances . . .	61 36 15	West of Greenwich.

JAMAICA.——In Fort Charles, Port Royal. Latitude 17° 56′ 06″ N.

DATE.	Time by the Chron. 423.	423 on Mean Time. Fast.	No. of Observations.	Arc passed through.	Apparent Distance of Limbs.	Apparent Altitudes. Moon.	Apparent Altitudes. Sun or Star.	Barom.	Therm.	True Distance.	Longitude.	PHENOMENA and REMARKS.
1822.	H. M. S.	H. M. S.		° ′ ″	° ′ ″	° ′ ″	° ′ ″	IN.	°	° ′ ″	° ′ ″	
Oct. 23	7 29 11.9	5 09 56.6	10	966 01 30	96 36 09	17 28 00	41 50 12	30.03	86	96 30 41	76 53 40.5	Moon from Sun W. of her.
,, ,,	7 42 55.3	5 09 56.6	10	966 52 45	96 41 16.5	20 17 01	39 06 44	30.03	86	96 37 11	76 54 43.5	,, ,,
,, ,,	7 51 38.4	5 09 56.6	10	967 23 00	96 44 18	22 03 09	37 20 20	30.03	86	96 41 15.5	76 53 30	,, ,,
,, ,,	8 00 33.3	5 09 56.6	10	967 54 00	96 47 24	23 47 52	35 30 15	30.03	86	96 45 28.7	76 53 54	,, ,,
,, ,,	8 10 25.9	5 09 56.6	10	968 27 50	96 50 47	25 44 19	33 25 56	30.03	86	96 50 14.5	76 57 05	,, ,,
,, ,,	8 19 39.6	5 09 56.6	10	968 56 45	96 53 40.5	27 32 15	31 28 33	30.03	86	96 54 30.2	76 54 18	,, ,,
,, 25	9 04 06.5	5 10 04.2	10	1206 50 30	120 41 03	22 33 46	21 27 36	30.03	87	120 38 11.5	76 53 45	,, ,,
,, ,,	9 13 43.7	5 10 04.2	10	1207 27 30	120 44 45	24 41 23	19 19 50	30.03	87	120 43 03	76 54 10.5	,, ,,
,, ,,	9 22 39.8	5 10 04.2	10	1207 59 55	120 47 59.5	26 39 16	17 20 02	30.03	87	120 47 28.6	76 51 54	,, ,,
,, 27	14 29 19.2	5 10 14	10	652 09 40	65 12 58	75 28 34	22 44 48	29.99	78	65 00 12	76 54 52	Moon from Aldebaran W. of her.
,, ,,	14 43 19.4	5 10 14	10	651 23 30	65 08 21	76 26 14	26 02 24	29.99	78	64 52 06.7	76 54 11	,, ,,
,, ,,	14 54 46.6	5 10 14	10	650 45 40	65 04 34	76 29 32	28 42 54	29.99	78	64 45 33.5	76 52 15	,, ,,
,, ,,	15 05 51.6	5 10 14	10	650 07 30	65 00 45	76 13 56	31 19 38	29.99	78	64 39 05	76 53 25	,, ,,
,, ,,	15 18 38.9	5 10 14	10	649 24 08	64 56 24.8	75 17 58	34 20 46	29.99	78	64 31 40.5	76 53 21	,, ,,
,, 28	14 23 09.2	5 10 18	10	509 41 52	50 58 11.2	70 02 55	22 19 20	29.98	78	51 00 12.6	76 54 24	Moon from Aldebaran E. of her.
,, ,,	14 36 35.3	5 10 18	10	508 55 00	50 53 30	73 00 57	25 28 08	29.98	78	50 52 15.3	76 54 32	,, ,,
,, ,,	14 47 32.8	5 10 18	10	508 16 20	50 49 38	75 22 16	28 02 50	29.98	76	50 45 47.7	76 53 54	,, ,,
,, ,,	14 59 50.3	5 10 18	10	507 33 25	50 45 20.5	77 53 38	30 56 34	29.98	76	50 38 30.7	76 54 09	,, ,,
Nov. 3	1 08 10.1	5 10 36	10	1223 20 35	122 20 03.5	31 24 46	25 18 23	30.03	85	122 05 11.8	76 52 19.5	Moon from Sun E. of her.
,, ,,	1 20 35.7	5 10 36	10	1222 24 10	122 14 25	28 43 44	27 54 08	30.03	85	121 58 17.7	76 51 10.5	,, ,,
,, ,,	1 32 50.2	5 10 36	10	1221 26 50	122 08 41	26 05 15	30 26 22	30.03	85	121 51 28.3	76 50 32	,, ,,
,, ,,	1 45 48.2	5 10 36	10	1220 22 45	122 02 16.5	23 18 43	33 04 15	30.03	85	121 44 10	76 52 00	,, ,,
,, 4	2 38 07.4	5 10 40	10	1082 46 00	108 16 36	24 31 53	42 44 35	30.04	85	108 00 44.7	76 53 18	,, ,,
,, ,,	2 47 22.6	5 10 40	10	1082 00 30	108 12 03	22 30 24	44 19 57	30.04	85	107 55 49.2	76 50 27	,, ,,
,, 5	2 30 12.7	5 13 43.5	10	953 23 30	95 20 21	38 01 17	41 07 26	30.03	85	95 12 10	76 52 21	,, ,,
,, ,,	2 39 01.1	5 10 43.5	10	952 47 20	95 16 44	36 00 41	42 40 28	30.03	85	95 07 34.6	76 50 33.5	,, ,,
,, ,,	2 47 34.2	5 10 43.5	10	952 09 30	95 12 57	34 02 44	44 08 00	30.03	85	95 02 57.6	76 53 00	,, ,,
MEAN of 270 Distances . . .											76 53 15	West of Greenwich.

The preceding tabular statement comprises the results of 1350 distances, divided into 123 sets, and distributed through seven stations. The following table collects in one view the mean results, and exhibits a summary of the differences of the individual sets on the general mean at each station.

STATIONS.	No. of Distances.	No. of Sets.	Individual Sets differing from the Mean							MEAN LONGITUDES.
			Less than 1 mile.	Less than 2 miles.	Less than 3 miles.	Less than 4 miles.	Less than 5 miles.	Less than 6 miles.	Less than 7 miles.	° ′ ″
Sierra Leone.	318	23	4	7	4	7	.	.	1	13 15 26.8 W.
St. Thomas .	150	11	2	4	1	2	2	.	.	6 45 00.4 E.
Ascension . .	164	16	5	2	2	5	2	.	.	14 23 35 W.
Bahia. . . .	128	14	5	4	3	2	.	.	.	38 32 39 W.
Maranham	158	16	12	3	1	.	.	.	.	44 21 25.5 W.
Trinidad . .	162	16	5	6	5	.	.	.	.	61 36 15 W.
Jamaica . .	270	27	15	7	4	1	.	.	.	76 53 15 W.
TOTAL . . .	1350	123	48	33	20	17	4	0	1	

Whence it may be inferred that in similar circumstances of observation, —i. e., on shore, and within the tropics, the observer being previously accustomed to lunar observation with sextants, and furnished with a correct knowledge of the time at the station,—it is about 2 to 1, that a single set composed of 10 or 12 distances observed with Mr. Dollond's circle, will give a result within two miles of the longitude, deduced from an extensive series, including the various states of the atmosphere occurring in such climates, and at different periods of the moon's age; that it is about 2 to 3 that the result will be within one mile; and that a difference, amounting to so much as between 4 and 5 miles, may not be expected to occur oftener than about once in 25 sets.

The improvement which practice will make in the habits of observation, (and consequently on the inferences that have been thus stated,) is evident on an inspection of the table; for if the three last stations only are regarded, the chances will appear more than equal, that the result of a single set is within one mile, and 4 to 1 that it is within two miles of a general mean; whilst the extreme difference, occurring only once in 59 sets, is under four miles.

No attempt has been made to correct the distances inserted in the Nautical Almanac, by a more exact knowledge of the moon's place derived from the Greenwich observations; because the design has been to afford a practical inference of the degree of accuracy which an observer may expect with the means with which he is furnished on the spot. It may be proper to mention also, that the table includes every set of distances observed at the stations to which it refers.

The conveniency of the circle in observation, and the facility with which it may be managed by those who will accustom themselves to its use, may be judged by the observations at Jamaica; where it may be seen that 60 distances of the sun and moon were observed within the hour, or one in each minute, including the observation and entry of the time of each distance, and the reading off at every tenth distance, and writing down the arc passed through. In the repetition of the same process with the moon and Aldebaran at night, the number of distances observed in an hour was 50, or, on an average, one in a minute and twelve seconds. Of the six sets into which the distances of the sun and moon under notice were divided, three are within one mile of the combined result of twenty-seven sets, and the two others within two miles; and of the five sets of the moon and Aldebaran, four, are within one mile, and the fifth within two miles: this is stated to shew that accuracy was not sacrificed to expedition; both the instances were without the advantage (in expedition certainly) of an assistant.

The observation of the angular distance of the moon from certain fixed stars, has long and universally been regarded as the best means of deducing the longitude of a vessel on the ocean from celestial phenomena; but it has not been so generally recognised as it deserves to be, as the most eligible of all the methods which present themselves to the choice of the geographer, or the practical astronomer, for determining positions on land, wherever time or the conveyance of instruments form a part of the consideration. It combines, in a degree far beyond comparison with any other method, the very important qualities of convenience, expedition, and accuracy. The whole apparatus which is required,—a circle, a chronometer, and an artificial horizon,—does not weigh twelve pounds; no temporary observatory is required for its protection, and all situations are equally convenient for its use; the latitude and longitude may both be determined in the first twenty-four hours after the arrival at a station, during three-fourths of every lunation; and as the observations by which the determinations are accomplished may be multiplied within that interval at the pleasure of the observer, so as to comprise, in respect to latitude, every important variety of circumstance, and almost every variety in regard to longitude, no sacrifice of accuracy to expedition is called for, but the precision will be proportionate to the labour which is bestowed.

There are occasions in which the qualities of convenience in portability, expedition and accuracy in determination, are almost equally essential. Such is the design which is understood to be entertained, of forming the basis of a survey of central India, by the celestial determination of the geographical position of stations, selected at proper intervals over that very extensive portion of the globe. Admitting the space to contain 130 or 140 square equatorial degrees, and the stations to be on an average 100 miles apart, above eighty such determinations must be made. Those who are acquainted with the apparatus which would be required, in any other mode of deducing the position from celestial observation than the one under notice, and will pursue in detail the consideration of the conveyance of such an apparatus over such an extent of country, independently of the accidents and interruptions to which it would be liable,—and who can appreciate the time which would be occupied in obtaining an equally precise determination at each station, as lunar distances would give in 24 hours,—will, I think, arrive at the conclusion, that it is only by lunar distances that the design is likely to receive its accomplishment.

With respect to the degree of accuracy to which the mean results of the lunar observations at the seven tropical stations may be considered to have been obtained, additional evidence may be afforded, by exhibiting their mutual accordance when connected with each other chronometrically. I have employed for that purpose the chronometer No. 357 of Messrs. Parkinson and Frodsham, which, having a small and convenient rate, and being wound weekly, was selected as a standard of comparison for the chronometers in the Pheasant. No. 357 was stationary on board, and was suspended in a cot from the deck, in a part of the cabin where the motion was least, and where it remained undisturbed, except for the purposes of comparison, and of being wound. Its treatment, therefore, was as favourable to the preservation of a steady and uniform rate, as that of No. 423, with which its weekly comparison is shown in the following table, may be esteemed to have been the reverse. The immediate purpose of the table is to enable the transfer to No. 357 of the comparisons of No. 423 with astronomical time, recorded in former pages, and thus to furnish the means of examining in detail, the chronometrical connexion of the lunar longitudes with each other, to those who may be disposed to take that trouble. The testimony, however, which the table incidentally bears to the excellence of the chronometers, so liberally lent by Messrs. Parkinson and Frodsham, and to which the accuracy of the observations contained in this volume is so essentially indebted, will probably be an object of more general interest. It can rarely have happened to any chronometer to have undergone so great a variety and such constant practical exposure, whilst its rate was submitted to so severe a scrutiny, as that of No. 423. It is known that the rates of chronometers are frequently found to vary on embarkation in vessels, insomuch as to have given rise to the distinctive terms of Land Rates, and Sea Rates*; no such variation can however be traced

* An opinion has lately prevailed, that the change in the rate of chronometers on embarkation, which used to be considered as a consequence of the motion of a ship, is principally occasioned by the magnetic influence of the iron which she contains; and it has been assumed by some of the writers, who have taken part in the recent discussions on the subject, that the effect so attributed is one of general experience. I believe, on the authority of others, rather than from my own observation, that a difference does sometimes, and even frequently, take place between the land and sea rates of chronometers; but from whatever cause the irregularity may arise, I must regard its occurrence as an evidence of the inferiority of the particular chronometer, to the advanced state to which the art of their construction has attained; because, amongst the many with which I have at different times been furnished by Messrs. Parkinson and Frodsham, and which I have frequently transferred from the ship to the shore, for two and three weeks at a time, for the purpose of trial, I have never been able to discover any systematic variation whatsoever, consequent on their removal.

With regard to the influence of the iron as a cause of the irregularity, a more decisive evidence can scarcely be imagined of its not being practically discovered under the most favourable circumstances for its exhibition, than took place in the four chronometers of Messrs. Parkinson and Frodsham, of which I have given an account, in the appendix to Capt. Parry's Voyage of Discovery, in 1819-1820, pages vii to xii, xviii, xix, and xx. On that occasion the Hecla was stationary and immoveable, being frozen up, for more than ten months in the vicinity of the magnetic pole, the dip being between 88 and 89 degrees; such is the situation and such the circumstances, which are supposed to be best adapted for the development of magnetism in the stanchions and other vertical iron of a ship; the chronometers were kept on board during the whole winter, and their rates, preparatory to the navigation of the following summer, were assigned from the average of the four months immediately preceding her extrication from the ice; at the expiration of an equal period of four months of navigation, the Hecla arrived at Leith, having experienced much bad weather in crossing the

in the going of No. 423, in any one of the six voyages in which it was embarked between April and November, and which alternated with nearly equal periods on land, when it was employed incessantly in observations, including those of magnetism. The uniformity of rate which it preserved from day to day, under every circumstance of change or exposure, was indeed admirable; and is deserving of regard, as an evidence of the high degree of perfection to which the mechanism and workmanship of chronometers have attained.

Atlantic, having been entirely dismasted on one occasion, and (which might have been expected to have had even a more prejudicial effect on the chronometers,) having sustained very frequent severe shocks from collision with ice; but on comparing the four chronometers at the Observatory at Leith, their Greenwich time, employing the Winter Harbour rates, proved less than two seconds in error. On the arrival of the Hecla in the Thames, the chronometers were returned to Messrs. Parkinson and Frodsham's house in London, where, after a month's interval, they were found to be still going at the same rates, as in the Hecla whilst in the harbour of Melville Island. These particulars are stated in detail in the pages referred to; but the circumstance is thus again generally noticed, because it appears to have been overlooked by many, whose ingenuity has been exerted in devising contrivances to remedy an evil which has no practical existence, where the common discretion of life is exercised, in obtaining the better article at an equal price. Had the especial purpose of the Hecla's voyage been to inquire whether the iron of a ship, in its ordinary distribution, would, under extreme circumstances, exert a sensible influence on the chronometers, better adapted arrangements could scarcely have been devised for the experiment, nor could a more decisive result in the negative have been obtained.

WEEKLY COMPARISON of the CHRONOMETERS 357 and 423 between February and November, 1822, with the average Daily Rate in each Week of 423 on 357.

DATE.	423 on 357.	423's Rate on 357.	STATION.	DATE.	423 on 357.	423's Rate on 357.	STATION.
1822.	M. S.			1822.	M. S.		
Feb. 23	Fa. 0 13			July 6	Fa. 0 07		
		+ 0.43				+ 0.71	
Mar. 2	,, 0 16			,, 13	,, 0 12		
		+ 0.29				+ 0.36	At Sea.
,, 9	,, 0 18			,, 20	,, 0 14.5		
		+ 0.14				+ 0.5	
,, 16	,, 0 19		Sierra Leone.	,, 27	,, 0 18		
		− 0.07				+ 0.43	Bahia.
,, 23	,, 0 18.5			Aug. 3	,, 0 21		
		− 0.21				+ 0.29	
,, 30	,, 0 16			,, 10	,, 0 23		
		0				− 0.14	At Sea.
April 6	,, 0 16			,, 17	,, 0 22		
		+ 0.5				423 down. and reset.	
,, 13	,, 0 19.5			,, 24	Sl. 0 08		
		+ 0.36				− 0.14	Maranham.
,, 20	,, 0 22			,, 31	,, 0 07		
		− 0.07				− 0.36	
,, 27	,, 0 21.5			Sept. 7	,, 0 09.5		
		423 down and reset.				0	
May 4	Sl. 0 08		At Sea.	,, 14	,, 0 09.5		At Sea.
		− 0.29				− 0.21	
,, 11	,, 0 10			,, 21	,, 0 11		
		− 0.29				− 0.21	
,, 18	,, 0 12			,, 28	,, 0 12.5		
		+ 0.43				0	Trinidad.
,, 25	,, 0 09			Oct. 5	,, 0 12.5		
		+ 0.14				+ 0.07	
June 1	,, 0 08			,, 12	,, 0 12		
		0	St. Thomas.			+ 0.5	At Sea.
,, 8	,, 0 08			,, 19	,, 0 08.5		
		+ 0.21				+ 0.29	
,, 15	,, 0 06.5			,, 26	,, 0 06.5		
		+ 0.5	At Sea.			+ 0.21	Jamaica.
,, 22	,, 0 03			Nov. 2	,, 0 05		
		+ 0.57				+ 0.43	
,, 29	Fa. 0 01			,, 9	,, 0 02		
		+ 0.86	Ascension.				
July 6	,, 0 07						

From the comparisons in the preceding table, and the transits and zenith distances observed at the different stations with No. 423 and detailed in former pages, the particulars are supplied which are arranged in the next table, and furnish the necessary data for conveying the lunar longitude of each station to the next on either side of it, and thus of comparing them with each other, as is done in the subsequent memoranda.

The 1st column of the tables states the earliest and latest days at each station, in which the correction of No. 423 to mean time was ascertained with precision, its amount being shewn in the 4th column; the 5th contains the corresponding correction of 357, and the 6th its difference from the mean Greenwich time obtained by lunar observations; in the 7th column is inserted the number of days included by the observations of rate at each station, and in the 8th the average daily rate of 357, deduced from the direct comparison of 423 with astronomical time, and transferred to 357.

The rates of both the chronometers appear to have accelerated in a very regular and uniform progression, which may not improbably be attributed to the gradual adaptation to each other of the several parts of the workmanship, in the process of wear, as both the chronometers were of recent construction. The acceleration being regular, an intermediate rate between that of the station on either side has been assumed for the intervals of passage from station to station.

A TABLE

Shewing the RATE and CORRECTIONS of No. 357 to MEAN TIME at the several TROPICAL STATIONS, and its Corrections to the Mean Greenwich Time obtained by the Lunar Observations.

DATE.	STATIONS.	Lunar Longitude.	On Mean Time at the Station. 423.	On Mean Time at the Station. 357.	357 on Mean Greenwich Time.	Intervals.	357's Daily Rate.
1822.		H. M. S.	H. M. S.	H. M. S.	M. S.	Days.	Gaining.
March 6	Sierra Leone	0 53 01.8 W.	Fa. 0 48 48	Fa. 0 48 31	Sl. 4 30.8	27	0.96
April 2			„ 0 49 12	„ 0 48 57	„ 4 04.8		
May 27	St. Thomas.	0 27 00 E.	Sl. 0 30 11.5	Sl. 0 30 03	„ 3 03	14	1.21
June 10			„ 0 29 54	„ 0 29 46	„ 2 46		
„ 26	Ascension .	0 57 34.3 W.	Fa. 0 55 08	Fa. 0 55 09	„ 2 25.3	13	1.54
July 9			„ 0 55 40	„ 0 55 29	„ 2 05.3		
„ 23	Bahia . . .	2 34 10.6 W.	„ 2 32 46.5	„ 2 32 31.2	„ 1 39.4	13	2.22
August 5			„ 2 33 22	„ 2 33 00	„ 1 10.6		
„ 24	Maranham.	2 57 25.7 W.	„ 2 56 48	„ 2 56 56	„ 0 29.7	11	2.63
Sept. 4			„ 2 57 17.5	„ 2 27 25	„ 0 00.7		
„ 23	Trinidad. .	4 06 25 W.	„ 4 07 00	„ 4 07 12.5	Fa. 0 47.5	17	3.06
Oct. 10			„ 4 07 52	„ 4 08 04.5	„ 1 39.5		
„ 22	Jamaica. .	5 07 33 W.	„ 5 09 51.5	„ 5 09 58.5	„ 2 25.5	14	3.54
Nov. 5			„ 5 10 44	„ 5 10 48	„ 3 15		

COMPARISONS OF THE LONGITUDE OBTAINED BY DIRECT LUNAR OBSERVATION AT EACH STATION, AND THE LUNAR LONGITUDE OF THE ADJOINING STATIONS, REFERRED BY MEANS OF THE CHRONOMETER No. 357.

Sierra Leone.—*In the West Bastion of Fort Thornton.*

	h. m. s.
By 318 Lunar distances at Sierra Leone	0 53 01.8 W.
By 150 Lunar distances at St. Thomas's referred, by 357	0 52 59.7
Final Longitude . . .	0 53 00.75=13° 15′ 11″ W.

In a letter received in 1823, from the late Thomas Stewart Buckle, Esq., Civil Engineer and Surveyor of the colony of Sierra Leone, the Geographical bearing of Fort Thornton from Cape Sierra Leone, is stated to be S. 83° E., and the distance about 7,600 yards, or 3.8 geographical miles. Whence the longitude of the Fort referred to the Cape, would make the latter in 13° 19′ 00″ West.

St. Thomas.—*At the Mansion-House of Fernandilla, Man-of-War Bay.*

	h. m. s.
By 318 Lunar distances at Sierra Leone, referred by 357	0 26 57.9 E.
By 150 Lunar distances at St. Thomas	0 27 00.
By 164 Lunar distances at Ascension, referred by 357 .	0 26 58.7
Final Longitude . .	0 26 58.9=6° 44′ 43.5″ E.

The Roadstead of Santa Anna de Chaves is about 4½ geographical miles east of the meridian of Man-of-War Bay; whence the longitude of the Roadstead is 6° 49′ 13″ E.

Ascension.—*In the Barrack-Square.*

	h. m. s.
By 150 Lunar distances at St. Thomas's, referred by 357	0 57 36.4 W.
By 164 Lunar distances at Ascension	0 57 34.3
By 128 Lunar distances at Bahia, referred by 357 . .	0 57 34.7
Final Longitude . . .	0 57 35.1=14° 23′ 46.5″ W.

By a recent trigonometrical survey of Ascension by the officers of the garrison, the particulars of which are in the Hydrographic Office of the Admiralty, the Barrack-Square bears from Cross-Hill about N.W. the distance being less than 4,000 feet; whence the longitude of Cross-Hill may be taken at 14° 23′ 20″ W.

BAHIA.—*At the House of William Pennell, Esq., British Consul; Vittoria.*

	h. m. s.
By 164 Lunar distances at Ascension, referred by 357 .	2 34 10.2 W.
By 128 Lunar distances at Bahia	2 34 10.6
By 158 Lunar distances at Maranham, referred by 357 .	2 34 15.8
Final Longitude . . .	2 34 12.2=38° 33′ 03″ W.

Mr. Pennell's house is situated about half a mile east of the meridian of Fort St. Antonio; whence the longitude of the fort may be inferred 38° 33′ 30″ W.

MARANHAM.—*At the House of Robert Hesketh, Esq., British Consul, adjoining the Cathedral.*

	h. m. s.
By 128 Lunar distances at Bahia, referred by 357 . .	2 57 20.5 W.
By 158 Lunar distances at Maranham	2 57 25.7
By 162 Lunar distances at Trinidad, referred by 357 .	2 57 31.5
Final Longitude . . .	2 57 25.9=44° 21′ 28.″5 W.

The longitude may be referred to the Cathedral without sensible error.

TRINIDAD.—*On the Second Ground Lot, West of the Protestant Church, Port Spain.*

	h. m. s.
By 158 Lunar distances at Maranham, referred by 357 .	4 06 19.2 W.
By 162 Lunar distances at Trinidad	4 06 25.
By 270 Lunar distances at Jamaica, referred by 357 .	4 06 18.6
Final Longitude . . .	4 06 20.9=61° 35′ 13.″5 W.

The longitude of the Protestant Church may be inferred 61° 35′ 00″ W.

JAMAICA.—*At the Governor's House, Fort Charles, Port Royal.*

	h. m. s.
By 162 Lunar distances at Trinidad, referred by 357 .	5 07 39.4 W.
By 270 Lunar distances at Jamaica	5 07 33.
Final Longitude . . .	5 07 36.2=76° 54′ 03″ W.

The spot at which the observations were made is marked by the position of the Flagstaff at Fort Charles.

I proceed to compare the longitudes of the stations thus obtained with their previously-received longitudes, as given in the Connaissance des Tems, (No. for 1823,) and in Professor Lax's Nautical tables, (edition of 1821,) which works are presumed to contain the tables of longitude of the most approved authority in Great Britain and France.

Cape Sierra Leone is not found in the table of the Connaissance des Tems; in Professor Lax's table, it is given on the authority of the late Hydrographer of the Navy, 13° 18′ 00″ W., and by the present observations is in 13° 19′ 00″ W.

The longitude of the Roadstead of Santa Anna de Chaves, in the Island of St. Thomas is stated, in the Connaissance des Tems, on chronometrical authority, to be 7° 32′ 22″ East of Greenwich Observatory; but by the present observations, it is only in 6° 49′ 13″ East. Man-of-War Bay is placed by Professor Lax, on the authority of the Hydrographic Office, in 6° 44′ 00″ East, and by the present observations is in 6° 44′ 43″.5 East.

In the Connaissance des Tems, the longitude of the Island of Ascension is inserted 13° 58′ 45″ W., but without specification of the part of the Island to which the geographical position refers. Professor Lax has repeated the longitude of the French table, with the same uncertainty of position; but has also given that of Cross-Hill from the Hydrographic Office, 14° 13′ 30″ W. It appears from the recent trigonometrical survey, the particulars of which are deposited in that office, that the eastern extremity of the island is less than 8 miles east of the meridian of the Barrack-Square, or Cross-Hill; whence it may be inferred that if 14° 23′ 46″.5, the present determination, be the true longitude of the Barrack-Square, no part of the Island is in a less Western longitude than 14° 15′ 00″; and that the authority for the position assigned by Professor Lax to Cross Hill is about ten miles, and that of the French table nearer twenty miles in error. The correct longitude of Ascension is of value, because the island is frequently made by vessels on the homeward passage from the East Indies, and a departure is taken from it; the error of former determinations is also on the side of danger.

The longitude of Fort St. Antonio is given in Professor Lax's table, 38° 28′ 00″ West, on the authority of the Hydrographic Office. The table in the Connaissance des Tems, does not notice Bahia or its environs; but in a memoir entitled "Navigation aux côtes du Bresil," published in Paris, in 1821, by M. Le Baron Roussin, then Capitaine de Vaisseau, and since Admiral, in the service of France, (who was employed in the command of a small squadron, in the survey of the coasts of Brazil, in 1819 and 1820, and subsequently in 1822, when the Pheasant was at Bahia,) Fort St. Antonio is placed provisionally in 38° 30′ 12″ W., dependant upon the longitude of Fort Santa Cruz d'Anhatomirim; the difference of meridians between the stations being inferred chronometrically, and the longitude of Fort Santa Cruz derived from lunar distances, which are stated in the memoir to require to be more rigidly computed, before their correct result should be known. In the additions to the Connaissance des Tems for 1826, a memoir is printed by M. Givry, in which the longitudes of M. Le Baron Roussin on the coasts of Brazil are discussed; in that memoir Fort St. Antonio is placed in 38° 31′ 35″ W., being a chronometrical inference from Rio Janeiro, and dependant upon the true longitude of Rio, which is assumed from the mean of various sources differing not less than 22 miles from each other. It will be observed that the longitudes of Fort St. Antonio, assigned in the memoirs of MM. Roussin and Givry, are in neither case from direct observation at the meridian itself, but in both from the provisional longitudes of distant meridians referred to Bahia by means of a chronometer; if, however, the lunar observations made at Bahia itself by M. Roussin and his officers, be regarded as a more direct authority for the longitude of Fort St. Antonio, the following would appear the result:—

By 57 series of eastern distances of the Moon from the Sun observed on the 11th and 12th of October at Bahia	38° 41′ 43″.2
By 51 series of western distances observed at the same spot on the 25th and 27th of October.	38° 31′ 38 .4
Longitude of Bahia, west of Greenwich	38 36 40.8

and that of Fort St. Antonio would differ not more than a few seconds to the westward. It has been seen that the lunar observations with Mr. Dolland's circle made Mr. Pennell's house at Vittoria in 38° 32′ 33″; those at Ascension chronometrically referred to the same spot 38° 32′ 39″; and those at Maranham similarly referred 38° 33′ 57″; whence Fort St. Antonio may be deduced from the mean of the three determinations to be in 38° 33′ 30″ W. Navigators will exercise their own judgment, in selecting the longitude amongst these various authorities which may appear most satisfactory; but it may be presumed that a mean of the six deductions, or 38° 33′ 15″, may be within one mile of the truth.

Maranham does not appear in the table in the Connaissance des Tems; but in that of Professor Lax the town of St. Luiz Maranham is placed in 44° 05′ 00″ W., on the authority of the Hydrographer of the Admiralty. This longitude however can scarcely be regarded as otherwise than very erroneous; the town itself is small, and the cathedral, to which the lunar observations at Bahia, Maranham, and Trinidad, have severally assigned with little variation the longitude of 44° 21′ 28″.5, is situated nearly in the middle of the town.

The longitude assigned to Port Spain in Trinidad in the Connaissance des Tems is 61° 38′ 00″ W. on chronometric deduction; and in Professor Lax's table is the same longitude, referring to the Connaissance des Tems as authority; the present determination is 61° 35′ 00″, referred to the meridian of the Protestant church, which nearly divides the town.

The Connaissance des Tems places Port Royal, Jamaica, in 76° 45′ 15″ W., from astronomical observations; Professor Lax's table in 76° 52′ 30″ W. from the Hydrographic office ; and the lunar observations at Trinidad and Jamaica in 76° 54′ 00″ W. The result of the 270 distances observed on the spot, being 76° 53′ 15″, is, perhaps, in this instance, a more satisfactory determination, than the mean between the lunars of Trinidad and Jamaica.

The advantage to navigation, of the very accurate determination of the longitude of places of frequent resort, consists in the means which it provides for the regulation of chronometers. In making a port, it is comparatively of little consequence that its longitude should be known nearer than to a few miles ; but since the use of chronometers has become so general, (and much has navigation benefited by their introduction, even more perhaps in the time that is saved, than in the dangers that are avoided,) it is a great desideratum to furnish a ready means, at the different ports which ships are accustomed to visit, of obtaining a correct comparison with Greenwich time. For that purpose, however, it is necessary that the longitude should have been *accurately* determined; because if an error exists, its amount will be charged against the previous going of the chronometer, and will occasion the assignment of an erroneous rate in continuance ; which may be productive of far more inconvenience than the original error itself, if the succeeding voyage should be of much longer duration than the preceding one was. The comparison of chronometers with Greenwich time, by means of established geographical positions, is so much more convenient and certain, than by celestial observation on board, that the full value of chronometers to navigation will not be derived, until the small number of stations which have been as yet determined with sufficient care for that purpose, are extensively increased; when lunar observations at sea will only be resorted to as a check in long passages, or on the approach to land. A revision of the geographical position of those stations, which are most frequently visited, would be a very important service, and is well worthy of accomplishment under the direction of the Board of Longitude of the first maritime nation. The revised tables should contain an additional column to those in the tables at present esteemed as of the best authority, for the purpose of specifying the spot to which the geographical position refers; without such specification, it is quite superfluous to insert the data, as is now done, to seconds of space. The spots should also be selected, as far as might be possible, with reference to their conveniency of access, with instruments, from vessels in the harbour.

Longitude of Columbia College, New York.

By 70 lunar distances, 40 of Pollux east, and 30 of Aldebaran west of the moon, the longitude of Columbia College appeared 74° 03′ 27″ West. The result alone is stated, because the observations are not considered as sufficiently entitled to confidence, to justify their publication in detail. The transition from a residence of several months within the tropics, to the severity of a New York winter, was too sudden for the requisite attention to ensure accuracy in night observation.

Longitude of Hammerfest.

The practical as well as theoretical merits of the reflecting circle on Mr. Dollond's construction having been thus manifested by extensive trial, it was conceived, that in order to derive the full benefit from the application of the principle of repetition, the diameter, and consequently the weight, of the circle would admit of reduction. Accordingly, on the return of the Pheasant to England, and whilst the Griper was fitting, Mr. Dollond consented to receive back the ten-inch circle, and to make in exchange one of six inches diameter only; which was completed a day or two before the Griper sailed. Whilst at Hammerfest, I observed with it about the usual number of lunar distances, but not having leisure to compute the results on the spot, the observations were put by until the passage between Hammerfest and Spitzbergen, when, in the course of reduction, I was surprised by finding discordances much beyond the ordinary occurrence. On a careful examination of the circle, their cause was traced to a connexion which had been established between the horizon-glass and the collar of the telescope, by a part of the frame-work to which the skreens of the index glass were attached. In consequence of this connexion, the pressure of the face against the eye tube of the telescope, which assists in steadying the instrument during the observation of lunar distances, deranged the verticality of the horizon glass; but as, on the pressure being removed, it instantly returned to its adjustment, the occasional derangement had escaped notice. The lunar observations at Hammerfest having been thus vitiated, the longitude of Mr. Crowe's establishment at Fugleness has been deduced by five chronometers of Messrs. Parkinson and Frodsham, dependant on the provisional longitude of the pendulum station at Spitzbergen, (11° 40 30″ East,) and employing a mean between the observed rates at Hammerfest and Spitzbergen, for the intermediate period of fourteen days from the 22nd of June to the 6th of July; this longitude is 23° 45′ 45″ East, the different chronometers varying from 23° 44′ 30″ to 23° 46′ 30″.

The six-inch circle was not employed subsequently in lunar distances; but the latitudes recorded in pages 323 and 324 were observed with it, being sufficiently steadied by the handle for the observation of altitudes. Mr. Dollond has recently made a circle on the same principle of construction for Mr. Renwick of New York, of eight inches in diameter, which weighs four pounds, and has the same telescope and glasses as the circle of ten inches. The eight-inch circle appears, and will probably prove, the most eligible of the three in respect to size.

LONGITUDE OF THE PENDULUM STATION AT SPITZBERGEN.

I. BY THE TRANSITS OF THE MOON AND REGULUS, OBSERVED WITH No. 649 ON THE 10th OF JULY, THE APPARENT MERIDIAN ALTITUDES OF THE MOON AND STAR BEING RESPECTIVELY 22° 26′ AND 23° 01′.

	TIMES OF TRANSIT BY THE CHRONOMETER, No. 649.					Mean by the Chronometer.
	1st Wire.	2d Wire.	Meridian Wire.	4th Wire.	5th Wire.	
	M. S.	M. S.	H. M. S.	M. S.	M. S.	H. M. S.
Moon's Western Limb	27 36.8	28 04.4	1 28 32.4	29 00.4	29 28.4	1 28 32.47
Regulus	02 30.8	02 57.2	2 03 24	03 50.8	04 17.6	2 03 24.07

DATE.	TRANSITS.	649 Slow of Mean Time by the Star's Transit.	Gaining per Diem.	Sidereal Interval between the Transits.	Moon's AR. at the Transit of her Limb.	Corresponding App. Time.		Difference of Meridians.
						Spitzbergen.	Greenwich.	
		H. M. S.	S.	H. M. S.	° ′ ″	H. M. S.	H. M. S.	H. M. S.
July 10	Moon's Western Limb, and Regulus.	0 44 48.7	0.7	0 34 57.3	141 17 07.45	2 28 30.74	1 21 49.82	0 46 40.92 E.

II. BY THE SOLAR ECLIPSE.

The termination of the solar eclipse, which took place on the 8th of July, was observed in front of the pendulum house, by Mr. Henry Foster and myself: by Mr. Foster with an achromatic telescope, made by Mr. Dollond, of 2 feet 6 inches focal distance, and $2\frac{3}{4}$ inch aperture, with a power of 51; and by me, with the telescope attached to the repeating circle, of 6 inches focal distance, and one inch aperture, with a power of forty; the eye in both instances being protected by a deep red glass.

The chronometers employed were No. 423 by Mr. Foster, and No. 649 by me; the corrections of both to mean time being derived, as follows, from the transits and zenith distances observed with No. 649, recorded in pages 150 and 158.

		m. s.		m. s.
July 7th	at the Sun's transit 649 slow	44 51.27 ;	whence at 7 A.M., on the 8th	44 50.49
,,	at the transit of η Ursæ . . .	44 50.67 ;	,, ,,	44 50.17
,,	,, Arcturus .	44 50.52 ;	,, ,,	44 50.03
,,	,, γ Draconis .	44 50.03 ;	,, ,,	44 49.7
,,	,, α Lyræ . . .	44 50.45 ;	,, ,,	44 50.15
8	,, η Ursæ . . .	44 49.7 ;	,, ,,	44 50.2
7 P.M., by the Repeating Circle		44 51.32 ;	,, ,,	44 49.96
8 A.M.		44 49.14 ;		
			Mean . . Slow	44 50.11
No. 423, Fast of 649 by comparisons made before and after the end of the Eclipse				3 15.3
			No. 423 . . Slow	41 34.8

		h. m. s.		h. m. s.	
Termination of the eclipse	by Mr. Foster	at 7 11 10.8	by 423, or at	7 52 45.61	A.M. mean time.
	by Capt. Sabine	at 7 07 56.4	by 649, or at	7 52 46.51	A.M. mean time.

Mr. Foster's observation may be regarded as preferable to a mean of both, on account of the superiority of the telescope with which he observed.

The longitude which this observation would assign for the observatory at Spitzbergen has been computed by Mr. Foster on the suppositions, that the moon's place is correctly given in the tables for that day,—that the ellipticity of the earth is $\frac{1}{306}$th,—and that the observation was not affected by irradiation or by the inflection of light: the result, under such circumstances, would be 11° 37′ 58″ 5, East.

Mr. Foster has permitted me to insert the following memorandum of a series of distances of the sun east of the moon, and of a second series of the sun west of the moon, observed by him at the same spot as the solar eclipse and lunar transit were observed; their mean result appears to confirm the accuracy of the deduction from the moon's transit, rather than that from the eclipse, according to the computation.

III. By Lunar Distances observed by Mr. Foster.

At the Observatory on the inner Norway Island, the following Lunar Observations were taken with a sextant of eight inches radius, made by Dollond; the highest power was applied to the telescope, and the same red-coloured screen glass was used in both Lunations.

July 3, P.M. 1823. Barometer 29.78. Thermometer 38.8 Sun East of Moon.

Apparent Time at the Place of Observation.			Observed Distance between Sun and Moon's nearest Limbs.			Error of Sextant.	True Distance of centres computed by the direct method.			Apparent Time at Greenwich interpolated by second differences.			Longitude of the Place of Observation.			
H.	M.	S.	°	′	″	″	°	′	″	H.	M.	S.	°	′	″	
4	43	13.2	63	38	41	+ 10	63	49	32	3	55	50.2	11	50	45	E.
4	48	9.2	63	35	38	”	63	46	36	4	1	16.8	11	43	6	
4	49	42	63	34	45	”	63	45	47	4	2	47.6	11	43	36	
4	51	16.5	63	33	57	”	63	45	3	4	4	9.5	11	46	45	
4	52	46.5	63	33	8	”	63	44	19	4	5	31.1	11	48	51	
4	54	17.7	63	32	15	”	63	43	29	4	7	3.9	11	48	27	
4	55	57.7	63	31	12	”	63	42	29	4	8	55.2	11	45	37	
4	57	49	63	30	7	”	63	41	25	4	10	54.4	11	43	39	
4	59	37	63	29	10	”	63	40	32	4	12	32.7	11	46	4	
5	1	20.7	63	28	13	”	63	39	38	4	14	13.3	11	46	51	
5	2	59.2	63	27	20	”	63	38	46	4	15	49.5	11	47	25	
5	4	36	63	26	23	”	63	37	54	4	17	26.4	11	47	24	
5	6	3.2	63	25	38	”	63	37	12	4	18	43.9	11	50	4	
5	7	28.7	63	24	40	”	63	36	14	4	20	31.6	11	44	16	
5	8	55.5	63	23	43	”	63	35	22	4	22	8.5	11	41	45	
5	9	40.5	63	23	10	”	63	34	48	4	23	11.1	11	37	21	

Longitude of Place of Observation by Lunars ☉ East of ☽ = 11 45 44.7 E.

July 11, P.M., 1823, Barometer 30.02, Thermometer 41.5, Sun West of Moon.					
Apparent Time at the Place of Observation.	Observed Distance between Sun and Moon's nearest Limbs.	Error of Sextant.	True Distance between the centres computed by the direct method.	Apparent Time at Greenwich interpolated by second differences.	Longitude of the Place of Observation.
H. M. S.	° ′ ″	″	° ′ ″	H. M. S.	° ′ ″
3 59 32.6	46 36 35	+ 10	46 48 25	3 12 52.7	11 39 59 E.
4 1 6.3	46 37 30	"	46 49 23	3 14 36.9	11 37 21
4 2 35.8	46 38 20	"	46 50 16	3 16 12	11 35 57
4 4 0.4	46 39 8	"	46 51 8	3 17 45.4	11 33 45
4 7 0.5	46 40 42	"	46 52 47	3 20 43	11 34 22
4 8 41.9	46 41 28	"	46 53 38	3 22 14.4	11 36 52
4 10 19.5	46 42 15	"	46 54 28	3 23 44	11 38 52
4 11 48.6	46 43 5	"	46 55 23	3 25 19.2	11 37 21
4 13 5.1	46 43 50	"	46 56 9	3 26 45.9	11 34 48
4 16 31.2	46 45 28	"	46 57 55	3 29 55.2	11 38 55
4 18 11.5	46 46 18	"	46 58 48	3 31 30.7	11 40 12
4 19 50.2	46 47 17	"	46 59 51	3 33 23.6	11 36 39
4 21 32.9	46 48 13	"	47 0 50	3 35 9.4	11 35 52
4 23 14.7	46 49 12	"	47 1 54	3 37 4.3	11 32 36
4 24 26.4	46 49 53	"	47 2 36	3 38 19.7	11 31 40
4 26 32.5	46 50 58	"	47 3 46	3 40 25.2	11 31 49
Longitude of Place of Observation by Lunars ☉ West of ☽ = 11 36 3.7 E.					

		° ′ ″
Longitude of the Observatory, Sun East of Moon.	=	11 45 44.7 E*.
Ditto ditto, Sun West of Moon	=	11 36 3.7 E*.
Mean Longitude by Lunar Distances	=	11 40 54.2 E.

RECAPITULATION.

	° ′ ″
By the Transits of the Moon and Regulus	11 40 13.8 E.
By Lunar Distances	11 40 54.2
By the Termination of the Solar Eclipse . . . . . . , ...	11 37 58.5

Longitude inferred . . . . ° ′ ″ 11 40 30 E.

---

* The difference which Mr. Foster found in the longitude derived from his Eastern and Western distances, amounting to 39 seconds of time, or to about 20 seconds of angular distance, may be adduced in illustration of the errors to which I have alluded, as occurring in sextants even of the best makers, and not compensated by the most careful ascertainment of the index correction at the zero end of the arc. It will be seen by the following memorandum, that nearly the same amount of error was found to obtain in the angles of altitude, measured with the same instrument, and read from nearly the same part of the arc.

DATE.	Sun's meridian double Altitude L.L.	Index Correction.	Barom.	Therm.	Latitude deduced.	
	° ′ ″	″	IN.	°	° ′ ″	° ′ ″
July 5 Midnight.	24 56 10	+10	30.00	38.5	79 50 13.2	79 49 59.9 N.
„ 7 Noon . .	65 13 00	+10	29.82	39.5	79 49 45.6	

The mean latitude differs only 2 seconds from the results with the repeating circle, in page 321; but the particular observations are 27.6 seconds apart, making an error uncompensated by the index correction, which was very carefully and repeatedly examined, of about 14 seconds in each angle, if divided equally between them. The error shewn by the distances and altitudes is of the same description; and indicates that the angles read between the 40th and 60th degrees of the arc, and corrected for the index error observed at the zero end, were in excess of the truth from 10 to 14 seconds.

LONGITUDE OF THE PENDULUM STATION IN GREENLAND.

I. By Lunar Transits.

DATE.	STARS.	TRANSITS BY THE CHRONOMETER 423.					Mean by the Chronometer.
		1st Wire.	2d Wire.	Meridian Wire.	4th Wire.	5th Wire.	
		M. S	M. S.	H. M. S.	M. S.	M. S.	H. M. S.
Aug. 21	α Aquilæ	07 17.6	07 44	11 08 10.4	08 36.8	09 03.2	11 08 10.4
,, ,,	Moon's E. Limb .	34 43.2	35 10.4	13 35 37.6	36 05.2	36 32.2	13 35 37.73
,, 22	α Cygni	56 14.8	56 51.2	11 57 28	58 04.8	58 41.6	11 57 28.07
,, ,,	Moon's E. Limb .	16 53.6	17 21.2	14 17 48.4	18 15.6	18 42.8	14 17 48.33
,, 23	Arcturus	25 53.2	26 21.2	5 26 49.2	27 17.2	27 44.8	5 26 49.13
,, ,,	Moon's E. Limb .	59 17.2	59 44	15 00 11.2	00 38.4	01 05.2	15 00 11.2
,, 24	α Andromedæ . . .	12 07.2	12 36.8	15 13 06.4	13 36	14 05.6	15 13 06.4
,, ,,	Moon's E. Limb .	42 56.8	43 24	15 43 51.2	44 18.4	44 45.6	15 43 51.2
,, 25	Moon's E. Limb .	28 54.8	29 22.8	16 29 50.8	30 18.4	30 46	16 29 50.6
,, ,,	α Arietis.	05 59.6	06 28	17 06 56.4	07 24.4	07 52.8	17 06 56.27
, 26	Moon's E. Limb .	18 13.6	18 42	17 19 10.4	19 38.8	20 07.2	17 19 10.4
,, ,,	Aldebaran . . .	30 23.2	30 50.4	19 31 17.6	31 44.8	32 12	19 31 17.6
,, 27	Moon's E. Limb .	11 34.4	12 03 6	18 12 32.8	13 06	13 31.2	18 12 32.8
,, ,,	Aldebaran . . .	26 34.4	27 01.6	19 27 28.8	27 56	28 23.2	19 27 28.8
,, 28	Moon's E. Limb SP.	39 50.4	40 20.4	6 40 50	41 20	41 50.4	6 40 50.2
,, ,,	Pollux SP. . . .	32 47.6	33 17.2	10 33 47.2	34 16.8	34 46.4	10 35 47.07
,, ,,	Moon's E. Limb .	09 04	09 34	19 10 04	10 34	11 04	19 10 04
,, ,,	Aldebaran . . .	22 44.8	23 12	19 23 39.2	24 06.4	24 33.6	19 23 39.2
,, 29	Moon's E. Limb SP.	39 09.6	39 40	7 40 10.4	40 40.8	41 01.2	7 40 10.4
,, ,,	Pollux SP. . . .	28 57.6	29 27.6	10 29 57.2	30 27.2	30 56.8	10 29 57.27
,, ,,	Moon's E. Limb .	09 52	10 22	20 10 52	11 22.4	11 52.4	20 10 52
,, ,,	α Orionis	38 32.8	38 59.2	20 39 25.6	39 52	40 18.4	20 39 25.6
,, 30	Moon's E. Limb SP.	40 56.8	41 27.2	8 41 57.6	42 28	42 58.4	8 41 57.6
,, ,,	Pollux SP. . . .	25 08.8	25 38.4	10 26 08	26 38	27 07.6	10 26 08.13

DATE.	TRANSITS.	Interval Sidereal.	423 Fast of Mean Time by the Star's Transit.	423's Daily Gain. Page 167.	Moon's AR. at the Transit of her Limb.	Corresponding Apparent Time. Greenland.	Greenwich.	Difference of Meridians.
1823.		H. M. S.	H. M. S.	S.	° ′ ″	H. M. S.	H. M. S.	H. M. S.
Aug. 21	Moon and α Aquilæ . .	2 27 50.86	1 23 46.6	6.61	332 15 41.23	12 08 55.63	13 24 09.2	1 15 13.6
„ 22	Moon and α Cygni . .	2 20 42.69	1 23 54.35	6.34	343 47 21.47	12 51 13.1	14 06 37.3	1 15 24.2
„ 23	Moon and Arcturus . .	9 34 53.87	1 23 57.53	6.0	355 22 34.2	13 33 46.74	14 49 03.94	1 15 17.2
„ 24	Moon and α Andromedæ.	0 30 49.72	1 24 05.89	5.96	7 16 40.76	14 17 36.82	15 33 01.4	1 15 24.6
„ 25	Moon and α Arietis . .	0 37 11.54	1 24 13.61	8.0	19 45 13.76	15 03 45.48	16 18 58.7	1 15 13.2
„ 26	Moon and Aldebaran .	2 12 28.2	1 24 22.33	7.72	33 03 43.67	15 53 14.13	17 08 31.63	1 15 17.5
„ 27	Moon and Aldebaran .	1 15 07.93	1 24 29.41	7.2	47 23 12.9	16 46 46.63	18 02 08.3	1 15 21.7
„ 28	Moon and Pollux SP. .	8 53 34.12	1 24 35.2	6.3	54 56 50.52	5 15 07.8	6 30 27.3	1 15 19.5
„ „	Moon and Aldebaran	0 13 37.37	1 24 35.69	6.2	62 45 18.05	17 44 29.16	18 59 45.7	1 15 16.5
„ 29	Moon and Pollux SP.	2 50 13.94	1 24 41.28	6.96	70 46 27.6	6 14 39.87	7 29 59.3	1 15 19.4
„ „	Moon and α Orionis .	0 28 37.93	1 24 42.54	6.67	78 56 57.85	18 45 28.9	20 00 47.6	1 15 18.7
„ 30	Moon and Pollux SP. .	1 44 27.16	1 24 48.19	6.7	87 12 58.69	7 16 38.36	8 31 57.8	1 15 19.4
						MEAN		1 15 18.8

II. By Lunar Distances.

The Distances were observed with a Sextant, of which the Index Correction was — 1′ 20″. The Moon's Altitudes were observed at convenient intervals with a Mercurial Horizon; those of the Sun were computed.

DATE.	Time by the Chron. 423.	423 Fast of Mean Time.	No. of Observations.	Barom.	Therm.	Observed Distance of Limbs.	Apparent Altitudes.		True Distances.	Difference of Meridians.	PHENOMENA.
							Moon's Centre.	Sun's Centre.			
1823.	H. M. S.	H. M. S.				Corrected for Index. — 1′ 20″	° ′ ″	° ′ ″	° ′ ″		
Aug. 28	10 10 02.92	1 24 31	10	30.03	39.5	100 39 04	29 54 48	19 52 34	100 47 32.8	1 15 12.98	Moon from Sun East.
" "	10 25 54.5	1 24 31.2	10	30.03	39.5	100 31 14.9	28 59 18	20 40 48	100 39 07.1	1 15 22.5	" "
" "	10 58 00.72	1 24 31.4	10	30.03	39.5	100 15 16.3	27 05 51	22 09 10	100 22 08.3	1 15 26.18	" "
" "	11 12 46.2	1 24 31.5	10	30 03	39.5	100 07 50	26 11 48	22 44 59	100 14 20.6	1 15 29.5	" "
" "	11 35 20.36	1 24 31.6	10	30.03	42	99 56 23.8	24 47 40	23 33 41	100 02 26.3	1 15 29.64	" "
" 29	10 37 07.2	1 24 38	10	30.15	37	87 31 32.9	33 55 03	20 52 57	87 45 52.1	1 15 22.3	" "
" "	10 54 16.24	1 24 38.2	10	30.15	37	87 23 02.8	32 59 23	21 39 14	87 36 37.9	1 15 21.86	" "
" "	11 13 52	1 24 38.4	10	30.15	38	87 13 18.5	31 52 23	22 27 08	87 26 08.7	1 15 13.7	" "
" "	11 35 36.68	1 24 38.5	10	30.15	38	87 02 22.5	30 34 56	23 13 22	87 14 28.5	1 15 08.23	" "
" "	0 07 05.4	1 24 38.7	10	30.15	38	86 46 07.2	28 39 04	24 06 55	86 57 26.8	1 15 14.3	" "
" "	0 33 44.48	1 24 39.8	10	30.15	39	86 32 08.7	26 58 35	24 38 54	86 42 54.6	1 15 28.03	" "
						Mean of 110 Distances . .			. . .	1 15 20.84	= 18° 50′ 12″.6 W. Longitude.

RECAPITULATION.

	° ′ ″	° ′ ″
Difference of Meridians, by 12 Lunar Transits	= 1 15 18.8	= 18 49 42
Difference of Meridians, by 110 Lunar Distances	= 1 15 20.84	= 18 50 12.6
Mean	1 15 19.82	= 18 49 57.4
		Or, 18 50.00 W.

Whence the Longitude of the Observatory on the Inner Pendulum Island, serving as the first Meridian of Captain Clavering's Chart of the East Coast of Greenland from 72° to 76° N. Latitude, is 18° 50′ 00″ W. of Greenwich, and the Latitude 74° 32′ 19″ N.

LONGITUDE OF THE PENDULUM STATION AT DRONTHEIM.

I. By Lunar Transits.

DATE.	STARS.	TIMES OF TRANSIT BY THE CHRONOMETER 649.					Mean by the Chronometer.	REMARKS.
		1st Wire.	2d Wire.	Meridian Wire.	4th Wire.	5th Wire.		
1823.		M. S.	M. S.	H. M. S.	M. S.	M. S.	H. M. S.	
Oct. 15	2 α Capricorni . .	53 58.8	54 25.6	5 54 52	55 18.4	55 45.2	5 54 52	
" "	α Cygni	20 40.8	21 17.6	6 21 54	22 30.4	23 07.2	6 21 54	
" "	Moon's Western Limb	58 49.2	59 16.4	7 59 43.6	00 10.8	00 38	7 59 43.6	
" 16	α Pegasi	37 18.8	37 45.6	8 38 12.8	38 39.6	39 06.8	8 38 12.73	
" "	Moon's Western Limb	41 19.6	41 46.4	8 42 13.2	42 40	43 06.8	8 42 13.2	
" 17	2 α Capricorni . .	46 07.6	46 34.8	5 47 01.6	47 28.4	47 55.2	5 47 01.53	
" "	α Cygni	12 50	13 26.8	6 14 03.6	14 40	15 16.8	6 14 03.47	
" "	α Cephei . . .	51 00.4	51 55.6	6 52 51.6	53 46.8	54 42.8	6 52 51.47	
" "	α Aquarii . . .	34 18.8	34 45.2	7 35 11.6	35 37.6	36 03.6	7 35 11.4	
" "	α Pegasi	33 22.4	33 49.6	8 34 16.4	34 43.6	35 10.4	8 34 16.47	
	Moon's Western Limb	24 38	25 05.2	9 25 32	25 59.2	26 26	9 25 32.07	
" 19	α Pegasi . . .	25 31.6	25 58.8	8 26 25.6	26 52.8	27 19.6	8 26 25.67	
" "	Moon's Western Limb	58 01.6	58 29.2	10 58 57.2	59 24.8	59 52.4	10 58 57.07	Moon less than one hour past the full.
" "	Moon's Eastern Limb	. . .	. . .	11 01 12.8	01 40.4	. . .	11 01 12.8	Observations indifferent from Clouds.
" 23	Capella	16 15.6	16 52.8	14 17 30.4	18 08	18 45.2	14 17 30.4	
" "	Rigel	19 02	19 28.4	14 19 54.8	20 21.6	20 48	14 19 54.93	
" "	β Tauri	27 58	28 27.6	14 28 57.2	29 26.8	29 56.8	14 28 57.27	
" "	Moon's Eastern Limb	49 32	50 02	14 50 32	51 01.6	51 31.6	14 50 31.87	

DEVIATION of the TRANSIT INSTRUMENT from the MERIDIAN, at DRONTHEIM, as shewn by the Interval between the Transits of Stars, differing considerably in Declination, but having nearly the same right Ascension.

DATE.	STARS.	DIFFERENCES.		Solar Interval between the Transits.	Chronometer's Daily Rate.	True Sidereal Interval.	Southern Star too late.	Deviation.
		In R.A.	In Decl.					
1823.		M. S.	° ′	M. S.	S.	M. S.	S.	
Oct. 15	2 α Capri. & α Cephei.	27 08.85	57 44.5	27 02	G. 1.4	27 06.41	2.44	4.47
„ 17	2 α Capri. & α Cygni.	27 08.83	57 44.5	27 01.94	Mean Time.	27 06.37	2.46	4.98
„ „	2 α Capri. & α Cephei.	1 06 05.72	74 55	1 05 49.94	Mean Time.	1 06 00.75	4.97	5.29
„ „	α Cephei. & α Aquarii.	42 22.43	63 00	42 19.93	Mean Time.	42 26.89	4.46	5.28
„ 23	Capella and Rigel.	2 22.33	54 13	2 24 53	Mean Time.	2 24.93	2.6	4.93
Mean Deviation of the Transit Instrument to the West, when pointed to the Southern Horizon . .								4.99

DATE.	TRANSITS.	649 Slow of Mean Time by the Star's Transit.	649's Rate.	Correction for Deviation.	Sidereal Interval.	Moon's AR. at the Transit of her Limb.	Corresponding Apparent Time. Drontheim.	Corresponding Apparent Time. Greenwich.	Difference of Meridians.
1823.		M. S.	S.	S.	H. M. S.	° ′ ″	H. M. S.	H. M. S.	M. S.
Oct. 15	Moon's W. Limb, and α Cygni .	39 41.08	G. 1.4	– 2.54	1 38 03.04	333 37 37.03	8 53 28.93	8 12 00	41 28.9 E.
„ 16	Moon's W. Limb, and α Pegasi.	39 39.84	Mean Time.	– 0.67	0 04 00.45	345 15 42.33	9 36 10.7	8 54 44.7	41 26
„ 17	Moon's W. Limb, and α Pegasi.	39 40.33	Mean Time.	– 0.45	0 51 23.57	357 06 43.74	10 19 42.8	9 38 13	41 29.8
„ 19	Moon's W. Limb, and α Pegasi.	39 39.21	Mean Time.	. . .	2 32 56.45	22 30 47.84	11 53 30.3	11 11 59.16	41 31.14
„ „	Moon's E. Limb, and α Pegasi .	39 39.21	Mean Time.	. . .	2 35 12.55	22 31 59.56	11 55 46.04	11 14 10.95	41 35.09
„ 23	Moon's E. Limb, and β Tauri .	39 31.83	G. 1.4	– 0.15	0 21 37.97	83 54 18.8	15 45 37.03	15 04 02	41 35.03
	Mean by Transits { of the Moon's Western Limb 41′ 29″; of the Moon's Eastern Limb 41′ 35″ }								41 32 E.

II. By Lunar Distances.

The Distances were observed by a Sextant, of which the Index Correction was – 1′ 18″; the Altitudes were calculated.

DATE.	Time by the Chron. 649.	649 on Mean Time.	No. of Observations.	Barom.	Therm.	Observed Distance corrected for Index.	Apparent Altitudes. Moon.	Apparent Altitudes. Star.	True Distances.	Difference of Meridians.	PHENOMENA.
1823.	H. M. S.	M. S.		IN.	°		° ′ ″	° ′ ″	° ′ ″	M. S.	
Oct. 17	9 55 44	39 40.3	10	29.52	36	69 23 04.5	29 23 47	26 27 50	68 52 25.1	41 26.1	Moon from Aldebaran E. of her.
„ 23	12 34 57.22	39 32	10	30.15	41	48 48 08	46 09 54	46 53 07.5	48 12 54.5	41 37.9	Moon from α Arietis W. of her.
	Mean by Lunar Distances . . .									41 32	

RECAPITULATION.

		M. S.
By Lunar Transits	difference of Meridians	41 32
By Lunar Distances	„ „	41 32
	Mean	41 32 = 10° 23′ 00″ East Longitude.

For the purpose of referring the latitude and longitude of the pendulum station at Mr. Wenzel's house to the Cathedral at Drontheim, a base of 684 feet was measured with a Gunter's chain, in an intermediate plain on the bank of the river, and a trigonometrical operation accomplished with a repeating circle. The bearing of the cathedral from the observatory was S. 85° 27′ E., and the distance 5010 feet, or 0.82 of a geographical mile; whence the difference of latitude is deduced 3″.6 S. and of longitude 1° 50′ E., making the Cathedral in N. latitude 63° 25′ 50″.4, and in E. longitude 10° 24′ 50″. The geographical position assigned in the Connaissance des Tems to Drontheim generally, is 63° 25′ 50″ N. latitude, and 10° 23′ 25″ E. longitude; but as no specification is made of the spot, to which the position, thus given to seconds, refers, no more particular comparison can be made with it.

MEMOIR OF A CHART OF THE EAST COAST OF GREENLAND, BETWEEN THE LATITUDES OF 72° AND 76°.

The longitude of the Lands contained in this chart are referred to, and rest on the longitude of the observatory on the inner pendulum island, considered as a first meridian; the difference of meridians between the observatory and Greenwich having been ascertained by the observations in page 411.

The chart may be divided into three sections, in respect to the mode in which the survey of each section was accomplished. The first comprises from the latitude of 76° to that of 74° 30′; the second, from 74° 30′ to 73° 40′; and the third from 73° 40′ to 72°.

The coast of the main land to the north of 74° 30′, and of the islands, with the exception of Shannon Island, has been delineated by astronomical bearings from two hills, one on the outermost, and the other on the innermost of the Pendulum Islands, aided by the view and bearings from a hill on Shannon Island.

The geographical distance between the two stations on the Pendulum Islands, being 9,010 miles, and their position relatively to each other and to the observatory was determined by a trigonometrical operation, in which the repeating circle was used in obtaining the value of the angles, and the extent of the levelled base was 2019.6 feet.

The positions of the Bluff headland to the north of Roseneath Inlet, and of the islands named from their appearance, Ailsa, and the Haystack, have been laid down by the intersection of bearings; the island placed in latitude 76° was seen only from the station on the Inner Pendulum Island, and was decidedly more distant than the other lands.

The position of the south eastern extremity of Shannon Island was determined by observations on the spot; the eastern side of the Island was coasted by the ship; the north-western side was viewed from the hill at the north-east extremity; and the southern side from the hills on the Pendulum Islands, and from Cape Philip Broke; the island itself consists of low land, with five eminences, which at a short distance resemble a group of islands; the whole is so low as not to have been discovered from the Griper, until she had passed to the northward of Cape Desbrowe.

The character of the main land, which throughout the chart is composed of rocks of the order usually termed the trap formation, is lofty, bold, and precipitous, with summits frequently tabled, and rising abruptly from the sea to 3 and 4000 feet of elevation. The most northern point to which the land was seen continuous, was by Capt. Clavering and myself, from the eminence at the northern extremity of Shannon Island, and in a direction N. 26° W. true; the coast was also visible from the same spot, in a direction N. 20° W., being seen indistinctly through clouds, but still high and bold, and apparently trending a few degrees to the westward of north; the Ice Horizon to the eastward of N. 20° W., was clear and uninterrupted.

Cape Desbrowe, which rises from the sea at a slope scarcely admitting of ascent, to a tabled summit nearly 3000 feet in height, is the land earliest seen, and most conspicuous in approaching the coast in the 74° parallel; the expectation, that it might prove the north-eastern extremity of Greenland, was not destroyed, until the Griper had passed to the northward of its parallel.

The middle section of the chart, from 74° 30′ to 73° 40′, or more exactly from Cape Wynn to Cape James, was surveyed by Captain Clavering with the boats of the Griper, in an expedition undertaken for that purpose, between the 19th and 30th of August, whilst the pendulum experiments were in progress; the track of the boats is marked by a faint dotted line on the chart.

The longitudes of this portion of the coast were determined by Capt. Clavering, with the chronometer No. 649, corrected to the mean time of the observatory; it is justly due to the makers of that excellent watch, as well as proper in regard to the accuracy of the longitudes deduced by it, to notice, that on the return of the boats after an absence of eleven days, during which period the chronometer was unavoidably exposed to very many circumstances and incidents unfavourable to its steady going, its error on the observatory time, after the allowance of its previous and ordinary rate, was less than two seconds.

The large bay, or fiord, the recesses of which were explored by Capt. Clavering, is without doubt the inlet of Gael Hamkes, the situation of which has been shifted at the caprice of modern chart-makers, between the latitudes of 73° and 75°, but which appears in its proper latitude of 74° in the charts of the period when the discovery was recent. In the oldest of these with which I am acquainted*, entitled "De Carten van Noorwegen, Finmarken, Laplandt, Spitzbergen, Jan Meyen, Eylandt, Yslandt, als mede Hitland," engraved at Amsterdam by Pieter Goos, in 1666, being only twelve years subsequent to the voyage of Gael Hamkes, the "Landt door Gael Hamkes, opgedaen in't Jaer 1654," forms an inlet corresponding so well, both in latitude and general outline, with the one to which the name of the old Dutch navigator has been in consequence preserved, that no doubt of their identity can be entertained.

The shores of the fiord are in general lofty, and the water deep; there is a good and well protected anchorage, on the northern side, above the spot marked in the chart as the summer residence of Esquimaux. The fiord was entirely clear of ice; nor was the progress of the boats

* In the possession of James Smith, Esq., of Jordan Hill, near Glascow.

embarrassed by the formation of bay ice, which took place to a very considerable thickness, in more exposed situations and in shallower water, during the hours in which the sun was low*.

The third section, from Cape James to the southward, was delineated during the progress of the Griper down the coast, in the track which is shewn in the chart. The exact position of the Cape Broer Ruys of the

* The following experiments on the amount of the cooling influence of radiation, are illustrative of the reason why bay ice did not form in the fiord, whilst in the open sea and in the roadstead in which the Griper was anchored, the surface of the water was covered for several hours in each day with a coating sometimes an inch and half thick :—

"August 25th. Prepared a circular piece of black wool 2 inches diameter and flattened, which weighed, when dry, very exactly 8 grains; at half past nine P.M., when the sun was obscured by the hills on the northern side of the island, and the bay ice began to form in the roadstead, I placed the piece of wool on a grass plat, and a thermometer with its bulb also covered with black wool, by its side. A similar thermometer freely suspended in the air, three feet from the ground, and protected from the effects of radiation by a linen cloth stretched horizontally two feet above it, shewed 30 degrees, and was itself dewed at that temperature, probably from the air at that height being chilled by its proximity to the surface of the ground, which was cooling so rapidly by radiation; the wooled thermometer on the grass fell in a few minutes to 20 degrees; as did also a black wooled register thermometer placed in the focus of a polished metallic mirror: after 4½ hours' exposure, the sheltered thermometer was at 29°.5; the thermometer on the grass 20°, and the register had been at 19°; the wool had increased in weight 3 grains, and on being carefully dried, recovered exactly its original weight; being again replaced on the grass plat, it gained rather less than a grain in the one hours' exposure which preceded the re-appearance of the sun on the eastern side of the hills.

"August 28th. The experiments of yesterday were repeated, with the difference, that the wool was allowed to remain on the grass during the whole period of the sun's obscuration, between 6 and 7 hours, when it was found to have gained 5½ grains, weighing at the close 13½ grains. The sheltered thermometer was 29°, the wooled thermometer on the grass 20°, and the register in the mirror 19°.

"August 29th. Experiments repeated; the thermometers shewed respectively 30°, 21°, and 19°.—The weather was always clear and calm."

It appears, therefore, that the surface of the water, in all situations of fair exposure to the heavens, was subject, when the sun's rays were shaded, to a cooling influence which would have sunk a thermometer to 20°; but that in the fiord, where the radiation into space was greatly reduced by reason of the elevated and nearly perpendicular banks, the temperature to which the surface water was exposed, was probably very little less at any time, than that of the air with which it was in contact.

old charts, was determined by observation on the Cape itself; that of the most prominent points, and the general outline of the great bay comprised between Capes Broer Ruys and Parry, by astronomical bearings from the former cape, and from an ascertained station on the ice, in the neighbourhood of Cape Parry. The Bontekoe Island of the Dutch charts was recognised by the correct relative position in which it is placed in them to Cape Broer Ruys, and to the coast generally.

The names inserted in previous charts have been preserved wherever it was possible to recognise with confidence the spot to which they were designed to belong; they are distinguished by a line beneath. The names not so distinguished, which have been assigned by Capt. Clavering to a few of the most prominent features of the land, will probably be sufficient for every practical purpose, on a coast so little visited. I have to acknowledge his kind attention, in having attached my name, in his manuscript chart, to the remarkable cape on the outer pendulum island, and in his having subsequently changed it, at my request, to Cape Desbrowe *.

So far as the experience of a single season may have influence, the voyage of the Griper does not encourage a hope, that the immediate neighbourhood of the land, in the latitudes of the chart, may prove a successful fishing ground; on one occasion only were whales seen, being five or six in number, near the northern entrance of Gael Hamkes Bay; otherwise, as far as its navigation is concerned, it is a coast remarkably free from dangers, and in which the nature of the soundings is invariably indicated by the character of the land.

It is probable that the east coast of Greenland is rarely, if ever, accessible directly from the westward, in a higher latitude than that in

* So named after the late Edward Desbrowe, Esq., M.P. for Windsor, and Vice Chamberlain to her late Majesty, Queen Charlotte; to which gentleman I was indebted for my entrance into the army.

which the barrier of ice was crossed by the Griper, namely, a few miles to the northward of 74 degrees; it was Captain Clavering's wish to have crossed it higher if possible, and he accordingly entered the ice in 77° 30′ on the 28th of July, its eastern boundary being found in that parallel, in 2° West longitude; his western progress was however almost immediately impeded by an unbroken field, the Eastern and continuous side of which was coasted for sixty miles, until it conducted again into the open sea, nearly in the same longitude in which the ice had been first entered, and a degree further to the southward. In latitude 76° the boundary of the ice receded considerably to the westward, and on the 2nd of August Capt. Clavering again entered it in 75° 30′, and in 8° West, and proceeded through sailing ice in a S.W. direction, along the margin of fields in which no lane was visible, to the latitude of 74° 05′ and longitude of 15° West, where the first practicable breach of continuity presented itself, by which a passage to the land was ultimately effected. The character of the field ice was heavier than that which occupies the middle of Davis's Strait, and Baffin's Sea in the early part of the season of navigation; but was not so heavy as the field ice in the Polar sea, in the neighbourhood of the North Georgian Islands. The barrier of fields, which in 77 degrees must have been about 200 miles across, was reduced to 60 miles in the latitude of 74°, and required five days of much exertion to cross; an attempt which would have been scarcely prudent in a ship less admirably strengthened and equipped than the Griper.

The circumstance of principal geographical interest, the knowledge of which was obtained by the Griper's visit to East Greenland, was, the non-existence of the current which has been stated to prevail, if not throughout the year, at least constantly in the summer season, and to carry the overflowing waters and the ice of the Polar Sea, with great velocity, down the coast of Greenland to the southward; on this current much stress has been laid in the recent discussions on the probability of a north-west passage. The desire which was felt to ascertain the velocity, extent, and depth of the supposed current, as well as the fear of being carried by it much to the southward of the latitude in which we were desirous of making the land, induced the comparison of observations with the reckoning, which in the Griper was kept with the utmost care, on every possible occasion, and could not have failed to have discovered a current of the tenth part of the velocity attributed to the one in question, even if an additional means of comparing the ship's daily position had not been furnished by the land, which was in sight from the first day on which she entered amongst the fields of ice. But neither in crossing the barrier, nor on any subsequent occasion in the seven weeks during which she remained on the coast, could the slightest indication of a southerly current be perceived, either by her own experience, or by the ice viewed from day to day from the summits of hills of 2 and 3000 feet of elevation. The only observable motion of the great body of the ice was to press occasionally upon the land, an effect which was probably caused by the influence of the wind or sea on its eastern or outward side; at other times, and especially when an off-shore wind prevailed, the ice retired, so as to leave a channel between the land and ice of one or two leagues in breadth; but there was no general set whatsoever to the southward; and it was quite remarkable to observe, at the close of the season, how much less disintegration the fields had undergone, than ice which has been itself in motion, or which has been subject, as in Davis's Strait, to the destructive agency of icebergs.

The fact, however, of the absence of current at, and within, the barrier, and the stationary condition of the fields which formed it during the season of 1823, is not irreconcilable with the occasional existence of a southerly current amongst the broken fields and packed ice on the eastern boundary, and in the vicinity of the open sea. Geographical investigation has traced the cause of currents generally, to the drift of the superficial waters of the ocean from the impulse of the wind; the agency of this cause is peculiarly operative and conspicuous within the Arctic Circle, where not only the surface of the sea is exposed to the action of the wind, but also a seventh portion of each of the masses of ice which float upon its surface. The drift of the loose ice in the direction of the wind commences immediately a breeze springs up, and if followed by a calm, continues for some time after the exciting cause has subsided; the motion of the six-sevenths of the ice which are below the surface must communicate an impulse, greater both in velocity and in depth, than the action of the wind upon the surface of the water alone would produce; it is not surprising, therefore, that currents in the direction, and in accompaniment with wind, should be more particularly experienced in seas encumbered with ice, than in the open ocean; and as easterly and north-easterly winds are prevalent in the summer months, in what may be called the Greenland and Spitzbergen sea, it cannot be doubted that a current in a southwest direction along the margin of the main body of the ice must frequently prevail. It is not impossible also that the main body of the ice itself in the parallels of 74°, 75°, and 76°, being certainly not fixed, may not be so stationary in all seasons as it was in 1823, which must certainly be regarded as a remarkable season in respect to the weather, the serenity of which was scarcely interrupted by more than a gentle breeze, from the beginning of August to the middle of September. The wind, however, rarely blows with strength over an extent of Ice, and it must be a north wind of both strength and continuance, to set in motion such extensive masses, the surfaces of which present to its action little more than an unbroken plain. The causes which have been thus noticed are, however, very distinct from that to which the existence of a constant current down the coast of Greenland has been attributed, and which would move the largest floating masses equally with the smallest; namely, the overflowing of the Polar Ocean; of such a current no trace was discernible.

The extraordinary distances at which land has been stated to have been seen in the Arctic Circle, rendered it an object of some curiosity, to ascertain correctly the distance at which the coast of Greenland should become first visible, for which observation the circumstances proved particularly favourable: the weather was remarkably clear when the Griper entered the ice on the 2nd of August, and continued uninterruptedly so, until long after the coast was seen; it was certainly at no time visible on the 2nd or 3rd of August, during which days the Griper neared the land from 180 miles to 60, by navigating amongst the sailing ice; and it was not until the forenoon of the 4th, that even the appearance of land was recognised; when distinctly made out, Cape Desbrowe was, both by observation of the ship's place, and by the angle subtended by stations subsequently ascertained, about 55 miles distant, which, for land of nearly 3000 feet of elevation, was by no means an extraordinary occurrence; nor was the mode of its appearance in any respect unusual, being first recognised, but not without doubt, by the most experienced eyes, and the outline becoming gradually more and more distinct as the ship made progress, until all on board admitted its reality. A still more favourable opportunity of investigating the visibility of distant land from the level of the ocean, presented itself during the progress of the experiments on the Pendulum Islands; on ascending the hills which rose to the northward of the observatory, the nearly insulated hill which forms Cape Broer Ruys, was seen in the horizon clear of the general line of the coast, and was then supposed to be an island; its bearing from the observatory being ascertained, a telescope was stationed accordingly, and the appearance of the supposed island was looked for both by the naked eye and with the telescope, at almost all hours of the twenty-four, in the following twelve days; twice during that period, it was decidedly seen, but much distorted by refraction, and was once besides doubtfully visible; the weather during the whole time was perfectly clear, and the sky almost without a cloud. The distance in a direct line was about 73 miles. The particulars of a Barometrical measurement of the height of Cape Broer Ruys, which was ascended by Captain Clavering and myself for the purpose, have been unfortunately mislaid, but according to a rough calculation from the particulars made on the spot, it was between 2700 and 3000 feet; the observatory was 30 feet above the sea.

It is by no means designed in this statement to imply a doubt of the extraordinary facts that have been alluded to, but merely to show that it was not from inattention that we are not able to corroborate them by a similar experience.

HYDROGRAPHICAL NOTICES.

Previously to my leaving England in 1821, I had had the great advantage of much conversation with Major Rennell, on the subject of the currents in the Northern and Southern Atlantic Oceans, and of having my attention directed by him to those points in particular, concerning their velocity, limits, and temperature, on which further inquiries might conduce to the advancement of hydrographical knowledge.

The method of ascertaining the existence, direction, and velocity of a current, where land is not in sight, and a ship cannot be rendered stationary by anchorage, is to compare her position at intervals of sufficient length, (generally of 24 hours,) by observation and by reckoning. By the former is learnt her real change of geographical position in the interval; by the latter, the course and distance that she has gone through the water; should the position by the reckoning not agree with the position by the observation, the difference (presuming both to be correct) is the indication and measure of current.

To determine a ship's position from day to day by observation, or rather her relative position on one day to the preceding, has become, since the introduction of chronometers, a matter of very simple accomplishment, and capable of much precision. It is far otherwise with the reckoning, however, when more is sought by it than such a rough approximation as may serve the ordinary purposes of navigation; it must in fact, require the most assiduous and unremitting attention, as well as considerable nautical experience and judgment, to estimate correctly the continually varying effects of the winds and sea, on a body that is also continually varying the measure of her exposure to their influence. It may be in the power of an individual in a vessel, to obtain, by his own exertions alone, that portion of the materials towards the evidence of currents, which depends on her real change of position; but the completion of the evidence by a sufficiently correct reckoning must be the result of an interest participated in by all the executive officers of a ship; or by the establishment of such habits of accuracy, under the authority of her com mander, as are not of usual practise, because they are not necessary for the general purposes of navigation; the employment of chronometers, by which the position of a ship is ascertained and a fresh departure taken on every day that the sun shines, has superseded the necessity of that vigilant and scrupulous regard, which the older navigators paid to all the details of the reckoning, on which alone they had to depend; and has tended to substitute general habits of loose and vague estimation, for the considerate and well-practised judgment with which allowances were formerly made for the incidental circumstances of steerage, leeway, making and shortening sail, &c., &c., on a due attention to which the accuracy of a reckoning so materially depends.

In ships of war especially, the reckoning is further embarrassed by a difficulty, less obvious but not less generally operative, by which, if not properly provided against, the knowledge of the true course which the ship has made is necessarily rendered very uncertain; it arises from the usual practice of directing the course by the binnacle compasses, which are two in number for the convenience of the helmsmen, and being placed one on the larboard and the other on the starboard side of the midship, with a space between them of greater or less extent according to the size of the vessel, can scarcely fail, and are in fact generally influenced differently by the ship's iron; and being subject to different *systems* of attraction, the compasses not only disagree, but their disagreement varies according to the direction of the ship's head, the amount of the dip of the needle, and the force of terrestrial magnetism. It is customary always to steer by the weather compass, and thus each is liable to become in its turn the directing compass for periods of more or less duration, and the corrections of the courses for the disturbing influence of the ship's iron become so various and complicated, as to render the deduction of a correct reckoning practically unattainable. For example, the binnacle compasses of the Iphigenia, on her passage from England to Madeira, were observed to differ from each other half a point in one direction when on south-westerly courses, and less than half a point in the opposite direction when on easterly courses, the indications of the compasses having crossed each other, and agreed at some intermediate point; it was requisite, therefore, that the correction to be allowed on every course by each of the two compasses should be ascertained, and that the compass by which each course was directed should be specially recorded, in order that the true course should be known.

The most obvious mode of preventing so much inconvenience and trouble, as well as the more correct practice, is to direct and note the ship's course by one compass only, stationed permanently in some convenient situation without reference to the helmsmen, and to use the binnacle compasses solely to steer by, on the point which may be noticed at the time to agree with the magnetic course of the standard compass; and by employing an azimuth compass for the latter purpose, the advantage is gained of enabling the variation to be observed directly with the compass by which the course is governed, and thus of avoiding intermediate comparisons, in which time is occupied, and errors frequently introduced. This arrangement of a standard compass was adopted by Captain Clavering in the Pheasant, and subsequently in the Griper, and was found to answer its purpose perfectly, and to be attended with no practical inconvenience whatsoever.

Although from the causes above noticed no satisfactory investigation of the direction or velocity of currents could be made in the Iphigenia, in her passage from England to the coast of Africa, a remarkable and very interesting evidence was obtained, by observations on the temperature of the sea, of the accidental presence in that year of the water of the gulf stream, in longitudes much to the eastward of its ordinary extension. The Iphigenia sailed from Plymouth on the 4th of January, after an almost continuous succession of very heavy westerly and south-westerly gales, by which she had been repeatedly driven back and detained in the ports of the channel; the following memorandum exhibits her position at noon on each day of her subsequent voyage from Plymouth to Madeira, and from thence to the Cape Verd Islands, the temperature of the air in the shade and to windward, and that of the surface of the sea; it also exhibits in comparison, the ordinary temperature of the ocean at that season, in the respective parallels, which Major Rennell has been so kind as to permit me to insert on his authority, as an approximation founded on his extensive inquiries; the last column shews the excess or defect in the temperature observed in the Iphigenia's passage.

DATE.		Latitude N.	Longitude W.	Air.	Surface Water.		Excess or Defect.
					Observed.	Usual.	
	1822.			°	°	°	
Plymouth to Madeira.	Jan. 5	47 30	7 30	47	49	50	− 1
	„ 6	44 20	9 30	52.5	55.7	52.5	+ 3.2
	„ 7	41 22	11 37	54	58.2	54	+ 4.2
	„ 8	38 54	13 20	54.2	61.7	55.7	+ 6
	„ 9	No Ob	servation.	56	63	58	+ 5
	„ 10	33 40	15 30	60.7	64	60	+ 4
Madeira to the Cape Verds.	„ 19	26 00	17 50	66	65.5	67	− 1.5
	„ 20	24 30	18 50	68	67	68.4	− 1.4
	„ 21	23 06	20 00	69	69	69 5	− 0.5
	„ 22	21 02	21 27	69.5	69.5	71.2	− 1.7
	„ 23	19 20	23 00	70.6	70.2	71.6	− 1.4

It is seen by the preceding memorandum, that in the passage from Plymouth to Madeira, the Iphigenia found the temperature of the sea, between the parallels of $44\frac{1}{3}°$ and $33\frac{2}{3}°$ several degrees warmer than its usual temperature in the same season; namely, 3°.2 in $44\frac{1}{3}°$, increasing to 6° in 39°, and again diminishing to 4° in $33\frac{2}{3}°$; whilst at the same period, the general temperature of the ocean in the adjoining parallels, both to the northward, and to the southward, even as far as the Cape Verd Islands in $19\frac{2}{3}°$, was colder by a degree and upwards than the usual average. The evidence of many careful observers at different seasons and in different years, whose observations have been collected and compared by Major Rennell, has satisfactorily shewn, that the water of the Gulf stream, distinguished by the high temperature which it brings from its origin in the Gulf of Mexico, is not usually found to extend to the eastward of the Azores. Vessels navigating the ocean between the Azores and the continent of Europe, find at all seasons a temperature progressively increasing as they approach the sun; the absolute amount varies according to the season, the maximum in summer being about 14 degrees warmer than the maximum in winter; but the progression in respect to latitude is regular, and is nearly the same in winter as in summer, being an increase of 3° of Fahrenheit for every 5° of Latitude. It is further observed, that the ordinary condition of the temperature, in the part of the ocean under notice, is little subject to disturbance, and that in any particular parallel and season, the limits of variation in different years are usually very small; after westerly winds of much strength or continuance, the sea in all the parallels is rather colder than the average temperature, on account of the increased velocity communicated to the general set of the waters of the North-eastern Atlantic towards the southward. To the heavy westerly gales which had prevailed almost without intermission in the last fortnight in November, and during the whole of December, may therefore be attributed the colder temperatures observed in the latitude of 47½°, and in those between 26° and 19⅓°.

If doubt could exist in regard to the higher temperatures between 44⅓° and 33⅔°, being a consequence of the extension in that year of the Gulf stream in the direction of its general course, it might be removed by a circumstance well deserving of notice, namely, that the greatest excess above the natural temperature of the ocean was found in or about the latitude of 39°, being the parallel where the middle of the stream, indicated by the warmest water, would arrive, by continuing to flow to the eastward of the Azores, in the prolongation of the great circle in which it is known to reach the mid Atlantic.

One previous and similar instance is on record, in which the water of the Gulf stream was traced by its temperature quite across the Atlantic to the coasts of Europe ; this was by Dr. Franklin, in a passage from the United States to France, in November, 1776 *. The latter part of his voyage, *i.e.*, from the meridian of 35° to the Bay of Biscay, was performed with little deviation in the latitude of 45° ; in this run, exceeding 1200 miles, in a parallel of which the usual temperature, towards the close of November, is about 55½°, he found 63° in the longitude of 35° W., diminishing to 60° in the Bay of Biscay ; and 61° in 10° West longitude, near the same spot where the Iphigenia found 55°.7 on the 6th of January, being about five weeks later in the season. At this spot then, where the Iphigenia crossed Dr. Franklin's track, the temperature in November, 1776, was 5½°, and in January, 1822, 3°.2 above the ordinary temperature of the season.

There can be little hesitation in attributing the unusual extension of the stream in particular years to its greater initial velocity, occasioned by a more than ordinary difference in the levels of the Gulf of Mexico and of the Atlantic; it has been computed by Major Rennell, from the

* Franklin's works, 8vo., London, 1806, Vol. II., pages 200, 201.

known velocity of the stream at various points of its course, that in the summer months, when its rapidity is greatest, the water requires about eleven weeks to run from the outlet of the Gulf of Mexico to the Azores, being about 3000 geographical miles; and he has further supposed, in the case of the water of which the temperature was examined by Dr. Franklin, that perhaps not less than three months were occupied in addition by its passage to the coasts of Europe, being altogether a course exceeding 4000 geographical miles. On this supposition, the water of the latter end of November, 1776, may have quitted the Gulf of Mexico, with a temperature of 83°, in June; and that of January, 1822, towards the end of July, with nearly the same temperature. The summer months, particularly July and August, are those of the greatest initial velocity of the stream, because it is the period when the level of the Caribbean sea and Gulf of Mexico is most deranged.

It is not difficult to imagine that the space between the Azores and the coasts of the old continent, being traversed by the stream, slowly as it must be, at a much colder season in the instance observed by the Iphigenia than in that by Dr. Franklin, its temperature may have been cooled thereby to a nearer approximation to the natural temperature of the ocean in the former than in the latter case; and that the difference between the excess of 5°.5 in November, and of 3°.2 in January, may be thus accounted for.

If the explanation of the apparently very unusual facts observed by Dr. Franklin in 1776, and by the Iphigenia in 1822, be correct, how highly curious is the connexion thus traced between a more than ordinary strength of the winds within the tropics in the summer, occasioning the derangement of the level of the Mexican and Caribbean seas, and the high temperature of the sea between the British channel and Madeira, in the following winter.

Nor is the probable meteorological influence undeserving of attention, of so considerable an increase in the temperature of the surface water over an extent of ocean exceeding 600 miles in latitude and 1000 in longitude, situated so importantly in relation to the western parts of Europe. It is at least a remarkable coincidence, that in November and December, 1821, and in January, 1822, the state of the weather was so unusual in the southern parts of Great Britain and in France, as to have excited general observation; in the meteorological journals of the period it is characterized as "most extraordinarily hot, damp, stormy, and oppressive;" it is stated "that an unusual quantity of rain fell both in November and December, but particularly in the latter;" that, "the gales from the W. and S.W. were almost without intermission," and that in December, the mercury in the barometer was lower than it had been known for 35 years before*.

* The following description of this very remarkable winter is extracted from Mr. Daniell's Essay on the climate of London, (*Meteorological Essays*, London, 1823,—pages 297 and 298) and becomes highly curious when viewed in connexion with the unusual temperature of the ocean in the direction from which the principal winds proceeded.

"November 1821 differed from the mean, and from both the preceding years, in a very extraordinary way. The average temperature was 5° above the usual amount; and although its dryness was in excess," [the relative dryness, in consequence of the increased temperature] "the quantity of rain exceeded the mean quantity by one half. The barometer on the whole was not below the mean. All the low lands were flooded, and the sowing of wheat very much interrupted by the wet.

"In December the quantity of rain was very nearly double its usual amount. The barometer averaged considerably below the mean, and descended lower than had been known for thirty-five years. Its range was from 30.27 inches to 28.12 inches. The temperature was still high for the season, and the weather continued, as in the last month, in an uninterrupted course of wind and rain; the former often approaching to an hurricane, and the latter inundating all the low grounds. The water-sodden state of the soil, in many parts, prevented wheat sowing, or fallowing the land at the regular season. The mild temperature pushed forward all the early sown wheats to an height and luxuriance scarcely ever before witnessed. The grass, and every green production increased in an equal proportion.

"January, 1822. This most extraordinary season still continued above the mean temperature, but the rain, as if exhausted in the preceding month, fell much below the usual quantity in this. There was not one day on which the frost lasted during the twenty-four hours.

"Serious apprehensions were entertained lest the wheats, drawn up as they had been by

On leaving the Cape Verd Islands, the Iphigenia proceeded to make the continent of Africa at Cape Verd. The distance between the Cape and the Islands is about 400 miles, both being in the same parallel of latitude. This passage afforded an interesting opportunity of observing, on the approach to land, the influence of its vicinity on the temperature of the sea. The general temperature of the surface in that parallel and at that season may be considered 71°.7, the observations made at sunrise, noon, and sunset, in the first 350 miles of the passage, varying from 71° to 72°.4: but at sunrise on the 31st of January, being then at the distance of 26 miles west of Cape Verd, with no land as yet in sight, the surface water had lowered to 69°.6. On approaching nearer it progressively diminished, until at one mile from the shore, it had fallen as low as 64 degrees, and continued from 64 to 65 degrees, between Cape Manoel and Goree. Cape Verd is situated nearly at equal distances, exceeding 70 miles, from the mouths of the Senegal and Gambia, the one being to the north and the other to the south. It is probable that the water of both these rivers is always colder at their entrance into the sea, than the ocean temperature of the parallel; that of the Gambia certainly was so at that season, but it was not so cold as the sea in the vicinity of Cape Verd, as on approaching the entrance of the Gambia, the temperature of the surface rose to 67°.5, and varied in the river itself at different hours from 66° to 67°.5; and at the depth of 36 feet, being within six feet of the bottom, a self registering thermometer indicated at high water less than a degree colder than the surface. The coast in the neighbourhood of Cape Verd is every where low and sandy, and is covered with trees to the water's

warm and moist weather, without the slightest check from frost, should be exhausted by excessive vegetation, and ultimately be more productive in straw than corn.

"The month of February, still five degrees above the mean temperature, ended a winter which has never been paralleled."

It would not be difficult to trace in detail, each of the effects described in the preceding extract, to the cause which has been thus placed in connexion with them.

edge. Such indeed is the general character of the shores of western Africa, with the exception of Cape Sierra Leone; but at no other part of the coast was the diminution of the temperature of the water, on approaching the land, so great, as in the instance which has been mentioned. Between the Gambia and Sierra Leone are a succession of rivers, originating in land of less elevation than the Senegal and Gambia, and much exceeding them in the temperature of the waters which they convey into the ocean; in the mid-channel of the Rio Grande, at a few miles from its mouth, the surface was never less than 74°, and occasionally as high as 77.°5, and at the depth of thirty or forty feet was less than a degree colder than the surface. At the entrance of the River Noonez the surface water was 77°.5, and at that of the Rokelle 80°. To the south of the Rokelle, and from thence to the extremity of the Gulf of Guinea, the coast is swept by a current of considerable rapidity, which renders the cooling effect of the land less apparent; but in the bays of the coast, where the current sweeps from point to point, and leaves still water in the inside, a difference is commonly found amounting to three and four degrees*

* The passage from the Cape Verd Islands to Cape Verd and the Gambia afforded a not less interesting opportunity of observing the difference in the hygrometrical state of the atmosphere at sea and in the vicinity of the continent, in the region of the trade winds. We had entered the N. E. Trade in the latitude of 24° N., nine degrees to the Northward of the Cape Verd Islands, and did not lose it until the afternoon of the day on which we quitted the Gambia, the strength declining on the approach to the continent, but the direction continuing unchanged. On the 28th, 29th, and 30th of January, in navigating the first 350 miles of the passage from the Islands to the continent, the air in the shade and to windward varied at different hours of the day from 70.2 to 71.2, and the dew point from 63 to 64.5. At sunrise on the 31st when at twenty-six miles West of Cape Verd, the Dew Point was 61.5, and lowered to 57.5 on nearing the land, the temperature of the air not being sensibly affected. Off the entrance of the Gambia on the 1st of February, and in the river on the 2nd, 3rd, and 4th, the Dew Point was never higher than 51°, and occasionally as low as 48.5, the air over the water and in the shade being generally during the day from 69° to 70°. When about to quit the Gambia on the morning of the 5th of February, we experienced, although in a very slight de-

The following summary account of the direction and force of the currents experienced in each day's navigation, commences with the appointment of the Pheasant to convey the clocks and pendulums from Sierra Leone to the subsequent stations. Captain Clavering entered with much interest into the inquiry, and by his judicious arrangements, and personal superintendence, until habits were established, the reckoning of his ship was rendered little inferior, as an element in the deduction of currents, to the observed difference of latitude and the chronometrical difference of longitude. On leaving England, I had obtained from the Admiralty a supply of the logs invented by Mr. Massey, which being towed at a sufficient distance astern to be clear of the back-water occasioned by a ship's progress, register her way by the revolutions of a spiral acted upon by the water through which it is drawn. The self-registering log was used as a check upon the estimated reckoning, and proved the value and efficacy of the attention paid to the latter, by its being a very rare circumstance to find a difference between them, amounting to a mile, in twenty-four hours. The comparison between the ship's run by observation and by reckoning was usually made by Capt. Clavering from forenoon to forenoon, and by myself from afternoon to afternoon; and the results being each reduced to noon and compared, served for the detec-

gree, the peculiar wind called the Harmattan, of which the season was nearly over; its direction was one or two points to the North of the trade wind, or about N.N.E.; the air during its influence fell to 66.5, and the Dew Point to 37.5; affording a reasonable inference, that in a genuine Harmattan, and before it reaches the sea, the constituent temperature of the vapour may be at least as low as 32°. In the progress to Cape Roxo, on the afternoon of the same day, we lost the Harmattan, and with it the continuance of the trade wind. The sea breeze which followed, raised the temperature of the air to 70°, and of the Dew Point to 61.5.

It appears, therefore, that when the North East wind first comes off the continent of Africa, it contains only 53 parts in 100 of the moisture which would be required for repletion at the existing temperature; that in blowing over the sea its proportion of moisture rapidly augments, until at fifty miles from the land, it has acquired 80 parts in 100; which proportion is not subsequently increased by its passage over 350 additional miles of ocean. In the Harmattan the air contained only 38 parts in 100 of the proportion of moisture required for its repletion.

tion and correction of errors, on either side. The table exhibits the ship's true position at noon on each day; the temperature of the surface water; and the direction and amount of the difference of her position, by observation and by reckoning, from noon to noon. On days when the sun was obscured, the direction of the apparent set is deduced from intervals of 48 hours instead of 24, but the rate is that due to each interval of 24 hours.

DATE.	Latitude.	Longitude.	Temperature of the Surface Water.	Apparent Set in each 24 hours.
	From Cape Mount to Cape Three Points.			
1822.	° ′	° ′	°	
Apr. 15	6 40 N.	11 48 W.	84	
„ 16	Sun obscured.		83	S. 53 E. 32 miles.
„ 17	4 53	9 04	83	
„ 18	4 28	8 18	..	S. 84 E. 24 „
„ 19	4 18	6 36	84.8	N. 79 E. 40 „
„ 20	4 37	3 48	84.5	N. 76 E. 51 „
	From Lagos to St. Thomas.			
May 8	5 22 N.	2 51 E.	83.5	°
„ 9	5 00	2 32	84.5	S. 45 E. 9 miles.
„ 10	4 46	2 49	84	S. 84 E. 17 „
„ 11	3 46	2 57	83.2	S. 24 E. 16 „
„ 12	Sun obscured.		83	S. 82 E. 22 „
„ 13	0 36	5 22	82.8	
„ 14	0 16	6 24	82.8	S. 81 E. 13 „

DATE.	Latitude.	Longitude.	Temperature of the Surface Water.	Apparent Set in each 24 hours.
	From the River Gaboon to Ascension.			
1822.	° ′	° ′	°	
June 15	0 03 N.	7 45 E.	..	
„ 16	0 44 S.	5 50	..	S. 80 W. 29 miles.
„ 17	Sun obscured.		73	West 48.5 „
„ 18	1 00	2 07	74	
„ 19	1 45	0 19	72.5	S. 86 W. 29 „
„ 20	2 34	1 55 W	72.8	N. 88 W. 37 „
„ 21	3 48	4 54	74.5	S. 81 W. 47 „
„ 22	5 10	7 50	77.5	S. 81 W. 32.5 „
„ 23	6 21	10 43	77.5	N. 63 W. 16 „
„ 24	7 27	13 22	78	N. 57 W. 18.25 „
	From Ascension to Bahia.			
July 10	7 57 S.	14 24 W.	. .	
„ 11	9 16	17 00	. .	N. 74 W. 11 miles.
„ 12	10 10	19 45	. .	North 2 „
„ 13	10 35.5	22 25	. .	N. 35 W. 6 „
„ 14	11 05	25 53	. .	West 16 „
„ 15	11 42	29 08	. .	S. 82 W. 14 „
„ 16	12 27	32 51	. .	S. 71 W. 14 „
„ 17	13 05	36 31	. .	N. 79 W. 11 „

DATE.	Latitude.	Longitude.	Temperature of the Surface Water.	Apparent Set in each 24 hours.
	From Bahia to Pernambuco.			
1822.	° ′	° ′	°	
Aug. 8	13 30 S.	38 22 W.	..	
„ 9	Sun obscured.		77.2	N. 69 W. 13 miles.
„ 10	13 48	37 59	77.1	N. 12 W. 2.5 „
„ 11	12 36.5	37 02	77.2	N. 31 W. 14 „
„ 12	11 03.5	36 20	78	N. 33 E. 13 „
„ 13	10 15	35 53.5	78	N. 27 W. 15 „
„ 14	9 33	35 13	78	
	From Pernambuco to Maranham.			
Aug. 15	8 04 S.	34 54 W.	78	North 22 miles.
„ 16	6 15	34 36	78.4	N. 44 W. 62 „
„ 17	3 22	36 45	78.3	N. 70 W. 41 „
„ 18	2 17.5	40 17	77.8	N. 66 W. 13 „
„ 19	1 55	43 06	77.8	
	From Maranham to Trinidad.			
Sept. 8	0 21 N.	45 58 W	79.8	N. 49 W. 48 miles.
„ 9	2 59	48 07	80.8	N. 54 W. 99 „
„ 10	5 18	50 39	81.8	N. 38 W. 68 „
„ 11	7 01	52 38	81.5	N. 41 E. 5 „
„ 12	7 05	53 32	83	S. 47 W. 18 „
„ 13	7 24	54 19	83.3	S. 87 W. 17 „
„ 14	7 43	55 55	84	N. 72 W. 28 „
„ 15	8 12.5	57 22	84	N. 33 W. 48 „
„ 16	9 29	59 30	84	N. 52 W. 57 „
„ 17	8 00	61 00	84	

DATE.	Latitude	Longitude	Temperature of the Surface Water.	Apparent Set in each 24 hours.
	From Trinidad to Jamaica.			
1822.	° ′	° ′	°	°
Oct. 10	10 55	61 56	. . . .	N. 52 W. 49 miles.
„ 11	12 24	63 43	. . 83	N. 53 W. 12 „
„ 12	13 16	65 56	. . 83	N. 79 W. 16 „
„ 13	13 53	67 59	. . 82.8	S. 83 W. 16 „
„ 14	15 02	70 45	. . 82.9	
„ 15	Sun obscured.		. . 83	N. 41 W. 19 „
„ 16	17 50	76 08	. . 83	
	From Havannah to New York.			
Nov. 27	23 09	82 23	. . . .	S. 85 E. 14 miles.
„ 28	23 52	81 42	. . 80.5	N. 31 E. 22.5 „
„ 29	25 20	79 47	. . 80.7	N. 4 W. 70 „
„ 30	28 38	79 32	8 A.M. 80.8 Noon. 80.5 9 P.M. 80.1	N. 17 E. 38 „
Dec. 1	32 02	78 33	8 A.M. 79.2 3 P.M. 80.1 8 P.M. 79.5	
„ 2	Sun obscured.		8 A.M. 78.2 Noon. 78.7 3 P.M. 78	N. 47 E. 44.5 „
„ 3	35 04	74 54	8 A.M. 77.5 Noon. 77.6 6 P.M. 77.3	N. 55 E. 77 „
„ 4	Sun obscured.		8 A.M. 77.5 Noon. 77.5	
„ 5	36 38	72 29	. . 62.4	West 16 „
„ 6	37 00	73 46	. . 60.6	S. 55 W. 10 „
„ 7	37 35	74 33	. . 59.5	S. 5 W. 15 „
„ 8	38 44	74 26	. . . .	S. 45 W. 6 „
„ 9	40 08	74 07	. . .	

REMARKS ON THE PRECEDING SUMMARY.

In the voyage between Cape Mount and Cape Three Points, the Pheasant's progress appears to have been accelerated about 180 miles, by the current, which, during the season when the S. W. winds prevail on that part of the coast of Western Africa, runs with considerable rapidity in the direction of the land, round Cape Palmas to the eastern parts of the Gulf of Guinea. The breadth of this current abreast of Cape Palmas varies with the season, and has been found as much as 180 miles; but, in its subsequent course to the eastward, it enlarges to nearly 300, and occupies the whole space between the land on one side, and the equatorial current running in an opposite direction on the other; the velocity abreast of Cape Palmas and Cape Three Points, and in the vicinity of the land, was in May, about two knots an hour; and further to the eastward, where the Pheasant crossed its breadth from Cape Formosa to St. Thomas, and where its velocity had been much diminished by the dissipation of its waters, it was found still to preserve a general rate of rather less than a mile an hour; and a direction, a few degrees to the southward of east. Between Cape Three Points and Lagos, the observations were suspended in consequence of the greater part of the officers and men being absent in the boats, examining merchant vessels anchored on the coast, and suspected of being engaged in the trade in slaves. The little effect of the current experienced between the 8th and 9th of May, was occasioned by the slack water in the Lagos bight, from which the Pheasant did not re-enter the fair stream until the morning of the 9th. There appears to have been a southerly deflection between the 10th and 11th, for which no very obvious reason presents itself. The general temperature of the stream in the mid-channel in the Gulf of Guinea, in April and May, exceeds 84 degrees, diminishing to 82 and 83, on its southern border, where it is in contact with the colder water of the equatorial current; and occasionally to 79°, and frequently to between 79° and 81°.5, on its northern side, in the proximity of land.

In the passage from the coast of Africa to the Island of Ascension, the Pheasant appears to have entered the equatorial current, almost immediately after her departure from the entrance of the River Gaboon; as she was decidedly under its influence when passing the southern extremity of the Island of St. Thomas. This current is formed by the drift water impelled by the trade winds in the southern Atlantic, (which in the neighbourhood of the continent of Africa are very much southwardly,) towards the eastern part or head of the Gulf of Guinea; where, being opposed by the waters brought to the same spot by the Guinea current, the drift water streams off in the direction of the equator and principally on its southern side; and being continually fed in its western progress by the drift from the S.E., (becoming more and more inclined to the meridian, as the influence of the continent on the regular direction of the trade wind lessens from distance,) the stream pursues its course quite across the Atlantic to the continent of South America, where one portion of it proceeds along the northern coast of Brazil to the Caribbean Sea and Gulf of Mexico, and contributes in part to raise the level of those seas, and thus to lay the foundation of the Gulf Stream.

The Pheasant's voyages from the coast of Africa, successively to Ascension, Bahia, Pernambuco, Maranham, Trinidad, and Jamaica, were performed principally in the current, the origin and progress of which have been thus stated.

The equatorial current is not usually met with so far to the northward, at its commencement on the coast of Africa, as it was found by the Pheasant in the month of June: but it is probable that at the season when the trade winds are strongest, and approach nearest the equator, the drift water may be impelled into a more northern parallel than at other seasons, before the opposition to its direct course becomes so strong, as to occasion it to stream off to the westward. Its more usual northern limit, in the meridian of the Island of St. Thomas, is considered by Major Rennell to be in the second or third degree of south latitude. The direction of the stream was as nearly west as could be inferred from the observations, and its rapidity between the meridians of 7½ East, and 7½ West, averaged forty miles a day. We appear to have passed out of the stream on the 22nd of June in latitude 5°+, S., and longitude 8°+, W., into the drift current from the S.E., which contributes to its supply and to preserve its velocity across the Atlantic; it may be seen that the drift water was pressing on the southern border of the stream with a force of 16 and 18 miles in 24 hours, in a direction oblique to and accelerative of its course.

In the passage between the River Gaboon and Ascension, being a distance of 1400 geographical miles, the Pheasant was aided by the current above 300 miles, in the direction of her course.

In consequence of the southing of the trade wind in the vicinity of the continent of Africa, the water impelled before it, which forms the commencement of the Equatorial Stream, arrives from a more remote southern parallel, and is therefore of a colder temperature than the drift water which successively falls into it from the S.E., impelled more obliquely to the meridian, and consequently arriving from latitudes less distant from the Equator. Thus the temperature of the stream varied from 72.5 to 74°, whilst that of the drift current was 77.5 and 78°. But the more important distinction, both in amount and in utility in navigation, is between the waters of the Equatorial and of the Guinea currents. These exhibit the remarkable phenomenon of parallel streams, in contact with each other, flowing with great velocity, in opposite directions, and having a difference of temperature amounting to ten and twelve degrees. Their courses continue thus parallel to each other and to the land for above a thousand miles; and according as a vessel, wishing to proceed along the coast in either direction, is placed in one or the other current, will her progress be aided from forty to fifty miles a day, or retarded to the same amount: the practical advantage, therefore, derivable from the difference of temperature, in enabling vessels to discriminate at all times in which current they are situated, is as great as it is obvious*

* The occasional advance of the cold water of the Equatorial Current to the Island of St. Thomas, may assist in explaining an apparent peculiarity in the climate of that island, when compared with the climate of the Coast of Western Africa generally. At all the British possessions, from the Gambia in 13° north latitude to the Forts on the Gold Coast, June, July, and August are accounted the unhealthy months; whilst at St. Thomas, on the contrary, they are the most healthy in the year to Europeans, although they are not so to the Negroes, who suffer much from colds and rheumatism during their continuance. It has been seen, that the water of the Equatorial Current is from 10 to 12 degrees colder than that of the Gulf of Guinea, and that its northern border, which at other seasons passes the meridian of St. Thomas at a distance from 120 to 180 miles south of its southern extremity, was found in June in contact, or very nearly so, with the island itself; and it is not improbable, from a consideration of the causes which occasion its advance towards the equator when the sun is in the northern signs, that in July it may extend so far, as even to include the whole island within its limits.

The temperature of the air is known to be immediately dependant on that of the surface water of the sea, and to be influenced nearly to the full extent of any alteration that may take place therein. In crossing the Gulf of Guinea from Cape Formosa to St. Thomas, the air, over the surface of the Guinea current, observed in the shade and to windward, at sunrise, noon, and sun-set, averaged 81°.5, the extremes being 79° and 83°.5; whilst in the passage from the river Gaboon to Ascension, over the Equatorial Current, the air averaged only 74°, the extremes being from 73°.5 to 74°.5; a part of the passage being, moreover, on the very edge of the two currents, and within sight of St. Thomas. The vicinity of the Equatorial Current, therefore, when the sun is in the northern signs, cannot fail materially to influence the temperature of the island, (particularly as the wind is always from the south), and thus to affect its climate. Situated on the equator, St. Thomas has naturally two cold seasons, or winters, in the year, the sun being equally distant in June and in December; but in June, July, and August, is superadded the influence of the surface water of the ocean several degrees colder than in November, December, and January; rendering the months of June, July, and August, pre-eminently the winter of St. Thomas; in which the natives complain of colds and rheumatism, and the health of Europeans is less affected than at other seasons, because the climate is then less dissimilar than usual to their own.

The comparative unhealthiness of Prince's Island to St. Thomas, and of both to Annabona, as the residence of Europeans, has been frequently and particularly noticed by Portuguese authorities, and is universally recognised at Prince's Island and at St. Thomas. It may be a sufficient explanation to remark, that Annabona is always surrounded by the Equatorial Current; Prince's always by the Guinea Current; and that the position of St. Thomas is intermediate,

The voyage from Ascension to Bahia commenced in the continuation of the same drift current from the S.E., in which the latter part of the passage to Ascension was performed; but on the 13th of July, the Pheasant appears to have re-entered the southern border of the equatorial current, in the longitude of 22½ W., and latitude of 10½ S. The evidence of many voyages in different years, the journals of which have been submitted to Major Rennell's examination, have led him to the conclusion, that it is the ordinary course of that stream, to divide into two branches about the twenty-third degree of west longitude, the

and its climate is occasionally influenced by both. In tropical climates a very few degrees of temperature constitute an essential difference in the feelings of the natives, and in the health of Europeans.

The point of deposition varied over the differently-heated surfaces of water, in correspondence with the difference in the temperature of the air; so that, although the quantity of moisture was diminished in the colder air over the Equatorial Current, the proportion of the quantity to that which would have been required for repletion, was as nearly as possible the same as over the Guinea Current, being on the average 84°.5 parts in 100° in both instances. The air, therefore, was equally moist over the Equatorial as over the Guinea Current, although in the one case the weight of vapour in a cubic foot (derived from the averages) was 10 grains, and in the other 7.93 grains only. The cold air incumbent on the Equatorial stream, being borne by the south wind over the surface of the Guinea Current, caused the deposition, which generally obscured the horizon to the north of St. Thomas, during the pendulum observations, as noticed in page 33; and which fell, as we understood, in heavy rain in the offing. The quantity of vapour in the atmosphere over the island being less than that over the nearly surrounding water of the Guinea current (an effect of the high land of which the island consists), no deposition took place on the island itself. The hygrometer indicated the temperature of its superincumbent vapour to be between the extremes of 71° and 74°.5, observed three times a day between the 26th of May and the 12th of June. The range in the Gulf of Guinea was from 76° to 80°

It is worthy of notice to what little distance the colder air, impelled by the constant south wind, attained over the Guinea current, before it became itself heated by the condensation of the vapour of higher constituent temperature. The great bodies of the air and of the vapour over the respective currents, though so dissimilar in temperature, were as little affected by their contiguity, as the surface waters of the currents themselves. By their mutual and opposite action, the air in condensing and thus reducing the temperature of the vapour, and the heat liberated in the condensation of the vapour in raising that of the air, the mixture speedily destroyed the differences; and the effects of the contiguity were thus limited to a very few miles within the border of either stream.

northern portion flowing in a N.W. direction, and diffusing its waters in the basin of the Atlantic, and the southern, which is the largest portion, proceeding in a direction to the southward of west, until it reaches the coast of the continent of South America; where it is again subdivided by the projecting part of the coast between Cape St. Roque and Cape St. Augustin, the northern branch coasting the north of Brazil and Guiana to the West Indies, and the southern branch proceeding down the eastern side of the continent towards Terra del Fuego. The Pheasant's experience corresponded in all respects with this general view. The direction of the southern part of the equatorial stream, into which she entered on the 13th of July, became gradually more and more to the southward of west on approaching the continent; being due west between the longitudes of 22°½ and 26°; S. 82 W. between 26° and 29°; and S. 71 W. between 29° and 33°; and the apparent set between the noons of the 16th and 17th of July is obviously compounded of the influence of the equatorial stream, (then probably become still more southwardly) during the first part of the twenty-four hours, and of the northerly current, during the latter part, which the observations between Bahia and Pernambuco shew to prevail in the vicinity of the coast included between those stations. The Pheasant may therefore be considered to have crossed the whole breadth of the branch of the stream which proceeds to the S.W., by having passed out on its western side between the longitudes of 33° and 36°, and to have ascertained its general velocity to have exceeded half a mile an hour, by the according observations of the 14th, 15th, and 16th of July.

From Pernambuco to Cape St. Roque, the northerly current rapidly accelerated, until in passing the Cape it may be considered that the Pheasant had entered the full stream of the other branch of the equatorial current; namely, of the one which pursues its way along the northern coast of Brazil and Guiana to the West Indies. Between the noons of the

16th and 17th, she was set 44.5 to the north, and 42.5 to the west, making a general effect in the twenty-four hours of N. 44 W., 62 miles; but as she did not round Cape St. Roque until midnight, the course having been altered for that purpose at half-past eleven p.m., it must be understood that the direction of the current was probably more northerly in the first part of the interval, and more westerly in the latter part, than the general effect; and that the velocity may in like manner have been less than the rate of 62 miles to the south of Cape St. Roque, and more than that amount after passing the Cape. The purpose of stopping at Maranham obliged the Pheasant to draw nearer the land on the following day, than would have been expedient, had she been bound direct to the West Indies, and been desirous of preserving the full advantage of the current in her favour; on examination of the tabular results, it will be obvious, that by thus nearing the land, she quitted the full strength of the stream, and that she did not re-enter it again until the day after her departure from Maranham, when it was found to be running with the astonishing rapidity of ninety-nine miles in twenty-four hours. It may also be seen that although in the space comprised between the direct course of the stream from Cape St. Roque to the West Indies, and the coast of Brazil, the velocity progressively diminished on approaching the land, no counter current was found to take place, but the westerly direction was still maintained, though at the reduced rate of less than half a mile an hour, when very near the land. It may be attributed to the rapidity with which the water is thus swept along the shore, that no change is perceptible in its temperature, on approaching a coast which is so remarkably shallow, as to have not more than seventeen fathoms water at thirty-six miles in the offing.

At 10 A.M. on the 10th of September, whilst proceeding in the full strength of the current, exceeding, as already noticed, four knots an hour, a sudden and very great discoloration in the surface water a-head was reported from the mast-head, and from the very rapid progress which the ship was making, was almost immediately afterwards visible from the deck. Her position in 5°.08′ north latitude, and 50°.28′ west longitude, both known by observation, sufficiently apprized us that the discoloured water which we were approaching could be no other than the stream of the river Amazon, preserving its original impulse at a distance of not less than 300 miles from the mouth of the river, and its waters being not yet wholly mingled with those of the ocean of greater specific gravity, over the surface of which it had pursued its course.

We had just time to secure some of the blue water of the ocean for subsequent examination, and to ascertain its temperature, before we crossed the line of its separation from the river water, the division being as distinctly preserved as if they had been different fluids.

The direction of the line of separation was N.W. by N.,,rather northerly; great numbers of gelatinous marine animals, species of the Genus Physalia, were floating on the edge of the river water, and many birds were fishing apparently on both sides of the boundary.

The temperature of the ocean water was 81°.1, and of the river water 81°.8, both within a short distance of the division line; the specific gravity of the former was 1.0262, and of the latter 1.0204, distilled water being unity: the ocean water had also been found 81° at seven A.M. on the same morning. At noon, having advanced considerably within the boundary, so that it was no longer in sight from the ship, the specific gravity of the surface water was 1.0185, and its temperature 81.°8.

Being desirous of ascertaining the depth at which the water of the ocean would be found unmixed with the river water, Dr. Marcet's very simple and practical apparatus was employed to bring up water from fifty fathoms, the specific gravity of which proved 1.0262; the bottle was then sent down a second time to twenty-one fathoms, at which depth the specific gravity was also 1.0262, limiting the depth of any admixture of the fresh water to less than 126 feet. Its superficiality was further evidenced by the colour of the water in the ship's wake, which was much more blue than that of the general surface. The temperature of the water from fifty fathoms was 77°.2, and from twenty-one fathoms, 80°.5; we had no bottom with 105 fathoms.

From noon on the 9th, till 10 A.M. on the 10th, we had found the current of the ocean running with an average velocity of four knots in a direction N. 54° W.; the ship's true course had been very nearly N. 45° W.; the division line of the streams trended about N. 33° W. It was obvious, by the general appearance of the respective surfaces, that the current of the river water was running with considerable rapidity, in a direction inclined to that of the ocean, and nearly coinciding with the line which marked their separation; the ship's course was, therefore, altered a point westerly. During the afternoon of the 10th, and morning of the 11th, the colour and specific gravity of the surface water indicated that we continued in the river stream; but that it was becoming latterly more and more mixed with the sea water. At noon, in latitude 7°.01′, and longitude 52° 38′.5, the specific gravity was 1.0248, temperature 81°.5; and from twenty fathoms, 1.0262. Between noon on the 10th and noon on the 11th the ship was set N. 38° W., sixty-eight miles, or rather less than three miles an hour; which may, therefore, be considered the general direction and rate at which the water of the Amazon was proceeding, at the distance of 300 miles and upwards from its natural banks. The original impulse at its discharge into the ocean is to the eastward of north; so much, therefore, had its course been deflected, by having to sustain the continual pressure of the current of the ocean on its eastern side. As the initial velocity must have greatly exceeded that which it had preserved after a course of 300 miles, and as the force of the current which presses on it is much less in the neighbourhood of the land than it subsequently becomes, it is probable that the deflection may have been scarcely sensible in the early part of the course, but much more rapid latterly than would be due to the whole effect divided by the distance; and that a further deflection of the 16 degrees, which measured the inclination of the streams where the Pheasant crossed the division line, might not require much more distance for its accomplishment; when the course of the streams being parallel, the obstacle to the diffusion of the river water on its eastern side would be removed, and the marked line of the separation of the streams would gradually cease to exist. In the early part of the river's marine course, as it may be termed, and where the force of the current of the ocean is comparatively weak, the greater obliquity of its direction may compensate for its want of force, in enabling it to oppose the diffusion of the river water. On the western side the fresh water is gradually and insensibly lost in that of the sea; at noon on the 12th, the specific gravity of the surface water was 1.0253, in latitude 7°.05', and longitude 53°½.

The effect which the stream of the Amazons produces on the current of the ocean in thus crossing its course, is to accumulate the water brought by the equatorial current, until it streams off with a rapidity which graqually deflects, and ultimately overpowers the obstacle, which opposes its more regular flow; it is to the accumulation from this cause, that the partial velocity of ninety-nine miles in twenty-four hours, much exceeding the average rate of the branch of the Equatorial current between Cape St. Roque and the West Indies, is to be attributed. The southern border of the current is also removed by it to a distance from the land, leaving a space of the ocean, bounded by the river water on the east, the land on the south, and the Equatorial current on the north, which is occupied by irregular streams of various and uncertain strength and direction, as shown by the Pheasant's experience between the 11th and the 14th of September. It is desirable that vessels bound from the Brazils to the

West Indies should, therefore, keep well off the land of Guiana, in order to preserve the strength of the Equatorial current in their favour; whilst others, endeavouring to make a passage along the coast to the eastward, should be especially cautious to keep in the space within the current. The Pheasant re-entered the current about the eighth degree of latitude, and in the fifty-seventh of longitude, and was subsequently indebted to its influence, between two miles and two miles and a half an hour, until her arrival in the Gulf of Paria*

The observations in the passage from Trinidad to Jamaica indicate a general set of the surface water across the Caribbean Sea towards the Gulf of Mexico, averaging about sixteen miles in each twenty-four hours. The northerly inflexion, on approaching Jamaica, was occasioned by the indraft between Cape Tiburon and Point Morant.

From Jamaica to the Havannah the Pheasant was engaged in conducting a convoy, which obliged a suspension of the observations.

In crossing the Caribbean Sea from Trinidad to Jamaica, between the 9th and 17th of October, the temperature of the surface water, observed always at 8 A.M., and occasionally at other hours, was never less than 82°.8, nor more than 83°; between Jamaica and Grand Cayman, on the 10th and 11th of November, the minimum was 83°, and the maximum at 3 P.M., 83°.8; from the Cayman Islands to the entrance into the Gulf of

* In the passage from Maranham to the West Indies, and in crossing the mouth of one of the largest rivers of the globe, the hygrometrical state of the atmosphere was the subject of very frequent and careful observation on each day; no effect of the river, however, on the state of the aqueous vapour was perceptible: the point of deposition varied only between 72°.5 and 74°, and the air between 79° and 82°, the higher temperatures of both taking place when we had arrived abreast of Surinam, and the surface water had increased to 83°. In the Gulf of Paria, where the general temperature of the surface is raised to 84°.5 by the admixture of the heated water from the smaller branches of the Orinoco, the air was further augmented to 84°, and the Point of Deposition to 75°.5. Between Point Galeotta and Port Spain, we crossed the stream of one of the branches of the Orinoco, the temperature of which was 85°.5, and the specific gravity not more than 1.0064; the general surface of the Gulf being 1.0204.

Mexico, between Yucatan and Cuba, and in the open part of the Gulf itself, the surface varied from 82° to 82°.5*; but on approaching Havan-

* The particular observations were as follows, and are accompanied by the temperatures of the air, and of the point of deposition, observed at the same hours.

Between Trinidad and Jamaica.					Port Royal, Jamaica.					Between Jamaica and Havannah.				
Oct.	Time.	Water.	Air.	Point of Depos.	Oct.	Time.	Water.	Air.	Point of Depos.	Nov.	Time.	Water.	Air.	Point of Depos.
11	8 A.M.	83°	83°.2	77°.5	20	10 A.M.	In the offing 83°; in the middle of the harbour 81°.5, at sunrise; 83°.5, at 2 P.M.	83°.7	76°	10	9 A.M.	83°.1	82°	°
„	2½ P.M.	83	83	78.5	22	noon.		83.5	76.5	„	3 P.M.	83.8	83.2	
12	8 A.M.	83	82	76.5	23	7 A.M.		78.8	76	11	8 A.M.	83	81.8	74
13	8 „	82.8	83	77.5	„	2 P.M.		82.5	76	13	8½ „	82.5	80.8	72
14	8 „	82.9	82	78	24	noon.		83	77	14	8½ „	82.2	79.7	72
15	8 „	83	82	rain.	25	9½ A.M.		81.7	75	„	3 P.M.	82	78	rain.
16	8 „	83	83.4	77.5	29	noon.		85.5	78	15	8½ A.M.	80	78.8	72.5
17	8 „	83	82		30	10 A.M.		84.5	78	17	8½ „	82	80.3	74.5
	noon.	....	82	77	31	10 „		83	76	„	2½ P.M.	82.1	80.2	71.5
					Nov. 1	10 „		82.5	75	18	8½ A.M.	80.5	78.9	73
					2	6 „		78	75					
					3	6 „		78.5	76					

The light rain which fell on the afternoon of the 14th of November, in the passage between Jamaica and Havannah, was a precipitation from an height above the earth's surface, as the air near the surface was very far from being replete with moisture at the time. It was produced by the commencement of a wind from the N.E. (the same, I believe, which is called at Havannah, El Norte), which almost instantly lowered the temperature of the air two degrees at the surface, and of course correspondingly in its ascending progression, whilst the dew point and its progression remained unaltered. The height, therefore, at which the temperatures of the air and vapour would coincide (by reason of the difference in their respective ratios of cooling), would at once descend a space equivalent to that required to diminish the temperature of the air two degrees in its ascending progression, and a precipitation would take place throughout that space too copious to be altogether re-dissolved by falling into a warmer atmosphere; and thus some portion of it would reach the surface, forming the light rain which we experienced. It was not, however, of long continuance, the superfluous moisture being disposed of, and the atmosphere speedily adapting itself to the new order of circumstances, by the processes which have been so well pointed out by Mr. Daniell, in his essay on the habitudes of an atmosphere of permanently elastic fluid mixed with aqueous vapour.

I am not able to assign with confidence the cause of the surface water being only 80° on the morning of the 15th; but I suspect that it evidenced the presence of a thread of the current which descends from the northern shores of the Gulf of Mexico along the coast of Florida; and of which a small portion from the western border is sometimes turned to the

nah, on the morning of the 18th, we were apprised by the colder temperature of 80°.5, that during the preceding night we had entered the current, which descends from the northern shores of the Gulf of Mexico along the coast of Florida, and forms the head of the Gulf stream. In the subsequent passage from Havannah to the Straits of Bahama, on the 27th, 28th, and 29th of November, we crossed the narrow sea formed by the northern shore of Cuba and the Florida reefs, in which the waters of the stream are comprised, previously to their discharge into the Atlantic: the surface water in this passage varied from 80°.5 to 80°.7, which may therefore be considered as the initial temperature of the gulf stream towards the end of November. The strait between the Bahama's and the eastern side of Florida, which forms the outlet of the stream, is rather less than 200 miles in length, and from 33 miles at the narrowest part of the water-way, to 50 miles at the widest, in breadth. The Pheasant was at the southern extremity of the strait at noon on the 29th, and at the northern extremity at noon on the 30th, with good observations of the latitude on both days, and with especial care given to the intermediate reckoning. The rate of three miles an hour (or more exactly seventy miles in twenty-four hours) may, therefore, be regarded, with confidence, as the initial velocity of the Gulf stream at that period.

The maximum of its temperature in the strait was 80°.8, and the minimum observed 80°.5; but the Pheasant did not approach the shore on either side, where the surface is known to be colder, by reason of the vicinity of land.

The diminution in the rapidity of the stream on the 1st, 2d, and 3d of December, is the consequence of its expansion after the outfall into the

westward by the northern coast of Cuba on which it impinges, and takes a course towards Cape St. Antonio. The charge of a convoy in a sea so much infested with pirates was incompatible with the measures which would have been necessary to have ascertained more particularly the cause of the decrease in temperature of the surface water.

Atlantic; it is probable, however, that on neither of the three days was the Pheasant in the full strength of the current, being nearer the inner border, where the velocity is checked, and the waters accumulated, by the direction of the coast of America between Charleston and Cape Hatteras; the consequence of the accumulation is seen in the increased rate on the 2d and 3d, in comparison with that on the 1st of December; and in the very remarkable circumstance, that after passing Cape Hatteras, the velocity experienced between the 3d and the 5th of December was actually greater than the initial velocity at the outlet, being 3.2 miles an hour on the average of the forty-eight hours, or seventy-seven miles in each twenty-four hours; and was, doubtless, considerably greater than the average during a part of the time. The accumulation of the water of the stream in the neighbourhood of Cape Hatteras, to such an extent as to occasion it to flow off with even greater rapidity than on its discharge into the ocean from the Gulf of Florida, is a fact which I believe had not been previously observed, but which may be explained by a brief notice of the different states, at different seasons, of the current, and of the ocean through which it pursues its course. In the summer months, the stream issues from the outlet with a velocity nearly one-third greater than at the period of the Pheasant's voyage; its original northerly direction, received from the Bahama channel, is turned considerably to the eastward of north, (about N. 50° E.) by the coasts of Georgia and South Carolina, in which new direction it passes Cape Hatteras, and pursues an unobstructed course, until it impinges upon the St. George's bank to the eastward of Nantucket, by which it is turned still more to the eastward; but as it strikes the bank very obliquely, it is deflected without material accumulation of its water, or increase of velocity. The St. George's bank is the last obstruction that the stream encounters, as it never afterwards approaches land. There is, therefore, no accumulation in the summer months in the neighbourhood of Cape Hatteras; but on the contrary the western border of the stream expands into the great Bay between Cape Hatteras and Nantucket, and occasions a diminution rather than an increase in the velocity at the surface; accordingly it is found that the force originally communicated at the outlet is progressively diminished from above eighty miles in twenty-four hours in the first 180 miles after its discharge into the Atlantic, usually to less than seventy miles when abreast of Cape Hatteras.

On the approach of winter, the disparity in the general level of the Gulf of Mexico and the Atlantic is diminished by the reduction in the level of the Gulf, and the impulse communicated to the stream at its fall into the Atlantic is proportionably lessened. At that season, also, an alteration takes place in the level of the part of the ocean towards which the course of the stream is directed. The heavy autumnal gales from the north and north-east impel before them the superficial waters of the north-western Atlantic into the space comprised between the coast of America and the Gulf stream: this space, which is of considerable width between Cape Race in Newfoundland and the northern border of the stream, narrows towards the westward, and has no outlet; the drift water consequently accumulates, and presses wholly against the northern and western borders of the current, and by raising the usual level of the ocean, prevents the surface water of the stream from reaching the Nantucket and St. George's banks, and opposes the expansion of the western border into the recession of the coast of the continent between Cape Hatteras and Nantucket; the accumulation of the Gulf water is thus occasioned, which streams off to the north east with the augmented velocity experienced by the Pheasant between the 3d and 5th of December. It is probable that the occasional effects thus noticed are very superficial, and that the great body of the water which issues from the Gulf of Florida, and is of considerable depth, is governed, both in direction and velocity, solely by the original impulse, and the banks on which it impinges; but navigation is more immediately concerned with the surface current only.

On the 5th of December, between 10 A.M. and noon, the Pheasant quitted the gulf stream, passing out on its northern side. At 8½ A.M. she was in longitude by observation 72° 25′ W., and in latitude, deduced from the subsequent noon, 36° 14′; the temperature of the surface water was 74°, and of the air 60°.5. At 10 A.M., the temperatures being still the same, the depression of the horizon, observed with a dip sector from the Pheasant's gangway, where the height of the eye was 15 feet 3 inches above the sea, was 4′ 56″.6, being an excess of 1′ 05″.6 above the usual computed and tabular depression. On repeating the observations at noon, it was found that a change of great magnitude had taken place intermediately; the horizon, viewed from the same height, making an angle, on the second occasion, of only 3′ 36″.6 with the horizontal line passing through the eye. As the ship was in both instances very steady, and the horizons perfectly clear, the observations were decided and certain; and the utmost error of which either might be suspected could not be more than 5″. So great an alteration in the refractive quality of the atmosphere led to the immediate suspicion, that the temperature of the surface-water of the sea must also have greatly altered, and that we must have passed from the warm water of the stream into the colder surface of the general ocean. This suspicion was confirmed on trial, the temperature having fallen from 74° at 10 A.M. to 62.°4 at noon, being a difference of 11°.6. As a measure of precaution on such a sudden and great decrease, Captain Clavering immediately sounded, but had no bottom with 120 fathoms: the temperature at 110 fathoms, indicated by a register thermometer attached to the line above the lead, was 51°.5. The distance from the nearest banks noticed in the charts was sixty-five miles.

The northern boundary of the stream, where we had thus quitted it, was between the latitudes of 36°26′ and 36°38′, and in the meridian of 72° 30′W. The surface water on which we entered was in motion to the westward, at the average rate of sixteen miles experienced in the following twenty-four hours, and generally to the west and south-west between the northern side of the stream and the banks on the coast of Maryland. This motion may be more properly characterized as a drift current, occasioned by the prevalence and strength of recent northerly gales, than as a counter-current. In approaching the banks, the surface water at 8 A.M. and at noon on the 7th of December was 59°.5; at 3 P.M. it had fallen to 54°.2, on which we immediately sounded, and found bottom in thirty-three fathoms: on the following morning, in thirty fathoms, the surface was 53°.5, and at 8 A.M. on the 9th in twelve fathoms, but still with no land in sight (being twenty miles off the coast), 49°.5. In the afternoon of the same day, when about two miles distant from Sandy Hook, the water had finally lowered to 45°. Thus in a space of the ocean scarcely exceeding 200 miles in direct distance, we found the heat of the surface progressively diminish from 74° to 45°.

On a general review of the influence of the currents which have been thus particularized, on the Pheasant's progress, in her voyage commencing at Sierra Leone, and terminating at New York, it may be seen, that she was indebted to their aid on the balance of the whole account, and in the direction of her course from port to port, not less than 1600 geographical miles, the whole distance being under 9000 miles; affording a very striking exemplification of the importance of a correct knowledge of the currents of the ocean to persons engaged in its navigation; and consequently of the value of the information, in the acquisition and arrangement of which Major Rennell has passed the later years of his most useful life. The publication of the charts of the currents in the most frequented parts of the ocean, which he has prepared with his accustomed and well-known indefatigable assiduity, and strict adherence to the evidence of facts,—as soon as he shall deem them sufficiently complete for the public guidance,—will be a most important service rendered to practical navigation.

On the Depth at which the Water of the Ocean within the Tropics is found at the temperature of its greatest density.

The greatest density of sea water resulting from its temperature takes place at 42°, or thereabouts: if heated above, or cooled below that amount, it is rendered specifically lighter, and in the natural progression must be found incumbent on water of 42°.

In the existing state of the ocean, the temperature of 42° may be considered as the mean heat of the surface of the sea in a parallel between the latitudes of 65° and 70°; from whence the influence of external causes renders the surface colder towards the pole, and warmer towards the equator, and in both cases specifically lighter, than water of 24°.

In approaching the equator, (or rather, more generally the space within the tropics, to every part of which the sun is periodically vertical,) the warmth of the surface water increases, and the heat penetrates to greater depths; in descending beneath the surface, the temperature progressively decreases and the density augments, until the term of 42° (or thereabouts) is reached; beneath which no further alteration of either takes place, dependent on influences operating on the surface. It was to ascertain the depth at which the term of this progression might be met with in the tropics, that the following experiment was made.

In the Caribbean sea, in latitude 20½ N. longitude, 83½ W., nearly midway between the Cayman Islands and Cape St. Antonio at the west end of Cuba, in the afternoon of the 13th of November, 1822, a

Six's self-registering thermometer, enclosed in an iron cylinder, having holes in the top and bottom to admit the free access of the water, was lowered to a depth exceeding 1000 fathoms. A weight of 75 lbs. was attached to the end of the line, and 11 coils of 113 fathoms each, and 3 fathoms of a 12th coil were veered, making altogether 1246 fathoms. The weather being very favourable, with light airs and little swell, the ship's drift was bodily to leeward, without either head or stern way. The 1246 fathoms were veered in rather more than twenty-five minutes, at the expiration of which time the line was fairly on the ship's qua ter. Under such circumstances, the depth to which the thermometer was sunk must have exceeded a thousand fathoms, as an allowance of 246 fathoms for stray line would be more than ample, if no bight of consequence existed in the rope, which, from the rapidity with which the weight drew out the line, appeared to be the case; 246 fathoms of stray line would be an equivalent to a drift of four-fifths of a mile in twenty-five minutes, whereas that of the ship did not exceed half a mile an hour. The line was hauled in in fifty-three minutes, when it appeared that the thermometer had registered 45°.5, the surface being 83°; whence it may be reasonably inferred that 100 fathoms more line would have sunk the thermometer to 42°, the rate of cooling being on the average of the whole depth about twenty-eight or twenty-nine fathoms to a degree of Fahrenheit; and thus that the sea water would have been found at its maximum of density, dependent on temperature, at about 1200 fathoms below the surface.

The thermometer used in this experiment was made expressly for the purpose to which it was applied; it was of the ordinary construction, except that the top of the tube, in which is contained the index of heat, was hermetically sealed instead of being closed by a cork, as is sometimes the case. I have since sunk the same thermometer in the same apparatus to 650 fathoms, accompanied by a similar thermometer enclosed in a strong iron cylinder without perforations, and of which the top screwed down upon leather, so as to exclude the access of the water to the interior of the cylinder, and thus to prevent any effect which might be supposed to be occasioned in the indications of the free thermometer, by the increased pressure of water at great depths upon its exterior surface. The two thermometers were suffered to remain below above an hour, to allow the air in the inside of the closed cylinder to adjust itself to the temperature of the surrounding water; and on their being drawn up, they were found to have registered precisely the same indication.

A notice of the preceding experiment in the Caribbean Sea was read before the Royal Society in April, 1823, and published in the *Philosophical Transactions* of that year. I have since learned that a similar experiment had been made in the ocean between the tropics, in 1816, by Capt. Wauchope of the Royal Navy, then commanding His Majesty's ship Eurydice; and as the interest and value of each of the results, separately considered, are greatly increased by the very remarkable corroboration which they afford to each other, I have obtained Captain Wauchope's permission to insert the following detailed account of his experiment.

The thermometer used was a common one of Fahrenheit; it was enclosed in the middle of six cases, all of tin except the outer one, which was of wood: each of the cases had valves at the top and bottom opening upwards, so that the valves might remain open whilst descending, but would close on being drawn up through the water; there was also a small spring to the upper valve, which prevented it from opening when once shut. The four inside cases were separated from each other about a quarter of an inch all round, allowing the water to pass freely between them; the fifth case was distant from the one next to it on the inside by about half an inch, which space was filled with tallow. The outer case was of wood an inch in thickness, and separated about an inch from the one next to it by a column of water. The size of the apparatus was two feet high, and ten inches diameter, having a weight of 72lbs fastened to the bottom, and the end of a coil of two inch rope to the top. Of this rope 779 fathoms were veered, then 390 of two and a half inch, and then 266 of three inch, making in all 1435 fathoms veered overboard. A 32lb shot was attached to every 200 fathoms, and the whole was run out twenty-two minutes. It was allowed to remain twelve minutes before the hauling in was commenced, that the whole might have time to sink; and it took an hour and twenty minutes to draw the thermometer to the surface, when it stood at 42°, the surface water being 73°. Latitude 3° 20′ S. and longitude 7° 39′ E.; date, September, 1816* Capt. Wauchope is of opinion that the thermometer must have sunk about 1300 fathoms, provided there was no curve in the rope, as the ship's drift was about five knots in twenty-four hours.

Both experiments, therefore, concur in assigning a depth of 1200 or 1300 fathoms as the term of the augmentation, occasioned by external influences within the tropics, of the temperature of the sea water above that of its greatest density.

In a previous experiment of the same kind, which Captain Wauchope made in latitude 10° N. longitude 25° W., the quantity of rope veered was 966 fathoms, the temperature of the surface water was 80°, and the enclosed thermometer shewed 51°; corresponding to a diminution of temperature averaging one degree in about 29 fathoms, being very nearly the same ratio as appeared by the experiment in the Caribbean Sea.

* From the geographical position, and the temperature of the surface water, Capt. Wauchope was in the equatorial current near its commencement. In his case the surface water was thus accidentally colder than is due to the parallel, whilst in mine it was warmer; the accidents being in both cases the effect of currents.

EXPERIMENTS FOR DETERMINING THE VARIATION IN THE INTENSITY OF TERRESTRIAL MAGNETISM.

The inquiry into the laws which regulate the phenomena of terrestrial magnetism has occupied, successively, the attention of many of the most eminent mathematicians and philosophers of the last and present centuries; whose endeavour it has been to elicit, from the facts collected in different parts of the globe, an experimental law, which, embracing the whole, should represent by calculation each individual observation, within such limits as might reasonably be ascribed to accidental causes of difference.

There are three distinct classes of phenomena, in which the influence of terrestrial magnetism is found by observation to vary in different parts of the globe: these are—1st. The direction which a needle, freely suspended, assumes in relation to the geographical meridian; 2d. the direction, in regard to the horizon; and 3d. the intensity of the attraction by which it is solicited. These are termed respectively the magnetic variation, dip, and force. The phenomena of the variation and dip have hitherto received the principal attention, and indeed almost the exclusive regard, of the experimentalists, as well as of those who have sought to submit the results of experiment to a systematic arrangement.

The necessity of ascertaining the direction of the compass needle in regard to the geographic meridian, for the purposes of navigation, has caused the variation to become known in almost all parts of the globe, and far more extensively than the dip, which is an object purely of physical research; it happens unfortunately that the class of phenomena with which we are thus best acquainted, is the one least susceptible of systematic reduction; the skill of the many eminent mathematicians, by whom the observed variation of the needle has been considered, has failed in deducing any general law by which the observations may be represented; and it may be concluded, in the words of one of the most distinguished existing authorities, that it is impossible to find any places for two, or even for a greater number of magnetic poles, which will correctly explain the direction of the needle (in regard to the meridian) in every part of the globe.

The prospect of a better success, founded on the phenomena of the dip, was entertained at the commencement of the present century; it had been learnt by observation that a needle suspended by its centre of gravity, and having perfect freedom of motion, assumed an horizontal position in certain parts of the globe which, in the absence of very precise observation, appeared to form a great circle of the sphere, inclined at about 12 degrees to the equator; that in receding from this circle in either direction, the horizontal position was departed from; one of the poles dipping beneath the horizon if the recession was towards the North, and the other if towards the South: and as the amount of the departure from horizontality appeared, as far as was then known, to bear a certain proportion to the distance from the circle, it was deemed probable that at its poles the position of the needle would be vertical. Reasoning then on certain hypothetical considerations, in which the circle of no dip was regarded as the magnetic equator, and its poles as the magnetic poles, of the earth, it was inferred, that between the equator and the poles the tangent of the dip should be equal to twice the tangent of the magnetic latitude. On submitting this law, partly experimental and partly hypothetical, to more extensive experience, it has not been found, however, so correct a representation of the facts as it had been expected to prove. M. Biot, to whose original suggestion it was substantially due, had inferred from the position on the globe of the circle (or rather, as it is now known to be, the inflected curve) of no dip, that the spot in the northern hemisphere, in which the needle would be vertical, would be found in 76° north latitude, and 25° west longitude of Paris. It was shewn, however, by the observations which I had an opportunity of making in the longitudes west of Baffin's Bay, in the years 1818, 1819, and 1820 (and which have been since confirmed by other observers with other instruments), that the dips of greatest magnitude are found in meridians nearly 90° distant from M. Biot's supposition; and (which is more directly to the point, because amidst the many irregularities of the dip it by no means follows that there might not be two places in the northern hemisphere, each with a dip of 90 degrees) I had a further opportunity, in 1823, of observing within a very few miles of the actual spot assigned by M. Biot as that of the maximum of dip, when it was found not to exceed 80 degrees and a few minutes. Nor is this instance of difference between the facts and the computation from the formula which was designed to represent them, an extreme case. M. Biot has attributed the failure of the formula, in its general application, to the existence of certain secondary centres of magnetic attraction, which are supposed to interfere with and disturb the regularity of the effects under the general law*; however that may be, it is certain that no two positions can be assigned to the magnetic poles, which shall enable a calculation of the dip by any function of the polar distance, in which differences from fact shall not be found of 10 degrees and upwards; and that whatever circle may be assumed as that of the magnetic equator, if the magnetic latitudes are parallels, stations will be found having the same magnetic latitude, (and not insulated stations only, but districts,) in which the dip may be shewn to differ from 10 to 15 degrees. It must be conceded, therefore, that the dip cannot be considered in strictness even as an approximate indication of the magnetic latitude, of the magnetic equator, or of the magnetic poles in the assumed hypothesis.

* Précis de Physique Expérimentale, Edition of 1821. Vol. II. chap. ix.

In tracing the progress of our experimental acquaintance with the principal phenomena of the earth's magnetism, it is impossible to avoid remarking the little attention that has been paid to the intensity of the force by which the magnetic needle is solicited, in comparison with the pains which have been bestowed to ascertain its direction.

We learn by the memoir drawn up by the Academy of Sciences at Paris, for the instruction and guidance of M. de la Perouse and his associates, that previously to the year 1782, observations had been made at Brest, at Cadiz, at Teneriffe, and at Goree, and subsequently at Brest and Guadaloupe, in which no sensible difference had appeared in the intensity of the magnetic force at those stations. We now know that observations by which a difference was thus not discovered, must have been defective; and such appears to have been the suspicion of the members of the Academy, by whom the memoir was drawn up, if a judgment may be formed from their recommendation that such experiments should be repeated, and especially that the comparative force of magnetism should be ascertained at those places on the globe, where the dip was most considerable, and at those where it was least so.

The recommendation thus made had little effective operation until the beginning of the present century, when it fortunately attracted the notice, and obtained the attention of M. de Humboldt, to whom almost every branch of natural knowledge has so much obligation: his comparative experiments in Europe and in South America, published in 1805, first made known a diminution of intensity on approaching those parts of the globe, where the dipping needle is horizontal; and further appeared to indicate, though with considerable irregularities, that the diminution was progressive, coincidently with the decrease of dip. M. de Humboldt's experiments, with a much fewer number made by M. Rossel, in the voyage of Admiral D'Entrecastreaux, (but published subsequently to those of M. de Humboldt, and inclining to the same indications, though less decidedly,) include, it is believed, the whole of our experimental knowledge in regard to the intensity, previously to the year 1818; when the determination of the British government to re-attempt the discovery of a North West Passage between the Atlantic and Pacific Oceans, opened a field of great interest for researches of every kind connected with magnetism, in countries to which the access had previously been extremely inconvenient.

The interest which I had felt in the perusal of M. de Humboldt's experiments, induced me, on my appointment to conduct the scientific operations of the Voyage of Discovery of 1818, to feel much solicitude in the preparation of the instruments by which the magnetic dip and force should be ascertained. The dipping-needles officially supplied by the government were of very inferior construction; but Mr. Browne was kind enough to allow me to employ a dipping-needle belonging to himself, the workmanship of the late Messrs. Nairne and Blunt, artists of deserved celebrity in the construction of such instruments: it has subsequently remained in my possession, and has accompanied me in three northern, and one equatorial voyage: the results obtained with it are now to be discussed.

In the voyages of 1818, and of 1819-20, the first to Baffin's Bay, and the second to Melville Island, the same needle served for both the purposes of the dip and force. It was furnished with a means of adjusting the centres of gravity and of motion to each other, by small screws on a cross of wires attached to the axis, as described in the *Philosophical Transactions* for 1772, article 35; and was adjusted with great care, and probably as well as that very difficult operation is ever performed, before its embarkation in 1818; since which period its magnetism has never been interfered with, nor has it undergone a change. The observations on the intensity of the force made during those two voyages, are consequently strictly comparable with each other; the dips observed in those voyages are probably also entitled to as much confidence as can attach to observations made with needles in which the influence of gravity is supposed to be wholly destroyed by an instrumental adjustment. It is scarcely possible that such an adjustment should be quite perfect, and there will always remain a liability to errors, amounting perhaps to a few minutes.

The magnetic observations made in the voyage of 1818, were published in detail in the *Philosophical Transactions* for 1819, and those of the voyage of 1819—20, (or at least a portion of them, as the space allotted for the purpose did not admit of the whole being so published,) in the appendix to the narrative of that voyage. As those on the intensity of the force will be employed on this occasion, in the discussion of the ratio of its variation in different parts of the earth, the following abstracts of them are now given.

ABSTRACT in the VOYAGE of 1818.

STATIONS.	Latitude.	Longitude.	Dip.	Time of 100 Vibrations in the Magnetic Meridian.
	° ′	° ′	° ′	M. S.
London	51 31 N.	0 08 W.	70 33 N.	8 02
Shetland	60 09.5	1 12	74 21 *	7 49.75
Davis Straits, on Ice	68 22	53 50	83 08	7 20
Hare Island	70 26	54 52	82 49	7 21.3
Baffin's three Islands	74 04	57 52	84 09	
Baffin's Sea, on Ice	75 05	60 23	84 25	7 25.5
Baffin's Sea, on Ice	75 51.5	63 06	84 44	7 23.25
Baffin's Sea, on Ice	75 59	64 47	84 52	. .
Baffin's Sea, on Ice	76 32	73 45	85 44	. .
Baffin's Sea, on Ice	76 45	76 00	86 09	7 15
Baffin's Sea, on Ice	76 08	78 21	86 00	7 16
Davis Strait, on Ice	70 35	66 55	84 39	7 16

ABSTRACT in the VOYAGE of 1819—1820.

STATIONS.	Latitude.	Longitude.	Dip.	Time of 100 Vibrations in the Magnetic Meridian.
	° ′	° ′	° ′	M. S.
London	51 31 N.	0 08	70 33	8 02
Davis Strait, on Ice	64 00 N.	61 50 W.	83 04 *	7 17.4 †
Baffin's Sea, on Ice	72 00	60 00	84 15	. .
Possession Bay	73 31	77 22	86 04 *	7 19.5 †
East Shore of Regent's Inlet	72 45	89 41	88 27	7 19 †
Regent's Inlet, on Ice	72 57	89 30	88 25 *	. .
North Shore of Barrow's Strait	73 33	88 18	87 36 *	. .
Byam Martin's Island	75 10	103 44	88 26	7 22.5 †
Polar Sea, on Ice	74 55	104 12	88 29	. .
Bay of the Hecla and Griper	74 47	110 34	88 30	. .
On Melville Island	74 27	111 42	88 37	7 24.3 †
Observatory, Winter Harbour	74 47	110 48	88 43	7 26.25
Davis Strait, on Ice	68 30	64 21	84 21.5	. .

In the seven observations of the dip which are marked with an asterisk in the foregoing abstracts, Captain Parry had the kindness to accompany and remain with me, whilst I was engaged in making the observations; and to confirm the correctness of my reading of the division of the arc, to which the needle settled at each repetition, by going over that part of the process himself. Our separate readings are recorded in the original detailed accounts, where it may be seen that in no one of the seven instances they differed more than one minute in a mean result; affording a satisfactory proof of the precision with which the divisions of the circle of Mr. Browne's dipping instrument are capable of being read. It is proper to add, that the unity of the observer, in all the respects in which it is of consequence that the same individual should conduct all the experiments of a suite, was equally preserved in those observations, as in the others in which I had not the pleasure of such an accompaniment. In recording so extensive a series of observations on the dip and force, covering nearly a fourth part of the northern magnetic hemisphere, it is not unimportant to notice such particulars, as might otherwise be thought, and more particularly by those who are themselves practical observers, to constitute an occasion of dissimilarity.

The time of 100 vibrations of the dipping-needle at the stations in the second abstract, marked with a †, are now first published.

On the return of the expedition of Arctic Discovery in 1820, and before my departure for Africa, in 1821, a conviction of the imperfection of the usual method of observing the dip, with needles in which it is a necessary condition to the accuracy of the result, that the axis of motion should pass through the centre of gravity, but which condition is probably never strictly fulfilled, induced me to cause a needle to be made on a principle recommended by Professor J. Tobias Mayer, in his treatise "*De usu Accuratiori acûs inclinatoriæ Magneticæ*," published in the *Transactions of the Royal Society of Sciences at Gottingen*, for 1814. In this needle, the centres of motion and gravity are designedly separated, so that inequalities of workmanship in the axis, or in the planes of suspension, are rendered of less effect, being opposed by the joint influences of gravity and magnetism; whilst by a peculiar process of observation, and an appropriate formula, the joint operation of the two forces is resolvable, and the position which the needle should assume from that of magnetism alone, deducible with precision. This needle, being found on trial to deserve the preference in practice, which had been inferred from its superiority in principle, was subsequently employed in the determinations of Dip; and Mr. Browne's needle was reserved solely for the indications of the Force.

In order that the observations made with the needle on Professor Mayer's construction, may be understood in the condensed form in which they are given in the following tables, it is necessary to prefix a description of the needle itself, of the mode of observation with it, and of deducing the results.

The needle was a parallelopipedon of eleven inches and a half in length, four-tenths in breadth, and one-twentieth in thickness; the ends were rounded; and a line marked on the face of the needle passed through the centre to the extremities, answering the purpose of an index line.

The cylindrical axis on which the needle revolved was of bell metal, terminated, where it rested on the agate planes, by cylinders of less diameter; the finer these terminations can be made, so long as they do not bend with the weight of the needle, the more accurate will be the oscillations; small grooves in the thicker part of the axis received the Y's which raised and lowered the needle on its supports, and ensured that the same parts of the axis rested on the planes in each observation.

A small brass sphere traversed on a steel screw, inserted in the lower edge of the needle as nearly as possible in the perpendicular to the index line passing through the axis of motion; by this mechanism the centre of gravity of the needle, screw, and sphere, may be made to fall more or less below the axis of motion, according as the sphere is screwed nearer or more distant from the needle, and according as spheres of greater or less diameter are employed.

The object proposed in thus separating the centres of motion and gravity, was to give to the needle a force arising from its own weight, to assist that of magnetism in overcoming the inequalities of the axis; and thus to cause the needle to return, after oscillation, with more certainty to the same point of the divided limb, than it would do were the centres strictly coincident.

The centres of motion and of gravity not coinciding, the position which the needle assumes, when placed in the magnetic meridian is not that of dip; but the dip is deducible, by an easy calculation, from observations made with such a needle, according to the following directions.

If the needle has been carefully made, and the screw inserted truly as described, the centres of motion and of gravity will be disposed as in the lever of a balance, when a right line joining them will be a perpendicular to the horizontal passing through the extremities, (or to the index line;) this condition is not indeed a necessary one, but it is desirable to be accomplished, because it shortens the observations, as well as the calculation, from whence the dip is deduced; its fulfilment may be ascertained with great precision, by placing the needle on the agate planes before magnetism is imparted to it, and observing whether it returns to a horizontal direction after oscillation, in each position of the axis; if it does not, it may be made to do so at this time with no great trouble.

With a needle in which this adjustment can be relied on, two observations made in the magnetic meridian are sufficient for the determination of the dip; the two faces of the needle are turned successively towards the observer, by reversing the position of the axis on its supports in such manner, that the edge of the needle which is uppermost in the one observation becomes lowermost in the other; the angles which the needle makes with the vertical in these two positions being read, the mean of the tangents of those angles is the co-tangent of the dip.

But when needles are used in which this adjustment has not been made, or where its accuracy cannot be relied on, four observations are required; two being those which are already directed; the two others are similar to them, but with the poles of the needles reversed; calling then the first arcs F and f, and those with the poles reversed G and g, and taking

$$\text{tang. F} + \text{tang. } f = \text{A}$$
$$\text{tang. F} - \text{tang. } f = \text{B}$$
$$\text{tang. G} + \text{tang. } g = \text{C}$$
$$\text{tang. G} - \text{tang. } g = \text{D}$$
$$\frac{\text{A . D}}{\text{B + D}} + \frac{\text{B . C}}{\text{B + D}} = \text{twice the co-tangent of the dip.}$$

In reversing the poles, it is not necessary that the magnetic force imparted to the needle should be the same in amount as it possessed previously to the operation.

By adopting the precaution of placing the needle in a groove to prevent its lateral motion, and by confining the sides of the magnet by parallel strips of wood, so that in moving along the needle they may preserve its direction, the poles may always be ensured to coincide with the extremities of the longitudinal axis.

If the distance between the centres of motion and of gravity be considerable, the arcs in the alternate observations will be on different sides of the vertical, especially when the dip is great; in such cases the arcs to the south of the vertical are read negatively.

The arcs in each of the four positions, forming the elements from whence the dip is deduced, are the arithmetical mean of (usually) six observations, half of which were with the face of the circle towards the east, and half with the face towards the West; the needle being lifted by the Y's and lowered gently on its supports between each observation; the arcs indicated by both ends of the needle were also read, to correct the errors arising from inequality in the divisions, or from the axis of the needle not passing correctly through the centre of the circle.

The observations, of which the following table presents an abstract, were made in the summer of 1821, for the double purpose of determining the dip in London, and of affording a satisfactory evidence of the consistency of the results obtained with Mayer's needle. The experiments were made with spheres differing considerably in magnitude, with a view of discovering if any limit of proportion between the respective forces of magnetism and gravity was desirable to be maintained; it is obvious that in proportion as the spheres are larger, the arcs read will deviate more and more from the position which the needle would assume from the force of magnetism alone; but it does not appear that the ultimate result, deducible from the readings by means of the formula, is affected even by differences which might be considered as excessive.

ABSTRACT of OBSERVATIONS with MAYER'S NEEDLE on the DIP, in the Regent's Park, London, August, 1821.

DATE.	The MARKED END of the NEEDLE being				Dip deduced.
	A North Pole.		*A South Pole.*		
	The Weight.		The Weight.		
	Uppermost.	Undermost.	Uppermost.	Undermost.	
	f.	F.	*c.*	C.	
1821.	° ′	° ′	° ′	° ′	
August 3	+ 9 22.3	31 08.2	+ 6 17.1	29 41.3	70 03.3 N.
„ 6	− 22 22	49 11.7	− 20 09	46 58.7	70 04 7 „
„ 6	+ 10 08.3	30 36.8	+ 7 46.6	28 47.7	70 01.4 „
„ 11	− 14 49.1	45 58	− 11 28.7	41 50.7	70 00.1 „
„ 13	+ 14 07.2	27 24.3	+ 13 21.7	24 14.2	70 05.9 „
„ 13	+ 10 17.2	30 36.2	+ 9 15.3	27 12.6	70 03.5 „
„ 15	+ 8 03.4	32 00.2	+ 7 40.9	28 57.4	70 05.2 „
„ 15	+ 17 34.1	24 14	+ 15 34.8	22 17.5	70 00.9 „
„ 20	− 19 25	48 24.7	− 17 19	44 57.1	70 00.3 „
„ 20	+ 17 38.5	24 27.6	+ 15 22	22 04	70 03.8 „
MEAN					70 02.9 „

At the same time that Mayer's needle was provided for the purpose of introducing greater accuracy in the observations of the dip, several other alterations were made in parts of the apparatus employed in the determinations both of the dip and intensity, which it may be proper to premise. To ensure the perfect horizontality of the agate planes which support the axis of the needle, a spirit level was attached to a circular brass plate, of the proper diameter to be placed upon the planes themselves, with adjustments to bring it parallel to the plate: the errors of the level are shewn by placing the plate in various positions horizontally, and of the planes by turning the whole instrument upon its horizontal centre; when these errors are adjusted, and the planes and plate perfectly horizontal, the apexes of two cones, which proceed perfectly at right angles from the plate uniting them at their base, and are equal to the diameter of the divided circle of the instrument, should coincide with the divisions 90° and 90° of the circle; if they do not, the cones afford the means of correcting the adjustment in that respect also.

Mr. Browne's needle appearing unnecessarily encumbered about the axis, having two adjusting weights on each wire of the cross, and the wires themselves being needlessly long, and consequently liable to derangement, the weights were removed altogether, and the needle re-adjusted by shortening all the wires one-half, and one of the pairs more than a half, and unequally. Having since had opportunities of observing the direction of this needle in dips of nearly every amount, I am able to say with confidence, from its correspondence with the results of Mayer's needle, that the re-adjustment of the centers of motion and gravity were very satisfactorily accomplished. The alteration occasioned a small difference in the time of vibration, making its oscillation in the two subsequent voyages not *directly* comparative with those in the two preceding: this circumstance, however, is of no moment whatsoever, because the whole suite of experiments are equally comparable with each other, by means of the station in London common to both, at which the time of vibration was very correctly ascertained before and after the alteration was made.

A moveable ring was fitted into the great ring of the instrument, and close to the back of the divided circle, for the purpose of retaining and releasing the needle at any arc at which it might be desired to commence the vibrations for determining the intensity of the force. To one part of the ring a double lever was attached, acted upon by a spring throwing one end of the lever within the divided circle, so as to support the needle at any division that might be required; a string was attached to the other end of the lever, passing through a hole in the great ring, which when drawn tight released the needle, by pulling the lever from under it. The needle was always retained, previously to oscillation, both in the meridian and in the perpendicular vibrations, at 70 degrees from its natural position; the account of the time of vibration was commenced when the arc had diminished to 60 degrees; and concluded when it had further diminished to less than 10 degrees. The number of vibrations which Mr. Browne's needle usually made between these arcs, was from 90 to 110; from which the average time of 10 vibrations was inferred as entered in the tabular abstracts; the times of commencement, and of every subsequent 10th vibration, were noted by a chronometer of small and steady rate, to the nearest beat, *i. e.*, to the nearest four-tenths of a second.

The following table presents an abstract of the observations on the Dip, made principally with Mayer's needle in the voyages of 1822 and 1823:—

ABSTRACT OF OBSERVATIONS ON THE DIP.—1822 and 1823.

STATION.	PLACE OF OBSERVATION.	Date.	The MARKED END of the NEEDLE being: A North Pole. THE WEIGHT. Undermost.	A North Pole. THE WEIGHT. Uppermost.	A South Pole. THE WEIGHT. Undermost.	A South Pole. THE WEIGHT. Uppermost.	DIP DEDUCED.		REMARKS.
			° F. ′	° f. ′	° G. ′	° g. ′	° ′		
Island of St. Thomas .	The Pendulum Station	1822. May	86 18.8	85 12.1	94 55.8	93 44.9	00 05.75 S.	00 04 S.	
			86 23.3	85 12.8	94 45.4	93 39.8	00 09.6 N.		
			86 21	84 04	95 42.5	93 17.5	00 15.5 S.		
Bahia	The Pendulum Station	July	. . . .	. . .	. . . .	. . .	. . .	04 12 N.	Mr. Browne's needle
Island of Ascension .	The Barrack Square	June	. . . .	. . .	. . .	. . . .	. . .	05 10 S.	Mr. Browne's needle
Maranham	A garden in the suburb	Aug.	65 59.5	61 30.5	72 44.2	68 15.5	23 11.5 N.	23 07.75 N.	
			67 58.1	58 50	73 51.25	65 05	23 04 "		
Sierra Leone . . .	300 yards south of Fort Thornton . . .	March	59 00.1	52 32.5	68 19	58 45.6	31 04 "	31 02.25 "	
			58 49.7	53 57.7	67 50.5	59 25.3	31 00.5 "		
Trinidad	Government-house, St. Ann's . . .	Sept.	50 58.3	46 32.2	54 59.2	51 06	38 58 "	39 02.5 "	
			54 49.4	42 09.4	57 37	45 46.1	39 07 "		
St. Mary's, R. Gambia	One mile north of the town	Feb.	62 30	5 33.7	68 50.1	11 28.3	40 23	40 23 "	
Port Praya, C. Verd Ids.	A palm grove, near the Watering Place .	Jan.	58 50.6	7 51.8	63 44.3	9 11.6	45 26.1	45 26.1 "	
Jamaica	The sea-beach, at Port Henderson . .	Oct.	43 33.2	38 32.3	47 55.5	42 13.75	47 01 "	46 58.25 "	
			46 35	36 43.1	49 53	38 06	46 55.5 "		
Grand Cayman . . .	The sea-beach	Oct.	. . .	. . .	. . . .	. . .	. . . .	48 48.3 "	Mr. Browne's needle
Havannah	A garden, near the city	Nov.	38 45.5	36 47	41 19.3	36 12.5	51 55.3	51 55.3 "	
Teneriffe . . .	The sea-beach, east of Santa Cruz . .	Jan.	49 09	—00 19	50 14.1	—01 15.75	59 50	59 50 "	
Madeira	The garden of the British Consul, Funchal	Jan.	. . .	.	. . . .	. . .	. . . .	62 12.8 "	Mr. Browne's needle
New York . . .	The garden of the Lunatic Asylum .	Dec.	19 21.4	13 40.3	20 26.9	13 57.5	73 05	73 05 "	
Drontheim	The Pendulum Station	1823. Oct.	19 42	9 54	20 54	10 47	74 35 "	74 43 "	
			38 40	—15 24	39 06	—14 18	74 51 "		
Hammerfest . .	The sea-beach, Fugleness . . .	June	16 35	8 40	17 02	7 46	77 24.3 "	77 15.7 "	
			34 31	—14 01	36 00	—14 40	77 17.3 "		
			13 09	12 23.5	13 18	12 43	77 05.7 "		
Greenland . . .	The Pendulum Station	Aug.	27 31	—10 41	27 30	— 8 50	80 06 "	80 11 "	
			14 49	4 00	14 34	5 11	80 16 "		
Spitzbergen . . .	The Inner Norway Island	July	31 46	—17 50	32 54	—18 13	81 16 "	81 11 "	
			24 16	— 8 54	22 45	— 5 09	81 06 "		
			9 44	7 51	9 12	8 30	81 10 "		

The next table contains an account of the average time of ten vibrations of Mr. Browne's needle in similar arcs at the several stations enumerated in the first column; the second column exhibits the time of vibration in the plane of the magnetic meridian, the squares of which numbers express the inverse ratio of the intensity of terrestrial magnetism; the third column contains the corresponding times of vibration in the plane perpendicular to the meridian; the squares of the numbers in this column should correspond with the squares of those in the preceding, divided by the sine of the dip; consequently the dip itself is deducible as an observation, from a comparison of the times of vibration in the meridian and perpendicular to it; the results so obtained are inserted in the 5th column, until the dip arrives at an amount when, from the reduced ratio of the increase of the sines, this method of deduction ceases to be practically useful. The fourth column shews the times of vibration of the same needle, suspended by an assemblage of untwisted silk fibres attached to one extremity of the axis, and limiting the needle to an horizontal motion; the vibrations were performed under the protection of a wooden cover with glazed windows, the silk suspension being fifteen inches in length: the squares of these times should correspond with the squares of the times in column 2, divided by the cosine of the dip; affording a third method of deducing the dip by observation, which becomes available when the preceding fails; the 6th column contains the dips so deduced; the seventh exhibits the results obtained by the direct method brought forward from the preceding tabular abstract, and placed in comparison with those in the 5th and 6th columns; in the eighth is inserted the Dip finally deduced from a mean of the methods thus compared.

STATION.	TIME OF TEN VIBRATIONS.			DIP DEDUCED.		Dip deduced in the ordinary mode. (Page 474.)	Dip finally deduced.
	In the magnetic meridian.	Perp. to the merid.	Suspended horizontally.	$\frac{M^2}{P^2} = \sin. D.$	$\frac{M^2}{H^2} = \cos. D.$		
1.	2.	3.	4.	5.	6.	7.	8.
	= M.	= P.	= H.				
	s.	s.	s.	° ′	° ′	° ′	° ′
Island of St. Thomas.	61.652	Stationary.	. .	. . .	. . .	00 04 S.	00 04 S.
Maranham	58.66	93.744	. .	23 03	. .	23 07.75	23 06 N.
Sierra Leone . . .	58.012	81.017	. .	30 50.5	. . .	31 02.25	30 57 „
Trinidad	53.262	67.1	. .	39 03.3	. . .	39 02.5	39 03 „
Jamaica	49.667	58.15	. .	46 51	. . .	46 58.25	46 55 „
Grand Cayman . . .	49.61	57.288	. .	48 35	. . .	48 48.3	48 42 „
Havannah . . .	48.177	54.2	. .	52 11	. . .	51 55.3	52 03 „
London	49.453	51	. .	70 05.8	. .	70 03.5	70 04.5„
New York . . .	44.667	45.687	. .	72 55	. .	73 05	73 00 „
Drontheim	49.643	50.552	96.75	74 40	77 44	74 43	74 42 „
Hammerfest	48.885	49.435	103.8	. . .	77 11	77 13.3	77 13.3„
Greenland	48.05	48.4	116.507	. . .	80 12.7	80 12	80 12 „
Spitzbergen	47.562	47.9	121.36	.	81 10	81 11	81 10 5„

In the Voyage of Discovery of 1819-1820, I had made experiments on the intensity of the force with needles limited by their mode of suspension to an horizontal motion, of which an account is given in the appendix to the narrative of that voyage; it has not been thought necessary to include those experiments in the present summary, because, on occasions where the dip so nearly approaches 90 degrees, the employment of horizontal needles is more *curious*, as evidencing to general apprehension the diminution of directive force which the compass needle undergoes in such situations, than *useful* towards a knowledge of the real intensity of magnetism*; the reverse, however, is the case in the parts of the earth where the horizontal differs little from the natural direction of the needle; in such situations they afford, perhaps, a more exact comparison of the relative intensity than the dipping needle, which by reason of the resistance of the planes supporting its axes, sooner arrives at rest than needles suspended by silken fibres; and the greater duration of the period through which the vibrations continue, enables the average time of vibration to be obtained with greater exactness. It is for this reason also that a silken suspension is preferable for the horizontal needles to an agate resting on a point.

The apparatus which I provided for the horizontal needles in the voyages of 1822 and 1823, was preferable in many respects to that which had been employed in 1819-20; and being simple, of little cost, and fully equal to its purpose, the following description may promote the further extension of experiments of the same nature: it consisted of a mahogany box, made, for convenience, in an octagonal shape, with a top of stout glass; the height was fifteen inches, and the diameter sufficient to allow a bar of seven inches in length to vibrate freely, when suspended by a silk line passing through a brass button inserted in a perforation in the middle of the glass top; a metal circle fixed in the bottom of the

* Their employment should cease, whenever the uncertainty to which the observation of the dip is liable induces a corresponding uncertainty in the reduction of the time of vibration dependant upon it, equalling in amount, and superadded to the probable mean error of observation with the horizontal needles themselves.

box, of rather more than seven inches diameter, marked the arc of vibration; the bar or needle was carried in a light stirrup, into which it slid until correctly balanced; the silk thread, of fifteen inches in length, consisted of a sufficient number of silk fibres to sustain the weight, and was always allowed to untwist itself, in the first instance, with a brass needle, of equal weight with the magnetic needles; and was so adjusted, by moving the button round, that the brass needle should settle, when at rest, in the magnetic direction. The box was usually placed on the ground, in a sheltered situation, far from buildings or other sources of local interference; the only adjustment required, (except that of the silk thread,) was to render the divided circle horizontal, which was accomplished by a pocket spirit level, and wooden wedges placed beneath the box; the silk being thus without twist, and one of the magnetic bars in the stirrup, and known to be horizontal by its accordance with the circle, the degree to which it settled was registered as the zero; it was then drawn about 40 degrees out of the meridian, and retained by a copper wire passing through the glass top, and capable of being moved in azimuth from the outside, and of being raised so as to release the needle at pleasure, to commence its oscillations; these were not noticed until the arc had diminished to 30 degrees, when the registry commenced, and was repeated at the close of every tenth vibration, until the arc had still further diminished to 10 degrees, when the experiment was concluded.

The six needles which were used in this apparatus differed from each other considerably both in rapidity of vibration, and in the duration of the interval of oscillation between 30 and 10 degrees; Nos. 1, 2, 3, and 5, were similar in shape; being bars 7 inches long, 0.25 broad, and 0.15 thick; No. 4 was a bar of the same magnitude in the middle, but gradually tapering to points at the extremities; No. 6 was cylindrical, flattened in the middle to fit the stirrup; they were all magnetised with very powerful magnets in the summer of 1821, and being tried in London, in 1823, on their return from the Equator, and in 1824 on their return from the Arctic Circle, were found to have preserved throughout the same average time of vibration as in 1821, with the exception of No. 2, which from some cause that was not obvious had changed its rate materially between 1821 and 1823, and was therefore discontinued in subsequent use, and the experiments made with it, although inserted with those of the other needles in the tabular abstracts, are rejected in the conclusions drawn. When not in use, the needles were kept in pairs in the customary manner, Nos. 1 and 3, 2 and 6, 4 and 5, being combined, with their opposite poles united, in separate boxes; and each needle was placed by itself in the direction of the meridian for two or three hours before its time of vibration was ascertained. The times were registered to fractional parts of a second by the beats of a chronometer, having a rate inappreciable in the interval. The detail of the experiments at a single station will suffice to explain more fully the manner in which they were proceeded with; at the remainder of the stations, the results only are collected, and are presented in one view in a tabular abstract.

TIMES of VIBRATION of the SEVERAL NEEDLES observed with No. 423, on the Beach at Man-of-War Bay, Island of St. Thomas, May, 1822.

Vibrations.	Needle 1.		Needle 2.		Needle 3.		Needle 4.		Needle 5.		Needle 6.	
	Time.	Interval.	Time.	Interval.	Time.	Interval.	Time.	Interval.	Time.	Interval	Time.	Interval.
	M. S.	S.	M. S.	S.	M. S.	S.	M. S.	S.	M. S.	S.	M. S.	S.
0	0 00		0 00		0 00		0 00		0 00		0 00	
		117.6		73.8		119.2		39		53.6		45.6
10	1 57.6		1 13.8		1 59.2		0 39		0 53.6		0 45.6	
		117.2		73.6		118		39		53.4		45.6
20	3 54.8		2 27.4		3 57.2		1 18		1 47		1 31.2	
		116.4		73.2		117.6		39		54		45.2
30	5 51.2		3 40.6		5 54.8		1 57		2 41		2 16.4	
		116.8		73.4		117.6		38		53		44.6
40	7 48		4 54		7 52.4		2 35		3 34		3 01	
		116		73		117		38		53.6		44.6
50	9 44		6 07		9 49.4		3 13		4 27.6		3 45.6	
		116		73		117		38		53.4		45.2
60	11 40		7 20		11 46.4		3 51		5 21		4 30.8	
		116.2		72.8		116.8		38		53.6		44.8
70	13 36.2		8 32.8		13 43.2		4 29		6 14.6		5 15.6	
		116.2		72.6		116.4		38		53.4		44.6
80	15 32.4		9 45.4		15 39.6		5 07		7 08		6 00.2	
		115.4		72.8		116.4		38		53		44.6
90	17 27.8		10 58.2		17 36		5 45		8 01		6 44.8	
		115.4		72.8		116.4		39		53		44.6
100	19 23.2		12 11		19 32.4		6 24		8 54		7 29.4	
		115.6		72.6		116.8		39		53		44 6
110	21 18.8		13 23.6		21 29.2		7 03		9 47		8 14	
		116		72.4		. .		. .		52.6		. .
120	23 14.8		14 36		.		. .		10 39.6		. .	
		115		72.6		. .		. .		52.8		. .
130	25 09.8		15 48.6		. . . .		.		11 32.4		. .	
		115.4		72.6		. .		. .		. .		. .
140	27 05.2		17 01.2		. . . .		. .		. .		. .	
		115.6		72.6		. .		. .		. .		.
150	29 00.8		18 13.8		. . . .		. .		.		. .	
Mean Time of Ten Vibrations.	116.05 Seconds.		72.92 Seconds.		117.2 Seconds.		38.45 Seconds.		53.26 Seconds.		44.9 Seconds.	

AN ABSTRACT of EXPERIMENTS on the INTENSITY of MAGNETISM, with Needles suspended Horizontally.

STATIONS.	Average Time of 10 Horizontal Vibrations between the Arcs of 30° and 10°.					
	No. 1.	No. 2.	No. 3.	No. 4.	No. 5.	No. 6.
	s.	s.	s.	s.	s.	s.
St. Thomas	116.65	72.92	117.2	38.45	53.26	44.98
Bahia	. . .	73.87	119.5	39.07	54.03	46.07
Ascension	. . .	. . .	. . .	38.75	53.87	.
Maranham	. . .	73.17	. . .	. . .	52.88	45.04
Sierra Leone	. .	74.08	119.76	38.93	53.56	45.67
Trinidad	116.43	73.75	117.44	38.73	52.96	45.24
St. Mary's River Gambia	122.26	. .	. . .	39.8	54.566	. .
Port Praya	125.49	. . .	. .	40.7	55.25	. . .
Jamaica	114.3	72.31	114.31	37.4	51.9	44.39
Grand Cayman	116.84	73.16	115.6	. .	. .	44.8
Havannah	117.5	73.41	118.07	38.41	. . .	45.27
Teneriffe	. .	84.136	.	45.6	62.366	53.2
Madeira	. .	. .	141.8	46.2	. .	. . .
London 1821	159	92.6	165.9	54	72.95	61.95
London 1823	161.33	103.6	164	53.24	74.37	63.42
London 1824	161.5	. .	163.2	52.4	73.9	62.74
New York	156.09	. . .	156	50.64	70.55	60.36
Drontheim	180.84	. . .	182.67	60.2	. .	70.46
Hammerfest	195.84	. . .	196.86	63.43	87.64	75.73
Greenland	220.8	. . .	221.42	71.78	. . .	84.76
Spitzbergen	229.26	. . .	231.81	75.11	10.39	88.17

In any application that it may be proposed to make, of the times of vibration of needles limited to an horizontal motion, towards a knowledge of the ratio of variation in the force of terrestrial magnetism, it is a preliminary step, alike necessary in all cases, to obtain from the observed times of *horizontal* vibration, the corresponding times in which an equal number of vibrations would have been performed, had the needles been free to have oscillated in the direction of the force itself. This purpose is accomplished by reducing the squares of the respective times of horizontal vibration in the proportion of the radius to the cosine of the Dip: the results so obtained are inserted in the following table; which exhibits in effect the times of vibration of the same needles, supposing that they had been fitted as dipping needles, and had been made to vibrate in the plane of the meridian: the errors only excepted, which may have been introduced by an inaccurate knowledge of the Dip, affecting the cosine used in the reduction; this consideration becomes of moment only where the Dip is great; it may be accordingly remarked in pages 474-476, that the endeavours to obtain that element correct were increased in reference to the occasion.

REDUCED TIMES in which **TEN VIBRATIONS** would have been performed in the direction of the **DIPPING NEEDLE**, by each of the Needles of which the Horizontal Vibration was observed.

STATIONS.		No. 1.	No. 2.	No. 3.	No. 4.	No. 5.	No. 6.
		s.	s.	s.	s.	s.	s.
St. Thomas		116.65	72.92	117.2	38.45	53.26	44.98
Bahia		.	73.77	119.3	38.96	53.96	46
Ascension		. .	. .	. .	38.67	53.75	. .
Maranham		. .	. .	. .	.	50.7	43.19
Sierra Leone		.	68.6	110.9	36.05	49.6	42.29
Trinidad		102 6	64.99	103.5	34.13	46.67	39.87
St. Mary's		106.7	. .	. .	34.74	47.63	. .
Port Praya		105.1	. .	. .	34.09	46.28	. .
Jamaica		94.46	59.76	94.47	30.91	42.9	36.69
Grand Cayman		94.92	59.44	93.91	. .	. .	36.4
Havannah		92.14	57.57	92.57	30.12	. .	35.5
Teneriffe		.	59.64	. .	32.32	44.21	37.71
Madeira		. .	.	96.88	31.56	. .	. .
London	1821	93	54.09	96.88	31.54	42.60	36.18
	1823	94.22	60.78	95.77	31.1	43.43	37.04
	1824	94.33	. .	95.31	30.6	43.16	36.64
New York		84.40	. .	84.35	27.38	38.15	32.64
Drontheim		92.89	. .	93.84	30.92	. .	36.19
Hammerfest		92.1	. .	92.58	29.83	41.22	35.61
Greenland		91.12	. .	91.35	29.62	. .	34.97
Spitzbergen		89.8	. .	90.8	29.42	42.7	34.54

The squares of the numbers in each column of the above table, are to each other, in the inverse ratio of the intensity of the magnetic force at the respective stations, as severally indicated by the different needles.

Having thus obtained the experimental relation of the magnetic force, at stations distributed over and comprehending an eighth of the surface of the globe, (pages 465, 466, 476, 481,) I proceed to employ the results in their designed application,—as facts collected towards the establishment of a general law, which shall represent by calculation the relative intensity in all parts of the globe.

It is assuredly deserving of remembrance, and highly creditable to the sagacity of the philosophers who drew up the admirable memoir for the guidance of M. de la Perouse, a memoir unparalleled in the annals of voyages of discovery, that at a period when so high an authority in physical knowledge as Mr. Cavendish believed and maintained the invariability of the magnetic force at all parts of the Earth's surface, and when that belief was apparently confirmed by experiments at various stations widely removed from each other (page 463), the authors of the memoir not only manifested doubt, but even implied the expectation that the force would be found to vary in accordance with the dip.

Since the publication of the experiments of M. de Humboldt in the present century, the latter opinion had gradually gained ground, but no suggestion of a specific relation was made until 1820, when Dr. Thomas Young, in one of the numbers of the Astronomical and Nautical Collections, published in the Journal of the Royal Institution, inferred from certain hypothetical considerations, that the force would probably be found to vary, inversely, as the square root of 4, less three times the square of the sine of the Dip. In founding this inference, however, Dr. Young had assumed as a fact, that the Dip itself varied in conformity with M. Biot's hypothesis, or at least that a sufficient approximation to the phenomena was furnished by it; an assumption, the correctness of which is far from being agreeable to experience; on the contrary, indeed, the computation and facts are so frequently and so much at variance, as to have pressed the alternative, either of giving up the hypothesis, or of rejecting the phenomena of the Dip as an evidence in its support or contradiction. The course pursued by M. Biot himself has been to preserve the hypothesis, by enlarging it so as to include the existence of secondary and local centres of attraction, interfering with and destroying the systematic correspondence of the Dip with the primary and general attraction; implying consequently the rejection of the phenomena of the Dip as a test of the hypothesis; and which indeed is expressly done in the admission by the same authority, that the irregularities to which the Dip is liable, and consequently its amount at any particular station, can only be known by actual experiment.

Under these circumstances, then, it may be proper to examine in the first instance, whether the variations of the Intensity are, in fact, in any correspondence with those of the Dip. For that purpose the relation between them suggested by Dr. Young may be assumed, as that which would subsist in the event of both the phenomena being in conformity with the original hypothesis. And it will be sufficient to exhibit the comparison of the experimental results with a single needle with the ratio so computed, because the differences between the several needles are insignificant, compared with those which will appear between the calculation and experiment.

The most exact method of examining the accordance of a body of observed results with a formula proposed to represent their differences, would doubtless be, by comparing the ratio at each station severally with that at all the others both by experiment and calculation; the process however is tedious, and would be needlessly precise in the present case; it may be preferable therefore to substitute a more ready mode of approximative comparison, by reducing the times of vibration observed at each station to the corresponding times which would be required agreeably to the formula where the dip should be $= 0$; if the formula and experiments agree, the reduced times should all be the same; otherwise, as the squares of the reduced times represent the magnetic force where there is no Dip, severally conformable to the observations from whence they are derived, the arithmetical mean of the several squares may be taken as the equivalent to unity; and being combined with the squares of the times actually observed at the several stations, will give the experimental ratio at each, to be compared with the computed ratio.

Thus if $T, \overset{''}{T}, \overset{'''}{T}$, *&c.*, be the times of vibration of a needle in the plane of the magnetic meridian at stations where the sine of the dip is respectively $\acute{S}, \overset{''}{S}, \overset{'''}{S}$, *&c.*, then will $\acute{T}^2, \overset{''}{T}^2, \overset{'''}{T}^3$, *&c.*, represent inversely the force of magnetism at those stations, and agreeably to the formula of Dr. Young

$$\frac{2\,\acute{T}^2}{\sqrt{4 - 3\,\acute{S}^2}}, \quad \frac{2\,\overset{''}{T}^2}{\sqrt{4 - 3\,\overset{''}{S}^2}}, \quad \frac{2\,\overset{'''}{T}^2}{\sqrt{4 - 3\,\overset{'''}{S}^2}}, \quad \&c.$$

should be equal to each other, and to T^2, which is the force where the horizontal direction is the natural position of the needle.

But should the several values of T^2, thus obtained, be found to differ, their arithmetical mean may be assumed to represent the force where there is no Dip, as deduced from, and corresponding suitably, with the whole of the experiments.

A TABLE exhibiting a COMPARISON of the RATIOS of the MAGNETIC INTENSITY, to unity when the Dip = 0, as deduced, 1st, by Computation from the Dip; and 2d, by the Experiments with Mr. Browne's Needle.

STATIONS.	Dip.	Ratios Computed.	Ratios Experimental.	Computation in Excess or Defect.	STATIONS.	Dip.	Ratios Computed.	Ratios Experimental.	Computation in Excess or Defect.
Dip = 0	° ′ 0 00	$T^2 =$ 1	$\overset{2}{61.3}$ / 1		Havannah . . .	° ′ 52 03	1.37	1.62	−0.25
St. Thomas . .	0 04	1	0.99	+0.01	London	70 03.5	1.72	1.54	+0.18
Maranham . .	23 06	1.06	1.09	−0.03	New York . . .	73 00	1.79	1.88	−0.09
Sierra Leone .	30 57	1.12	1.12	−0.00	Drontheim . .	74 42	1.82	1.52	+0.30
Trinidad . . .	39 03	1.19	1.33	−0.14	Hammerfest . .	77 13	1.87	1.57	+0.30
Jamaica . . .	46 55	1.29	1.52	−0.23	Greenland . . .	80 12	1.92	1.62	+0.30
Cayman . .	48.12	1.32	1.53	−0.21	Spitzbergen . .	81 10.5	1.93	1.66	+0.27

The differences between the experimental and computed ratios shewn in the preceding table, are obviously far greater than can be attributed to any probable errors of observation; the defect of the computation at all the West India stations, for example, and its excess at the northern stations, are so great and so systematic, as to be decisive against the supposed relation of the Force to the observed Dip, and equally so against any other relation whatsoever, in which the respective phenomena might be supposed to vary in correspondence with each other.

It is further observable, that if an attempt be made to compute the dip at the several stations of experiment, on the supposition that its tangent is equal to twice the tangent of the magnetic latitude, the differences between the Intensities, as evidenced by experiment, and as computed from the observed Dip, will be found to take place at the same stations and nearly in the same degree, as the observed Dips will differ from those of computation; that in fact the Intensity does not correspond with the Dip, when the variations of the latter deviate from the general law of the hypothesis; and that, in consequence, the secondary local attractions, conceived to be influencial on the direction of the needle, must be further conceived to have little or no sensible effect on the general Intensity of magnetism.

It is desirable therefore, in the next place, to examine, whether the original supposition of two magnetic centres, infinitely near to each other and to the earth's centre, acting on all points of its surface in the inverse ratio of the squares of the distances, a supposition strongly supported by the analogous magnetism of an iron sphere, may not afford a general law of magnetic Intensity, capable of representing within certain small limits, incidental to the experiments themselves, all the variations of Intensity which have been thus observed.

In the supposed magnetic sphere, the Intensity would be at a minimum in a great circle representing the magnetic equator, and at a maximum at each of the poles of that circle; and by pursuing a similar course of demonstration to that which has been adopted by mathematicians in regard to the supposed variations of the Dip, the force between the equator and the poles should vary in the proportion of 1 to 2, and intermediately as $\sqrt{1 + 3 \cos.^2 i}$, i being the itinerary distance from either of the magnetic poles.

To refer this arrangement to the terrestrial sphere, in order to examine its conformity or otherwise with the actual phenomena, the magnetic equator must be sought by connecting those points on the Earth's surface, where the Intensity is observed to correspond, and to be at a minimum in regard to all other points; or, the geographical position of the maximum of Intensity in either hemisphere must be determined, to fix the place of the magnetic pole: the latter operation requires the less extensive experimental inquiry, and on examination the observations in the northern hemisphere recorded in the preceding pages will be found sufficient for the purpose.

The experiments in the two voyages of North-western Discovery (pages 465-466), will alone furnish the means of assigning an approximate position for the maximum of Intensity; as first, in regard to its geographical *longitude*, the force was observed to increase, in sailing to the westward on or near a parallel of latitude in Baffin's and the adjoining seas, until about the meridian of 80° West longitude; but, in proceeding still further to the westward, it was found to diminish: now, as the amount of the force is supposed to be in inverse proportion to the itinerary distance, the point of any particular parallel where the force will be greatest must be at the intersection of the geographical meridian passing through the maximum in the hemisphere, or, in other words, through the magnetic pole; whence the situation of the pole may be inferred to be in or about 80° West longitude, whatever may be its latitude. Second, in regard to its geographical *latitude*, it was further noticed, that in ascending Davis Strait on a meridian or nearly so, and, generally, wheresoever in the countries or seas adjacent opportunities presented themselves of comparing the magnetic force at stations nearly in the same meridian, but in different parallels, the intensity diminished as the latitude increased; indicating that even the most southerly stations (between 60° and 70° N. latitude) were to the North of the parallel in which the magnetic pole was situated.

Having this approximate position, it was not difficult to fix the more precise spot which should best correspond with the general body of observations collected in the four voyages. After a few trials, it was found that the latitude of 60° N., and longitude of 80° W., would fulfil the purpose decidedly better, than when either 59° or 61° of latitude were substituted, or than when 85° W. longitude was employed instead of 80°; a greater precision than to degrees of latitude, and to 5° of longitude (150 miles), might have appeared a refinement beyond the occasion, otherwise 78° of longitude might have been preferred to 80°. From the spot thus indicated, therefore, *i. e.*, in 60° N. latitude, and 80° W. longitude, the itinerary distances of the several stations of experiment were computed, as entered in the fourth columns of the subjoined Tables; as well as the computed ratios of the force to unity at the equator, varying in the direct proportion of $\sqrt{1 + 3 \cos.^2 i}$., and inserted in the last column but one of each of the three tables. In obtaining the corresponding experimental ratios, a similar process has been followed to that already described in page 486; thus $T'^2 . \sqrt{1 + 3 \cos.^2 i'}$, $T''^2 . \sqrt{1 + 3 \cos.^2 i''}$, $T'''^2 . \sqrt{1 + 3 \cos.^2 i'''}$, *&c.*, is the force at the magnetic equator corresponding to the several experiments, the arithmetical mean of which gives the values of T^2 for each of the needles, as inserted at the head of their respective columns; these values being again compared with the experiments at each station, the inverse proportion which the experiments severally bear, to the force at the equator regarded as unity, is inserted in the columns opposite to the stations to which the experiments belong. In the final column of each Table, the ratios by experiment and calculation at the several stations are compared, and their differences stated.

TABLE. I.——COMPARISON of the EXPERIMENTAL and COMPUTED RATIOS of the MAGNETIC INTENSITY, at the Stations visited in the voyages of 1822 and 1823.

STATIONS.	Geographical Position.		Computed itinerary Distance.	Ratio of Intensity by Experiment.							Ratio.		Experimental Ratio in excess or defect.
				Horizontal Needles.					Mean by the Horizon. Needles.	Dipping Needle.			
	Latitude.	Longitude.		1	3	4	5	6			by Experim.	by Computat.	
Magnetic Equator .	. .	. . . .	$T^2=$	120.27^2	120.81^2	39.24^2	54.05^2	46.41^2		63.4^2			
	°	°	° ′										
St. Thomas . .	00.5 N.	6.75 E.	87 58	1.06	1.06	1.04	1.03	1.06	1.05	1.04	1.045	1.005	+0.04
Ascension . .	8. S.	14.5 W.	85 08	. .	.	1.03	1.01	. .	1.02	. .	1.02	1.01	+0.01
Bahia . .	13. „	38.5 „	80 16	. .	1.03	1.02	1.	1.02	1.02	. .	1.02	1.04	−0.02
Sierra Leone .	8.5 N.	13.5 „	71 02	. .	1.19	1.18	1.19	1.2	1.19	1.19	1.19	1.15	+0.04
Maranham .	2.5 S.	44. „	68 31	. .	. .	. .	1.14	1.16	1.15	1.17.	1.16	1.18	−0.02
Gambia . . .	13.5 N.	16.75 „	65 07	1.27	. .	1.28	1.29	.	1.28	. .	1.28	1.24	+0.04
Port Praya . .	15. „	23.5 „	60 48	1.31	. .	1.32	1.36	. .	1 33	. .	1.33	1.31	+0.02
Teneriffe . .	28.5 „	16.25 „	52 36	. .	. .	1.47	1.49	1.51	1.49	. .	1.49	1.45	+0.04
Trinidad . .	10.5 „	61.5 „	51 23	1.37	1.36	1.32	1.34	1.36	1.36	1.42	1.39	1.47	−0.08
Madeira . .	32.5 „	17. „	48 52	. .	1.55	1.55	. .	. .	1.55	. .	1.55	1.52	+0.03
London .	51.5 „	. . .	42 57	1.64	1.59	1.59	1.58	1.61	1.60	1.64	1.62	1.62	. .
Jamaica .	18. „	77. „	42 03	1.62	1.64	1.61	1.59	1.6	1.61	1.63	1.62	1.63	−0.01
Cayman .	19.25 „	81.5 „	40 43	1.61	1.65	. .	. .	1.63	1.63	1.63	1.63	1.65	−0.02
Drontheim . .	63.5 „	10. E.	39 14	1.68	1.66	1.61	. .	1.65	1.65	1.63	1.64	1.67	−0.03
Hammerfest .	70.5 „	24. „	39 01	1.71	1.7	1.73	1.72	1.7	1.71	1.68	1.69	1.68	+0.01
Havannah . .	23. „	82.5 W.	36 53	1.70	1.70	1.70	. .	1.71	1.705	1.73	1.72	1.71	+0.01
Spitzbergen . .	80. „	11.5 E.	31 46	1.79	1.77	1.78	1.76	1.80	1.78	1.78	1.78	1.78	. . .
Greenland . .	74.5 „	19. W.	26 09	1.74	1.75	1.76	. .	1.76	1.75	1.74	1.75	1.85	−0.10
New York . .	40.5 „	74. „	19 40	2.03	2.05	2.05	2.01	2.02	2.03	1.96	1.99	1.91	+0.08

TABLE II.——COMPARISON of the EXPERIMENTAL and COMPUTED RATIOS of the MAGNETIC INTENSITY, at the Stations visited in the voyage of 1818.

STATIONS.	GEOGRAPHICAL POSITION.		Computed itinerary Distance.	Time of Ten Vibrations, (page 465.)	RATIOS OF INTENSITY.		Experimental Ratio in excess or defect.	REMARKS.
	Latitude North.	Longitude West.			by Experim.	by Calculation.		
Magnetic Equator . . .	. . .	. . .	° ′ 90 00			1.		
	° ′	° ′	° ′	s.				
1. London	51 31	0 08	42 57	48.2	1.62	1.62	. . .	
2. Shetland	60 09	1 12	37 01	46.975	1.70	1.70	. . .	
3. Davis Strait . . .	68 22	54 00	13 53	44.	1.94	1.95	−0.01	On an iceberg.
4. Hare Island . . .	70 26	55 00	14 36	44.13	1.94	1.95	−0.01	
5. Davis Strait . . .	70 35	67 00	11 50	43.6	1.98	1.97	+0.01	On an iceberg of immense size.
6. Baffin's Sea . . .	75 05	60 30	16 17	44.55	1.90	1.94	−0.04	On field ice.
7. Baffin's Sea . . .	75 51	63 00	16 56	44.325	1.92	1.94	−0.02	On field ice.
8. Baffin's Sea . . .	76 08	78 21	16 08	43.6	1.98	1.94	+0.04	On an iceberg.
9. Baffin's Sea . . .	76 45	76 00	16 49	43.5	1.99	1.94	+0.05	On an iceberg.

TABLE III.——COMPARISON of the EXPERIMENTAL and COMPUTED RATIOS of the MAGNETIC INTENSITY, at the Stations visited in the voyage of 1819—1820.

STATIONS.	GEOGRAPHICAL POSITION.		Computed itinerary Distance.	Time of Ten Vibrations. (page 466.)	RATIOS OF INTENSITY.		Experimental Ratio in excess or defect.	REMARKS.
	Latitude North.	Longitude West,			by Experim.	by Calculation.		
Magnetic Equator . . .	. .	.	° ′ 90 00			1.		
	° ′	° ′	° ′	s.				
London	51 31	0 08	42 57	48.2	1.63	1.62	+0.01	
Davis Strait . . .	64 00	61 50	9 22	43.74	1.98	1.98	. . .	On ice.
Possession Bay . .	73 31	77 22	14 20	43.95	1.95	1.96	−0.01	
Regent's Inlet . . .	72 45	89 41	13 17	43.9	1.96	1.96	. .	
Byam Martin's Island .	75 10	103 44	17 22	44.25	1.93	1.93	. . .	
Melville Island . .	74 27	111 42	18 30	44.43	1.92	1.92	. .	
Winter Harbour . .	74 47	110 48	18 30	44.625	1.90	1.92	−0.02	

On viewing the differences between the calculation and experiments, contained in the final columns of the three preceding Tables, and on duly considering the delicate nature of such experiments, and their liability to various sources of error of observation,—as well as the possible local and accidental disturbing attractions which may occasionally have been encountered in the course of so extensive a series,—the accordance of the experimental results with the general law proposed for their representation cannot be contemplated as otherwise than most striking and remarkable.

The general applicability of the law to all the stations of the present experiments, is further strengthened by a more attentive consideration of the differences themselves, and of the causes which may seem to have occasioned them; there are three of principal note in Table I., in two of which, at Trinidad and Greenland, the calculation is in excess, and in the third, at New York, in defect. The differences at Trinidad and New York are produced principally by the results with the horizontal needles, which differ more from the results with the dipping-needle in those instances than upon any other occasion, and more in fact than the latter does from the calculation: now a difference between the results of the two methods of experiment will be occasioned by, and is indicative of, the presence of some partial or accidental attraction at the station in question, the effect of which will necessarily be more conspicuous in the vibrations of the horizontal needles, on which only a portion of the regular terrestrial force acts, than in those of the dipping-needle, on which the whole unresolved force operates. It is, indeed, this very difference that furnishes the best means with which I am acquainted of detecting the presence of a disturbing force; and accordingly the agreement of the two methods at all the other stations may be with propriety referred to, as evidencing the general success of the caution which was at all times observed, in selecting a situation for the experiments sufficiently distant from iron.

That the discrepancy at Trinidad was occasioned by some such accidental cause, may also be inferred from the correspondence of the experiments and calculation at the neighbouring stations on opposite sides, and nearly at equal distances, of Maranham and Jamaica.

With respect to the correctness of the experiments themselves at Greenland and Trinidad, the opportunities were excellent; the place of observation at Trinidad was on the lawn of the Government House, far distant from buildings, and above a mile from Port Spain; at Greenland there were no houses to remove from, and nothing indeed to suspect, except the soil, which was everywhere strongly impregnated with iron. The circumstances were not so favourable at New York, the weather being extremely cold, with snow falling during the time that the needles were in use; so that the observations on the Dip with Mayer's needle were not repeated, as at the other northern stations, where, from the considerable amount of the Dip, it became an important element in the reduction of the times of horizontal vibration, and where consequently its correct knowledge was materially conducive to the accuracy of the inferences which might be drawn from the horizontal needles, in regard to the intensity of the force in the direction of the dipping needle: for this reason, the results at the horizontal needles at New York are not perhaps entitled to the same consideration as elsewhere, and the Intensity deduced by the dipping needle alone, may with more propriety be regarded as furnishing the experimental ratio at that station.

The differences next in rank in Table I., and the only remaining ones in that table worthy of notice, are at St. Thomas, Sierra Leone, the Gambia, and Teneriffe, at all of which the calculation is in defect nearly to the same small amount; these stations are situated nearly alike in respect to the assumed position of the magnetic pole, of which it has already been remarked, that preserving the latitude of 60°, the longitude of 78° west would have accorded better with the general body of the experiments than that of 80°; it is particularly in regard to these four stations that the longitude of 78° would have been preferable, as the removal of the pole two degrees to the eastward of its assumed position would reconcile the calculation and experiments at them, with comparatively very little effect on the itinerary distances of any of the other stations.

Of the nine stations comprised in Table II., the experiments made on land are entitled to principal confidence; those on ice were on occasions when the vessels were detained by circumstances of weather, and may have been affected either by the presence of the ship, anchored to the iceberg on which they were [necessarily] made,—by a vibrating motion of the ice from the impulse of the waves, affecting the horizontality of the planes,—or, if on field or floe ice, by a circular motion of the whole mass, only sensible to the observer, by perceiving, at the close of the observations, that the instrument had moved in azimuth perhaps several degrees from the meridian during their course. One of the ice stations should be excepted from this remark, the experiments having been made on an iceberg of immense size very securely grounded in Davis Strait, on which I remained during great part of a day, whilst the ships were distant under sail. It is the 5th station of the Table; and with the land stations, being the 1st, 2d, and 4th, will be found to present the best accord between the experiments and calculation; whilst at the other stations the differences are not greater than might very reasonably have been expected from the causes above noticed.

The experiments comprised in Table III. were made generally under more favourable circumstances than those of the preceding voyage, and with an attention to particulars, suggested by experience, and conducing to greater accuracy in the mode of observation; their results present, therefore, as might be expected, a closer accordance with the calculation, than those of the preceding voyage: several of the stations in this table are peculiarly interesting, from their proximity and relative bearing to the assumed position of the magnetic pole.

In taking a general review of the experiments on the intensity of magnetism contained in the preceding pages, and of the inferences that have been drawn from them, it appears, that if the earth be considered a magnetic sphere, with poles analogous to those of the induced magnetism of an iron ball, and if in the year 1822, or thereabouts, the geographical position of the pole in the northern hemisphere be assumed in 60° N. latitude, and 80° (or more exactly in 78°) W. longitude,—and if the magnetic force be supposed to vary between the pole and the equator in the proportion of two to one, and intermediately, as the square root of one increased by three times the square of the cosine of the distance from the pole, measured on a great circle of the sphere,—the relation of the intensities actually observed at thirty-three stations, distributed over, and comprehending a space equal, or nearly so, to a fourth part of the surface of an hemisphere, will be represented within such small limits as may reasonably be ascribed to the unavoidable uncertainties of experiment.

The annexed map of the northern magnetic hemisphere has been constructed for the purpose of producing a more distinct impression than can be conveyed by verbal description, first, of the arrangement of the intensities under the supposed law; second, of the portion of the hemisphere in which the phenomena have been proved in conformity to it; and third, of that portion which yet remains for a more extended experience. The map is a correct delineation, on a polar projection, of the land in an hemisphere, having its pole in the latitude of 60° north, and in the longitude of 80° west; the parallels are those of equal magnetic intensity, and are drawn at the proper intervals, to express the ratios of the force under the respective parallels of 1.1, 1.2, 1.3, *&c.*, to the force at the equator considered as unity; the space comprehended by the experiments, and over which they are distributed, is shewn by the insertion of the names of the stations. This map furnishes also a ready means to travellers in the northern hemisphere, of perceiving the relation of the magnetic force according to the law which the present experiments have suggested, at any two or more stations which they may design to visit; and by thus facilitating comparison, may conduce towards the obtainment of further experimental testimony.

Meanwhile, the evidenced agreement over so considerable a portion of the hemisphere affords a reasonable ground of expectation, that the phenomena may equally be found in correspondence with the law at other parts of the same hemisphere; and that the law may even prove, still more extensively, one of general application over the whole surface of the globe, the intensities in the southern hemisphere being computed from the southern magnetic pole. It would assuredly be an highly interesting subject of physical research, to ascertain by direct experiment, whether the magnetic force varies in the southern hemisphere in the same ratio as it has been shewn to do in the northern; and if so, to determine, by a sufficiently extensive series of comparative experiments, the geographical position of the southern magnetic pole, as that of the northern has now been deduced. Presuming the supposed regularity of the phenomena, it would be an additional advantage, that the situation of the two poles should be ascertained as nearly contemporaneously as possible; and if the same instruments which have already traversed so large a portion of the one hemisphere were employed for a similar purpose in the other hemisphere, and if on their return to England, they should be found to have preserved their magnetism unchanged, as on former occasions, their employment would further shew whether or not the general intensity of magnetism is the same in each hemisphere.

Experience has shown, in all countries where sufficiently precise observations have been made, that both the dip and variation undergo an apparently systematic periodical oscillation, sensible in its annual progression, but of which the extent and period have not yet been determined at any station on the globe. It may be inferred that the intensity is subject to an analogous variation, but the evidence of experiment is yet wanting. Were it ascertained that the absolute intensity at any particular place had undergone an alteration in a certain number of years, a circumstance far more difficult of experimental proof than the changes in the dip and variation, the cause might be ascribed either, to a fluctuation in the general magnetic intensity of the globe,—or to such an alteration in the system of terrestrial magnetism in regard to its geographical relations, as the changes of the dip and variation are usually ascribed to; namely, to a change of position of the magnetic poles. The existence of the latter cause, however, as producing the effect, may become the subject of a distinct and decisive experiment, if the inferences which have been drawn in the preceding pages shall be established by more extensive experience; as by the repetition of a similar series of experiments in a future year, the position of the maximum of intensity in the northern hemisphere may be shewn either to have been stationary in the interim, or to have advanced to a spot, of which the geographical relation may be determined in the same manner as on the present occasion; when, if the number of years elapsed have been sufficient, the difference of position and the interval may become the elements of ascertaining the nature of the progression to which the magnetic pole is subject; and to which the alteration of intensity, at any particular station, should have been conformable, if thegeneral magnetic intensity of the globe is a constant force.

The experiments recorded in the preceding pages have placed beyond question the fact, that the variation in the intensity of the magnetic force in different parts of the globe cannot be represented by any function of the known dip; consequently, whenever it may be desired to trace with precision the compound effect of the forces acting on the compass needle in ships, the intensity must be regarded as an essential element of the computation, distinct from the dip, and necessary to be known by observation, until that necessity shall be superseded, by the law of its variation having been thoroughly ascertained.

There are two forces which act on the compass needle of ships; the natural force of terrestrial magnetism, and the disturbing force of the ship's iron. The latter is usually considered constant in different parts of the globe; the former varies in its influence on the horizontal needle, inversely as the cosine of the dip, and directly as the intensity. Now as the intensity doubles between the equator and the pole, it is obvious that its variation must by no means be omitted in the computation; and a reference to the experiments with the horizontal needles at New York and at London will suffice to shew, that the magnitude of the horizontal force cannot be assumed to vary as a function alone of the observed dip, as is done in Dr. Young's "Table of corrections for clearing the compass of the regular effect of a ship's permanent attraction," published in the Journal of the Royal Institution, vol. ix., page 375, without incurring occasionally very considerable error. The dip at New York being 73°, and at London 70°, the natural force acting on the horizontal needle should be reduced at New York, according to that computation, in the ratio of 0 85, to 1 in London, and the influence of the disturbing force proportionally increased; whereas, the experiments shew that the natural horizontal force is actually greater at New York than in London (in the proportion of 1.1 to 1), notwithstanding the increase of three degrees in the inclination of the dip; and it is so because the absolute force of magnetic attraction is greater at New York in the proportion of 1.96 to 1.62 in London, as shewn by experiment, instead of being in the proportion of 1.78 to 1.72, in which it would be inferred to vary by the author of the article in the Journal of the Institution. The effect of the increased, instead of diminished, magnitude of the natural force acting on the compass needle at New York, in comparison with London and its vicinity, was further obvious in the amount of the disturbing influence of the iron in the Pheasant, which was observed by Captain Clavering to be less at New York than in the River Thames, notwithstanding the difference in the dip. In fact, if the maximum disturbance at the Nore were 16°, it should be augmented to upwards of 18° by the table of corrections which has been referred to, but under the actual magnetic circumstances was really reduced to little more than 15°. A much stronger practical example of error produced by neglecting the consideration of the actual intensity might doubtless be furnished by situations of equal dip, chosen in the North Sea and in the Gulf or River St. Lawrence; and is the reason why comparatively so much fewer complaints have been made of the errors of the compasses in the latter navigation, than might have been expected from the amount of the dip in the vicinity of the St. Lawrence.

It is indeed a fortunate circumstance for navigation generally, that the amount of the dip of the needle is not always commensurate with that of the intensity of the force; and that the dips of greatest magnitude in both hemispheres are confined to regions, which, from other natural causes, are rarely traversed. Had the dip, for instance, in the West Indies, in the homeward passage from thence by the course of the Gulf Stream, and generally on the coasts of the United States and of the British North American possessions, been in correspondence with the intensity, the irregularities in the direction of the compass needles would have proved a most serious embarrassment, instead of being a consideration of very little practical importance, in the navigation of those extensive and greatly-frequented districts of the ocean.

Observations on the Diurnal Oscillation of the Horizontal Needle at Hammerfest and Spitzbergen.

The few satisfactory observations which time and opportunity enabled me to make on the diurnal oscillation of the needle, are appended to the preceding memoir, in the hope that, from the localities in which they were made, and I may venture to add from the extreme care that was taken to obtain results worthy of confidence, they may prove of service to those persons who are engaged in the investigation of the nature and causes of that phenomenon.

The observations were made with a very complete and delicate apparatus, the property of Mr. Browne, made by Mr. Dollond. The needle is suspended by a silk line of several inches in length, passing over a pulley, and having a weight attached to the other extremity, which may be either a counterpoise for the whole weight of the needle, or for a portion of it, so that the weight may be either partially or entirely relieved from the central pivot on which the needle is otherwise supported; the graduated circle is of seven inches diameter, and is divided into spaces of ten minutes, which are again sub-divided to single seconds by micrometer wires in the field of two compound microscopes, one of which is fixed to see the north end, and the other the south end of the needle, at the same time with the nearest divisions of the circle. The adjustments required are,—of the horizontality of the circle, and of the needle when suspended,—the freedom from tension in the silk suspension,—and the coincidence of the micrometer wires, when at zero, with the nearest primary division to the indication of the needle. The whole apparatus requires to stand on an insulated and very firm support, and to be approached, and the micrometer screws touched, with great caution. It was protected from the weather by a circular canvass tent, of which the wood work was copper fastened; and situations were selected for the tent, where the needle might be undisturbed by accidental visitors. Both ends of the needle were observed, but the movements of the north end alone recorded, as those of the south end corresponded, in consequence of the counterpoise being rather less than the weight of the needle. The primary division with which the micrometer wire coincided when at zero, was to the west of the north end of the needle, both at Hammerfest and Spitzbergen; so that the higher numbers of the registry indicate its greatest oscillation to the eastward, and the lower numbers the limit to the westward. The following tables comprise the observations.

OBSERVATIONS on the DIURNAL OSCILLATION of a NEEDLE SUSPENDED HORIZONTALLY.—Hammerfest, June, 1823.

Latitude 70° 40′ N. Dip 77° 13′ N. Variation 11° 26′ W.

June	12th	13th	14th	15th	16th	17th	18th	19th	20th	21st	22d	No. of Days of Observation.	Mean Place of the Needle at the respective hours.	Deviation of the North end of the Needle from its average Mean Place.
H.	′ ″	′ ″	′ ″	′ ″	′ ″	′ ″	′ ″	′ ″	′ ″	′ ″	′ ″		′ ″	′ ″
6 A.M.	..	..	..	..	..	12 00	..	10 30	12 10	12 10	10 00	5	11 11	2 14 Ey.
7½ „	..	..	..	..	..	12 00	..	..	..	..	10 30	2	11 15	2 18
9 „	..	..	..	..	13 50	12 00	..	..	10 55	11 20	10 00	5	11 37	2 41
10½ „	.	12 25	12 00	12 00	11 25	10 40	..	9 10	9 15	11 00	..	8	10 59	2 02
Noon.	..	3 00	4 45	9 30	8 50	5 00	9 15	5 23	6 43	6 16	5 08	10	6 23	2 34 Wy.
1½ P.M.	..	3 20	..	8 40	8 20	7 40	..	2 50	6 43	5 45	..	7	6 11	2 46
3 „	..	5 00	4 45	8 00	8 20	7 45	5 30	3 40	6 43	6 14	..	9	6 15	2 42
4½ „	..	6 30	4 00	7 30	8 50	7 00	6 30	5 25	6 43	6 30	..	9	6 34	2 23
6 „	..	8 10	4 40	8 20	7 00	8 25	8 20	8 15	7 34	6 40	..	9	7 30	1 27
7½ „	8 20	7 00	5 00	9 10	6 30	8 25	7 30	8 40	7 34	6 40	..	10	7 29	1 28
9 „	8 20	5 00	5 40	8 20	8 00	8 25	7 30	8 40	7 31	6 40	..	10	7 25	1 52
10½ „	..	9 30	8 30	8 20	8 00	..	8 30	9 00	8 40	6 40	..	8	8 24	0 33
Midnight.	12 45	12 20	21 30*	8 40	8 00	..	9 40	9 30	8 40	6 40	..	8	9 32	0 35 Ey
1½ A.M.	..	12 00	..	..	10 30	..	9 40	10 07	8 40	..	..	5	10 11	1 14
3 „	..	12 00	..	10 50	10 30	..	..	..	..	..	..	3	11 07	2 10
4½ „	..	12 10	..	..	11 00	..	..	12 00	11 00	9 25	..	5	11 05	2 08
Mean Place of the Needle													8 57	

* There appears no reason to doubt, that the deviation of the needle at midnight of the 14th 10 or 12 minutes to the eastward of its usual direction at the same hour, arose from some natural cause. No one had been near the Observatory Tent since the preceding observation, and when I entered it with the usual care to avoid disturbance, I found the needle perfectly steady at the recorded division; and on the following day, at 10½ A.M., it had returned of its own accord to its ordinary indication. This observation, being the only irregularity of amount which occurred in the ten days, has been omitted in the Mean.

OBSERVATIONS on the DIURNAL OSCILLATION of a NEEDLE SUSPENDED HORIZONTALLY.—SPITZBERGEN, July, 1823.

Lat. 79° 50′ N. Dip 80° 10′ N. Variation 25° 12′ W.

JULY.	4th	5th	6th	7th	8th *	9th	10th	11th	No. of Days of Observation.	Mean Place of the Needle at the respective hours.	Deviation of the North end of the Needle from its average Mean Place.
	′ ″	′ ″	′ ″	′ ″	′ ″	′ ″	′ ″	′ ″		′ ″	′ ″
6 A.M.	. . .	10 34	10 20	. .	10 15	. .	. . .	.	3	10 23	2 42 Ey.
7½ ,,	. .	10 00	10 20	. . .	10 18	. . .	. . .	. . .	3	10 13	2 32 ,,
9 ,,	. .	9 30	8 56	. . .	9 30	. . .	. . .	9 00	4	9 14	1 33 ,,
10½ ,,	. . .	8 30	7 40	8 08	8 30	7 28	. . .	. .	5	8 03	0 22 ,,
Noon.	. . .	7 30	6 00	7 35	. . .	. . .	8 00	7 58	5	7 25	0 16 Wy.
1½ P.M.	. . .	4 44	6 00	. . .	6 25	6 30	5 48	. . .	5	5 53	1 48 ,,
3 ,,	. .	4 35	6 26	6 30	. . .	5 44	5 12	. .	5	5 41	2 00 ,,
4½ ,,	. . .	3 35	5 30	6 30	5 58	. . .	5 08	5 42	6	5 23	2 18 ,,
6 ,,	. . .	5 08	5 07	5 43	5 53	4 54	4 53	5 31	7	5 18	2 23 ,,
7½ ,,	. . .	4 13	5 07	3 40	6 12	. . .	5 27	. . .	5	4 56	2 45 ,,
9 ,,	. . .	4 13	. . .	4 35	6 50	. . .	5 44	. . .	4	5 20	2 21 ,,
10½ ,,	. . .	6 30	6 33	6 30	7 20	6 14	6 20	6 52	7	6 37	1 04 ,,
Midnight.	7 37	7 20	8 50	8 00	. . .	. . .	8 00	. . .	5	7 57	0 16 Ey.
½ A.M.	8 50	8 45	9 34	. . .	. . .	. . .	. . .	. . .	3	9 03	1 22 ,,
3 ,,	11 36	11 37	10 46	. . .	10 37	10 30	10 15	9 14	7	10 40	1 59 ,,
4½ ,,	11 50	10 50	11 30	10 15	9 31	12 00	. . .	10 20	7	10 54	2 13 ,,
MEAN Place of the Needle . .										7 41	

* Sun eclipsed at 7 A.M.

ATMOSPHERICAL NOTICES.

On the Depression of the Horizon of the Sea over the Gulf Stream.

In estimating the depression of the horizon of the sea, corresponding to the different heights of an observer's eye, the horizon is supposed to be raised by terrestrial refraction one-fourteenth part of the depression due to the spherical figure of the earth; and the corrections for different heights, rigorously computed from the dimensions of the earth, are reduced, accordingly, in that proportion, in the tables of the most approved authorities. Experience has shewn that, in general, when the temperature of the air is colder than that of the surface of the sea, the tabular depressions, so computed and reduced, are in error, in defect,—and when the air is warmer than the sea, in excess,—of the true depression: the proportion of the error to the difference of temperature being, however, too irregular, and the inferences themselves subject to exceptions of too decided a character, to allow any practical rule to be established for a corresponding allowance in correction. So long as the error of the tables is confined to a few seconds in amount, its occurrence may be safely disregarded in all the ordinary purposes of navigation; but it was a question, only to be solved by experiment, whether in cases of an extreme difference between the temperatures of the air and water, the amount of error might not be so considerable as to require attention, especially in deducing a ship's place by chronometrical observations within three hours of noon It was for the purpose of having this question tried in the Gulf Stream, where the sea is frequently many degrees warmer than the air, that Dr. Wollaston contrived the dip sector, an instrument now too well known to need description, but which, from accidental circumstances, had not been applied in its original design until the present occasion.

The dip sector which I employed was the property of the Admiralty, and was one of those originally made for the northern expeditions; previously to my leaving England in 1821, I had it fitted with a telescope of much larger field than before, so that the spaces of the opposite horizons to be brought into contact were greatly augmented, and the observations rendered thereby much more exact; with the instrument thus improved, used under proper circumstances, and with due repetition, the results may be confided into less than five seconds.

The following table presents an abstract of the observations, by which it will be seen, that so far as their evidence can determine, a navigator may be right nine times in ten, in apprehending a tabular error in defect when the sea is warmer than the air; but that with differences in the temperature of the air and water, frequently amounting to between 10 and 20 degrees, and once even as great as 29 degrees (the sea being always the warmer), the error of the tables was not found even in a single instance so great as two minutes.

OBSERVATIONS on the DEPRESSION of the HORIZON of the SEA, made principally over the GULF STREAM.

DATE.	Geographical Position.	Latitude.	Longitude.	Height of the Eye.	Temperature.		Depression.		Tabular in excess or defect.	Temp. of the air in excess or defect.	REMARKS.
					Air.	Sea.	Observed.	Tabular.			
1822.		° ′	° ′	Feet In.	°	°	′ ″	′ ″	′ ″	°	
Aug. 9	Atlantic . .	13 20 S.	37 45 W.	19 2	76.1	77.1	4 25.4	4 19	−0 06.4	− 1	Light airs.
Nov. 10	Caribbean Sea	18 33 N.	79 36 „	15 3	82	83.1	4 03.4	3 51	−0 12.4	− 1.1	Wind, E.N.E.; pleasant breeze, with fine weather.
„ 10	Caribbean Sea	18 33 „	80 01 „	15 3	83.2	83.8	3 56.2	3 51	−0 05.2	− 0.6	do. do. do.
„ 11	Caribbean Sea	19 20 „	81 40 „	15 3	81.8	83	3 53.3	3 51	−0 02.3	− 1.2	Wind, E.N.E.; sunshine, with occasional clouds.
„ 13	Caribbean Sea	20 25 „	83 30 „	15 3	80.8	82.5	4 17.2	3 51	−0 26.2	− 1.7	Wind, N.E.b.E. do. do.
„ 14	Caribbean Sea	20 40 „	88 45 „	15 3.5	79.5	82.2	4 21.2	3 51.3	−0 29.9	− 2.7	do. do. do.
„ 14	Caribbean Sea	21 00 „	84 20 „	15 10	78	82	4 18.7	3 55.5	−0 23.2	− 4.	Wind, N.E.; with light rain (page 450, note).
„ 15	Caribbean Sea	21 30 „	84 55 „	15 3.5	78.8	80	4 21.2	3 51.3	−0 29.9	− 1.2	Fresh N.Easterly breeze, with occasional squalls.
„ 17	Gulf of Mexico	22 49 „	84 57 „	15 3.5	80.3	82	3 51.3	3 51.3	. . .	− 1.7	do. do. do.
„ 17	Gulf of Mexico	22 51 „	81 20 „	15 3.5	80.2	82.1	3 48.7	3 51.3	+0 02.6	− 1.9	Little wind, with sunshine.
„ 30	Gulf Stream .	29 12 „	79 30 „	15 3.5	78.8	81.1	4 14.2	3 51.3	−0 22.9	− 2.3	Sunshine.
Dec. 4	Gulf Stream .	35 40 „	73 30 „	18 3	47	76	4 50	4 13	−0 37	−29.	Becoming calm, after a northerly gale.
„ 5	Gulf Stream .	36 24 „	72 40 „	15 3	60.5	74	4 56.6	3 51	−1 05.6	−13.5	Wind light, southerly.
„ 5	Atlantic . .	36 39 „	72 30 „	15 3	61.5	62.4	3 36.6	3 51	+0 11.4	− 0.9	Wind freshening.
„ 6	Atlantic . .	36 58 „	73 40 „	15 3	43	60.6	4 57.8	3 51	−1 06.8	−17.6	Wind, N.W.; fresh.
„ 7	Atlantic .	37 35 „	71 33 „	15 6	49.5	59.5	5 36.2	3 55.2	−1 41	−10.	Wind light, easterly.
„ 7	Atlantic . .	37 45 „	74 33 „	15 3	50.8	54.2	4 37	3 51	−0 46	− 3.4	Soundings in thirty-three fathoms.
„ 8	Atlantic . .	38 30 „	74 26 „	15 1	43.8	53.5	5 15	3 50	−1 25	− 9.7	Wind, N.W.; fresh; soundings in thirty fathoms.
„ 9	Atlantic . .	40 00 „	74 00 „	15 1	37.8	49.5	5 45.6	3 50	−1 55.6	−11.7	Wind faint N.W.; clear weather; horiz. very distinct, but with the appearance of the inversion of a ripple.
1823.											
Jan. 8	Gulf Stream .			15 10	52.5	69.2	5 16.6	3 55.6	−1 21	−16.7	Steady breeze, E.N.E.
„ 8	Gulf Stream .			15 10	56	69	5 00.8	3 55.6	−1 05.2	−13.	do. do.
„ 9	Gulf Stream .	In the passage from New York to the British Channel.		16 1	64.2	67	4 19.2	3 57.6	−0 21.6	− 2.8	Wind south, fresh.
„ 10	Gulf Stream .			15 9	61.8	67.8	4 04.4	3 55	−0 09.4	− 6.	Wind S.W.; with slight squalls.
„ 16	Gulf Stream			15 00	56.5	60	3 32.6	3 49	+0 16.4	− 3.5	Wind east; fresh, with light rain.
„ 17	Gulf Stream .			16 0.5	53.5	59	4 00	4 00	0 00	− 5.5	do. do.
„ 17	Gulf Stream .			16 0.5	53	59	4 00	4 00	0 00	− 6.	do. do.

On the Intensity in Effect of the Radiation of Heat in the Atmosphere, at Heights, and at the Level of the Sea.

The following attempt to compare the heat produced by the radiant power of the sun at the level of the sea and at a considerable elevation,—and, conversely, the cooling influence of nocturnal radiation in a calm and clear sky,—was made at Jamaica. The stations selected for the experiment were on the Glacis of Fort Charles at Port Royal at an elevation of 8 feet, and on a plateau near the summit of the ridge of the Port Royal Mountains, 4080 feet above the sea.

The thermometers employed to measure the opposite extremes of the radiating influences, were a mercurial thermometer, having an index registering the maximum, and a spirit thermometer, having an index registering the minimum of heat; the stems only were attached to scales, and the bulbs were coated with lamp-black and covered with black wool. The same thermometers were used at both stations, and were placed for exposure on thick vegetation, by the filaments of which they were supported horizontally, without being screened; the vegetation on the mountain was meadow grass, and at Port Royal the plant called Tibullus Maximus, by which the ground between Fort Charles and the extremity of the Point is over run. The exposure to the heavens was equally perfect in both cases; the height above the ground, at which the thermometers were supported by the vegetation on the mountain, was between three and four inches, and at Port Royal about ten inches.

The extremes of the true atmospheric temperature, the maximum in the day and the minimum at night, were registered by a thermometer suspended at about five feet above the ground, in a situation shaded from the heavens by a roof, but open in all other directions; the thermometer was enclosed in an highly-polished metallic cylinder of eight inches diameter, protecting it from the influence of radiation from surrounding bodies, and pierced with large holes in the top and bottom to admit a thorough draft.

The observations, of which the particulars are collected in a table, shew, that whilst a blackened thermometer, exposed during six days to the sun at the level of the sea, did not rise higher on any occasion than 36.5 above the temperature of the surrounding atmosphere, the same thermometer similarly exposed at an elevation of 4080 feet, rose on a single day of experiment 59 degrees higher than the surrounding atmosphere; and that, notwithstanding the reduction at 4080 feet of 13 degrees in the temperature of the atmospheric medium in which the thermometer was exposed, and which medium was constantly operating in counteraction of the measure of heat produced by the absorption of the rays, the exposed thermometer was actually nine degrees higher on the one day in the mountains than its maximum had been in any one of the six days at the level of the sea. It is also shewn, that whilst a thermometer, fairly exposed to the heavens at night, at the level of the sea, fell on the average of seven nights nine degrees below the temperature shown by a thermometer protected from radiation into space (the greatest partial effect being 11.5 degrees), the same thermometer similarly exposed on a single night at 4080 feet, fell eighteen degrees below the protected thermometer. And thus, that the vegetation on which the thermometer was placed was respectively subject,—at Port Royal, to a difference of 55°.5,—and in the mountains to a difference of 77°,—of temperature in each twenty-four hours; evidencing a far greater intensity in the action of radiation at the elevation, than at the level of the sea.

The circumstances of the weather were favourable for the experiments at both stations; that is to say, the weather was clear and calm during a portion of each day and of each night.

EXPERIMENTS upon SOLAR and TERRESTRIAL RADIATION at Jamaica.

DATE.	DAY OBSERVATIONS. Maximum of Heat.			NIGHT OBSERVATIONS. Minimum of Heat.			Extreme Differences in each 24 hours.	
	Blackened Therm. in the Sun.	Atmosph. Temp.	Difference.	Blackened Therm. exposed.	Atmosph. Temp.	Difference.	Of the exposed Therm.	Of the Atmosph.
On the Glacis of Fort Charles at Port Royal, 8 feet above the Sea.								
	°	°	°	°	°	°	°	°
Oct. 25	122	86	36	72	76	4	50	10
„ 26	123	87	36	69	76	7	54	11
„ 27	122	86	36	65	76	11	57	10
„ 28	122	86	36	66	76	10	56	10
„ 29	123	86.5	36.5	65	76.5	11.5	58	10
„ 30	123	86.5	36.5	65	76	11	58	10
Nov. 3				67	76	9		
Means	122.5	86.3*	36.2	67	76	9	55.5	10.25
On a plateau near the summit of the Port Royal Mountains, 4080 feet above the Sea.								
Nov. 1	132	73	59	45	63	18	77	10

* A mercurial thermometer suspended freely in the air, about five feet above the ground and in the sun, was carefully observed at intervals of the fore and afternoon, from the 25th to the 30th of Oct., and was never seen to rise higher than 92°, being a difference of 6° Faht. above the shaded thermometer, occasioned by the absorption of the rays in the bulb, which was naked and not blackened. This thermometer usually attained 92° at 10 A.M., before the sea breeze set in: on the commencement of the breeze it fell, but regained the same height in the afternoon, although the breeze had freshened intermediately.

The evidence which the preceding experiments furnish, of the greater intensity in the effect of the sun's rays at an elevation than at the level of the sea, was further strengthened by the indications of a differential thermometer in vacuo exposed to the sun at both stations. The sentient-ball of this thermometer was of dark-coloured glass, designed to absorb the rays; the other ball was pellucid, and was protected by a double case of polished silver, with no part of which it was in contact: the whole instrument was enclosed in a glass cylinder hermetically sealed at the lower end, and containing a tolerable vacuum. The degrees of the thermometric scale were millesimal, the interval between the boiling and freezing of water being divided into 1000 degrees. On placing the thermometer in a fair exposure to the sun, and on removing the wooden cover by which it was usually protected, the fluid in the stem adjoining the sentient ball fell rapidly, until it reached an amount proportioned to the influence to which the thermometer was subjected, when it remained stationary for some minutes. This amount being registered, the cover was restored and the instrument re-placed in a secure and shaded situation.

EXPERIMENTS upon SOLAR RADIATION with a DIFFERENTIAL THERMOMETER in VACUO at Jamaica.

AT PORT ROYAL.					AT AN ELEVATION OF 4080 FEET.				
Date.	Hours.	Therm.	Effect produced in	Remarks.	Date.	Hours.	Therm.	Effect produced in	Remarks.
Oct. 24	noon.	88°	. .	Strong breeze, very clear.	Nov. 1	8 50A.M.	74°	1½ min.	Faint haze.
„ 24	1½ P.M.	88	. . .	do. do.	„ 1	9 50 „	100	1½ min.	Clear.
„ 26	9½ A.M.	82	.	Calm, do.	„ 1	10 45 „	84	1½ min.	Light clouds.
„ 26	noon.	88	2 min.	Fresh sea-breeze.	„ 1	11 45 „	110	1½ min.	Very clear.
„ 29	noon.	90	2 min	Almost calm, clear.					
„ 29	2 P.M.	86	2 min.	Freshening, clear.					
„ 30	10 A.M.	88	2 min.	Calm, very clear.					
„ 30	½ P.M.	91	2 min.	Very strong sea-breeze.					
„ 30	3½ „	71	2 min.	do. do.					
Nov. 3	8 A.M.	68	. .	Calm and clear.					
„ 3	9 „	82	. . .	do.					
„ 7	7 „	48	. . .	do.					

POSTSCRIPT.

June, 1825.

An intention having been expressed in page 186, of appending, at the close of this volume, a notice of a comparison of the thermometer employed in the pendulum experiments, with a standard thermometer which was understood to be in preparation under the superintendence of the members of a committee of the Royal Society, in the expectation that the completion of the standard would be accomplished before the publication of the experiments,—it is necessary to state, in explanation of the non-fulfilment of that intention, that no such standard thermometer has yet been completed.

LONDON
PRINTED BY W: CLOWES
Northumberland-court.

www.ingramcontent.com/pod-product-compliance
Ingram Content Group UK Ltd.
Pitfield, Milton Keynes, MK11 3LW, UK
UKHW050730090726
473066UK00013B/1093

* 9 7 8 1 1 0 8 0 6 2 0 7 7 *